Advances in Hydroinformatics

Advances in Hydroinformatics

Edited by Roman Morris

MURPHY & MOORE
www.murphy-moorepublishing.com

Published by Murphy & Moore Publishing,
1 Rockefeller Plaza,
New York City, NY 10020, USA
www.murphy-moorepublishing.com

Advances in Hydroinformatics
Edited by Roman Morris

International Standard Book Number: 978-1-63987-018-9 (Hardback)

Cataloging-in-Publication Data

Advances in hydroinformatics / edited by Roman Morris.
p. cm.
Includes bibliographical references and index.
ISBN 978-1-63987-018-9
1. Hydrology--Data processing. 2. Hydrology--Computer simulation.
3. Hydrologic models. I. Morris, Roman.
GB656.2.E43 A38 2022
551.480 285--dc23

TABLE OF CONTENTS

PREFACE

Over the recent decade, advancements and applications have progressed exponentially. This has led to the increased interest in this field and projects are being conducted to enhance knowledge. The main objective of this book is to present some of the critical challenges and provide insights into possible solutions. This book will answer the varied questions that arise in the field and also provide an increased scope for furthering studies.

Hydroinformatics is the branch of informatics that applies information and communication to address the crucial problems of the efficient use of water. Computational hydraulics is one of the major disciplines that fall under the domain of hydroinformatics. It primarily focuses on the numerical simulation of water flows and related processes. Different techniques of artificial intelligence are applied in the field of hydroinformatics. A few of them are artificial neural networks, support vector machines and genetic programming. Hydroinformatics draws from the diverse fields of hydrology, hydraulics, data mining and environmental engineering. This book provides comprehensive insights into the field of hydroinformatics. It will help the readers in keeping pace with the rapid changes in this field. The aim of this book is to present researches that have transformed this discipline and aided its advancement.

I hope that this book, with its visionary approach, will be a valuable addition and will promote interest among readers. Each of the authors has provided their extraordinary competence in their specific fields by providing different perspectives as they come from diverse nations and regions. I thank them for their contributions.

Editor

Precipitation variability within an urban monitoring network via microcanonical cascade generators

P. Licznar[1], C. De Michele[2], and W. Adamowski[3]

[1]Faculty of Environmental Engineering, Wroclaw University of Technology, Wrocław, Poland
[2]Department of Civil and Environmental Engineering, Politecnico di Milano, Milan, Italy
[3]Institute of Environmental Engineering, John Paul II Catholic University of Lublin, Stalowa Wola, Poland

Correspondence to: P. Licznar (pawel.licznar@pwr.edu.pl)

Abstract. Understanding the variability of precipitation at small scales is fundamental in urban hydrology. Here we consider the case study of Warsaw, Poland, characterized by a precipitation-monitoring network of 25 gauges and microcanonical cascade models as the instrument of investigation.

We address the following issues partially investigated in literature: (1) the calibration of microcanonical cascade model generators in conditions of short time series (i.e., 2.5–5 years), (2) the identification of the probability distribution of breakdown coefficients (BDCs) through ranking criteria and (3) the variability among the gauges of the monitoring network of the empirical distribution of BDCs.

In particular, (1) we introduce an overlapping moving window algorithm to determine the histogram of BDCs and compare it with the classic non-overlapping moving window algorithm; (2) we compare the 2N–B distribution, a mixed distribution composed of two normal (N) and one beta (B), with the classic B distribution to represent the BDCs using the Akaike information criterion; and (3) we use the cluster analysis to identify patterns of BDC histograms among gauges and timescales.

The scarce representation of the BDCs at large timescales, due to the short period of observation (~ 2.5 years), is solved through the overlapping moving window algorithm. BDC histograms are described by a 2N–B distribution. A clear evolution of this distribution is observed, in all gauges, from 2N–B for small timescales, N–B for intermediate timescales and B distribution for large timescales.

The performance of the microcanonical cascades is evaluated for the considered gauges. Synthetic time series are analyzed with respect to the intermittency and the variability of intensity and compared to observed series. BDC histograms for each timescale are compared with the 25 gauges in Warsaw and with other gauges located in Poland and Germany.

1 Introduction

Urban hydrology requires access to very precise information about the precipitation variability over small spatial and temporal scales. Widespread use of surface runoff models coupled to urban drainage networks increases the common request for rainfall data inputs at high temporal and spatial resolutions. As already estimated a decade ago by Berne et al. (2004), the necessary resolution of rainfall data as the input in hydrological models in Mediterranean regions was about 5 min in time and 3 km in space for urban catchments of ~ 1000 ha. For smaller urban catchments of ~ 100 ha, even higher resolutions of 3 min and 2 km were required. Results obtained with the application of operational semi-distributed urban hydrology models fully confirmed earlier observations from select study cases in England and France (Gires et al., 2012, 2013). These authors strongly recommend the use of radar data in urban hydrology, especially in the context of real-time control of urban drainage systems. In particular, they opted for X band radars (whose resolution is hectometric), as opposed to the more common C band radars, because they are affected by less uncertainty. Additionally, Gires et al. (2012) stated that small-scale rainfall variability under 1 km resolution cannot be neglected and should be accounted for in probabilistic ways in the real-time management of urban drainage systems. As a matter of fact, the implementation of radar techniques gained a rising popularity in major cities across the EU (for details refer to the Thames Tideway Tunnel (TTT), 2010).

Despite the obvious benefits of radar instruments, radar data are not always available for practical applications. Thus, current versions of even the most advanced computer rainfall-runoff urban drainage models do not consider radar data as rainfall input. Therefore the only possibility of accounting for spatial rainfall variability is to consider different point time series for each subcatchment (Gires et al., 2012). The vast majority of engineering practical calculations and modeling of drainage systems is still associated with point rainfall time series or their elaborations, such as intensity–duration–frequency (IDF) curves, depth–duration–frequency (DDF) relations or simplified design hyetographs. This explains the necessity of high temporal resolution of point rainfall measurements in urban catchments. It should be noted that time series at high temporal resolution (1–10 min) with a considerable record length (at least 20–30 years) are nowadays required, especially from the European perspective with respect to the probabilistic assessment of the urban drainage network functioning (Schmitt, 2000; BS EN 752-3, 1997) or the probabilistic assessment of retention volumes at hydraulic-overloaded storm-water systems (Deutsche Vereinigung für Wasserwirtschaft, Abwasser und Abfall e.V., 2006).

The strategy of using local precipitation time series as the basis of the probabilistic assessment of urban drainage systems has two important shortcomings. In the case of local precipitation data shortage, this strategy fails completely. Whereas in all other situations when some local precipitation data sets are accessible, questions and doubts about the representativeness and reliability of data arise. First of all, we consider the doubts regarding the temporal representativeness of data: short data sets could not describe (Willems, 2013) the multi-decadal oscillatory behavior of rainfall extremes in storm-water outflow modeling. Other doubts regarding the spatial representativeness of data include the recording of rainfall time series only in a limited number of gauges installed in selected subcatchments. This results in assigning the same time series to a group of neighboring subcatchments or, in critical but not rare cases, one time series for the whole urban drainage system, habitually collected by a gauge installed near the airport. Sometimes in situations of local precipitation shortage, time series from other locations are allowed by technical guidelines (Schmitt, 2000) only if there is compatibility in terms of annual precipitation totals and IDF values.

Finally, since most of the modeling activity is oriented to predict the future behavior (e.g., in the next 50 years) of drainage systems, the mere use of historical precipitation time series of the last 20–30 years could not be significant to represent the future scenarios. Alternatively, the generation of synthetic time series from precipitation models could represent probable precipitation scenarios to feed hydrodynamic urban drainage models and take into account the uncertainty associated with the discharge. However, it should be pointed out that the information content of historical precipitation records is not increased by precipitation models and synthetic data generation, which just provide an operational basis for the extraction of such information.

Thus, there is a strong motivation for the development of local precipitation models at high temporal resolutions. Many of them are based on the idea of precipitation disaggregation in time. Disaggregation refers to a technique generating consistent rainfall time series at some desired fine timescale (e.g., 5 min resolution) starting from the precipitation at a coarser scale (e.g., daily resolution). At the same time, as stressed by Lombardo et al. (2012), the downscaling techniques aim at producing fine-scale rain time series with statistics consistent with those of observed data. A general overview of rainfall disaggregation methods is given by Koutsoyiannis (2003). Among an ensemble of known techniques, random cascade models, especially microcanonical cascade models (MCMs), are quite often used. The popularity of the latter could be explained by their appeal to engineering applications, the assumption of mass conservation (i.e., rainfall depth conservation) across cascade levels and straight rules for the extraction of cascade generators from local precipitation time series (Cârsteanu and Foufoula-Georgiou, 1996). Olsson (1998), Menabde and Sivapalan (2000), Ahrens (2003) and Paulson and Baxter (2007) provide contributions demonstrating the potentiality of MCMs in rainfall downscaling. Molnar and Burlando (2005) and Hingray and Ben Haha (2005) highlight the application of MCMs in urban hydrology. Hingray and Ben Haha (2005) apply a continuous hydrological simulation to produce from synthetic rainfall series continuous discharge series used afterwards for the retention design. Recently, Licznar (2013) illustrates the possibility of substituting synthetic time series generated from MCMs for observed time series of the probabilistic design of storm-water retention facilities.

Two decades of random cascade applications to precipitation disaggregation has progressed the construction of generators. Quite soon the assumption of independence and identical distribution of the cascade weight generators at all timescales was questioned and found suitable only for a limited, rather narrow range of analyzed scales (Olsson, 1998; Harris et al., 1998). As an alternative, Marshak et al. (1994), Menabde et al. (1997) and Harris et al. (1998) promote the use of the so-called "bounded" random cascade; its weights distribution systematically evolves, decreasing the weights variance with the reduction of timescale. In addition, Rupp et al. (2009) suggest that microcanonical cascade weights should not only be timescale-dependent but also intensity-dependent. The common practice of assuming the beta distribution for MCM generators is questioned by Licznar (2011a, b), especially for sub-hourly timescales. Alternatively, MCM generators are assumed normal–beta (N–B) distributed, with

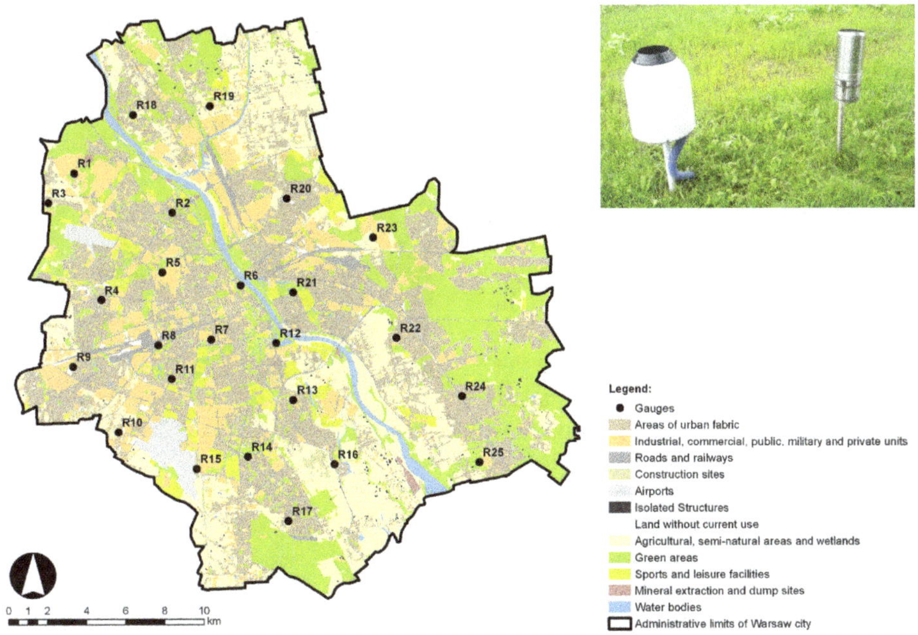

Figure 1. Map of 25 gauges composing the precipitation-monitoring network in Warsaw. Administrative limits of Warsaw city were marked in black. The land use classification was made with the Urban Atlas, which provides pan-European comparable land use and land cover data for large urban zones with more than 100 000 inhabitants (http://www.eea.europa.eu/data-and-maps/data/urban-atlas#tab-metadata). The average density of the network is one instrument for 20.7 km^2. MPS weighing-type TRwS 200E gauges were accompanied by standard Hellman gauges for the routine control of daily precipitation totals.

the atom at 0.5, or 3N–B distributed, composed of three N and one B distribution. For the sake of clarity, it should be stressed that B refers solely to the distribution of MCM generators and has nothing in common with the β model, the simplest cascade model often known as the monofractal model (for details refer to Over and Gupta, 1996).

Molnar and Burlando (2008) explore the variability of MCM generators on a large data set of 10 min time resolution, including 62 stations across Switzerland. These authors investigate seasonal and spatial variability in breakdown distributions to give indications concerning the parameters' estimation of MCM in ungauged locations. To our knowledge, there are only studies considering the large-scale variability (i.e., among different urban areas) of MCM generators, and there is a lack of knowledge concerning the small-scale variability (i.e., within an urban area).

It should be stressed that the fitting of cascade generators was relatively simple but extremely data-demanding. Observational precipitation time series in high resolution usually exceeding 20 years were unavoidable for the fit of the cascade parameters. This resulted in the prevailing practice of comparing the statistics of synthetic and observed time series. In the majority of studies, data originating from old-type manual gauges were subject to obvious uncertainty related to the precision of measurements as well as the resolution of records digitization. Simultaneously, the fitting of theoretical distributions to breakdown coefficients (BDCs) was, in almost all cases, not supported by statistical tests confirming the correctness of achieved results or by the use of some information criteria to rank the theoretical distributions.

Keeping in mind the above discussed needs of urban hydrology, the current state of MCMs and severe limitations of this rainfall disaggregation technique, the goals of our study were the following:

1. Propose a methodology to calibrate microcanonical cascade generators in conditions of short time series;

2. Identify the probability distribution of BDCs through the use of information criterion;

3. Investigate the variability of empirical BDC distributions among a group of gauges;

4. Address the following questions of interest in urban hydrology: is it sufficient to use a single time series for the probabilistic assessment of the entire urban drainage system? Is it sufficient to fit just one MCM for the analysis of the whole city area? Could we continue the practice of supplying urban rainfall-runoff models by time series recorded outside city center by gauges located at the airport or over rural areas?

2 Data and methodology

2.1 Data

We used data belonging to a precipitation network of 25 gauges distributed across 517.24 km^2 of Warsaw, Poland (Fig. 1). The data set was the same used by Rupp et al. (2012) and consisted of a 1 min precipitation (both liquid and solid) time series recorded by electronic weighing-type gauges. All stations, TRwS 200E of MPS system Ltd (Fig. 2), were installed and operated by the Municipal Water Supply and Sewerage Company (MWSSC) in Warsaw. Prior to the network installation, studies about the location of the stations were done by the MWSSC to identify the configuration most representative of the precipitation variability within the urban area (Oke, 2006). Finding good places for the installation of gauges was possible due to the fact that the MWSSC in Warsaw operates a vast number of local water intakes and water- and sewage-pumping stations. Due to sanitary standards, all these installations had to occupy terrain with green areas to serve as buffers, e.g., for the spread of odors. In addition, all facilities were fenced and guarded for safety reasons. Therefore all instruments were placed on grass and the neighborhood met at least the requirements of class 2 or 3 as recommended by the WMO (2012). The majority of gauges (i.e., R1, R3, R5, R7, R8, R10, R12, R17, R18 and R19) were able to be installed on flat, horizontal surfaces, surrounded by an open area, thus meeting even requirements for class 1 instruments. In addition, gauge R15 was installed in perfect conditions on the ground at the Warsaw Fryderyk Chopin Airport.

Since the installation of the precipitation network in Warsaw was mainly motivated by the real-time control of the drainage system, all gauges (Fig. 1) were connected to a single data acquisition system. The accuracy of gauge measurements as claimed by the manufacturer was 0.1 %, and the data resolution was 0.001 mm for depth and 1 min for time. As previously mentioned by Rupp et al. (2012), field tests conducted prior to the operational use of the precipitation network have shown good agreement between simulated and recorded totals and have revealed a dampening/broadening of the input signal evident over the range of a few minutes. The last phenomenon – known as a "step response error" – was studied in detail in laboratory conditions for different gauge types by Lanza et al. (2005). They found that the step error of TRwS gauge was quite small in comparison to other gauges and equal to 3 min in laboratory conditions. Our short 15 min field test (as displayed on Fig. 2) suggested a dampening of the gauge-recorded signal for the first 3 min initial phase of the generated hyetograph and the slightly longer 5 min broadening at the final phase of hyetograph. Detailed discussion of the origins of gauge step response errors is beyond the scope of this paper and is in fact hard to realize since it is introduced by the gauge's inner microprocessor algorithm for data processing. This algorithm is known only by the gauge manufacturer and is not reported in the technical documentation.

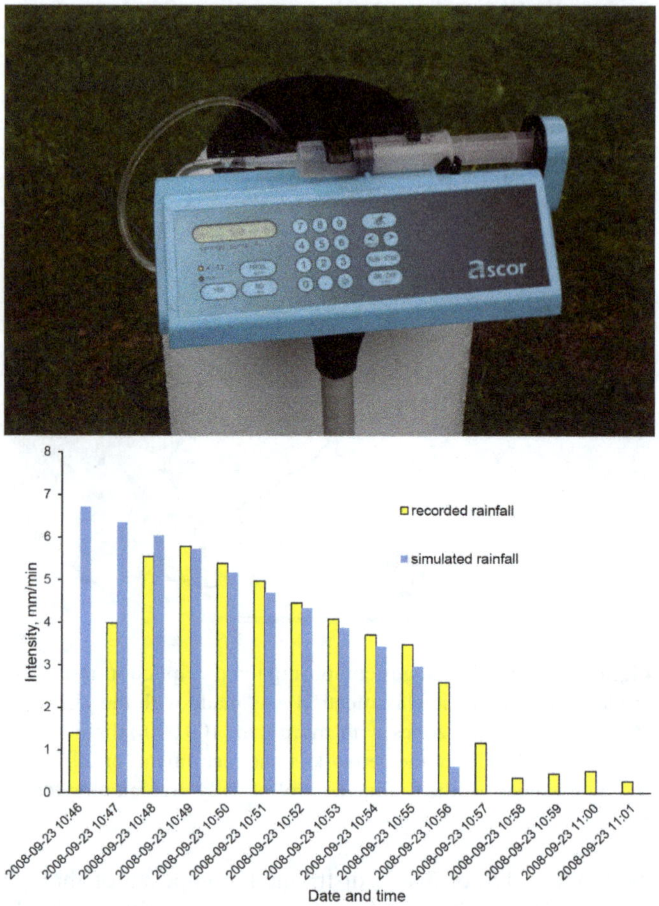

Figure 2. Weighing-type TRwS 200E gauge during some tests (upper panel). Rainfall is simulated by means of a precise medical pump. Sample of test results reporting simulated and recorded rainfall depths (lower panel).

In general, it could only be stated that in weighing-type electronic gauges, the weight of deposed precipitation was sampled by some electronic (often piezometric) sensor with some high temporal resolution at presumably kHz rate. Afterwards all samples were averaged over longer time windows, unknown to the user. This process was repeated for overlapping time windows, and the difference of the rainfall total of adjacent windows was calculated to obtain the temporal rainfall rate reported as instrument output at its recording time resolution. In addition, rainfall rates were always rounded regardless of the magnitude of real precipitation (resulting in additional rounding errors discussed afterwards). This procedure allowed for satisfactory smoothing of electronic sensor signal fluctuation due to wind effects and temperature changes. It allows for the introduction of some additional filters, cutting sudden signal jumps due to foreign-object deposition inside the open orifice of the gauge inner tank (e.g., falling leaves or acts of vandalism by throwing small stones or garbage).

Figure 3. Location of Polish and German precipitation gauges used during the comparison of Warsaw results with other studies.

In light of our personal experiences and test results of the WMO (Lanza et al., 2005), it could be stated that reliable precipitation recording at single minute scales by commercially available gauges is still the goal to be achieved and not a current reality. Having this in mind, as well as timescales of previous microcanonical cascade studies concerning urban hydrology realized on time series recorded by old-type gauges, we decided to work with the aggregated precipitation time series at 5 min resolution. The technique used to aggregate original 1 min data into 5 min time series is discussed afterwards; here we only mention that this operation was opposite to the rainfall total differentiation for adjacent time windows operated by the gauge microprocessor.

Despite the limited timespan of available data covering the period from the 38th week of 2008 to the 49th week of 2010, we believe that the Warsaw precipitation network might support good probing ground for the variability study in the microcanonical cascade parameters over small-scale urban areas. In fact, the Warsaw precipitation-monitoring network belongs to the biggest European urban gauge network. Its size can be compared only with similar networks of 25 gauges in Vienna (414.87 km^2) or with 24 gauges spread throughout Marseille (240.62 km^2) and Barcelona (100.4 km^2) (see TTT, 2010).

We compare the results of our study with those related to other Polish and German gauges. We limit our comparison to results previously published by Licznar et al. (2011a, b) for four gauges in Germany (gauges A, B, C and D represent local climates of different parts of western Germany) and one gauge in Wroclaw, Poland and yet unpublished results by Górski (2013) for a rain gauge in Kielce, Poland (Fig. 3). Our choice is motivated by the similarity of the used methodology and the investigated range of timescales, as well as by the indispensable accessibility to precise recordings of the BDC histograms.

Finally, to investigate the existence of possible statistical bias induced by the calculation of BDCs on short precipitation records, we use additional data recorded by an old-type pluviograph gauge installed previously at the current location of gauge R7 on the ground of Lindley's Filters station. This pluviograph gauge was operated only in summer months from 1 May to 31 October. Data are in the form of 15 min rainfall time series read off the original paper strips with the resolution of 0.1 mm for depth, covering a period of 25 years (1983–2007).

2.2 Microcanonical cascade models

We use MCMs as in Licznar et al. (2011a, b). We consider the disaggregation of precipitation totals from 1280 min (quasi daily) into 5 min time series, assuming the branching number b equal to 2, and constructing cascades assembled from only nine levels ($n = 8, \ldots, 1, 0$) corresponding to timescales $\lambda = 2^n$ from $\lambda = 256$ to $\lambda = 1$ (Fig. 4). Precipitation-depth time series generated by such cascades are the products of the original precipitation total R_0 at timescale $\lambda = 256$, multiplied by the sequence of weights at the descending cascade levels:

$$R_{j,k} = R_0 \prod_{i=1}^{k} W_{f(i,j),i}, \tag{1}$$

where $j = 1, 2, \ldots 2^k - 1, 2^k$ marks the position in the time series at the kth cascade step. The sequence of randomly

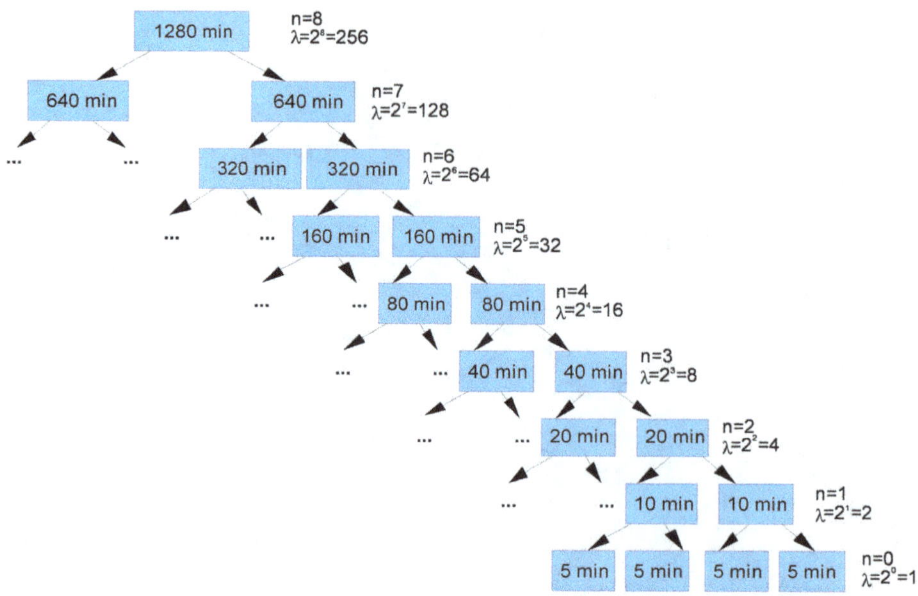

Figure 4. Schematic diagram of a developed microcanonical cascade model with branching number $b = 2$.

generated weights $W_{f(i,j),i}$ is steered at the following ith cascade step by the function $f(i, j)$, which rounds up $j/2^{k-i}$ to the closest integer. The weights in the microcanonical cascades are forced to sum to 1 so their pairs are always equal to W and $1-W$, where W is a two-sided truncated random variable from 0 to 1. The microcanonical assumption conserves the mass (precipitation depth in our case) at each branch and eliminates the risk of cascade degeneration. From an engineering perspective, this means that the downscaling process can be seen as opposite to precipitation summation realized by Hellman gauges, recording daily totals only, and a pragmatic solution for the generation of synthetic precipitation time series at 5 min resolution.

In our study we do not focus our attention on the disaggregation capabilities of microcanonical cascades already discussed in numerous papers. We concentrate on the small-scale variability of their generators W among gauges constituting the urban precipitation network. The obvious attractiveness of MCMs arises from the possibility of extracting the distribution of W from data on the basis of BDC studies (Cârsteanu and Foufoula-Georgiou, 1996). By definition, BDCs are generally calculated using non-overlapping adjacent pairs of precipitation time series:

$$\mathrm{BDC}_{j,\tau} = \frac{R_{j,\tau}}{R_{j,\tau} + R_{j+1,\tau}} \qquad j = 1, 3, 5, \ldots, N_\tau - 1, \tag{2}$$

where $R_{j,\tau}$ is the precipitation amount for the time interval of length τ at position j in the time series, and N_τ is the length of time series at timescale τ. The calculation of BDCs with respect to Eq. (2) for Warsaw gauges is conducted only for non-zero pairs of R_j and R_{j+1}. Calculations are executed at aggregated intervals of length $2^n \tau_{\mathrm{org}}$, where τ_{org} is the original time step equal to 5 min and n is a cascade level, increasing from 0 to 8 with increasing cascade timescales λ from 1 to 256 (Fig. 4). Simultaneously for all analyzed timescales, BDC couples equal to 0/1 or 1/0 (when only one between R_j and R_{j+1} is 0) are separated from resulting data sets and their occurrence probabilities, p_0(LEFT) and p_0(RIGHT), respectively, are used to estimate intermittency probability p_0:

$$\begin{aligned} &\Pr(\mathrm{BDC}_n(j) = 0 \text{ or } \mathrm{BDC}_n(j+1) = 0) \\ &= p_0(\mathrm{LEFT}) + p_0(\mathrm{RIGHT}) = p_0. \end{aligned} \tag{3}$$

The probability p_0 is used within a MCM generator to take into account the intermittency characteristic of precipitation, forcing some portion of random weights W to be equal to 0.

The preliminary results have revealed an over-representation of BDC values equal to 1/2 or 1/3, 2/5 and 1/4 (2/3, 3/5 and 3/4, respectively), especially for small timescales, i.e., λ =1 and λ =2. Figure 5 (left panel) shows an example of BDC histogram for timescale $\lambda = 1$ with evident artificial spikes. Similar phenomenon was already reported by Rupp et al. (2009) and Licznar et al. (2011b) and explained as the result of the instrument or the recording precision of precipitation gauges. The magnitude of observed rounding errors for Warsaw gauges is, however, smaller than in the case of German gauges (Licznar et al., 2011b); the precipitation depths recorded with a better resolution of 0.001 mm still result in irregularity of BDC distribution, induced by sharp peaks at discrete BDC values, and hinder the identification of the theoretical distribution. In order to correct the rounding errors, a randomization procedure originally proposed by Licznar et al. (2011b) is applied. This type of procedure, also known as jittering,

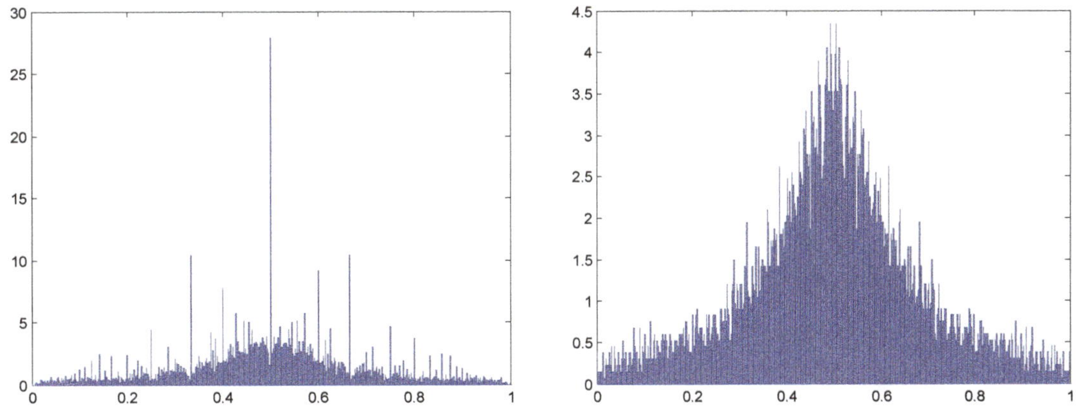

Figure 5. Comparison of BDC histograms for gauge R7 and timescale $\lambda = 1$, calculated according to the non-overlapping moving window algorithm and using original (left panel) and randomized (right panel) non-zero precipitation data. Horizontal axes show BDC range and vertical axes show the respective frequency values.

is fundamental to the analysis of data characterized by the presence of ties (De Michele et al., 2013). Thus, the original 1 min time series are slightly modified by adding some random corrections to the precipitation depths exceeding zero. Random correction values are sampled from the uniform distribution in the range $[-0.0005, 0.0005]$ mm, resulting in visible BDC histogram smoothing (Fig. 5, right panel). Note that the uniform distribution is used for the randomization of the rounding errors because, in the absence of information, it is the most intuitive distribution requiring less assumption (for more details, please see Licznar et al., 2011b).

Irregularities in BDC histograms are observed for timescales $\lambda > 8$. These are due to the decreasing sample size, calculated from a limited timespan of accessible data slightly exceeding 2 years. This issue is rather irrelevant in former studies (Molnar and Burlando 2005, 2008; Licznar et al., 2011a, b) realized on data series 10 or even 20 times longer. To solve this issue, we apply the overlapping moving window algorithm as an alternative to the classical non-overlapping moving window algorithm for the calculation of BDC values. Figure 6 shows the differences between the two algorithms for $\lambda = 1$. Switching from non-overlapping to overlapping moving window algorithm leads to an increase of the number of time segments for the calculation of BDC values. For time series of n data and a time window of size $m \leq n$, the number of non-overlapping windows is $\lfloor n/m \rfloor$, where the symbol $\lfloor \cdot \rfloor$ represents the integer part while the number of overlapping windows is $(n - m + 1)$. For large $n \gg m$, the overlapping moving window algorithm leads to almost m times the number of time segments available in the overlapping moving window algorithm. It should be stressed that the real strength of the overlapping moving window algorithm in analyzing distributions of BDC values can be observed for the largest timescales. The reason is that for small timescales most of the time segments are characterized by zero precipitation and thus not involved in the calculation of BDCs, whereas for larger timescales, time segments are becoming larger and rarely characterized by zero precipitation. This phenomenon arises from the fractal properties of rainfall time series, and similar conclusions result from the "box-counting" analysis.

It is clear that the overlapping moving window algorithm is especially desired for limited observational data sets. However, its implementation for short time series may be characterized by a poor representativeness of BDC distributions due to multi-decadal oscillations of precipitation totals and extremes (Willems, 2013). To investigate the magnitude of the oscillations in the BDC distributions, we use historical time series from former old-type gauge R7 covering a 25-year period from 1983 to 2007 at 15 min resolution. For each year, there are only 6 months of data from May to October available. For this data set, we make the calculations of BDCs in seven time periods. First, we calculate BDCs for the 5-year periods 1983–1987, 1988–1992, 1993–1997, 1998–2002 and 2003–2007 using the overlapping moving window algorithm. We consider this temporal size (5 years $\times$ 6 months $=$ 30 months) because it is comparable to the one available for electronic gauges. Afterwards, we repeat the same calculation with a 25-year size using both non-overlapping and overlapping moving window algorithms. As we work here with a coarser resolution (15 min instead of 5 min of electronic gauges), we perform the analysis with a smaller hierarchy of sub-daily timescales λ' from 1 to 32 and breakdown times from 15 to 30 min up to 480 to 960 min. For all calculations we perform the randomization of non-zero values. Since their reading precision was set to 0.1 mm, we introduce a random correction belonging to the uniform distribution in the range of $[-0.05, 0.05]$ mm.

To compare BDC histograms obtained for all analyzed timescales λ and λ', with theoretical functions, a probability distribution assembling two truncated (with truncation points at 0 and 1) N distributions (Robert, 1995) and one B

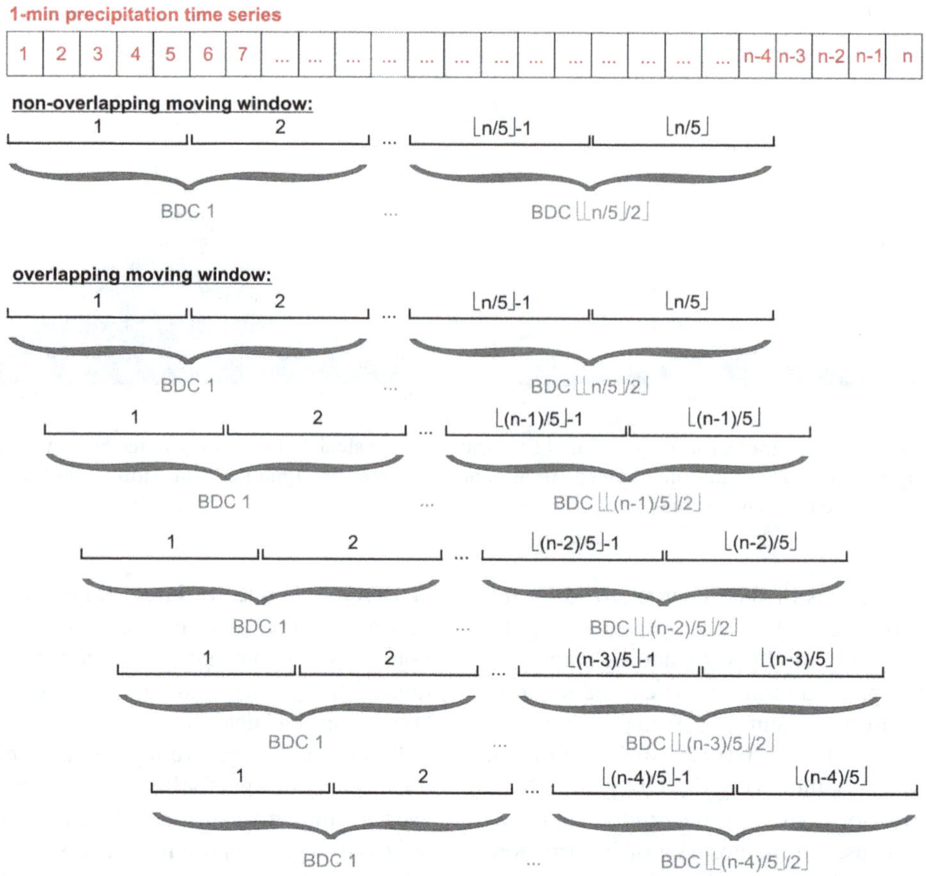

Figure 6. Example showing differences between non-overlapping and overlapping moving window algorithms for the calculation of BDCs in the case of 1 min precipitation time series and breakdown time of 5–10 min. Note that $\lfloor n \rfloor$ means the integer part of n, where n is the total length of 1 min precipitation time series.

symmetrical distribution is implemented. This distribution, indicated as 2N–B distribution, has the following density function:

$$p(w) = p_1 \left\{ \frac{1}{\sigma_1\sqrt{2\pi}} e^{\frac{-(w-0.5)^2}{2\sigma_1^2}} \right\} + (1-p_1)\left\{ p_2 \left\{ \frac{1}{B(a)} w^{a-1}(1-w)^{a-1} \right\} + (1-p_2)\left\{ \frac{1}{\sigma_2\sqrt{2\pi}} e^{\frac{-(w-0.5)^2}{2\sigma_2^2}} \right\} \right\}, \tag{4}$$

where p_1 and p_2 are weights characterizing the contribution of the individual distributions within the 2N–B distribution, σ_1 and σ_2 are the scale parameters of truncated N distributions and $B(a)$ is the symmetrical B function parameterized by a.

The fitting of 2N–B distribution parameters is performed numerically by means of maximum likelihood estimation. It is very likely that the use of the model given in Eq. (4), governed by five parameters, can suffer from overparameterization in comparison to the most commonly used B symmetrical distribution with only one parameter. Note that the application of goodness-of-fit tests (namely the Kolmogorov–Smirnov test or χ^2 test) at 1 or 5 % levels of significance gives negative results for both B and 2N–B distributions. This is because the large sample size of empirical BDCs leads to the rejection of the hypothesis, even in the case of very small differences between observed and theoretical distributions, as also pointed out in Licznar et al. (2011a). Here, we use the Akaike information criterion (AIC) as a measure of the relative quality of 2N–B and B models for given sets of empirical BDCs. AIC is the maximized value of the log-likelihood function (LL) penalized by the number of model parameters k:

$$\mathrm{AIC} = 2k - 2\mathrm{LL} \tag{5}$$

The preferred distribution is the one with the minimum value of AIC.

2.3 Cluster analysis

To our knowledge, until now, the variability of MCM generators among a group of gauges was investigated comparing the value of the parameter of B distribution (Molnar and Burlando, 2008). Here, we prefer to compare directly the empirical distribution of BDCs instead of the parameters of the theoretical distribution, which are possibly biased by fitting errors. We encounter the same problems found in the implementation of statistical tests due to the large sample size. For this we use the cluster analysis to compare the shape of BDC histograms among the stations of the monitoring network in Warsaw and with other Polish and German gauges.

In particular, a hierarchical clustering is used. This is a data-mining tool applied to segment data into relatively homogeneous subgroups, or clusters, where the similarity of the records within the cluster is maximized (Larose, 2005). Prior to the application of the cluster analysis for each timescale and each site, the BDC histogram is sampled in 100 points selected at equal distance one from each other. These 100 values are the components of a vector representing the empirical BDC distribution. Note that a basic requirement of cluster analysis is the comparison of records of equal length. As all BDC distributions are left and right truncated in the interval (0, 1), sampling their histograms with a resolution of 0.01 produces vectors that describe well the shape of histograms. The clustering of these vectors (searching similar sites) is operated using the Euclidean distance. It is computed as

$$d_{\text{Euclidean}}(\boldsymbol{X}, \boldsymbol{Y}) = \sqrt{\sum_i (x_i - y_i)^2}, \tag{6}$$

where x_i and y_i with $i = 1, \ldots, 100$ represent the ith component of $\boldsymbol{X}$ and $\boldsymbol{Y}$ vectors, respectively.

The Euclidean distance is a measure of similarity, not having, in general, a physical interpretation. Initially, in hierarchical clustering analysis, each vector is considered to be a tiny cluster of its own. Then, in following steps, the two closest clusters are aggregated into a new combined cluster. By replication of this operation, the number of clusters is reduced by one at each step and eventually sites are combined into a single huge cluster. During the agglomerative process, the distance between clusters is determined based on single-linkage criterion. In this case, the distance between clusters A and B is defined as the minimum distance between any element in cluster A and any element in cluster B. This single linkage is often termed the nearest-neighbor approach and tends to form long, slender clusters, clearly indicating similarities among clustered elements. As a final result of agglomerative clustering, a treelike cluster structure (named dendrogram) is created.

Dendrograms show similarities as well as dissimilarities between BDC distributions among the considered sites and they are prepared separately for all analyzed timescales. In addition, the cluster analysis is also applied to the intermittency parameter, in this case comparing vectors of eight components: the p_0 value for the eight timescales $\lambda = 1, 2, 4, 8, 16, 32, 64, 128$.

3 Results and Discussion

Results are presented relative to gauge R7 for brevity. This station has been selected because of its localization in the strict city center, its installation in perfect meteorological conditions on the ground and the existence of former historical rainfall records. Results for the other gauges are qualitatively similar to those shown for R7.

3.1 Empirical BDC distributions

BDC histograms are calculated using the non-overlapping moving window algorithm and plotted in Fig. 7 for gauge R7 and a sequence of analyzed breakdown times. It is clearly visible that, despite the randomization procedure that removes pronounced peaks of histograms at certain specific BDC values, like 0.5 or $1/3, 2/5, 1/4$ and $2/3, 3/5, 3/4$, respectively (Fig. 5), the plots remain irregular, especially for timescales exceeding $\lambda = 8$, reducing the possibility of identifying the proper theoretical distribution. Visible irregularities of BDC histograms increase with increasing timescales, which is an obvious effect of decreasing data sets and thus decreasing populations of calculated BDC values do not allow the production of histograms of fine bins resolution (referring to the populations). Similarly, Fig. 8 reports the distributions of BDC calculated through the overlapping moving window algorithm. The comparison between Figs. 7 and 8 shows how the change of algorithm from a non-overlapping to an overlapping moving window brings evident smoothing of BDC histograms, occurring not only at larger timescales but also at small timescales. Note that the smoothness of BDC histograms in Fig. 8 is comparable with the quality of BDC histograms showed by Licznar et al. (2011b) for German gauges, derived using non-overlapping moving window algorithm for much longer precipitation time series ranging from 27 to 46 years of continuous records. The introduction of the overlapping moving window algorithm allows for the fitting of MCM parameters with the availability of extremely short time series (i.e., 2 years long) in the case of Warsaw gauges. The overall acceptance of overlapping moving window algorithm implementation, including for short rainfall time series, is discussed in Sect. 3.3.

3.2 Theoretical BDC distributions and their evolution along timescales

In Fig. 8 we also report the fitted theoretical distributions (2N–B distribution in solid red curves and B distribution in blue dashed lines) for each timescale considered. The visual comparison clearly indicates a better fit of 2N–B (or N–B in

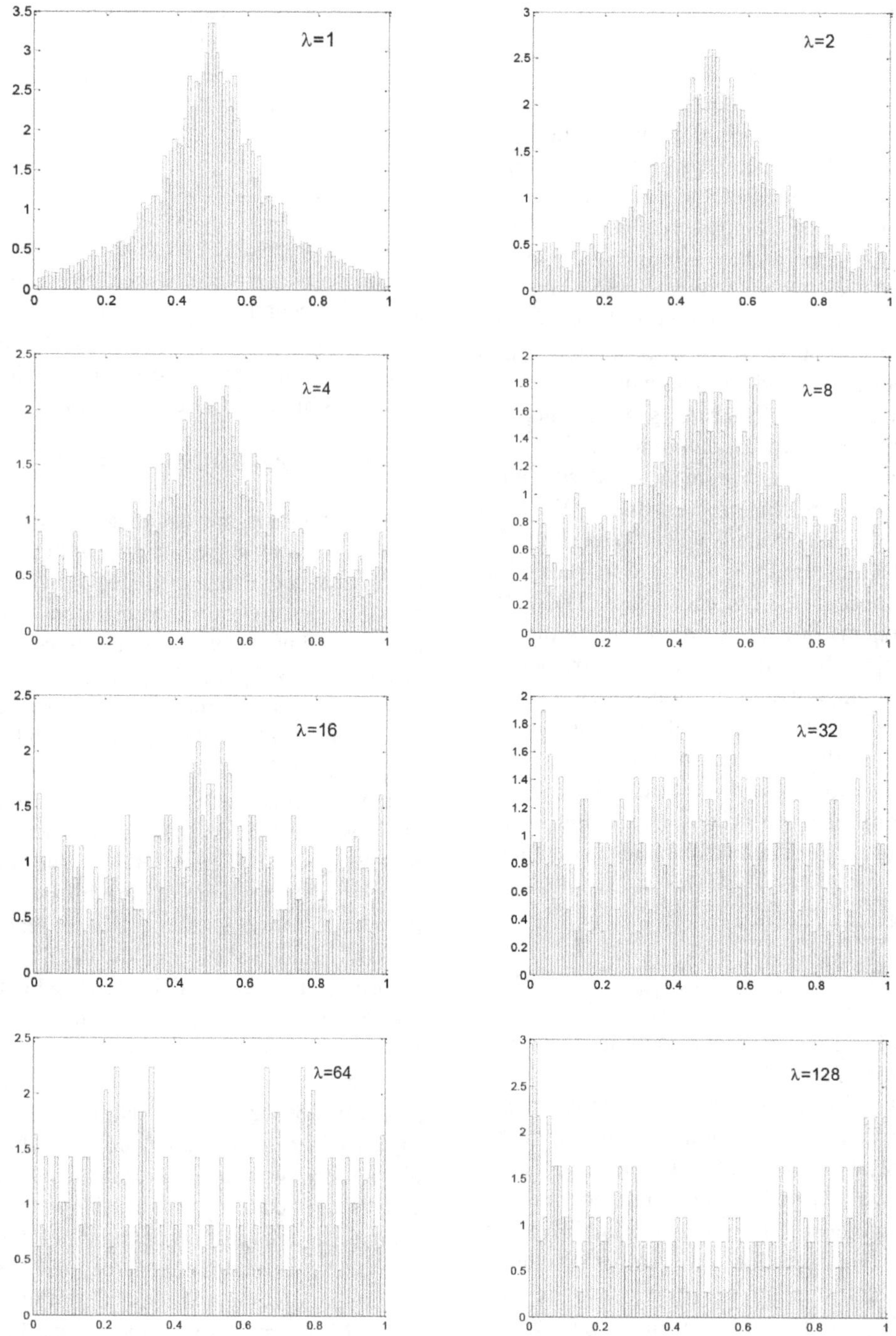

Figure 7. Histograms of BDC values for gauge R7 calculated according to the non-overlapping moving window algorithm and based on randomized precipitation time series. Horizontal axes show BDC range and vertical axes show the respective frequency values.

some cases) distribution for timescales smaller than $\lambda = 64$. In Fig. 8 it is possible to see how the distribution with the best fit changes from a B distribution at $\lambda = 128$ to a joined 2N–B for the smallest value of λ through a N–B distributions. This is in agreement with previous studies by Licznar et al. (2011a, b). This observation is supported by higher values of log-likelihood for 2N–B distribution (or the simplified N–B) in comparison to the B distribution (Table 2). These differences are in the range of thousands, and even after accounting for the number of model parameters, the AIC for 2N–B (or the simplified N–B) distributions are much smaller (or equal) than that of B distributions, confirming the visual result given in Fig. 8. Based on this, we prefer the 2N–B distribution with respect to the B distribution except for the case

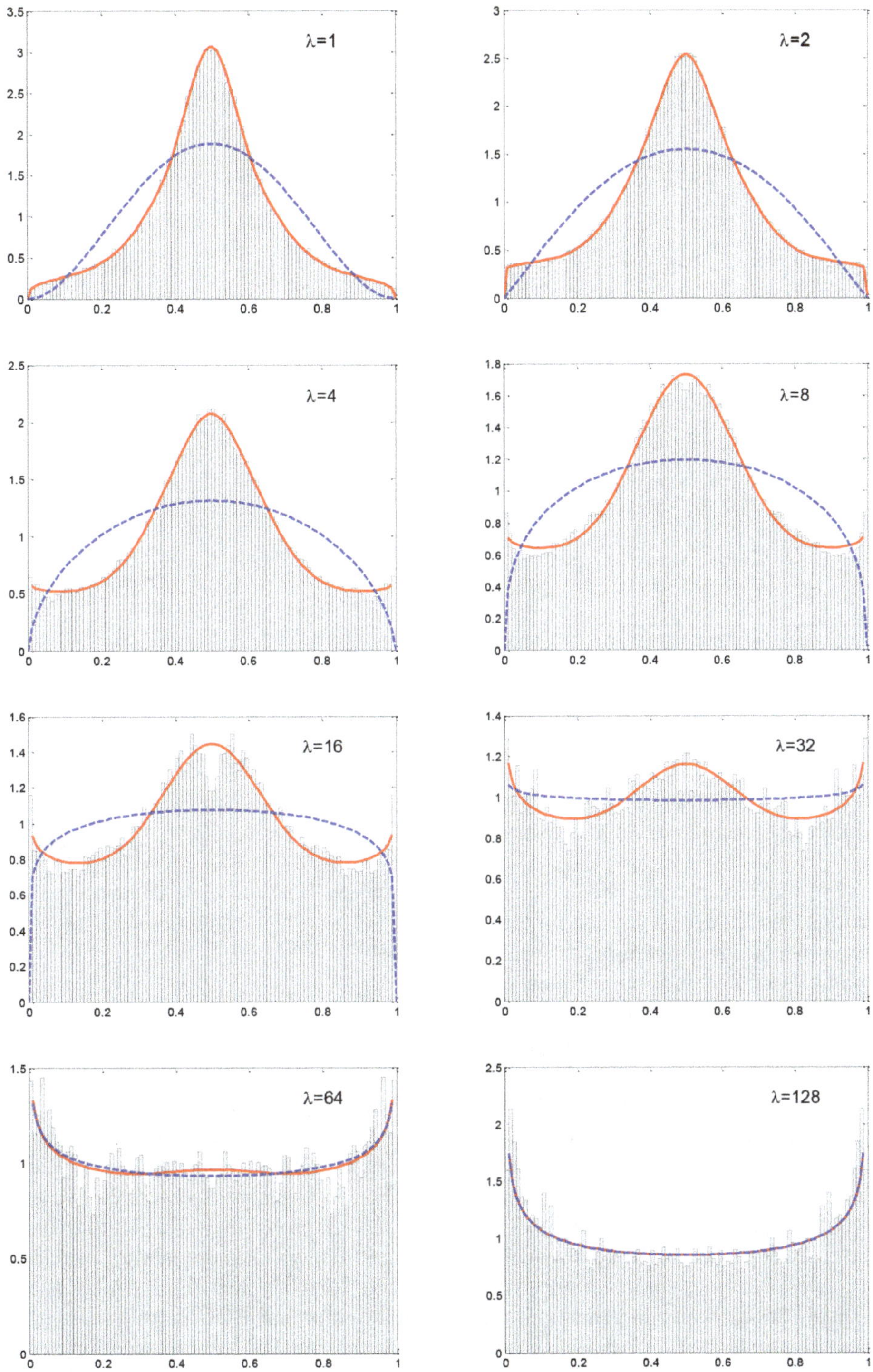

Figure 8. Histograms of BDC values calculated according to overlapping moving window algorithm and based on randomized gauge R7 precipitation times series. Horizontal axes show BDC range and vertical axes show the respective frequency values. The solid red curves represent the 2N–B probability density function, whereas the blue dashed curves represent the B probability density function.

Table 1. Values of p_1, p_2, a, σ_1 and σ_2 parameters at different timescales for gauge R7. The values of parameters are reported in bold, whereas their 95 % confidence limits are in italic.

Breakdown times	Timescale	p_1	p_2	a	σ_1	σ_2
		0.1541	**0.3479**	**1.3350**	**0.0559**	**0.1341**
5–10 min	$\lambda = 1$	*0.1474*	*0.3377*	*1.3097*	*0.0523*	*0.1300*
		0.1608	*0.3580*	*1.3604*	*0.0595*	*0.1383*
		0.0706	**0.4036**	**1.0632**	**0.0559**	**0.1341**
10–20 min	$\lambda = 2$	*0.0644*	*0.3950*	*1.0474*	*0.0523*	*0.1300*
		0.0768	*0.4121*	*1.0789*	*0.0595*	*0.1383*
		0.0212	**0.5036**	**0.9437**	**0.0559**	**0.1341**
20–40 min	$\lambda = 4$	*0.0155*	*0.4954*	*0.9325*	*0.0523*	*0.1300*
		0.0270	*0.5118*	*0.9548*	*0.0595*	*0.1383*
		–	**0.6175**	**0.9484**	–	**0.1341**
40–80 min	$\lambda = 8$	–	*0.6091*	*0.9390*	–	*0.1300*
		–	*0.6259*	*0.9579*	–	*0.1383*
		–	**0.7548**	**0.9170**	–	**0.1341**
80–160 min	$\lambda = 16$	–	*0.7494*	*0.9098*	–	*0.1300*
		–	*0.7601*	*0.9242*	–	*0.1383*
		–	**0.8873**	**0.8929**	–	**0.1341**
160–320 min	$\lambda = 32$	–	*0.8827*	*0.8873*	–	*0.1300*
		–	*0.8919*	*0.8985*	–	*0.1383*
		–	**0.9797**	**0.8799**	–	**0.1341**
320–640 min	$\lambda = 64$	–	*0.9758*	*0.8754*	–	*0.1300*
		–	*0.9835*	*0.8843*	–	*0.1383*
		–	**1.0000**	**0.7783**	–	**0.1341**
640–1280 min	$\lambda = 128$	–	*0.9973*	*0.7754*	–	*0.1300*
		–	*1.0027*	*0.7813*	–	*0.1383*

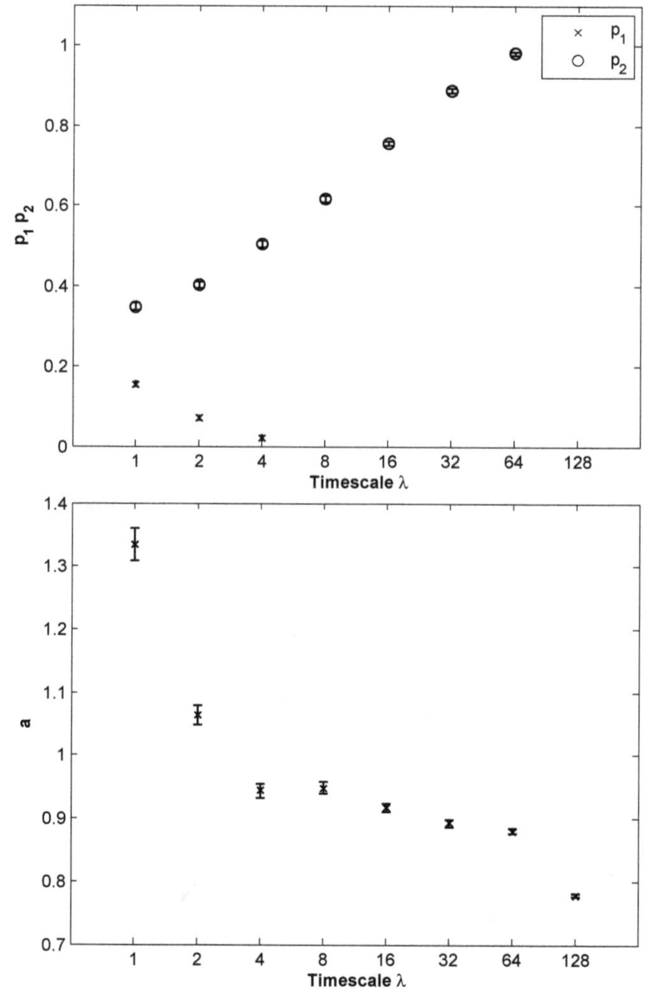

Figure 9. Value and 95 % confidence intervals of parameters of p_1, p_2 and a with λ for gauge R7. Horizontal axes are plotted at binary logarithm scale $\log_2$.

$\lambda = 128$. Analyzing the data reported in Table 2, it is worth noting the systematic increase of sample size n increasing the timescale.

From a practical point of view, a rapid increase in the number of BDCs, equal or close to 0.5, decreasing the timescale should be expected as a symptom of enclosing a limit of the precipitation temporal variability in a point by accessible instruments. The precipitation averaging over some small area of orifice and time intervals is inevitable for gauges; thus, for small timescales, most of the small-scale precipitation variability remains undetected and smoothed, leading to an over-representation of constant precipitation time intervals. From a theoretical point of view, it should be noted that bounded cascades allow the multiplicative weights (or precisely their distributions) to depend on the cascade level and converge to unity as the cascade proceeds. As a consequence, the simulated random process becomes smoother on smaller timescales (Lombardo et al., 2012), which in general mimics the dynamics of precipitation collected by gauges. In other words, as postulated by Marshak et al. (1994), Menabde et al. (1997) and Harris et al. (1998), the variance of weights reduces with every descending cascade level. As a simple extension of this rule, the increasing frequency of weights at the central part of their distribution plots has to be observed. The increase in the number of BDCs equal or close to 0.5 with decreasing timescale is well illustrated by empirical histograms in well-known pioneering contributions to MCM applications for rainfall time series disaggregation published by Olsson (1998), Menabde and Sivapalan (2000) and Güntner et al. (2001). Quite recently, this behavior was also proved to be rainfall-intensity-dependent by Rupp et al. (2009).

For each analyzed timescale, we estimate the parameters of 2N–B probability distribution (or its simplifications N–B and B): p_1, p_2, a, σ_1 and σ_2. Table 1 gives the values for gauge R7 with their 95 % confidence limits. A good visual fit of empirical BDC distributions in Fig. 8 corresponds to quite narrow 95 % confidence limits of the fitted parameters (mostly invisible in Fig. 9 plots). The 95 % confidence limits do not exceed more than a few percent of the estimated values, with the sole exception of parameter p_1 for $\lambda = 4$, where the differences range up to 27 %. Additionally, the scale parameters of N distributions, σ_1 and σ_2, appear to be constant among analyzed timescales not only for gauge R7 but also for the other Warsaw gauges.

Table 2. Values of the Akaike information criterion (AIC) for the 2N–B distribution (model 1) – or its simplifications, N–B and B – and the B distribution (model 2) and the hierarchy of analyzed timescales λ at gauge R7. Calculations were based on estimates of the maximized value of the log-likelihood function (LL), known sample size (n) and number of model parameters (k).

Breakdown times	Timescale	n	Model 1 Distr.	k	LL	AIC(M1)	Model 2 Distr.	k	LL	AIC(M2)	Δ =AIC(M2) – AIC(M1)
5–10 min	$\lambda = 1$	132 940	2N–B	5	48 480	−96 950	B	1	36 307	−72 612	24 338
10–20 min	$\lambda = 2$	136 968	2N–B	5	32 272	−64 534	B	1	19 798	−39 593	24 941
20–40 min	$\lambda = 4$	144 778	2N–B	5	19 071	−38 132	B	1	8794	−17 585	20 547
40–80 min	$\lambda = 8$	159 272	N–B	3	11 119	−22 232	B	1	4464	−8927	13 305
80–160 min	$\lambda = 16$	185 014	N–B	3	4591.9	−9178	B	1	925	−1848	7330
160–320 min	$\lambda = 32$	230 716	N–B	3	1167.3	−2329	B	1	46	−91	2238
320–640 min	$\lambda = 64$	315 360	N–B	3	1543.70	−3081	B	1	1491	−2979	102
640–1280 min	$\lambda = 128$	501 092	B	1	12 614.40	−25 227	B	1	12 614	−25 227	0

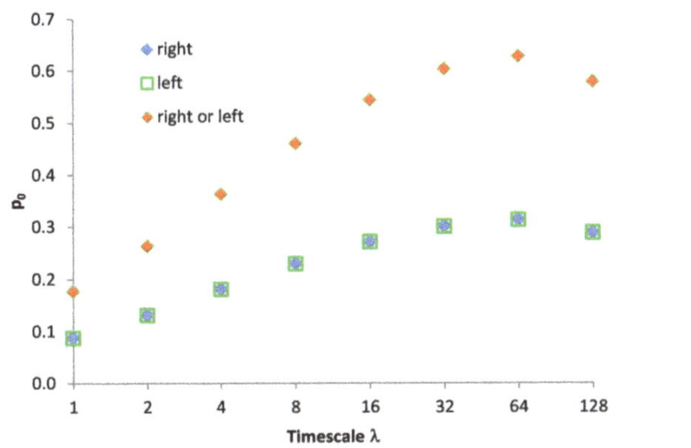

Figure 10. Variability of the intermittency parameter p_0 with λ for gauge R7. Horizontal axis is plotted at binary logarithm scale $\log_2$.

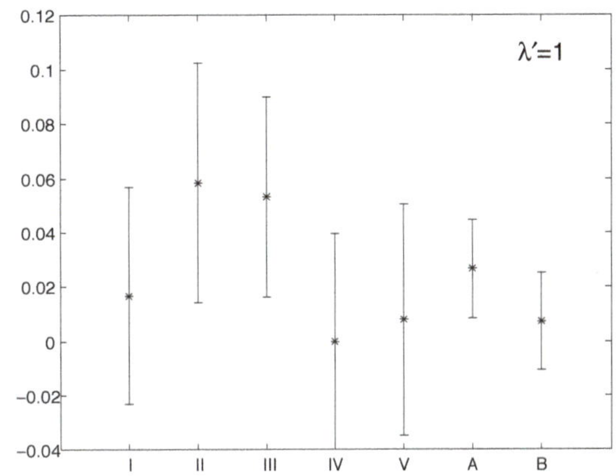

Figure 11. Value and 95 % confidence intervals of parameter p_1 at timescale $\lambda' = 1$ for gauge R7. Roman numerals I–V on horizontal axes indicate the 5-year ranges: 1983–1987, 1988–1992, 1993–1997, 1998–2002 and 2003–2007. Uppercase letters A and B indicate values calculated using the entire 25-year range of 1983–2007 and non-overlapping (A) and overlapping (B) moving window algorithm.

The variability of p_1, p_2 and a with λ is presented in Fig. 9 for gauge R7. A systematic decrease of p_1 down to 0 increasing the timescale is observed, denoting a decreasing importance of the first N within the 2N–B distribution. An opposite systematic increase of p_2 up to 1 increasing the timescale is observed, denoting a decreasing importance of the second N within the 2N–B distribution. The evolution of the B parameter a shows a fast reduction, with values below 1 noticed for the smallest scales, changing the B distribution shape from convex to concave. At larger timescales the reduction of a is hardly visible, with the sole exception of $\lambda = 128$. Figure 10 shows the variability of intermittency parameters p_0 with timescale λ. For all of them, the values of p_0(LEFT) match the values of p_0(RIGHT), which is in good agreement with previous studies by Molnar and Burlando (2005) and Licznar et al. (2011a, b). This could be interpreted as the proof of the fully random occurrence of intermittency in the precipitation time series. A systematical increase of p_0 with λ is observed, with the sole exception of a small drop at $\lambda = 128$. General increase of p_0 with timescale is a natural outcome of the fractal properties of the geometric support of rainfall occurrence.

3.3 Performance of the overlapping moving window algorithm

The performance of the overlapping moving window algorithm was investigated in detail at gauge R7, where a 25-year time series at 15 min resolution was available. We calculated the parameters of 2N–B distribution for the hierarchy of subdaily timescales λ' relative to the 5-year periods of 1983–1987, 1988–1992, 1993–1997, 1998–2002 and 2003–2007 (indicated afterwards with the Roman numerals I, II,..., V, respectively) and the whole 25-year data set (indicated in the next as case A) using the overlapping moving window algorithm. In addition, we calculate the parameters of 2N–B

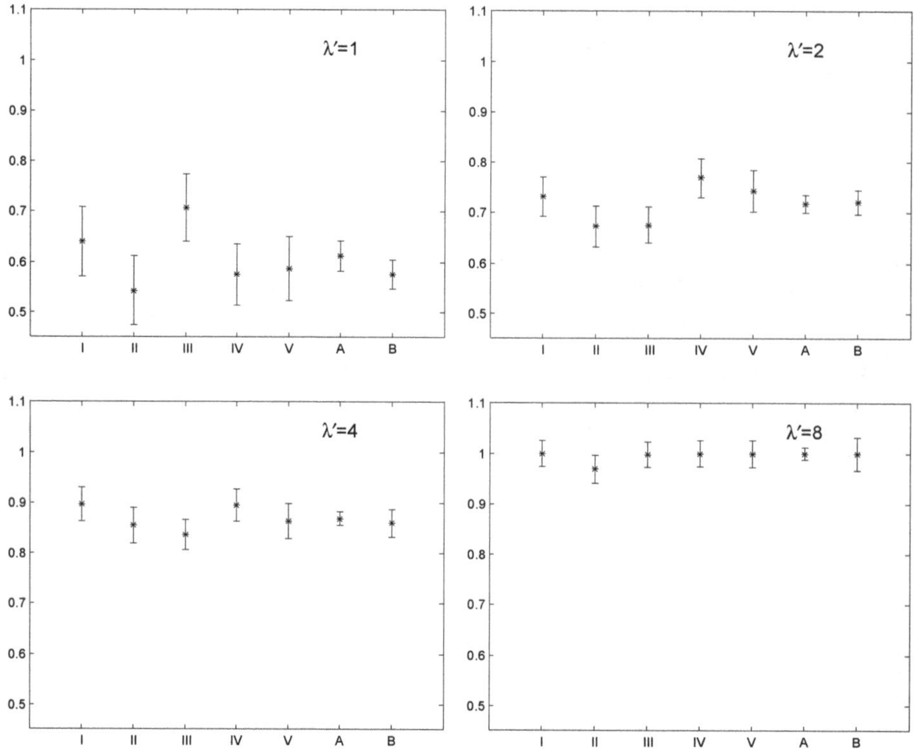

Figure 12. Value and 95 % confidence intervals of parameter p_2 at timescales $\lambda' = 1, 2, 4, 8$ for gauge R7. Roman numerals I–V on horizontal axes indicate respectively the 5-year ranges: 1983–1987, 1988–1992, 1993–1997, 1998–2002 and 2003–2007. Uppercase letters A and B indicate values calculated using the entire 25-year range of 1983–2007 and non-overlapping (A) and overlapping (B) moving window algorithm.

distribution also using the classical non-overlapping moving window algorithm over the whole 25-year data set (indicated in the next as case B). The results are shown in Figs. 11–13.

In general, the selected probability distribution is a B distribution for the largest timescales ($\lambda' = 16, 32$), a N–B distribution for $\lambda' = 2, 4, 8$ and a 2N–B distribution for $\lambda' = 1$ (the only exception is the period IV). The above listed timescales λ' are not compatible with timescales λ; however, transposing them on a coherent time axis leads to the conclusion that characteristic transitions from B to N–B and 2N–B distributions occur at approximately the same time ranges. The estimated parameters σ_1 and σ_2 appear to be constant among analyzed timescales and equal to 0.0646 and 0.1363, respectively. These values are very close to those reported in Table 1. Figure 11 shows the estimates of p_1 for $\lambda' = 1$, with a variability in the range 0–0.058 for the 5-year periods I–V. At the same time, the 95 % confidence limits of p_1 overlap each other partially and values estimated for cases A and B. Confidence limits for periods I–V are rather wide and are only reduced by 50 % for cases A and B. Note that here we work with 15 min time series and not 1 min time series as before.

A better agreement is observed for larger timescales, as illustrated in Figs. 12 and 13, with visibly narrow 95 % confidence limits; however, they still partial overlap one another. For smaller timescales, larger oscillations of p_2 parameter can be observed over the periods I–V but, due to wider 95 % confidence limits, they overlap one another and those relative to cases A and B. The only exception is found for the period III at timescale $\lambda' = 1$.

For parameter a and $\lambda' = 1$, 95 % confidence limits for all calculations overlap except period V, which has slightly lower values. For $\lambda' = 2$ and $\lambda' = 4$, mutual overlay of 95 % confidence limits is noticed. Passing to $\lambda' = 8$ and $\lambda' = 16$, the overlapping among all pairs of periods from I to V is not always present; however, it is present with 95 % confidence limits drawn for case B. For $\lambda' = 32$, 95 % confidence limits for periods I–V and case A are extremely narrow.

Results reported above suggest good repeatability of BDC distributions calculated during all periods; this is graphically confirmed in Fig. 14, with the exception of period II and timescale $\lambda' = 1$. Probably this could be explained by the poor performance of the newly proposed overlapping moving window algorithm applied to low time resolution of the original time series. Our observations support the use of the overlapping moving window algorithm for BDC calculations in situations of short (about 2 years) precipitation time series access, while in previous microcanonical cascade studies (e.g., Molnar and Burlando, 2005, 2008), longer (e.g., about 20–30 years) time series are indispensable. In addi-

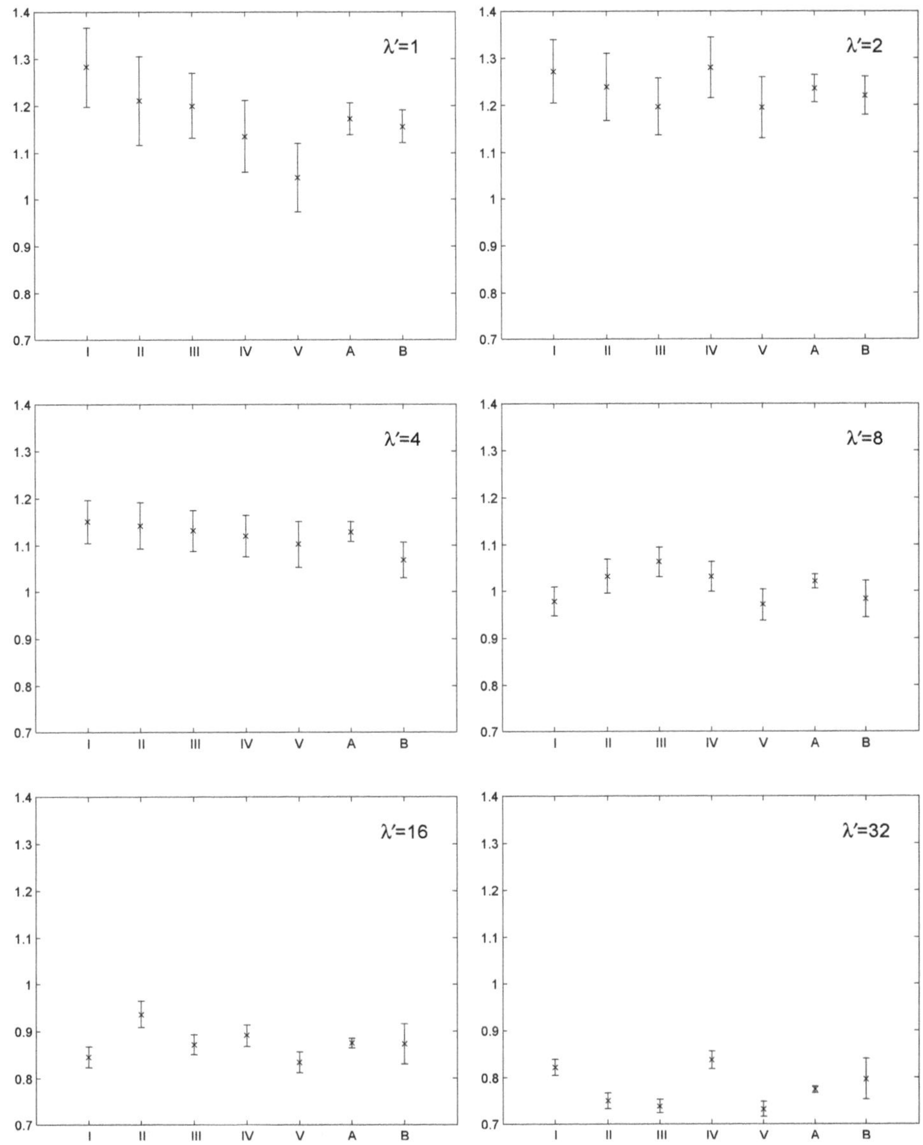

Figure 13. Value and 95 % confidence intervals of parameter a at timescales $\lambda' = 1, 2, 4, 8, 16, 32$ for gauge R7. Roman numerals I–IV on horizontal axes indicate the 5-year ranges: 1983–1987, 1988–1992, 1993–1997, 1998–2002 and 2003–2007. Uppercase letters A and B indicate values calculated using all 25-year range of 1983–2007 and non-overlapping (A) and overlapping (B) moving window algorithm.

tion, even in situations of longer precipitation time series access, BDC calculations by means of the proposed algorithm should be favored over the old non-overlapping moving windowtechnique because the new algorithm leads to narrowed 95 % confidence intervals of fitted BDC distributions parameters.

We do not claim here that the moving window technique combined with MCMs solves the problem of local precipitation time series shortage. It is obvious that rainfall statistics derived from short periods may be biased against long-term statistics (e.g., due to climate oscillations). Until now, to our best knowledge, there have been no attempts made to assess the possible bias of MCM generators due to precipitation oscillations driven by climate change. Hitherto contributions of MCM generators are mostly based on precipitation series that are not too long, presumably displaying only very weak, if any, oscillations and are always treated as a single data set.

Possible bias of MCM generators due to precipitation oscillations undoubtedly should be verified on other, much longer time series of better resolution, such as the 10 min time series collected in Uccle, Belgium (Willems, 2013). Simultaneously, only a detailed analysis based on long and complete precipitation time series covering at least a few decades could deliver us the answer to the question of whether the climate change effects could be retrieved via the temporal evaluation of microcanonical cascade generators.

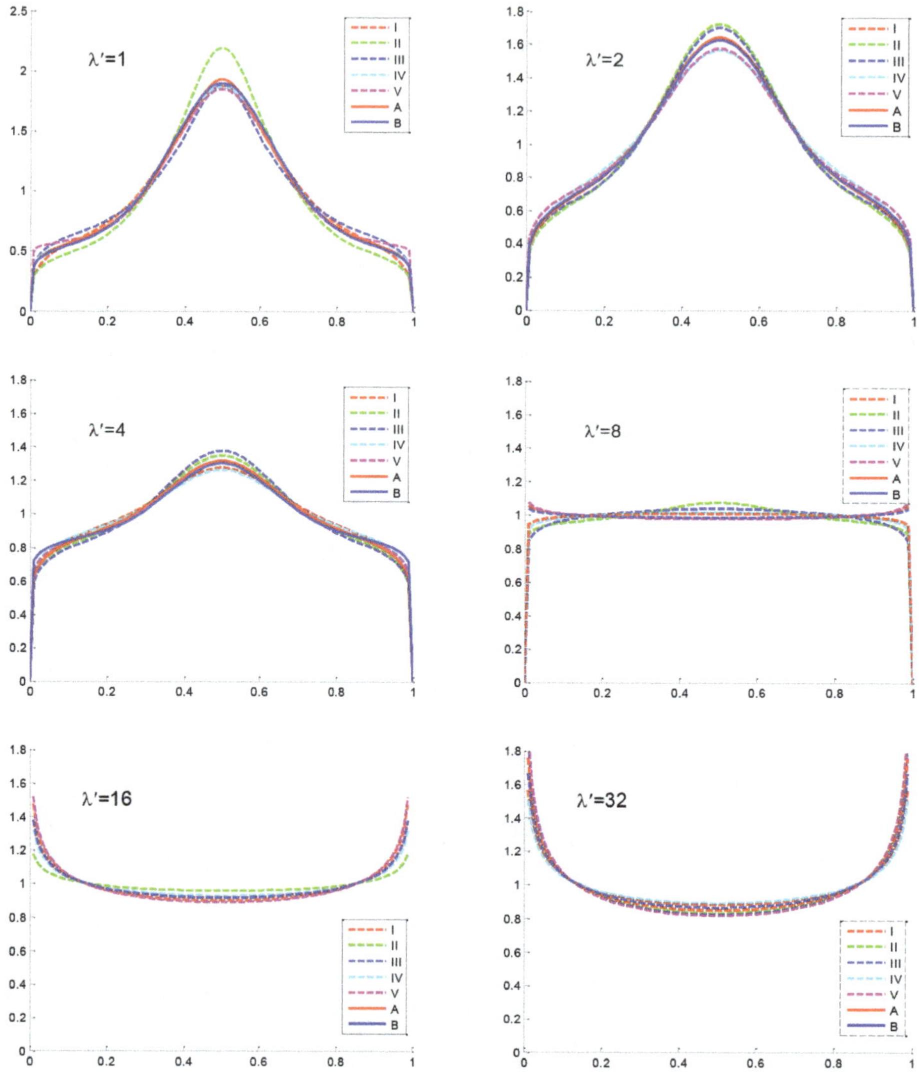

Figure 14. Variability of fitted theoretical BDC distributions histograms at timescales $\lambda' = 1, 2, 4, 8, 16, 32$ for gauge R7. Roman numerals I–V in legend indicate the 5-year ranges: 1983–1987, 1988–1992, 1993–1997, 1998–2002 and 2003–2007. Uppercase letters A and B indicate results calculated using all 25-year range of 1983–2007 and non-overlapping (A) and overlapping (B) moving window algorithm. In all plots, horizontal axes show BDC ranges and vertical axes show the frequency values.

From this perspective, the moving window technique could be of considerable usefulness in BDC distributions fitting for periods corresponding to 11-year solar spot cycles.

3.4 Performance of microcanonical cascade in disaggregation

As an additional check of the overall performance of the applied techniques (i.e., the randomization procedure, the overlapping moving window algorithm and the 2N–B probability distribution), we test the performance of microcanonical cascade in disaggregating the precipitation at the analyzed gauges. The MCM is used to generate 100 synthetic time series at 5 min resolution on the basis of the observed 1280 min precipitation totals (similar to Molnar and Burlando, 2005; Licznar et al., 2011a, b). To evaluate the goodness of disaggregation, we compare the probability of zero precipitation at synthetic and observed time series for all analyzed timescales. Moreover, we calculate the survival probability function of non-zero synthetic precipitation amounts and compare it to the survival probability function of observed precipitation amounts. This analysis is limited to 5 min data, i.e., terminal results of the disaggregation most suitable for urban hydrology application. Special attention to the 5 min synthetic time series was also paid by other researchers (see e.g., Molnar and Burlando, 2005, 2008; Licznar et al., 2011a, b). An example of 56.3 mm event disaggregation is plotted in Fig. 15 for gauge R7. It should be stressed that the structure of the synthetic time series is composed

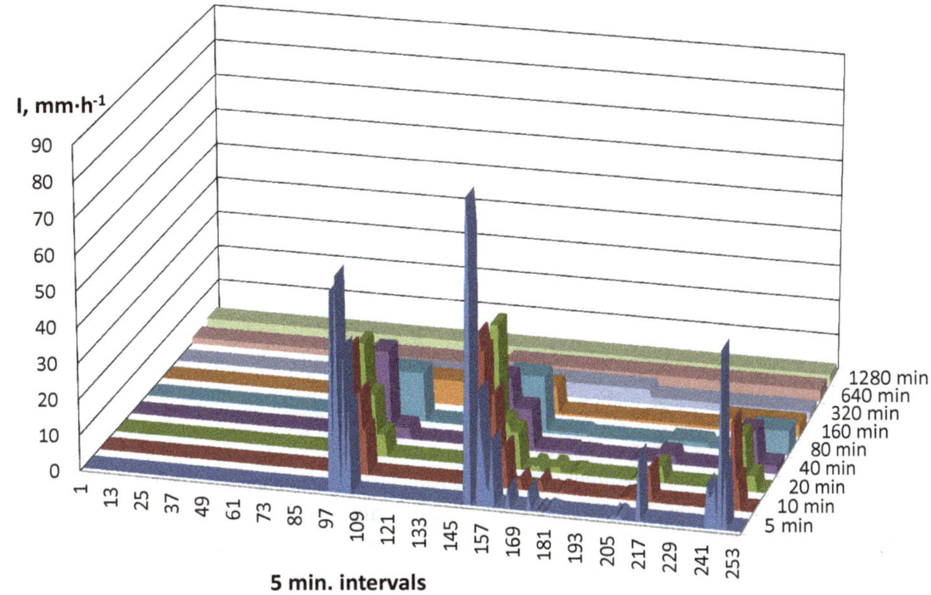

Figure 15. An example of precipitation disaggregation of a 56.3 mm event from 1280 to 5 min for gauge R7.

of uncorrelated segments like the one presented in Fig. 15. Thus, the synthetic time series is missing the correct autocorrelation structure of natural precipitation (for detail discussion see Lombardo et al., 2012). The expected value of the zero-precipitation probability $E(p_0)$ for observed and generated series is given in Fig. 16 for gauge R7. The synthetic values of $E(p_0)$ are calculated as the average over 100 MCM disaggregations. The differences in terms of $E(p_0)$ between observed and simulated are negligible (see Fig. 16). In addition, for comparison we also give the synthetic values of $E(p_0)$ for gauges R15 and R25.

Figure 17 shows the comparison between observed and simulated survival probability function of rainfall amount at 5 min for gauge R7. In Fig. 17, for gauge R7, we report the empirical survival probability function for a synthetic series out of 100 and the averaged function using all the generated series. In addition, for comparison, we give the averaged survival functions for gauges R15 and R25. At first glance, highest rainfall intensities drawn in Fig. 17 show strange behavior manifested by constant exceedance probability above a given precipitation threshold. This is especially pronounced for observed or synthetic series from a single MCM run. This is due to the very short rainfall time series used for the calculation of survival probability functions. According to multifractal theory, singularities in a small data set are very rare. Highest rainfall intensities as singularities are very rare in 2-year series. The behavior of both the synthetic functions for gauge R7 in Fig. 17 is very similar, with the sole exception of the extended and smoothed tail of the averaged function plot. Both the synthetic functions are placed above the observed function. This displacement reveals overprediction of 5 min precipitation depths, particularly at the range of intensities

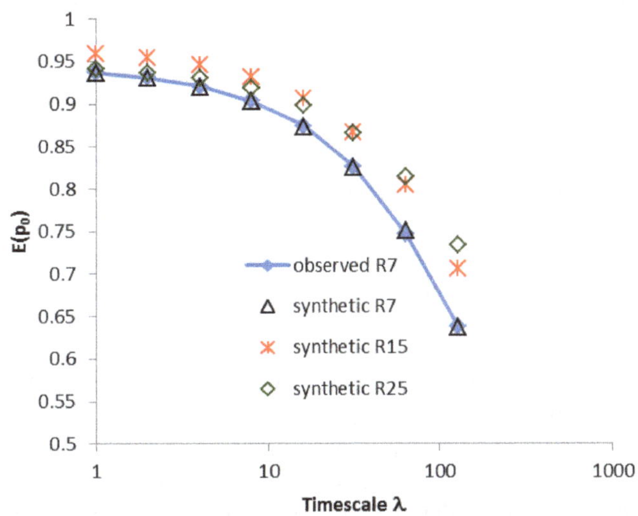

Figure 16. Comparison between observed for gauge R7 and synthetic series for gauges R7, R15 and R25 in terms of intermittency E(p_0) for the considered timescales. The values for the generated data are calculated as the average of 100 disaggregation runs. The variability between runs was negligible and thus is not shown here.

from 0.3 to about 2.0 mm/5 min. It should be noted that the magnitude of dissimilarities between synthetic and observed survival functions for gauge R7 did not exceed the ones reported in other works, e.g., Molnar and Burlando (2005) and Licznar et al. (2011a, b). In comparison, the magnitude of dissimilarities between observed survival probability for gauge R7 and synthetic (average) survival probability function for other gauges R15 and R25 is much more pronounced.

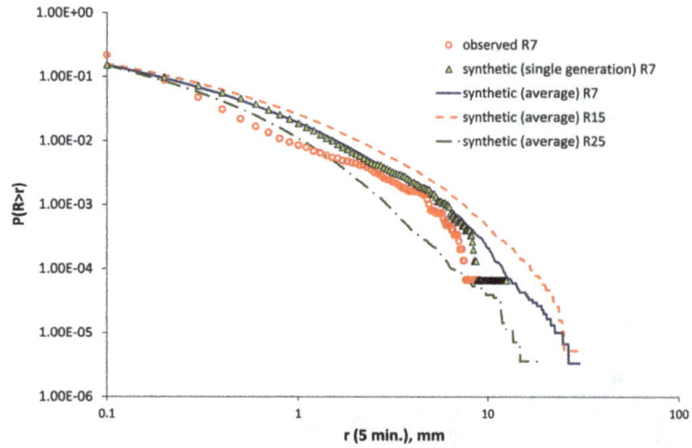

Figure 17. The survival probability function of 5 min precipitation amounts for the observed time series (circles) and the synthetic time series (triangles) generated by the disaggregation of 1280 precipitation amounts for gauge R7. The lines represent the average distributions calculated over the generation of 100 synthetic time series for gauge R7 and for comparison for gauges R15 and R25.

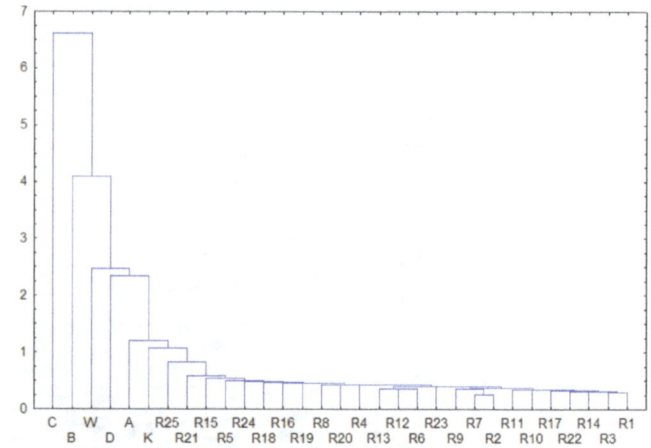

Figure 18. Dendrogram resulting from the cluster analysis of BDC histograms for $\lambda = 1$. The vertical scale shows binding distance, whereas names of gauges are given on the horizontal scale ("K" stands for Kielce gauge and "W" stands for Wroclaw).

3.5 Cluster analysis results and their interpretation

Dendrograms summarizing the results of the cluster analysis for BDC histograms are produced for each timescale and reported in Figs. 18 and 19 for $\lambda = 1$ and $\lambda = 128$, respectively. Results for the first four timescales, i.e., $\lambda = 1, 2, 4, 8$, are unsurprising and easily interpreted. All Warsaw gauges are grouped in a single cluster with similar shapes of BDC histograms; their interconnection on the dendrogram is placed at the level of binding distance equal to about 0.5. Only R25 seems to be characterized by a slightly different pattern of BDC histogram. However, gauge R25 has a behavior that is still much closer to other Warsaw gauges than to the other cities considered. For example, at $\lambda = 1$ gauge R25 is merged into the Warsaw gauges cluster at an Euclidean distance equal to 0.81, whereas the same occurs for the Kielce (the closest considered Polish city) gauge at the Euclidean distance equal to 1.07. For timescales $\lambda = 2, 4, 8$, gauge R25 merges the cluster of Warsaw gauges at quite similar Euclidean distances: 0.89, 0.83 and 0.81, respectively.

The dendrogram for $\lambda = 128$ is given in Fig. 19, being representative of timescales $\lambda = 16, 32, 64, 128$. From Fig. 19 it is possible to see the departure of gauge R15 from the cluster of other Warsaw gauges. The position of gauge R15 is isolated from other Warsaw gauges, and its Euclidean distance from the closest one is large and increases with greater timescale; it is equal to 1.8, 3.19, 3.88 and 8.03 for $\lambda = 16, 32, 64$ and 128, respectively. Simultaneously, the Euclidean distance from the cluster of Warsaw gauges to the nearest neighbor does not exceed 0.9, 1, 1.4 and 1.89 for $\lambda = 16, 32, 64$ and 128, respectively.

This last observation proves that the variability of BDC shapes among Warsaw gauges generally increases with a greater timescale. It may partly be explained by the already mentioned evolution of histogram shapes, the replacement of 2N–B distribution by less centered N–B and B-distribution characterized by a higher variance of BDC. In the specific case of gauge R15, the BDC histograms for the largest timescales are boldly concave (not shown for brevity) and their shapes are becoming similar to B symmetrical distributions parameterized by very small values of a: 0.76, 0.64, 0.54 and 0.45 for $\lambda = 16, 32, 64$ and 128, respectively.

In the last step we used the cluster analysis to investigate the variability among the gauges in terms of the intermittency parameter p_0, which is considered a vector with eight components as values corresponding to the considered timescales. Results are given in the form of a dendrogram in Fig. 20. With respect to p_0, all Warsaw gauges form one single chain-like cluster. Three gauges in the cluster, namely R14, R25 and R15, are characterized by the largest distances from the nearest neighbor with Euclidean distances equal to 0.079, 0.064 and 0.0614, respectively. The distance of gauges R15 and R25 from the other stations in the cluster is similar to observations made for Figs. 18 and 19. A possible, but not certain, explanation for gauge R14 could be its location close to gauge R15 in a weakly developed part of the city.

Unfortunately, we do not have access to other meteorological data to compare our results with other local climate conditions. To our knowledge, studies about microclimate or local turbulence have not been conducted for Warsaw. However, in our opinion the anomalous behavior of gauges R15 and R25 does not originate from random errors due to gauges installation. As previously mentioned, all gauges were installed in very good conditions, and R15 was an airport gauge. A plausible explanation of the anomalous behavior of gauges R25 and R15 could be found in its location. Gauge R25 was

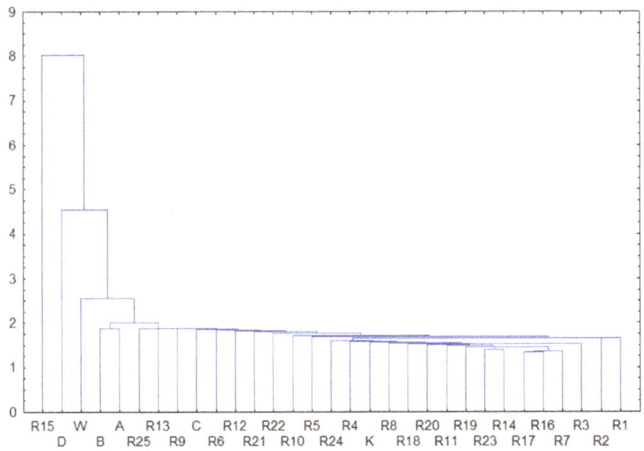

Figure 19. Dendrogram resulting from the cluster analysis of BDC histograms for the timescale $\lambda = 128$. The vertical scale shows binding distance, whereas names of gauges are given on the horizontal scale ("K" stands for Kielce gauge and "W" stands for Wroclaw).

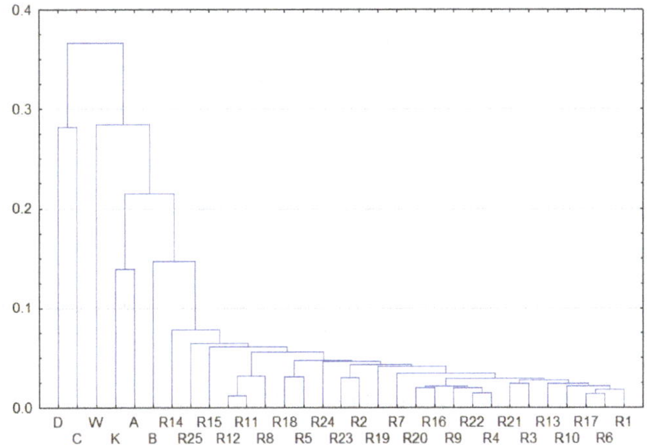

Figure 20. Dendrogram resulting from the cluster analysis of the intermittency parameter p_0. The vertical scale shows binding distance, whereas the names of gauges are given on the horizontal scale ("K" stands for Kielce gauge and "W" stands for Wroclaw).

located in a southeast suburban area in the close vicinity of a forested area and the Vistula river valley. This specific suburban area is also most frequently a place for the development of local convection processes (Prof. S. Malinowski, personal communication, 2013). The anomalous behavior of gauge R15 seems to arise from its specific location on the ground at the Warsaw airport. In the neighborhood of the instrument there are no high buildings and trees and the ground is covered only by short-cut grass. The local atmospheric turbulence conditions, additionally influenced by taking off and landing aircrafts, could have favored the different behavior of this station. In general, gauges R15 and R25 are the only instruments installed outside the areas of urban fabric (Fig. 1) in rather rural conditions of surrounding green areas. The suburban location of these gauges combined with the direct green surrounding reduces, or even minimalizes to zero, urban heat island effects. Peng et al. (2012) investigated the surface urban heat island intensity across 419 global big cities (including Warsaw). These authors showed that the distribution of daytime surface urban heat island intensity correlates negatively across cities with the difference of vegetation fractional cover and of vegetation activity between urban and suburban areas. Kłysik and Fortuniak (1999) found the occurrence of urban heat in the second biggest city in Poland, Łódź (about 120 km southwest), which is comparable to Warsaw flat topography. According to statistics calculated over many years at two stations – one in the center and one at the airport – over 80 % of nights were characterized by a surplus heat in town amounting generally to 2–4 °C and sporadically to 8 °C and more. Once more for Łódź, Fortuniak et al. (2006) investigated the data from two automatic stations: one urban and one rural. They found the relative humidity to be lower in the town, sometimes by more than 40 %, and water vapor pressure differences to be either positive (up to 5 hPa) or negative (up to −4 hPa). Air-temperature differences between the urban and rural station exceeded 8 °C. It could be that similar processes occurred in Warsaw and affected local precipitation dynamics and thus gauges R7, R15 and R25. As a consequence, statistics of synthetic time series vary visibly in Figs. 16 and 17. However, the significance of these differences should be studied in more details in the future.

4 Conclusions

Keeping in mind the simplicity of the retrieval of microcanonical cascade generators from observational data, we proposed to use this technique for the local variability of very short precipitation time series within an urban monitoring network.

We considered a network of 25 gauges deployed in Warsaw over an area of 517.2 km^2. An attempt was made to define the generators of a MCM able to produce 5 min time series, as requested by urban hydrologists, through the disaggregation of quasi-daily precipitation totals. We showed that smooth distributions of BDC are possible for all analyzed timescales, even in the case of a limited length of time series, which in our case slightly exceeded 2 years only. This was made possible by the implementation of a randomization procedure and the use of an overlapping moving window algorithm for the calculation of BDCs.

The correctness of the overlapping moving window algorithm is checked using additional 15 min rainfall time series, 25 years long, at gauge R7. The algorithm is implemented for a hierarchy of sub-daily timescales and separate 5-year periods. The results of BDC calculations are compared to those obtained using all 25 years of data with both overlapping and non-overlapping moving window algorithms. Despite the coarse resolution of data and winter time gaps in the series, the results show a good agreement of BDC distributions calculated over the different periods, suggesting the correctness of the overlapping moving window algorithm, at least in central Poland.

To adequately describe the shapes of BDC histograms, we have implemented a special joined probability distribution, 2N–B, assembled from two N distributions and one B symmetrical distribution. A systematical evolution of BDC histograms from joined 2N–B through joined N–B up to B distributions was observed increasing the timescale. To test the use of more complicated models alternative to the classical B distribution, we suggested the AIC.

To check all the applied techniques (i.e., the randomization procedure, the overlapping moving window scheme and the 2N–B distribution), MCMs were used to disaggregate 1280 min precipitation into 5 min time series. The quality of the generated series was checked, comparing the statistical properties of these with those of observed series. In particular, we compared probabilities of zero precipitation and the survival probability functions of non-zero 5 min precipitation amounts for the considered timescales with agreement comparable to previous studies done in Switzerland, Germany and Poland.

As a main part of this study, we conducted an intercomparison of BDC histograms among the 25 Warsaw gauges and considered as a term of reference another six gauges located in Poland and Germany. The intercomparison was made, scale-by-scale, by means of cluster analysis. Resulting dendrograms for small timescales (i.e., $\lambda = 1, 2, 4, 8$) revealed rather little variability of BDC histograms among all Warsaw gauges in comparison to the variability exhibited with respect to the other external gauges. Only gauge R25 seemed to be characterized by a slightly different pattern. It might originate from the specific gauge location on the city boundary in the vicinity of forested areas and Vistula river valley.

Dendrograms obtained for large timescales (i.e., $\lambda = 16, 32, 64, 128$) also delivered a general picture of similarity among Warsaw gauges with the very clear exception of gauge R15. To our best knowledge, a possible explanation of this was its installation on the ground at the Warsaw airport, which was strongly man-modified and with turbulent local conditions. In addition, R25, R15 and R14 were also identified as gauges presenting slightly different behavior in terms of the intermittency parameter p_0.

As final remarks we can affirm that MCMs combined with cluster analysis could be used as a tool for the assessment of the spatial variability of local precipitation patterns among a group of gauges. This framework could be effectively implemented even in the case of very short observational series thanks to the proposed overlapping moving window algorithm. We believe that the use of this algorithm could increase the development and use of MCMs in urban hydrology. At the same time, we are fully aware of the inherent MCM limitations in the quality of rainfall disaggregation and the necessity of additional verifications of the overlapping moving window algorithm for other gauges with longer and higher quality observational time series.

Returning to questions of interest in urban hydrology addressed at the end of the Introduction, we can formulate following answers:

1. Small precipitation variability within gauges located in city centers, as measured via microcanonical cascade generators, justifies the practice of a single time series use for the probabilistic assessment of the entire urban drainage system.

2. From current engineering needs in urban hydrology, it is enough to use only one fitted MCM for the precipitation time series disaggregation in Warsaw city. We suppose that this result could be valid even in larger urban areas, but the verification is necessary. We dissuade from the cascade generation fitted on precipitation time series collected at instruments located out of the city center in unrepresentative sites like, in our case, the ground at the airport.

3. We question the practice of using gauges from airports for urban hydrology.

Finally, we recommend further research to assess the influence of the local conditions on BDC histograms to find clearer explanations of observed anomalies. We also recognize the necessity of further tests in other cities and precipitation monitoring networks, especially in the case of cities with complicated orography and the presence of hydrological networks.

Acknowledgements. This project was financed by Polish National Science Center (NCN) funds allocated on the basis of decision no. 2011/03/B/ST10/06338. It was realized as a part of the scientific project "Spatiotemporal analysis and modeling of urban precipitation field." Precipitation data were provided by the Municipal Water Supply and Sewerage Company (MWSSC) in Warsaw, Poland. Authors also acknowledge the three anonymous reviewers for their suggestions and comments.

Edited by: R. Uijlenhoet

References

Ahrens, B.: Rainfall downscaling in an alpine watershed applying a multiresolution approach, J. Geophys. Res., 108, 8388, doi:10.1029/2001JD001485, 2003.

Berne, A., Delrieu, G., Creutin, J. D. and Obled, C.: Temporal and spatial resolution of rainfall measurements required for urban hydrology, J. Hydrol., 299, 166–179, doi:10.1016/j.jhydrol.2004.08.002, 2004.

BS EN 752-3: Drain and sewer systems outside buildings. Planning, British Standards Institution, ISBN 0580267822, 34 pp., 1997.

Cârsteanu, A. and Foufoula-Georgiou, E.: Assessing dependence among weights in a multiplicative cascade model of temporal rainfall, J. Geophys. Res., 101, 26363–26370, doi:10.1029/96JD01657, 1996.

De Michele, C., Salvadori, G., Vezzoli, R., and Pecora, S.: Multivariate assessment of droughts: Frequency analysis and dynamic return period, Water Resour. Res., 49, 6985–6994, doi:10.1002/wrcr.20551, 2013.

Deutsche Vereinigung für Wasserwirtschaft, Abwasser und Abfall e.V. (DWA): Arbeitsblatt DWA-A 117: Bemessung von Regenrückhalteräumen, Hennef, 2006.

Fortuniak, K., Klysik, K., and Wibig, J.: Urban–rural contrasts of meteorological parameters in Lodz, Theor. Appl. Climatol., 84, 91–101, doi:10.1007/s00704-005-0147-y, 2006.

Gires, A., Onof C., Maksimovic C., Schertzer D., Tchiguirinskaia I., and Simoes N.: Quantifying the impact of small scale unmeasured rainfall variability on urban hydrology through multifractal downscaling: a case study, J. Hydrol., 442–443, 117–128, doi:10.1016/j.jhydrol.2012.04.005, 2012.

Gires, A., Tchiguirinskaia, I., Schertzer, D., and Lovejoy, S.: Multifractal analysis of an urban hydrological model on a Seine-Saint-Denis study case, Urban Water J., 10, 195–208, doi:10.1080/1573062X.2012.716447, 2013.

Górski J.: Analysis of Precipitation Time Series for Needs of Urban Hydrology on Example of Kielce City. PhD thesis, Kielce University of Technology, Poland, 2013 (in Polish).

Güntner, A., Olsson, J., Calver, A., and Gannon, B.: Cascade-based disaggregation of continuous rainfall time series: the influence of climate, Hydrol. Earth Syst. Sci., 5, 145–164, doi:10.5194/hess-5-145-2001, 2001.

Harris, D., Menabde, M., Seed, A., and Austin, G.: Breakdown coefficients and scaling properties of rain fields, Nonlin. Processes Geophys., 5, 93–104, doi:10.5194/npg-5-93-1998, 1998.

Hingray, B. and Ben Haha, M.: Statistical performances of various deterministic and stochastic models for rainfall series disaggregation, Atmos. Res., 77, 152–175, 2005.

Kłysik, K. and Fortuniak, K.: Temporal and spatial characteristics of the urban heat island of Łódź, Poland, Atmos. Environ., 33, 3885–3895, 1999.

Koutsoyiannis, D.: Rainfall disaggregation methods: Theory and applications, in: Proc. Workshop on Statistical and Mathematical Methods for Hydrological Analysis, edited by: Piccolo, D. and Ubertini, L., Universitá di Roma "La Sapienza", Rome, http://itia.ntua.gr/en/docinfo/570, 2003.

Lanza, L., Leroy, M., Alexandropoulos, C., Stagi, L., and Wauben, W.: WMO laboratory intercomparison of rainfall intensity gauges, Final report, IOM Report No. 84, WMO/TD No. 1304, 2005.

Larose, D. T.: Discovering knowledge in data: an introduction to data mining, John Wiley & Sons, Inc., Hoboken, New Jersey, USA, 222 pp., 2005.

Licznar, P.: Stormwater reservoir dimensioning based on synthetic rainfall time series, Ochrona Srodowiska, 35, 27–32, 2013.

Licznar, P., Łomotowski, J., and Rupp, D. E.: Random cascade driven rainfall disaggregation for urban hydrology: An evaluation of six models and a new generator, Atmos. Res., 99, 563–578, doi:10.1016/j.atmosres.2010.12.014, 2011a.

Licznar, P., Schmitt, T. G., and Rupp, D. E.: Distributions of microcanonical cascade weights of rainfall at small timescales, Acta Geophysica, 59, 1013–1043, doi:10.2478/s11600-011-0014-4, 2011b.

Lombardo, F., Volpi, E., and Koutsoyiannis, D.: Rainfall downscaling in time: theoretical and empirical comparison between multifractal and Hurst-Kolmogorov discrete random cascades, Hydrolog. Sci. J., 57, 1052–1066, doi:10.1080/02626667.2012.695872, 2012.

Marshak, A., Davis, A., Cahalan, R., and Wiscombe, W.: Bounded cascade models as nonstationary multifractals, Phys. Rev. E, 49, 55–69, 1994.

Menabde, M. and Sivapalan, M.: Modeling of rainfall time series and extremes using bounded random cascades and Levy-stable distributions, Water Resour. Res., 36, 3293–3300, doi:10.1029/2000WR900197, 2000.

Menabde, M., Harris, D., Seed, A., Austin, G., and Stow, D.: Multiscaling properties of rainfall and bounded random cascades, Water Resour. Res., 33, 2823–2830, doi:10.1029/97WR02006, 1997.

Molnar, P. and Burlando, P.: Preservation of rainfall properties in stochastic disaggregation by a simple random cascade model, Atmos. Res., 77, 137–151, doi:10.1016/j.atmosres.2004.10.024, 2005.

Molnar, P. and Burlando, P.: Variability in the scale properties of high-resolution precipitation data in the Alpine climate of Switzerland, Wat. Resour. Res., 44, W10404, doi:10.1029/2007WR006142, 2008.

Oke, T.: Initial guidance to obtain representative meteorological observations at urban sites, Instruments and Observing Methods, Report no. 81, World Meteorological Organization, WMO/TD-No. 1250, 2006.

Olsson, J.: Evaluation of a scaling cascade model for temporal rain-fall disaggregation, Hydrol. Earth Syst. Sci., 2, 19–30, doi:10.5194/hess-2-19-1998, 1998.

Over, T. M. and Gupta, V. K.: A space-time theory of mesoscale rainfall using random cascades, J. Geophys. Res., 101, 26319–26331, 1996.

Paulson, K. S. and Baxter, P. D.: Downscaling of rain gauge time series by multiplicative beta cascade, J. Geophys. Res., 112, D09105, doi:10.1029/2006JD007333, 2007.

Peng, S., Piao, S., Ciais, P., Friedlingstein, P., Ottle, C., Bréon, F.-M., Nan, H., Zhou, L., and Myneni, R. B.: Surface urban heat island across 419 global big cities, Environ. Sci. Technol., 46, 696–703, doi:10.1021/es2030438, 2012.

Robert, C. P.: Simulation of truncated normal variables, Stat. Comput., 5, 121–125, 1995.

Rupp, D. E., Keim, R. F., Ossiander, M., Brugnach, M., and Selker, J. S.: Time scale and intensity dependency in multiplicative cascades for temporal rainfall disaggregation, Water Resour. Res., 45, W07409, doi:10.1029/2008WR007321, 2009.

Rupp, D. E., Licznar, P., Adamowski, W., and Leśniewski, M.: Multiplicative cascade models for fine spatial downscaling of rainfall: parameterization with rain gauge data, Hydrol. Earth Syst. Sci., 16, 671–684, doi:10.5194/hess-16-671-2012, 2012.

Schmitt, T. G.: ATV-DVWK Kommentar, ATV-A 118 Hydraulische Berechnung von Entwässerungssystemen, DWA, Hennef, 2000.

Thames Tideway Tunnel (TTT): Needs Report, Appendix B, Report on Approaches to UWWTD Compliance in Relation to CSO's in major cities across the EU, available at: http://aim.prepared-fp7.eu/viewer/doc.aspx?id=28 (last access: 6 January 2015), 2010.

Willems, P.: Multidecadal oscillatory behaviour of rainfall extremes, Clim. Chang., 120, 931–944, doi:10.1007/s10584-013-0837-x, 2013.

World Meteorological Organization (WMO): WMO-No. 8, Guide to Meteorological Instruments and Methods of Observation, 7 edition, World Meteorological Organization, Geneva, Switzerland , 2012.

2

Statistical forecast of seasonal discharge in Central Asia using observational records: development of a generic linear modelling tool for operational water resource management

Heiko Apel[1], **Zharkinay Abdykerimova**[2], **Marina Agalhanova**[3], **Azamat Baimaganbetov**[4], **Nadejda Gavrilenko**[5], **Lars Gerlitz**[1], **Olga Kalashnikova**[6], **Katy Unger-Shayesteh**[1], **Sergiy Vorogushyn**[1], **and Abror Gafurov**[1]

[1]GFZ German Research Centre for Geoscience, Section 5.4 Hydrology, Potsdam, Germany
[2]Hydro-Meteorological Service of Kyrgyzstan, Bishkek, Kyrgyzstan
[3]Hydro-Meteorological Service of Turkmenistan, Ashgabat, Turkmenistan
[4]Hydro-Meteorological Service of Kazakhstan, Almaty, Kazakhstan
[5]Hydro-Meteorological Service of Uzbekistan, Tashkent, Uzbekistan
[6]CAIAG Central Asian Institute for Applied Geoscience, Bishkek, Kyrgyzstan

Correspondence: Heiko Apel (heiko.apel@gfz-potsdam.de)

Abstract. The semi-arid regions of Central Asia crucially depend on the water resources supplied by the mountainous areas of the Tien Shan and Pamir and Altai mountains. During the summer months the snow-melt- and glacier-melt-dominated river discharge originating in the mountains provides the main water resource available for agricultural production, but also for storage in reservoirs for energy generation during the winter months. Thus a reliable seasonal forecast of the water resources is crucial for sustainable management and planning of water resources. In fact, seasonal forecasts are mandatory tasks of all national hydro-meteorological services in the region. In order to support the operational seasonal forecast procedures of hydro-meteorological services, this study aims to develop a generic tool for deriving statistical forecast models of seasonal river discharge based solely on observational records. The generic model structure is kept as simple as possible in order to be driven by meteorological and hydrological data readily available at the hydro-meteorological services, and to be applicable for all catchments in the region. As snow melt dominates summer runoff, the main meteorological predictors for the forecast models are monthly values of winter precipitation and temperature, satellite-based snow cover data, and antecedent discharge. This basic predictor set was further extended by multi-monthly means of the individual predictors, as well as composites of the predictors. Forecast models are derived based on these predictors as linear combinations of up to four predictors. A user-selectable number of the best models is extracted automatically by the developed model fitting algorithm, which includes a test for robustness by a leave-one-out cross-validation. Based on the cross-validation the predictive uncertainty was quantified for every prediction model. Forecasts of the mean seasonal discharge of the period April to September are derived every month from January until June. The application of the model for several catchments in Central Asia – ranging from small to the largest rivers (240 to 290 000 km^2 catchment area) – for the period 2000–2015 provided skilful forecasts for most catchments already in January, with adjusted R^2 values of the best model in the range of 0.6–0.8 for most of the catchments. The skill of the prediction increased every following month, i.e. with reduced lead time, with adjusted R^2 values usually in the range 0.8–0.9 for the best and 0.7–0.8 on average for the set of models in April just before the prediction period. The later forecasts in May and June improve further due to the high predictive power of the discharge in the first 2 months of the snow melt period. The improved skill of the set of forecast models with decreasing lead time resulted in narrow predictive uncertainty bands at the beginning of the snow melt period. In summary, the proposed generic automatic forecast model development tool provides robust predictions for seasonal water availability in Central Asia, which will be tested against the official forecasts in the upcoming years, with the vision of operational implementation.

---

1 Introduction

The Central Asia (CA) region, encompassing the five countries Kazakhstan, Kyrgyzstan, Tajikistan, Turkmenistan and Uzbekistan as well as northern parts of Afghanistan and north-western regions of China, is characterized by the presence of two major mountain systems. The Tien Shan and Pamir Mountains are drained by a number of endorheic river systems such as the Amudarya, Syrdarya, Ili, Tarim, and a few smaller rivers. The CA river basins are characterized by semi-arid climate with strong seasonal variation of precipitation. Most precipitation falls as snow during winter and spring months in the western and northern Tien Shan (Aizen et al., 1995, 1996; Sorg et al., 2012). In contrast, parts of the central Tien Shan and the eastern Tien Shan receive their largest precipitation input during the summer months. The Pamir Mountains receive the highest portion of precipitation during winter and spring months, with the minimum in summer (Schiemann et al., 2008; Sorg et al., 2012).

Precipitation also exhibits a high spatial variation, ranging from less than 50 mm yr^{-1} in the desert areas of Tarim and around 100 mm yr^{-1} on leeward slopes of the central Pamir, to more than 1000 mm yr^{-1} in the mountain regions, which are exposed to the westerly air flows. These flows are a major moisture source in the region (Aizen et al., 1996; Bothe et al., 2012; Hagg et al., 2013; Schiemann et al., 2008). The combination of the low precipitation in the CA lowlands with high precipitation in the mountains highlight the Tien Shan and Pamir Mountains as the most important regional water source (the so called "water towers of Central Asia"). Snow melt in the mountains is the dominant water source for the lowlands during spring and summer, i.e. for most of the vegetation period. During summer, glacier melt and liquid precipitation gain some importance depending on the basin location and degree of glacierization (Aizen et al., 1996). The Tien Shan and Pamir Mountains exhibit particularly high relative water yield compared to the lowland parts of these catchments (Viviroli et al., 2007). Related to the economic water demands in the lowland plains primarily for irrigated agriculture, the Tien Shan and Pamir Mountains are among the most important contributors of stream water worldwide (Viviroli et al., 2007). These mountains also have a very high fraction of glacier meltwater in summer, particularly in drought years (Pritchard, 2017). Within the Aral Sea basin, to which the Amudarya and Syrdarya rivers drain, the irrigated area amounts to approximately 8.2–8.4 million ha (Conrad et al., 2016; FAO, 2013). Additionally, considerable irrigation areas are located in the Aksu–Tarim basin, where agricultural land doubled in the period 1989–2011 and land use for cotton production even increased sixfold (Feike et al., 2015). Irrigated agriculture in CA is mainly fed by the stream water diversion with only a small portion of groundwater withdrawal (FAO, 2013; Siebert et al., 2010). Hence, reliable prediction of seasonal runoff during the vegetation period (April–September) is crucial for agricultural planning and yield estimation in the low-lying countries in the Aral Sea basin, as well as for the management of reservoir capacities including dam safety operations in the upper parts of the catchments. Seasonal forecasts are one of the major responsibilities of the hydro-meteorological (hydromet) services of the CA countries and are regularly released starting in January and continuing until June, with the primary forecast issued at the end of March–beginning of April for the upcoming 6-month period. In some post-Soviet countries, these forecasts are typically developed based on empirical relationships for individual basins relating precipitation, temperature and snow depth and snow water equivalent (SWE) records to seasonal discharge, partly available only in analogue form as look-up tables or graphs (hydromet services, unpublished questionnaire survey undertaken within the CAWa project, http://www.cawa-project.net). Particularly, point measurements of snow depth and/or snow water equivalent, which have been carried out by helicopter flights or footpath surveys in mountain regions in recent decades, are costly or not feasible due to access problems nowadays. Other hydromet services apply the hydrological forecast model AISHF (Agaltseva et al., 1997), developed at the Uzbek hydro-meteorological service (Uzhydromet), which computes discharge hydrographs by considering temperature, snow accumulation, and snow melt. Snow pack is accumulated in winter and temperature and precipitation are taken from an analogous year to drive the model in the forecast mode. The hydro-meteorological services rely on meteorological and hydrological data acquired by the network of climate and discharge stations, which, however, strongly diminished during the 1990s (Unger-Shayesteh et al., 2013). Fortunately the density of the monitoring network recovers nowadays, partly with substantial international support (e.g. Schöne et al., 2013; CAHMP Programme by World Bank; previous programmes by SDC and USAID), but at a slow rate. In any case, the hydro-meteorological services need timely to near-real-time data and simple methodologies capable of utilizing available information in order to fulfil their mandatory tasks.

Some research activities were undertaken towards the establishment of simple forecast methods in the past, mainly trying to establish a relationship between large-scale precipitation records and seasonal discharge. Schär et al. (2004) showed the potential of the ERA-15 precipitation data from the December–April period to explain about 85 % of the seasonal runoff variability in May–September in the large-scale Syrdarya river basin. The explained variance for the Amudarya River amounted, however, to only about 25 %, presumably due to poor precipitation modelling in the ERA data set, strong influence of glacier melt, and water abstraction for irrigation purposes. Similarly, Barlow and Tippett (2008) explored the predictive power of NCEP–NCAR cold-season (November–March) precipitation for warm-season (April–August) discharge forecasts using canonical correlation analysis. Though for some of the 24 CA gauges, no skilful prediction could be achieved, for a few catchments 20 to 50 % explained variance could be attained. Archer and Fowler (2008) utilized temperature and discharge records as well as precipitation for spring and summer seasonal flow forecasts on the southern slopes of Himalaya in northern Pakistan using multiple linear regression (MLR) models. Despite good predictions of spring and early summer flows, late summer discharges were poorly forecasted due to the strong influence of the summer monsoon. Recently, Dixon and Wilby (2015) demonstrated the skill of a linear regression model for the Naryn basin, Kyrgyzstan, based on TRMM precipitation from October to March to explain 65 % of the seasonal flow variance in the vegetation period. The authors selected specific TRMM pixels in the catchments showing the highest correlation to seasonal discharge. They also explored the predictive skill of multiple linear regression models additionally including temperature and antecedent discharge and testing different lead times from 1 to 3 months. They showed that forecasts based on multiple linear regression models are always superior to zero-order forecasts, i.e. the mean flow.

The fact that substantial snow accumulation in CA mountain regions during the winter and spring months significantly governs runoff in the vegetation period can be effectively utilized for seasonal forecasts. For a similar climatic setting, Pal et al. (2013) included the measurements of snow water equivalent at point locations into multiple linear regression models along with precipitation, antecedent discharge and temperature-based predictors. Linear models with multiple predictor combinations achieved skilful forecasts of the spring (March–June and April–June) seasonal flow in northern India on the southern Himalayan slopes. Point snow measurements are, however, rarely available, and remotely sensed snow cover extent can provide a viable alternative. Based on the monitored snow cover extent, e.g. using optical satellite imagery, and additionally considering temperature and precipitation to implicitly approximate SWE, a solid basis for seasonal discharge forecast can be assembled. The MODIS snow cover product proved to be highly accurate for the CA region (Gafurov et al., 2013). Methodologies to remove cloud obstruction of optical imagery have matured over the past decade (Gafurov and Bárdossy, 2009; Gafurov et al., 2016) and tools for automated image acquisition and processing reached the operational level (Gafurov et al., 2016). MODIS snow cover data were, for example, used for runoff forecast in the Argentinian Andes in the high-flow season (September–April), though no cloud elimination algorithms were applied (Delbart et al., 2015). Snow cover in September–October could explain about 60 % of the high-flow season discharge variance. However, no skilful forecasts with lead times greater than zero were possible. Rosenberg et al. (2011) proposed a hybrid (statistical–hydrological model) framework for seasonal flow prediction in Californian catchments using accumulated precipitation in antecedent periods and SWE modelled by a distributed hydrological model. These two predictors were linked to seasonal discharge by principal component and Z-score regression (Rosenberg et al., 2011). The hybrid approach was found to be comparable and in some cases superior to a purely statistical approach, but this required substantial efforts for hydrological simulation of the SWE dynamics.

Based on the finding of the studies listed above, we propose a simple methodology for the operational forecast of seasonal runoff for the vegetation period (April–September) for all CA catchments, whose individual drainage areas range over 3 orders of magnitude. The method is based on multiple linear regression models with automatic predictor selection. The predictors are based on the readily available precipitation, temperature, and discharge gauge records, augmented by the operationally processed cloud-free MODIS snow cover product (Gafurov et al., 2016). Based on theoretical considerations, it is argued that in linear modelling the use of meteorological data from a single gauging station for a large catchment is justified, as long as the variability of the station records are representative for the variability within the whole catchment. The validity of this assumption can be verified by the achieved model performance. We demonstrate the model predictive skill and robustness in a cross-validation and discuss the relative importance of the automatically selected predictors depending on the prediction lead time.

2 Study sites and data

For the testing of the forecast models, 13 catchments were selected. The catchments cover a wide range of geographical regions, ranging from catchments along the western slopes of the Altai mountains in eastern Kazakhstan (Uba, Ulba), the western and northern rim of the Tien Shan (Chirchik, Talas, Ala-Archa, Shu (also known as Chu), Chilik, Charyn) and central Tien Shan (Karadarya, Naryn; cf. Aizen et al., 2007), to the northern and central Pamir (Amudarya) and the northern Hindu Kush (Murgap). The size of the catchments varies by over 3 orders of magnitude, from 239 to 288 000 km^2. Figure 1 provides an overview of the location and size of the catchments, while Table 1 additionally lists the discharge and meteorological gauging stations used for the seasonal flow forecast. Note that the Naryn catchments are nested. The upper Naryn represents a high alpine catchment and is the headwater catchment of the larger Naryn catchment draining into the Toktogul reservoir. This separation was undertaken in order to test the proposed method for a high alpine catchment with a comparably high degree of glacialization. The wide range of catchment locations, climatic conditions, and sizes enable a testing of the proposed forecast models under different boundary conditions, and thus provides an indication

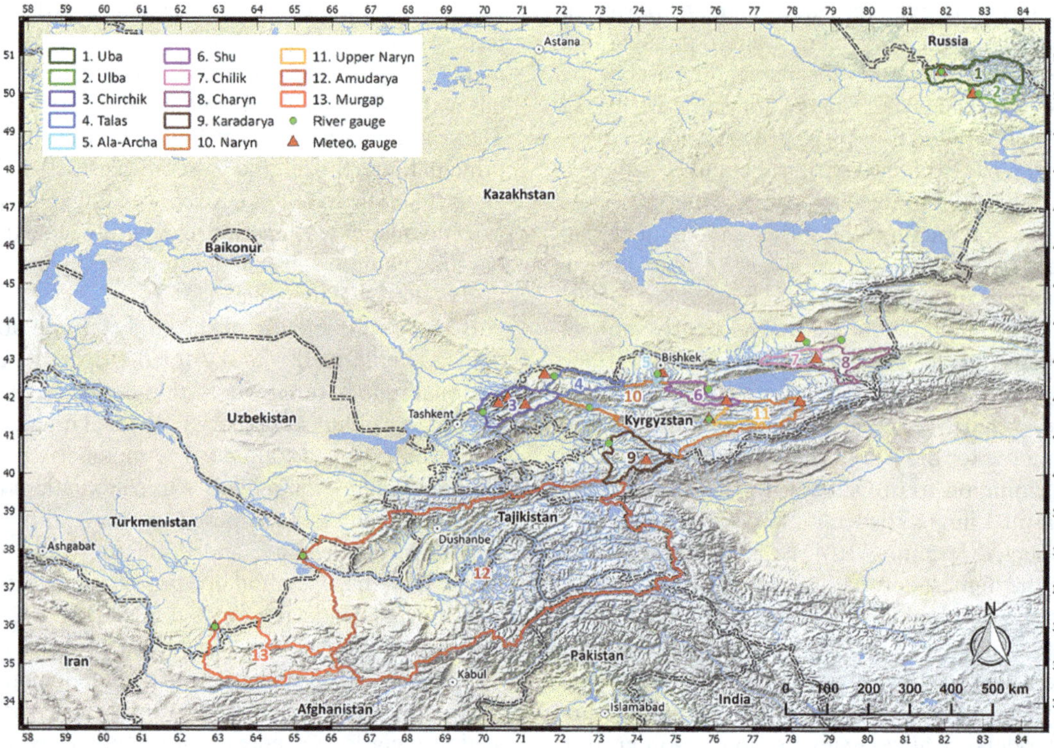

Figure 1. Overview of the catchments for which prediction models were established, with locations of discharge and meteorological gauging stations used (coordinates in latitude–longitude).

of the applicability, robustness, and transferability of the approach.

The catchment boundaries are derived to map the catchment area draining to the selected discharge stations. For the meteorological data (temperature and precipitation) meteorological stations run by the individual hydromet services were selected. Ideally those are located in the catchment area and have sufficient data coverage of at least 16 years (starting in 2000 in order to be consistent with the MODIS temporal coverage). However, for two catchments meteorological stations fulfilling these criteria were not available (Talas, Chilik). For those catchments, meteorological stations nearby were selected for the prediction. For both catchments the meteorological stations are located in the same river catchment downstream of the discharge station.

Monthly values of discharge and meteorological data were obtained for the stations listed in Table 1, i.e. monthly mean discharges, monthly mean temperatures, and total monthly precipitation. For the present study meteorological station data were used, because of its operational availability to the CA hydromet services. Gridded reanalysis products like ERA-Interim typically have a latency of weeks to months, and thus cannot be used for operational forecasts to fulfil the mandatory regulations. A limitation of station temperature and precipitation data is that they are likely not representative for basin-average values. However, it is assumed that the variability of the catchment averages and the variability of station data are similar. This, in turn, enables the use of the station data in the statistical forecast using multiple linear regressions.

In addition to the station data, mean monthly snow coverage values for the individual catchments were calculated using daily snow cover data derived by the MODSNOW-Tool (Gafurov and Bárdossy, 2009; Gafurov et al., 2016). MODSNOW uses the MODIS satellite snow cover product and applies a sophisticated cloud elimination algorithm (Gafurov and Bárdossy, 2009; Gafurov et al., 2016) to obtain cloud-free daily snow cover images. The MODSNOW-Tool runs operationally in most of the CA hydromet services, thus enabling the use of snow cover information for operational forecasts.

Due to the use of MODIS snow cover data, which are available for March 2000 onwards, the time series of the data used for the construction of the forecast models had to be limited to post-2000. The time period for the model development and testing was thus set to 2000–2015, for which mostly complete continuous time series for all data and stations were available.

The seasonal discharge, i.e. the predictand of the forecasts, is calculated as the mean monthly discharge for the period April to September.

Table 1. List of the catchments for which prediction models are derived with discharge (Q) and meteorological gauging stations used for the prediction. Note that Charvak, Andijan and Toktogul are reservoir inflows summing several tributary inflows. For the Charvak reservoir the mean temperature and precipitation data of three meteorological stations located in the catchment were used. Latitudes and longitudes are in decimal degrees (WGS84). Q mean seas. is the multiannual mean seasonal discharge from April to September for the period 2000–2015. Mean ann. P is the mean total annual precipitation of the meteorological station for the period 2000–2015. Mean ann. T is the mean annual mean temperature of the meteorological station for the period 2000–2015. Mean winter SC is the mean of the mean daily snow coverage from January to February for the period 2000–2015.

	Catchment	Discharge station	Q deg. lat.	Q deg. long.	Meteo. station	Meteo. deg. lat	Meteo. deg. long	Meteo. altitude (m)	Catchment area (km^2)	Q mean seas. (m^3 s^{-1})	Mean altitude (m)	Mean ann. P (mm)	Mean ann. T (°C)	Mean winter SC (%)
1	Uba	Shemonaikha	50.620	81.880	Shemonaikha	50.620	81.880	300	9324	269.2	740	460	3.6	69.2
2	Ulba	Perevalochnaya	50.033	82.843	Oskemen	50.030	82.700	375	5080	151.4	950	483	3.8	87.7
3	Chirchik	Charvak	41.626	69.969	Chatkal	41.822	71.097	2300	10903	346.21	2575	708	5.5	97.3
					Oygaing	42.000	70.633	1620	10 903					
					Pskem	41.861	70.384	2220	10 903					
4	Talas	Kluchevka	42.581	71.836	Kyzyl-Adyr	42.616	71.586	1764	6663	19.62	2424	327	9.0	72.1
5	Ala-Archa	Kashka-Suu	42.650	74.500	Baytik	42.670	74.630	1579	239	8.83	3288	559	3.2	79.6
6	Shu	Kochkor	42.250	75.833	Kara Kuzhur	41.930	76.300	855	4961	34.53	2934	253	1.1	59.4
7	Chilik	Malybai	43.494	78.392	Shelek	43.597	78.249	600	3964	70.67	2603	274	11.0	74.5
8	Charyn	Sarytogai	43.553	79.293	Zhalanash	43.043	78.642	1690	7921	59.06	2260	507	6.1	82.4
9	Karadarya	Andijan	40.814	73.257	Ak-Terek	40.365	74.222	1190	11670	186.21	2663	913	9.5	82.4
10	Naryn	Toktogul	41.760	72.750	Naryn city	41.460	75.850	2040	51926	653.13	2850	374	4.4	88.0
11	Upper Naryn	Naryn city	41.460	75.85	Tien Shan	41.910	78.210	3614	10343	168.64	3546	345	−5.8	91.0
12	Amudarya	Kerki	37.842	65.23	Kerki	37.842	65.230	237	287714	2551.02	2578	173	17.9	56.7
13	Murgap	Takhta Bazar	35.966	62.907	Takhta Bazar	35.966	62.907	354	35767	40.13	1707	217	18.2	37.5

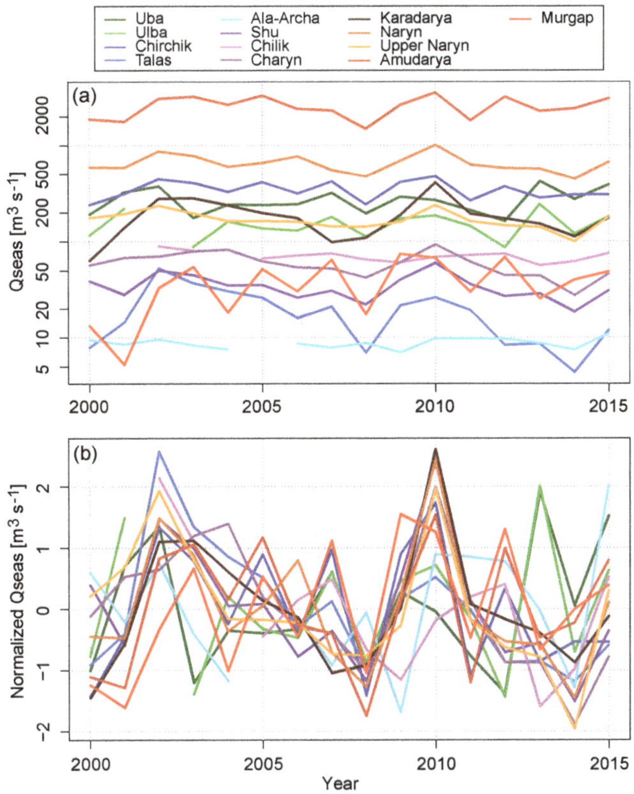

Figure 2. Seasonal discharge (mean monthly discharge for the period April–September) for the catchments under study **(a)**. Panel **(b)** shows the seasonal discharge normalized to zero mean and standard deviation of 1.

Seasonal discharge variability

Figure 2 shows the seasonal discharges for all catchments considered in this study. The top panel highlights the differences in the magnitude of the seasonal discharge, spanning almost 3 orders of magnitude (see also Table 1). Discontinuous lines indicate data gaps. In order to illustrate differences in the inter-annual variability of the seasonal discharge the lower panel of Fig. 2 plots the seasonal discharges normalized to zero mean and standard deviation of 1. This plot indicates similar but also different inter-annual variability patterns of the different catchments. In order to distinguish between similar and different inter-annual variabilities, cross-correlations of the seasonal discharges are calculated and hierarchically clustered (Fig. 3). Cluster memberships were established using the Ward algorithm, clustering the catchments based on the dissimilarities of the correlation between the seasonal discharge time series of the different catchments. The correlation matrix in Fig. 3 shows that the seasonal discharges mainly cluster according to their geographical location. The variability of the seasonal discharge of the two catchments in the Altai region (Uba, Ulba) is distinctively different to all the others. Also, the two most

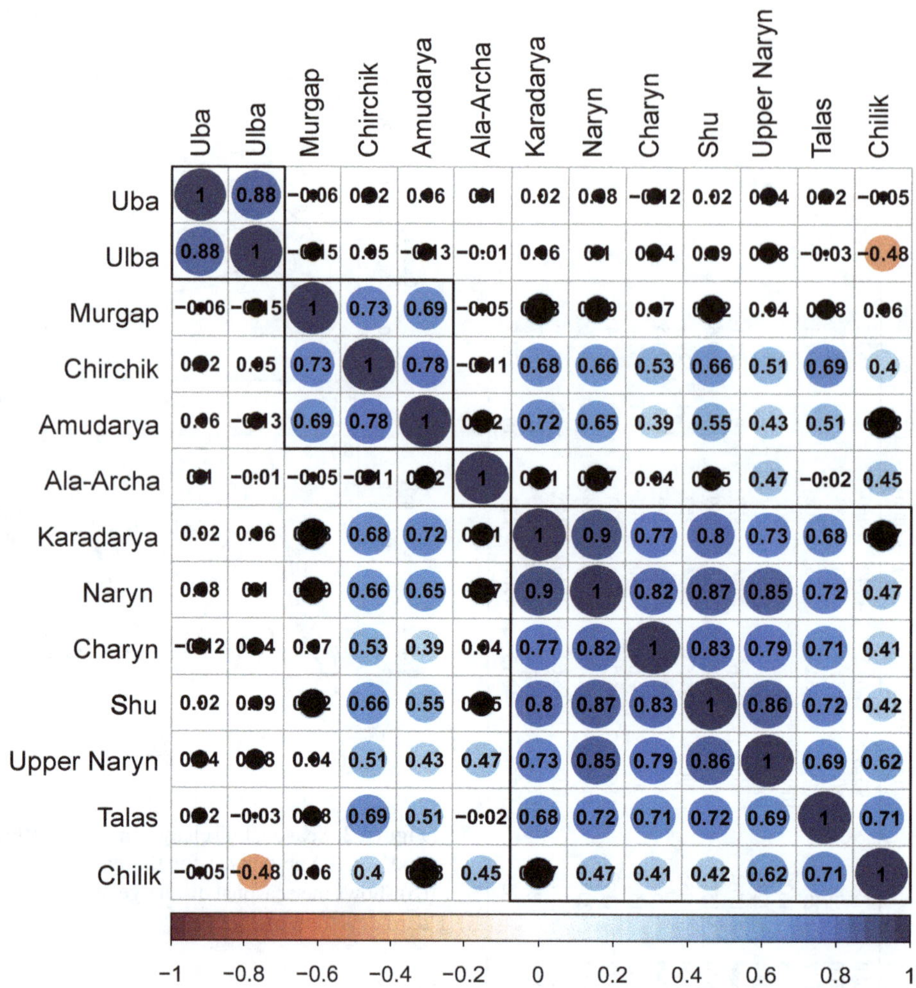

Figure 3. Correlation matrix of the seasonal discharges of the catchments under study. The catchments are hierarchically clustered using the Ward algorithm. The colour and size of the circles indicate the direction and strength of the correlations, with blue colours indicating positive and red colours indicating negative correlations. The numbers provide the actual linear correlation coefficient. The coloured circles indicate significant correlation at a significance level of $p = 0.05$.

southern catchments (Amudarya and Murgap) form a distinct cluster that is joined by the most western catchment of the northern Tien Shan, Chirchik. However, Chirchik is also well correlated to the largest group, the catchments in the Tien Shan, which all show similar inter-annual variability of the seasonal discharge. An exception to this is the smallest catchment in the study, Ala-Archa, which is not correlated to any of the other catchments, presumably due to the strong influence of local small-scale meteorology and glacier-melt-dominated discharge formation in the summer months.

The analysis of the inter-annual variability broadly maps the geographical and climatic differences of the catchments considered in this study. These differences in variability, but also in the magnitude of the discharges and catchment size, imply that the forecast methods can be tested against a wide range of boundary conditions and seasonal variabilities. If skilful forecasts are obtained for all catchments, it can be argued that the approach delivers robust forecasts that are not obtained by chance or due to similar variabilities in all catchments. If successful, it could also be inferred that the approach can be transferred to other regions with similar streamflow generation characteristics.

3 Method

As mentioned in the introduction, the seasonal discharge during the vegetation period of April to September in CA is dominated by snow melt in the mountains. Therefore a good estimation of the snow accumulation and snow water equivalent in the catchments during the winter months may provide reliable forecasts of the discharge during the vegetation period. However, snow depth and snow water equivalent are not regularly monitored except for some dedicated research sites. Thus, alternative data containing proxy information about the snow depth and snow water equivalent must be used. Based on these considerations predictors for the forecast models were derived from mean monthly temperature records, total monthly precipitation and monthly mean snow coverage of the catchments. It is argued that combinations of these factors are able to serve as proxy data for snow depth and water equivalent. While the precipitation directly contains information about the snow fall amount and thus accumulation, temperature may contain information on the wetness of the snow pack. In combination with snow coverage, temperature and precipitation may thus provide information about the snow volume and water content. In addition to the climate data, monthly antecedent discharge can serve as an indicator of the magnitude of the snow melt process and groundwater storage state and release and is used as predictor, too. The usefulness of antecedent discharge for flow prediction has been shown, e.g. in Slater et al. (2017). It should be noted that in this study only direct observations are used. Forecasted climate data are not used for the predictions.

For some regions, particular the Altai catchments, early summer (May–July) precipitation plays a larger role for the seasonal discharge generation. This precipitation is partly considered as observations in the late forecasts presented here. However, reliable information on the May–July precipitation some months in advance could possibly improve the early forecasts. But due to the low predictability of the typically convection-type summer precipitation (Gerlitz et al., 2016), this is not considered in the predictor set.

Evaporative losses in the presented mountain catchments are considered to be low due to the low summer temperatures, fast catchment response, and high water flow velocities in the rivers. Higher losses can occur in reservoir lakes, but with the exception of the large Amudarya basin no reservoirs are present in the selected catchments. The Nurek reservoir lake exists on the Vakhsh river In the Amudarya catchment, but evaporative losses from the lake surface area of 98 km^2 can be considered negligible in comparison to the large catchment size. Therefore evaporation is not directly considered as a predictor for the forecasts.

The catchment area of the Vaksh river at the conjunction with the Panj river amounts to 31 415 km^2, equivalent to about 11 % of the Amudarya catchment at Kerky considered here. Assuming further that the reservoir can manage only a fraction of the total discharge of the Vakhsh river, and that the effects of the water retention are further buffered by the seasonal mean discharge spanning 6 months, it can be assumed that the regulating effect of the Nurek dam on the overall seasonal discharge of the Amudarya at Kerky is rather low. Additionally, the dam is operational since 1980, and therefore a discontinuity in the time series 2000–2015 can be ruled out. We thus argue that the anthropogenic influence of the seasonal discharge time series of the Amudarya is negligible for the presented study.

3.1 Generation of the predictor set

The core set of predictors consists of the monthly values preceding the prediction date. According to the operational forecast schemes of the CA hydromet services a series of different prediction dates were defined. The first prediction of the seasonal mean discharge for the vegetation period (April to September) is issued on 1 January, followed by predictions on 1 February, 1 March, 1 April, 1 May, and 1 June. The predictions from January to March are preliminary forecasts, while the prediction on 1 April is the most important for the water resource planning in the CA states. The following operational forecasts serve as corrections of the April forecast. However, if the discharge of the whole vegetation season is predicted in these late forecasts, the predictors for the late forecasts are not fully independent from the predictand, because the discharge of April and May is included in the seasonal discharge, but at the same time used as predictor. This violates formal requirements of a linear regression and biases the performance evaluation of the late predictions. Therefore we deviate from the official procedures of the CA hydromet services for the late forecasts. In this study the mean discharge of the remaining vegetation season is predicted in the late forecasts (i.e. May–September for the May forecast, and June–September for the June forecast), and the observed discharge in April and May is then added to the forecasts in order to obtain values for the whole vegetation period. By this procedure the formal requirements of the MLR are fulfilled, and the results of the forecasts are in line with the formal procedures of the hydromet services. This is necessary to obtain acceptance of the proposed method in the services and their use in the official forecast procedures, because the water regulation procedures and, for example, agricultural yield estimations are traditionally based on bulk numbers for the entire period. If these procedures are not followed, the obtained results, which are better than the forecasts issued with the existing procedures, might not be implemented or come into practise, and thus a chance would be missed to bring research results into application. In this context it should be noted that sub-seasonal discharges are highly correlated to the full seasonal discharge (see Appendix A1). This means in consequence that regression results obtained with sub-seasonal discharges will be very similar to results obtained using the full discharge time series.

The selection of the predictor set to be used in the MRL follows this rationale, which is based on a combination of principle hydrological considerations, preliminary correlation tests, and expert judgement: for the prediction up to 1 April the monthly values over the whole winter period, i.e. from October to the prediction date are used. For later predictions (i.e. in May and June) the set was limited to data of the prediction year, i.e. from January onwards. This restriction was implemented in order to keep the number of predictor combinations and thus the calculation time within reasonable limits. It can be justified by the typically low correlation of the October–December data with the seasonal discharge compared to later values. The monthly predictor values were accompanied by multi-monthly means, spanning over 2 and 3 months prior to the prediction date, and mean values for the whole predictor period defined above, i.e. either from October to the prediction month, or from January to the prediction month, respectively. An exception for this is the snow coverage, which is typically invariant in the deep winter months (100 % coverage). Therefore multi-monthly means are only calculated for the whole predictor period for the early forecasts in January and February. For later forecasts, multi-monthly means from January onwards are used. The early winter predictors are not used in the March to June forecasts because of their low predictive power, which is illustrated by the comparatively low performance of the early forecasts (see Sect. 4).

Furthermore, composites were calculated from the climatological data in order to extend the predictor set. They are introduced in order to explore their potential to reflect snow wetness. It is argued that composites can improve the prediction by linear models, as some non-linear interactions might be reflected better by composites compared to the raw data (as shown in Hall et al., 2017, for example). Analogously to the original data, monthly and multi-monthly composites were derived. For the composites, products of "temperature and precipitation", "temperature and snow coverage", "precipitation, snow coverage and temperature", and "precipitation and snow coverage" were used. Antecedent discharge was not included in the composites, because this should not influence the snow cover characteristic.

3.2 Statistical modelling

For the development of the statistical forecast models, standard multiple linear regression was applied. It is argued that the discharge generation from snow melt over whole catchments and on a seasonal timescale can be approximated by linear models. In fact, this was shown by a large number of studies using hydrological models based on linear concepts like linear storage, e.g. Duethmann et al. (2014, 2015) in CA. Additionally a number of studies have shown that linear regression is a valid approach for seasonal forecasts (e.g. Delbart et al., 2015; Dixon and Wilby, 2015; Seibert et al., 2017). A linear modelling approach is thus seen as a valid approach for seasonal forecasts in the study region from a general point of view. However, in order to statistically support the assumption that the runoff-generating processes can be approximated by linear models, the formal assumptions of MLR were also tested: the assumption of normal distribution of the residuals was tested by the Shapiro–Wilk test, the independence of the residuals was tested by calculating the autocorrelation with lag 1, and the heteroscedasticity of the residuals was tested by the Breusch–Pagan test.

In the model selection procedure all possible predictor combinations, which are different for every prediction month as described in 3.1, are used in the MLR for the construction of forecast models. However, some restrictions were put on the predictor combinations in order to avoid overfitting and thus spurious regression results:

1. The predictors are grouped into 8 groups: snow cover, temperature, precipitation, antecedent discharge, and the four composite types.
2. The maximum number of predictors in a regression is limited to four.
3. Only one predictor from each group of predictors can be used in an individual regression model.

This resulted in 7728 predictor combinations, i.e. multiple linear models to be tested in January, and increased to 155 690 possible models in April. A complete list of the predictors for the different prediction months is provided in Appendix A2. The coefficients for all these linear models were automatically fitted during the MLR by the least-squares method. Only models with all predictors statistically significant at $p = 0.1$ and with an overall model significance of at least $p = 0.1$ were retained. From the remaining models the best models were selected based on the lowest predicted residual error mean of squares (PREMS) value obtained by a leave-one-out cross-validation (LOOCV). In the LOOCV, 1 year of the seasonal discharge time series is removed from the data set for fitting the MLR. The missing data point is then estimated by the model fitted to the remaining data. The PREMS value is the mean of squared errors of all seasonal discharges left out and the associated predicted LOOCV values. PREMS is thus defined as follows.

$$\text{PREMS} = \frac{1}{n}\sum_{i=1}^{n} e_{(i)}^2,$$

with $e_{(i)}$ being the residuals of the LOOCV:

$$e_{(i)} = |y_i - \hat{y}_{(i)}|,$$

where $\hat{y}_{(i)}$ is the regression estimate of y_i based on a regression equation computed leaving out the ith observation of the overall number of n observations. The PREMS was used in this study instead of the usual PRESS (predictive residual error sum of squares) in order to avoid biases possibly introduced by missing predictor or predictand data. Using the sum of squares could favour models with missing data compared to models providing predictions for all 16 years. Using the mean of the squares can avoid this to a large extent.

The LOOCV is testing the MLR for robustness and can avoid overfitting and incidental good MLR results valid for the whole data set only. In order to avoid an overestimation of the forecast skill the seasonal discharge time series were tested for auto-correlation, which could lead to spurious estimation of model robustness in the LOOCV.

Model skill was evaluated by a number of measures: adjusted R^2, root mean square error RMSE, and mean absolute error MAE. The robustness of the model set was quantified as the ratio between the adjusted R^2 based on the LOOCV residuals and the adjusted R^2 of the complete model residuals. The reliability of the model was analysed by probability integral transform (PIT) diagrams (e.g. Crochemore et al., 2017) and quantified as PIT scores (Renard et al., 2010).

In the presented study, not only the single-best model according to PREMS of the LOOCV–MLR was selected as prediction model, but rather the best 20 models, if more than 20 models pass the significance tests. This selection aims to analyse the differences between the best models in terms of performance and predictors, but also serves as a set of models for the forecast of the seasonal discharge. The distribution of the residuals of the best 20 forecast models was evaluated to provide 80 % predictive uncertainty bounds for every forecast. However, it should be noted that the choice to use the best 20 models is subjective, and this number can be increased or reduced. Moreover, a distinct set of specific models from the best models can be selected according to their performance measures and temporal dynamics by experts knowledgeable of the individual catchments. A sufficient amount of freedom was left for the selection of the number and individual selection of the best models to be used for the forecasts, in order to enable an expert selection of models by the forecasters of the CA hydromet services. The forecasters can check every model retained for their performances (quantitatively and qualitatively) and select the models to be used for the prediction accordingly.

3.3 Predictor importance

The predictors of the selected best models were analysed for their importance, i.e. their share of the overall explained variance (R^2) of the individual models. This was achieved by the *lmg* algorithm implemented in the R-package *relaimpo* (Grömping, 2006). The *lmg* algorithm is based on sequential R^2 values, but explicitly eliminates the dependence on predictor orderings by averaging over orderings using simple unweighted averages. In sequential R^2 calculations, the model is re-run with a single predictor only and the explained variance is calculated. Then the next predictor is added and the gain in explained variance is calculated. By this procedure the variance explained by individual predictors can be quantified. However, in this procedure the sequence of predictors added influences the share of explained variance associated with the individual predictors. Therefore the *lmg* algorithm tests all possible predictor sequences and calculates the mean importance of every sequence in order to overcome the problem of predictor ordering in sequential R^2 values. The predictor importance calculation yields information about the importance of the individual predictors at different forecast points in time for the catchments under study. This can be used for a discussion of the factors responsible for the winter snow accumulation and snow water content in the catchments.

However, such a discussion is complicated by the use of the composite predictors. Therefore the importance of composite predictors is divided into equal proportions to the components of the composites. If more than one composite is used in a model, the proportions associated with the component factors (snow cover, precipitation, temperature) are summed up and displayed as parts of the composite importance in the figures presented in Sect. 4.2. This analysis is not meant to provide a quantitative estimation for the component importance of the composite predictors, but rather to enhance the discussion and interpretation of the predictors of the selected forecast models.

In addition to the importance for an individual model (here the best LOOCV model), the mean importance over the best 20 LOOCV models is calculated. This is achieved by calculating the fractions of the sum of importance of an individual predictor for all 20 models to the sum of the R^2 values of all 20 models for each catchment and month. These fractions are then multiplied by the mean R^2 values of the best 20 models. This mean predictor importance can be compared to the predictor importance of the best model in order to analyse the stability of the predictor selection within the best 20 LOOCV models.

4 Results

In order to test the suitability of the LOOCV for the seasonal streamflow forecast the autocorrelation and partial autocorrelation of the streamflow time series was calculated and plotted (Appendix A2). Any autocorrelation in the discharge time series could lead to artificial overestimation of the forecast's skill by the LOOCV. Hardly any autocorrelation at $\alpha = 0.05$ could be detected. Only for the Ulba some significant autocorrelation for lag 1 and 2 is shown just above $\alpha = 0.05$, but by the partial autocorrelation only. No autocorrelation was found at $\alpha = 0.01$. Therefore it can be stated that autocorrelation does not exist in the discharge time series of all catchments, and thus the proposed LOOCV is an appropriate validation method.

The MLR fitting with LOOCV (see Sect. 3.2) was applied for different forecast dates ranging from 1 January to 1 June for all catchments. Out of the models that passed the significance tests after fitting, the best 20 models according to the PREMS resulting from the LOOCV were retained for the forecasts. The tests for possible violations of the formal MLR assumptions showed that 89.5 % of all selected models for all catchments and prediction months fulfilled the criteria of normal distributed residuals, 95.8 % of the selected models passed the test for independence of the residuals, and 99.5 % of the selected models have homoscedastic residuals (see Appendix A3). In summary, the formal requirements of MLRs are fulfilled by almost all models, and the use of linear models for seasonal discharge forecasts is also justified from a formal point of view.

Table 2. Adjusted R^2 values of the best-performing prediction models from the LOOCV for all catchments and prediction months. "Best" indicates the single best model according to the LOOCV, and "Mean" indicates the mean percentage over the best 20 models according to the LOOCV. The adjusted R^2 values are associated with indicators for significance levels. Additionally, for the prediction months May and June the adjusted R^2 values for the prediction of the remaining season (i.e. for the mean seasonal discharge May–September for prediction month May, and June–September for prediction month June) are given in italics.

		January		February		March		April		May		June	
		Best	Mean	Best	Mean	Best	Mean	Best	Mean	Best	Mean	Best	Mean
		0.678	0.511	0.824	0.714	0.842	0.743	0.811	0.790	0.868	0.828	0.968	0.958
1	Uba	++	++	+++	+++	+++	+++	+++	+++	*0.775*	*0.723*	*0.836*	*0.793*
										+++	+++	+++	+++
		0.624	0.429	0.714	0.444	0.781	0.672	0.869	0.811	0.886	0.919	0.987	0.979
2	Ulba	o	+	+++	+	+++	++	+++	+++	*0.858*	*0.875*	*0.921*	*0.862*
										+++	+++	+++	+++
		0.253	0.278	0.594	0.556	0.650	0.593	0.891	0.884	0.959	0.940	0.972	0.965
3	Chirchik	++	–	+++	++	+++	++	+++	+++	*0.931*	*0.918*	*0.948*	*0.932*
										+++	+++	+++	+++
		0.669	0.408	0.794	0.703	0.808	0.728	0.823	0.787	0.921	0.872	0.966	0.958
4	Talas	+++	+	+++	+++	+++	+++	+++	+++	*0.894*	*0.831*	*0.940*	*0.925*
										+++	+++	+++	+++
		0.393	0.353	0.597	0.431	0.758	0.524	0.761	0.623	0.742	0.635	0.820	0.778
5	Ala-Archa	+	o	++	o	+++	+	+++	++	*0.733*	*0.603*	*0.772*	*0.719*
										+++	++	+++	++
		0.274	0.260	0.709	0.440	0.903	0.729	0.680	0.569	0.812	0.754	0.895	0.858
6	Shu	+	–	+++	o	+++	+++	+++	++	*0.805*	*0.745*	*0.882*	*0.841*
										+++	+++	+++	+++
		0.865	0.818	0.856	0.787	0.910	0.873	0.757	0.770	0.884	0.800	0.934	0.829
7	Chilik*	+++	++	+++	++	+++	+++	+++	++	*0.843*	*0.770*	*0.934*	*0.831*
										+++	+++	+++	+++
		0.643	0.503	0.844	0.786	0.792	0.765	0.873	0.810	0.943	0.947	0.989	0.986
8	Charyn	+++	+	+++	+++	+++	+++	+++	+++	*0.918*	*0.919*	*0.964*	*0.951*
										+++	+++	+++	+++
		0.573	0.449	0.589	0.411	0.880	0.845	0.976	0.968	0.994	0.984	0.983	0.978
9	Karadarya	++	+	+++	++	+++	+++	+++	+++	*0.988*	*0.970*	*0.947*	*0.930*
										+++	+++	+++	+++
		0.782	0.679	0.657	0.657	0.844	0.800	0.853	0.819	0.824	0.824	0.932	0.898
10	Naryn	+++	+++	+++	+++	+++	+++	+++	+++	*0.811*	*0.806*	*0.896*	*0.847*
										+++	+++	+++	+++
		0.832	0.810	0.898	0.850	0.916	0.897	0.947	0.923	0.862	0.863	0.946	0.944
11	Upper Naryn	+++	+++	+++	+++	+++	+++	+++	+++	*0.847*	*0.839*	*0.917*	*0.916*
										+++	+++	+++	+++
		0.213	0.304	0.841	0.691	0.857	0.840	0.878	0.839	0.897	0.893	0.982	0.977
12	Amudarya	+	+	+++	+++	+++	+++	+++	+++	*0.869*	*0.845*	*0.961*	*0.949*
										+++	+++	+++	+++
		0.465	0.367	0.757	0.551	0.802	0.642	0.807	0.700	0.985	0.970	0.997	0.996
13	Murgap	++	o	+++	+	+++	++	+++	++	*0.965*	*0.934*	*0.985*	*0.981*
										+++	+++	+++	+++

* The performance of Chilik is not representative and comparable to the other catchments due to too many missing discharge and predictor data. Significance level p: +++ = 0.01, ++ = 0.05, + = 0.1, o = 0.2, – = > 0.2; for mean the lowest significance of the model set is used.

In general the performance of the linear models increases from January to June, with the best models reaching adjusted R^2 (adj. R^2) values in the range of 0.68–0.97 in April and 0.89–0.99 in June. For most of the catchments high adj. R^2 values in the range of 0.57–0.83 were already obtained in January. Only for Ala-Archa, Amudarya, Shu, and Chirchik is the performance unsatisfyingly low in January, but increases to adj. $R^2 > 0.59$ already 1 month later in February. Table 2 lists the adj. R^2 values of the best LOOCV models for all catchments and forecast months. Note that the adj. R^2 in the table is calculated using the coefficients of the linear models fitted to the whole data set, i.e. they are not cross-validated adj. R^2. For the months May and June the performance is (a) calculated for the whole seasonal discharge by adding the forecast values of the sub-season to the observed discharge, and (b) for the sub-season only. As expected, the adj. R^2 values are higher for the whole seasonal discharge, because in this case parts of the seasonal discharge is explained by actual observations. However, the performance is just slightly lower for the sub-seasonal discharge predictions, typically in the range of 0.01–0.05, less than the adj. R^2 for the whole season. Only for Uba is a larger difference of about 0.1 in adj. R^2 in May observed. This good performance of the sub-seasonal forecasts can be explained by the general high correlation of the sub-seasonal discharges with the full seasonal discharge (Appendix A1), and the high predictive power of antecedent discharge for the later forecasts (see Sect. 4.2).

While for most of the catchments, the performance of the models gradually increases with decreasing lead time including the sub-seasonal discharge forecasts, the performance for Chilik shows significant decreases and increases. This is mainly caused by a comparatively large number of missing discharge and predictor data, but possibly also by the fact that the meteorological station used for this catchment is located outside of the catchment. The automatic fitting algorithm takes advantage of this by finding models able to explain the fewer data points better compared to the full time series despite the use of PREMS instead of PRESS. However, these models can already represent an overfitting and are thus less reliable or less stable in time compared to models fitted to longer time periods.

In order to get a more comprehensive picture of the model performance, Fig. 4 shows the temporal evolution of the adj. R^2 evaluated for the complete time period of the single best LOOCV model, the minimum adj. R^2 of the best 20 models, the mean adj. R^2 of the best 20 models, the root mean square error (RMSE) of the single best LOOCV model calculated for the full data set normalized to the mean seasonal discharge (see Table 1), the normalized mean absolute error MAE, and the PREMS value of the best model. Note that the highest adj. R^2 value is not necessarily the adj. R^2 of the single best model, because the best model is selected according to the lowest PREMS in the LOOCV, and not the best adj. R^2 evaluated using the whole time series. Therefore the mean adj. R^2 in January is occasionally higher than the adj. R^2 of the best LOOCV model, i.e. the most robust model. In general, Fig. 4 shows that the different adj. R^2, RMSE, MAE, and PREMS values are similar in their evolution in time, i.e. increase (adj. R^2) or decrease (RMSE, MAE, PREMS) with later forecast months. This indicates that for all best 20 models the performance is improving with later forecasts.

Furthermore, the difference between minimum adj. R^2 and mean adj. R^2 to the adj. R^2 of the single best LOOCV model is typically larger in the early prediction months. This indicates a wider spread of model performance within the selected 20 models for the predictions with longer lead times. This difference decreases with shorter lead times, meaning that more models with similar high performance can be found, and thus uncertainty of the model set is reduced. To a certain extent this is likely caused by the larger number of possible predictors for later prediction months, but it is also well justified to assume that the later predictors have more predictive power: data from the late winter months can better describe the snow coverage and water content compared to predictors from the previous autumn–early winter. This issue will be discussed further in Sect. 4.3.

Figure 4 shows that the RMSE as well as the MAE of the best model of the LOOCV is at its maximum at about 35 % of the long-term seasonal mean discharge. (Talas and Murgap in January). However, for most catchments the normalized RMSE and MAE is already below 20 % in January. For the important April forecast they generally fall below 10 %, except for Talas and Murgap, where it remains at 20 %. These values state the high performance of the linear forecast models in terms of actual discharge and are thus a useful source of information for practitioners to assess the value of the forecasts.

Figure 4 also shows the PREMS values of the best models and the performance development with the forecast months. The PREMS values generally decrease (i.e. improve) with decreasing lead time. However, occasionally increases can be observed for later forecast months. This can be also seen in the adj. R^2 values, but is less pronounced because of the scale of the left y axis. This phenomenon is caused by the changing predictor sets from forecast month to forecast month. In particular, multi-monthly predictors change for each prediction date according to the parameter selection outlined in Sect. 3.1. As this phenomenon of increasing PREMS values usually occurs in April or May, it can be hypothesized that the information of the late winter–early spring months used in the later forecasts does not contain better information about the snow cover as the previous months. With respect to a practical application, the better performing forecasts from the previous months can be used, which is equivalent to an extension of the predictor set by the predictors of the previous month.

This general reduction of PREMS also means that the models become more robust for later prediction months. To illustrate this more clearly, Fig. 4 also shows the relation between the mean adj. R^2 of the LOOCV for all 20 models to the mean adj. R^2 of the full model fit. The mean adj. R^2 of the LOOCV is calculated from the LOOCV residuals used to calculate the PREMS. According to the rationale of the LOOCV, a model is more robust and less prone to overfitting, if the LOOCV R^2 is very close to the overall R^2. Figure 4 shows that this is generally the case for the catchments with very high adj. R^2 values, and also for later prediction months. This means that the selection of the predictors is likely stable even if additional data are added to the time series in future. However, there are some catchments for which comparably

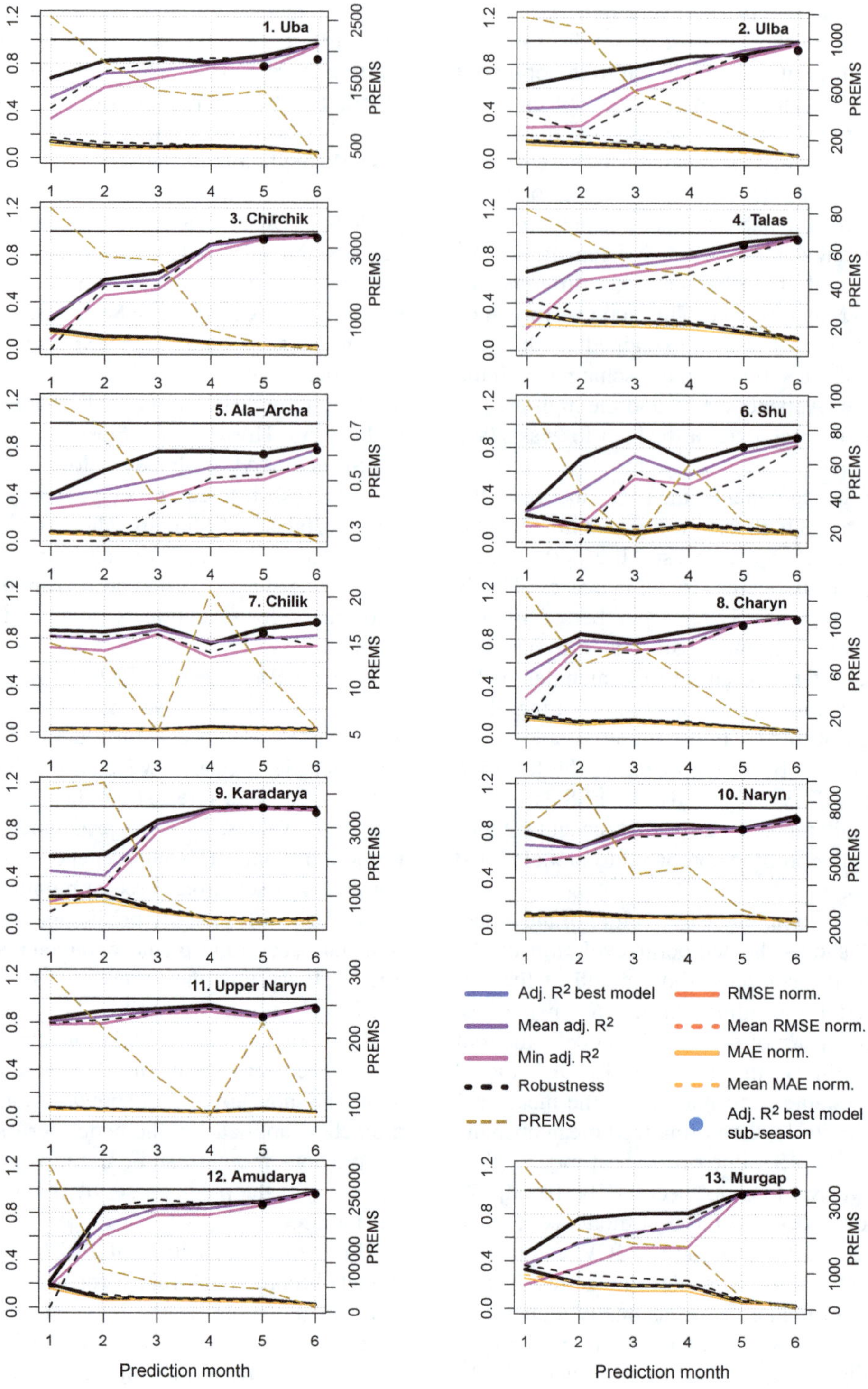

Figure 4. Performance of the prediction models for the different catchments and prediction months. Adj. R^2 best model is the adjusted R^2 of the single best LOOCV model given for the prediction of the complete seasonal discharge (lines) and the sub-seasonal discharge prediction in May and June (dots); mean adj. R^2 is the mean adj. R^2 of the best 20 LOOCV models; min adj. R^2 is minimum adj. R^2 of the best 20 LOOCV models; robustness is mean LOOCV–adj. R^2 of the best 20 models divided by the mean adj. R^2; RMSE (MAE) norm. is the root mean squared error (mean absolute error) of the single best model normalized to mean multi-annual seasonal discharge; mean RMSE (MAE) norm. is the mean root mean square error (mean absolute error) of the best 20 LOOCV models normalized to the multi-annual mean seasonal discharge; PREMS is the predictive residual sum of squares (PRESS) of the single best model, divided by the number of years for which prediction could be made (i.e. predictive residual mean of squares). Prediction months 1–6 refer to predictions made in January to June.

Table 3. Number of times the models yield acceptable prediction according to the criteria of the CA hydromet services for all catchments and prediction months. Numbers indicate percentage of the years of the period 2000–2015 for which the criteria for an acceptable forecast is fulfilled. "Best" indicates the best model according to the LOOCV, and "Mean" indicates the mean percentage over the best 20 models according to the LOOCV.

		January		February		March		April		May		June	
		Best	Mean	Best	Mean	Best	Mean	Best	Mean	Best	Mean	Best	Mean
1	Uba	69 %	70 %	88 %	85 %	88 %	85 %	88 %	83 %	93 %	93 %	100 %	100 %
2	Ulba	80 %	62 %	87 %	71 %	87 %	77 %	93 %	87 %	100 %	99 %	100 %	100 %
3	Chirchik	50 %	54 %	75 %	73 %	75 %	75 %	88 %	93 %	100 %	100 %	100 %	100 %
4	Talas	81 %	67 %	94 %	82 %	88 %	81 %	88 %	88 %	100 %	99 %	100 %	100 %
5	Ala-Archa	67 %	59 %	73 %	63 %	80 %	69 %	87 %	75 %	85 %	81 %	92 %	87 %
6	Shu	69 %	55 %	81 %	70 %	88 %	81 %	81 %	77 %	93 %	90 %	100 %	96 %
7	Chilik	85 %	83 %	85 %	82 %	85 %	93 %	92 %	87 %	92 %	92 %	100 %	93 %
8	Charyn	75 %	67 %	88 %	84 %	88 %	83 %	94 %	88 %	100 %	100 %	100 %	100 %
9	Karadarya	75 %	70 %	69 %	69 %	88 %	84 %	88 %	88 %	100 %	100 %	100 %	100 %
10	Naryn	88 %	79 %	75 %	79 %	88 %	84 %	88 %	87 %	100 %	99 %	100 %	100 %
11	Upper Naryn	88 %	86 %	88 %	87 %	88 %	90 %	94 %	92 %	100 %	96 %	100 %	100 %
12	Amudarya	44 %	51 %	81 %	70 %	75 %	79 %	81 %	82 %	94 %	96 %	100 %	100 %
13	Murgap	75 %	66 %	88 %	76 %	88 %	78 %	88 %	86 %	100 %	100 %	100 %	100 %

less robust models could be derived even for later prediction months (Ala-Archa, Shu). For these catchments it is likely that the predictor selection will change with additional data.

In addition to the performance metrics, Fig. 5 plots the temporal dynamics of the best LOOCV models for all six prediction months. It can be seen that the models can map the high variability of the observed seasonal discharges very well, often already in January or February. This graphically corroborates the findings derived from the performance metrics and underlines that the good performance of the models is not a statistical artefact.

In order to set the performance of the presented models in the context of the routines and guidelines of the CA hydromet services, the performance of the models was also estimated according to the performance criteria used by the hydromet services. This is defined by the following:

$$S_\sigma = \frac{|\mathrm{res}|}{\sigma_{Q_\mathrm{s}}}, \quad (1)$$

with |res| denoting the absolute value of the residual of an individual forecast, and σ_{Q_s} the standard deviation of the seasonal discharge (here calculated for the discharge time series used, i.e. for the period 2000–2015). According to the protocols of the hydromet services an acceptable ("good") forecast is defined by $S_\sigma < 0.675$. Table 3 shows how often this criteria was fulfilled during the analysis period 2000–2015 for the best model, and on average by the best 20 models. For the critical forecast month April the criteria was fulfilled for at least 81 % of the years (13 out of 16 years) by the best model for all catchments. For all catchments the percentages increase further for the later forecast months. These findings are also valid for the set of the best 20 models, as the very similar percentages of the mean of all models compared to the best model indicate. This means that the developed models would provide acceptable forecasts for the hydromet services in the range of 80–90 % for the important forecast month April.

4.1 Predictive uncertainty

In order to quantify the predictive uncertainty the empirical 10 and 90 % percentiles were calculated for every prediction month based on the residuals of the forecast model sets consisting of the best models according to PREMS and fulfilling the formal selection criteria. Note that for the early prediction months, occasionally less than 20 models passed the significance test. The tables in Appendix A3 indicate when this was the case. The quantiles of the residuals were then added to the median of the predictions of the model set, thus providing an 80 % predictive uncertainty band, i.e. an interval in which the true value of the seasonal discharge should lie with a probability of at least 80 %. Figure 6 shows the predictive uncertainty bands for every catchment along with the observed seasonal discharge. The predictive uncertainty for the different prediction months is shown in shades of orange. In general it can be seen that the predictive uncertainty bands narrow with later prediction months, illustrating the better prediction during later prediction months described above. While this is perfectly visible for most catchments (e.g. Chirchik, Karadarya), it is not the case for some others (Ala-Archa, Shu, Chilik, Naryn). For Chilik and Naryn this is a consequence of the already high performance of the early forecasts, which results in similar uncertainty bands for the different prediction months. For Ala-Archa and Shu, however, this is caused by the larger difference between the predictions and performance of the best 20 models compared to the other catchments, as indicated by the difference between the best and mean adj. R^2 shown and listed in Fig. 4 and

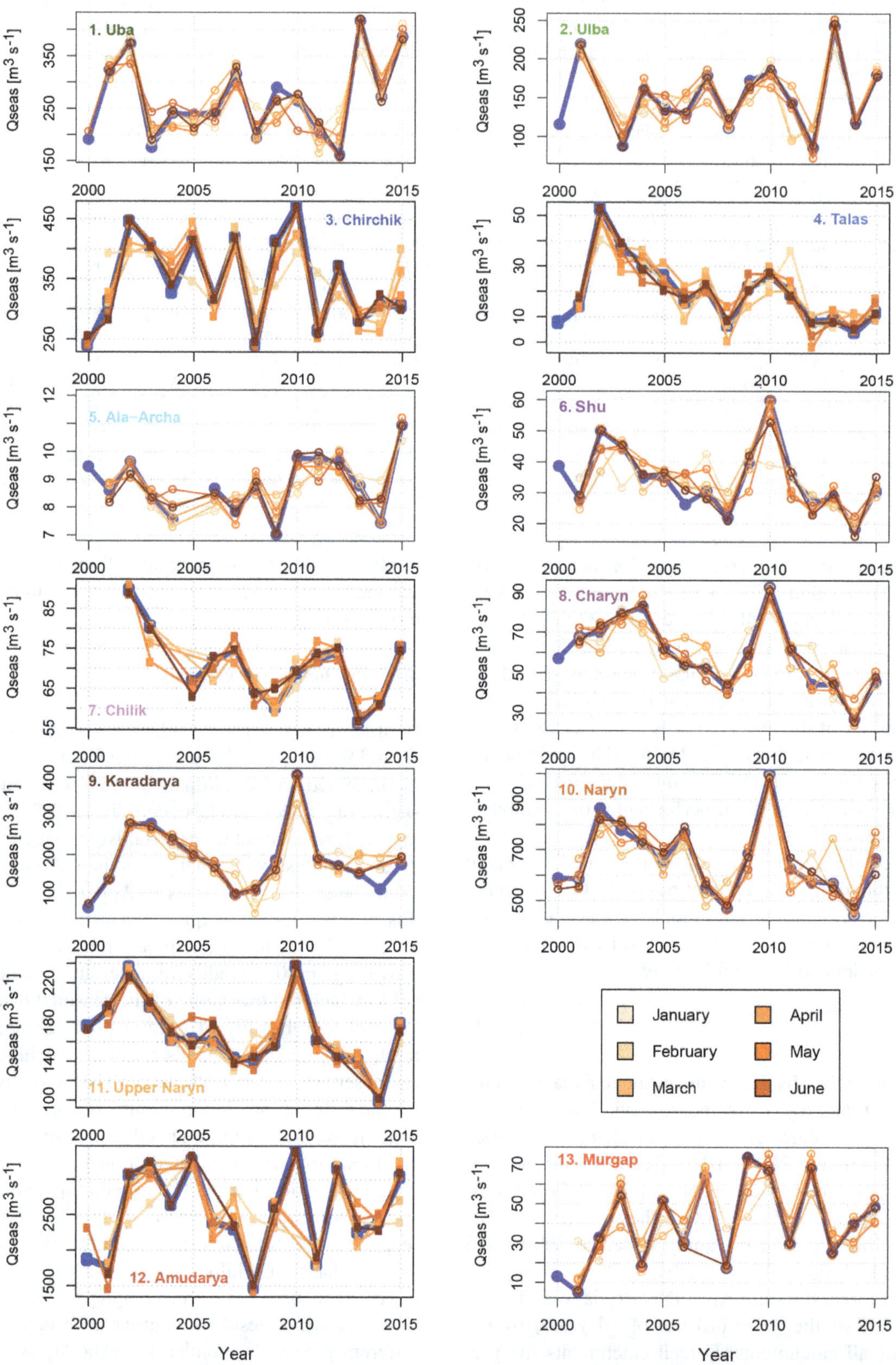

Figure 5. Forecasts of the seasonal discharge by the single best model selected by the LOOCV for the individual catchments and all prediction months. The blue lines show the observed seasonal discharges. Note that some models do not provide forecasts for every year due to missing predictor data.

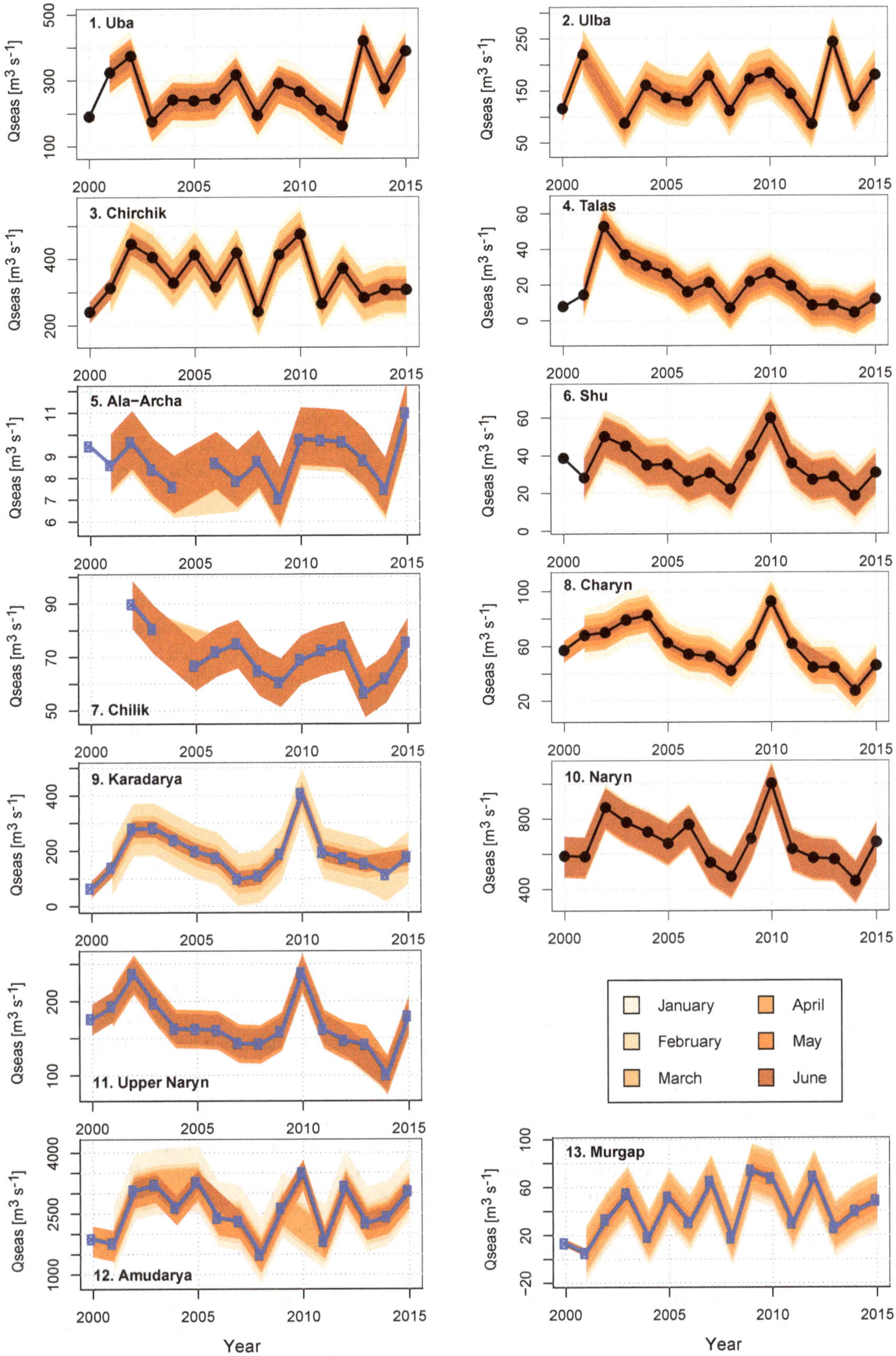

Figure 6. The 80 % predictive uncertainty bands for all catchments and forecasts months. The blue lines indicate the observed seasonal discharges.

Table 2, respectively. This causes a wider distribution of the residuals of the best 20 models and thus higher predictive uncertainty. However, if only the best or a smaller selection of the best 20 models are used for a forecast, the predictive uncertainty would also be reduced. This means that the uncertainty bands derived depend on the subjective choice of the number of models to be kept in the model set. Another reason for wider predictive uncertainty bands for later months is the observed decline in performance during later months in some catchments due to the changed predictor set (e.g. for Shu). This causes again higher predictive uncertainty bands, which overlay the narrower band from the previous month.

From a formal point of view the uncertainty bands correctly include at least 80 % of the observed seasonal discharges, even for very narrow bands (e.g. in June for Chirchik or Karadarya). This indicates that the uncertainty estimation derived from the regression residuals provides a reliable estimation of the uncertainty associated with model selection, and can be used to derive decisions based on the forecasts given by the MLR model sets. However, it must be noted that the derived uncertainty bands represent the predictive uncertainty of the MLR models fitted to the available time series. They do not account for any uncertainty stemming from a possible lack of representativeness of the rather short time series used. Longer discharge time series might show a different variability of seasonal discharge, which would then not be covered by the derived models. However, as the models can be updated every year in future, this potential problem is expected to decrease with further use of the approach in the CA hydromet services.

Moreover, it should be noted that the estimated uncertainty cover only the model selection uncertainty. Other uncertainty sources are the following:

- model structure, which is assumed to be rather low given the high explained variances;
- data sources, which are not quantifiable, but might be high, particularly the discharge data;
- the performance criteria for selecting the best models.

The last aspect has been tested. Using other performance criteria as PREMS can result in a slightly different selection of the best models, but more often just in a different order of the best models. The best PREMS model is not necessarily the best cross-validated R^2 model, or the best MAE or RMSE model. However, as this mainly affects the ordering of the best models, the results in terms of predictions and predictive uncertainty of the model set, if unweighted as presented, would be very similar.

In addition to the predictive uncertainty, the reliability of the forecasts was also quantified by PIT diagrams and PIT scores. Figure 7 shows the PIT diagrams for every catchment and all forecast months using the forecasts of the selected set of models. The PIT diagrams show that the model set predictions are in most cases close to the 1 : 1 line, i.e. provide reliable forecasts. However, in some cases the predictive uncertainty is underestimated (PIT diagram lines with pronounced vertical component around the 50 % quantile). This means that some of the predictive uncertainty bands presented in Fig. 6 are too narrow to reliably quantify the predictive uncertainty. This is mostly the case for the late forecasts with high skill, where the models in the set often produce very similar forecasts. In addition to the diagrams a PIT score was calculated as the area between the PIT curve and the 1 : 1 line as a summarizing indicator for the reliability (Renard et al., 2010). The theoretically least reliable model has a score of 0.5, a perfect model a score of 0. The highest score, i.e. the lowest reliability, of all models is 0.2, with the majority of the models being in the range of 0.07–0.15. Interpreting the scores with the curves in the PIT diagram it can be deduced that the reliability of model sets with PIT scores $\leq$ 0.1–0.14 is acceptable. For higher scores the predictive uncertainty derived from the model set is likely to be underestimated.

4.2 Predictor importance (is there some hydrological process information in linear models?)

Figure 8 illustrates the importance of the predictors of the selected MLR models as absolute fractions of the R^2 values, whereas it is not differentiated between individual predictors, but rather between predictor classes described in Sect. 3.1. The left panels of Fig. 8 show the importance for the single best LOOCV model, while the right panels show the average importance of the predictors for the best 20 LOOCV models. A comparison of the left and right panels shows that the predictor selection and importance for the different catchments and prediction months of the best model is quite similar to the mean of the best 20 models. This indicates that the predictor selection for the models in the set is quite stable, and hence that the predictor selection is not random, but rather has some hydrological meaning. However, an interpretation of the contributions of the different factors is complicated by the use of the composites, which are almost always selected as one or more predictors in the MLR models. Nevertheless, some general features can be identified from Fig. 7:

- Typically there is no single factor dominating the explained variance, with the exception of Karadarya, where the composites have an exceptionally large share of the explained variance. But as the composites are comprised of the other predictors (except antecedent discharge), this statement is actually valid for all catchments. This indicates that the winter snow accumulation providing the bulk of the seasonal discharge is best described by a combination of the factors determining the extent and water equivalent of the snow pack in the catchments (precipitation, temperature, snow coverage). Omitting one of these predictors leads in fact to a reduction in model performance.

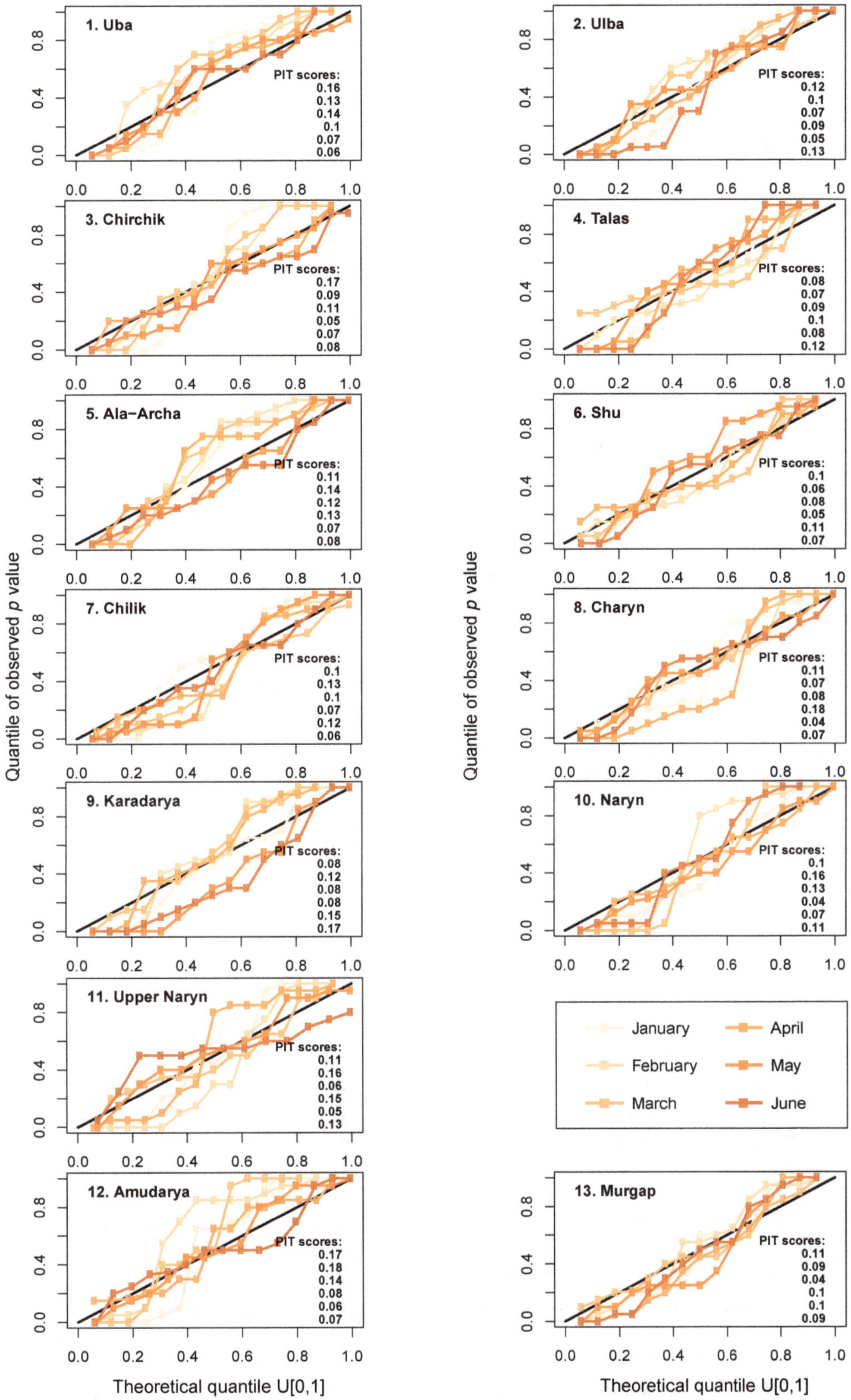

Figure 7. PIT reliability diagrams for every catchment and forecast month. The PIT score is calculated as the area between the reliability plots and the 1 : 1 line as suggested in Renard et al. (2010). The lower the PIT score, the higher the reliability. The theoretically least-reliability score is 0.5, the best 0. The colour codes of the PIT scores indicate the forecast month as in the legend.

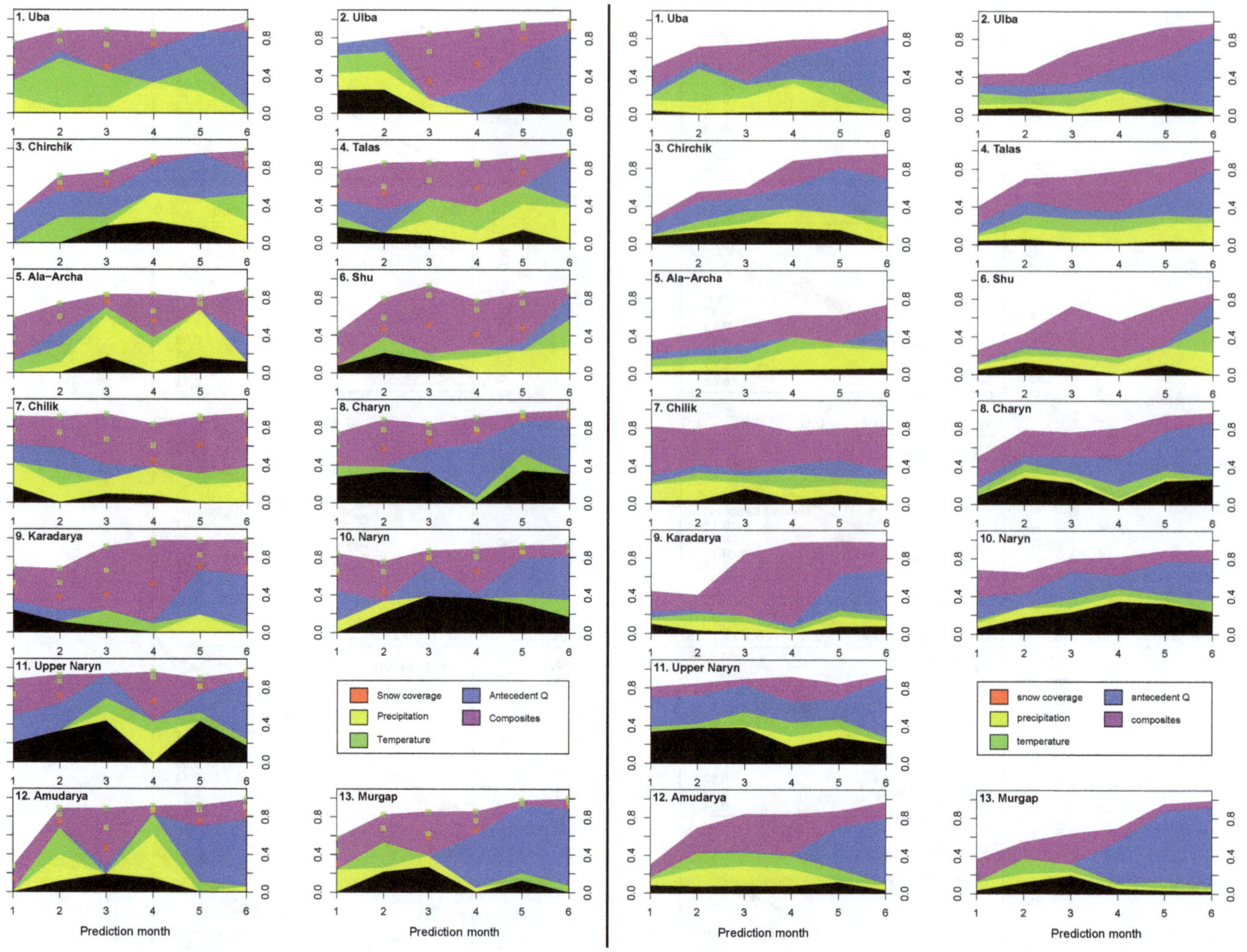

Figure 8. Importance of the predictors in the linear models as absolute contribution to the explained variance (R^2) for all catchments and prediction months. Left panels: of the best LOOCV model; right panels: on average for the best 20 LOOCV models. Squares in the left panels indicate the presence of the different predictors used in the composites: snow cover, precipitation, and temperature, using the same colour codes as for the individual predictors.

- There is a general and plausible trend for higher importance of antecedent discharge in the later prediction months. In this period it can be expected that antecedent discharge has higher predictive power of the seasonal discharge compared to the winter months, i.e. during the accumulation phase, because it directly indicates the magnitude of the discharge generation from snow melt. This finding is valid for most catchments except Chirchik, Ala-Archa, and Chilik. For Chirchik the importance of antecedent discharge is almost constant throughout the prediction months, both for the best model and on average. Contrary to this, antecedent discharge has very little importance for Ala-Archa and Chilik. For Ala-Archa this observation can be explained by the very small catchment size and thus the quick response of discharge to precipitation events and snow melt, i.e. lower transit times, but also with the high proportion of glacier melt during the summer months. The high importance of precipitation, which is higher than in any other catchment particularly in the later prediction months, also supports this reasoning. For Chirchik and Chilik, however, no plausible explanation can be derived from the basic catchment characteristics presented here.
- The importance of the snow coverage predictors indicate a regional differentiation in the predictor importance. For the two catchments in the Altai region (Uba, Ulba, cluster 1 in Fig. 3) snow coverage as an individual factor is of less importance compared to the other regions. This observation can be attributed to different snow cover characteristics in these catchments, which have lower altitudes compared to other catchments in

this study. Therefore, snow accumulation in these catchments is comparably low and quickly responds to increasing temperature already in the first months of the year. Seasonal snow cover variations obtained from the MODSNOW-Tool (Gafurov et al., 2016) for these catchments also illustrate sudden snow cover depletion in the month of April for both catchments, and also for Uba with multiple depletions in winter months until April (analysis not shown in this study). Thus, snow melt is not important in these catchments for seasonal summer discharge, although it may be of high importance for spring discharge, which is beyond the focus of this study. The reverse line of argument can be applied for the relatively high importance of snow coverage for the high-altitude central Tien Shan catchments (Naryn and upper Naryn) with mean annual temperatures below zero (see Table 1), where snow coverage alone explains up to almost 40 % of the explained variance by the MLR models, to which the share of snow coverage contained in the composites has to be added. For these catchments snow coverage alone is thus already a good indicator of the seasonal discharge.

- In terms of predictor importance no obvious differences can be detected on average (right panels in Fig. 8) between the Tien Shan and Pamir discharge regimes identified in the cluster analysis (Fig. 3), with the exception of the Naryn catchments as stated above. The mean predictor importance figures for those catchments are all very similar. This can indicate that although the variability of seasonal discharges varies with geographical location, the runoff-generating processes seem to be similar.

This general interpretation of the predictor importance shows that the selection of the predictors, particularly the change of predictors with prediction months and geographic region, has some hydrological meaning. Due to the simplicity of the approach and the simple linear relationship between the predictors, it is unlikely that more hydrological process information and understanding can be extracted from the MLR results. If this can be achieved at all, then it is on an individual catchment basis only and by the interpretation of the exact predictors, i.e. not aggregated by classes as above. This is, however, beyond the scope of this study. But nevertheless, the observations described above indicate that the general runoff-generation processes can be described by linear models, and that the presented forecast results are unlikely to be obtained by pure chance only.

4.3 Potential of operational application

A lot of management and strategic decisions are based on seasonal forecasts of water availability in CA. The main consumer of water resources in the Aral Sea basin is the agricultural sector, which is based on one of the world's largest irrigation systems (Dukhovny and de Schutter, 2011). The planning of agricultural production crop types and water allocation through the irrigation network are important decisions that are made based on water availability forecasts. Also the estimation of agricultural yield is related to water availability and is needed for country income planning, which depends heavily on agricultural export in some countries. Therefore reliable forecasts of seasonal water availability are essential for the economies of CA states.

In order to design a generic and readily applicable forecast tool, the presented method was designed according to the needs and data availability of the CA hydromet services, which are responsible for the seasonal forecasts. The method is based on station data readily available to the state agencies, thus fulfilling a core prerequisite for an operational implementation of the method. Moreover, the procedure for deriving forecast models is fairly simple and implemented in the open-source software R. Therefore no limitations due to licence issues exist. The model development is automated requiring only some basic definitions, such as the formatting and provision of the predictor data as ASCII text files, and the specification of the prediction month. Therefore the code can be applied by the staff of the hydromet services after a short training period. However, it should be noted that the provided predictor data should be as complete as possible in order to avoid spurious model fitting results (overfitting). Due to the automatic model fitting the algorithm may find the best-performing models fitted to a few years only, if too many predictor data are missing. The chances of overfitting are then greatly increased as the degree of freedom of the linear models, i.e. the ratio of the years used for fitting to the variables in the prediction models, decreases.

The presented model system can also be run with alternative predictor data. For example, it has been tested using gridded ERA-Interim reanalysis data for precipitation and temperature, averaged monthly over the individual catchment areas. Results similar to, if not better than, those presented were obtained. However, due to the latency of at least 2 months until the data are released, an operational use of the model system with ERA-Interim data is not feasible at the moment.

5 Discussion and conclusions

The presented study aimed to develop a flexible and generic forecast model system for the prediction of the seasonal (April–September) discharge in CA river basins, with the final goal of operational use by the CA hydromet services. In order to achieve this the data requirements were kept as low as possible, using only monthly precipitation and temperature data from a single station in the individual catchments, accompanied by operationally processed monthly MODIS snow coverage data and monthly antecedent discharge. Based on this core predictor data set, a variety of monthly, multi-monthly, and composite predictors were automatically derived for different prediction dates. The predictors were then used for predicting the seasonal discharge with multiple linear regression models. In order to avoid overfitting, restrictions were set on the selection and number of predictors in each MLR, and the models were tested for significance and for robustness by a leave-one-out cross-validation. A set of significant prediction models was then selected based on the predictive residual error mean of squares of the LOOCV.

The prediction model system was tested for the period 2000–2015 on a selection of 13 different river basins in different geographic and climatic regions, and with different catchment characteristics. It could be shown that the models provided good to excellent predictions for all catchments and for all defined prediction dates and lead times. For the first prediction on 1 January, i.e. for a lead time of 3 months, the explained variance (expressed as adjusted R^2) is already high in the range of 0.46–0.86 for 9 catchments. For the following prediction on 1 February the explained variance is above 0.59 for all catchments and increases further with the following months. For the important prediction date for the planning of water resources in the region on 1 April just before the high-flow season, adj. R^2 values of the best models for each catchment are in the range 0.68–0.97, indicating exceptional high performance for a seasonal forecast.

The automatic selection of the predictors and their importance revealed some geographic or temporal patterns. Geographically the northern Altai catchments differ in the predictor selection of the best LOOCV–MLR models from the other regions, as snow melt in this region has less contribution to seasonal discharge (April–September), with snow cover often reduced to zero already in early spring. For all catchments the importance of antecedent discharge is increasing with progressing prediction dates. This is plausible from a hydrological perspective: while during the winter months the discharge is dominated by groundwater contribution, the discharge in April and later contains information about the snow melt process and thus has predictive power. This means in summary that the selected predictors and their importance have some hydrological meaning, thus supporting the validity of the forecast models derived by the model system. However, it should be noted that specific features of runoff generation in the catchments cannot be detected and discovered by the rather abstract level of predictors, predictor importance, and the very basic catchment characteristics presented here. Overall, the presented simple forecast system proved to be able to provide robust, very skilful, and reliable forecast models for CA.

The reason for the high performance is surely the temporal separation of most of the annual precipitation (snow in winter), and the runoff generation (snow melt in spring and summer). The forecast is thus based on an estimation of the snow pack accumulation in winter and its snow water equivalent, for which the predictors and their combinations provide proxy information. Moreover, the proxy information is not forecasted, but measured, thus providing more reliable information compared to forecasted predictors. An additional incorporation of climate predictions from dynamical (Kim et al., 2012) and statistical (Gerlitz et al., 2016) seasonal prediction models is unlikely to further increase the forecast skill. The contribution of spring and summer rainfall to discharge variations appears to be dispensable, as represented by the high forecast performance of the applied predictor variables. Furthermore, the prediction of seasonal climate anomalies is highly uncertain, particularly for the rather dry summer season, which impedes its application for seasonal runoff forecasts (Gerlitz et al., 2016). Potential improvements could be achieved using gravity-based water storage variations, as provided by the GRACE mission for example. The total water storage variation monitored by GRACE should actually map the snow accumulation and the snow water content over the whole winter period. This information could be used as predictor for catchments large enough to match the spatial resolution of GRACE.

As the timely separation of precipitation and runoff is a unifying feature of all CA headwater catchments encompassing high-mountain ranges, the model system is able to perform exceptionally well for all tested catchments. It is thus also very likely that the model system will also work well in the CA catchments not included in this study, with some limitations for very small catchments. Thus, the proposed methodology provides a generic and flexible tool for the development of seasonal discharge forecast models for CA rivers. This tool can be used by the responsible hydromet services without the need for larger investments in hardware, software, and education and training of staff. In fact, the model system is already tested in four CA national hydromet services. The only prerequisite for the application of the model system is the availability of meteorological data from stations within the catchments. If this is not given, stations nearby can be used as an alternative. However, they need to be representative for the meteorological variability in the catchment. As this study has shown, this might work, for example, for stations located downstream of the discharge stations within the same but larger catchment. Moreover, a comparable forecast skill can be expected in other regions with similar climatic settings, e.g. the South American dry Andes or the western US (e.g. the Sierra Nevada). The provided information of seasonal water availability could also be used in dam operation and dam safety procedures, as well as strategic flood hazard management plans.

Appendix A

A1 Correlation of seasonal discharge to sub-seasonal discharge

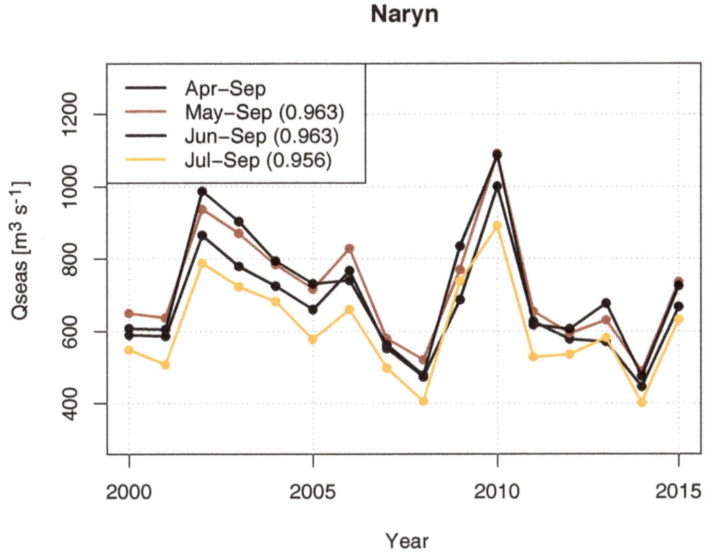

Figure A1. Comparison of seasonal discharge for the whole vegetation period April to September to sub-seasonal discharge time series, taking the Naryn basin as example. The sub-seasonal series are highly correlated to the seasonal time series. Numbers in the legend provide the linear correlation coefficient of the sub-seasonal discharges to the seasonal discharge of the whole vegetation period.

A2 Predictors used for the different prediction dates

The following sections list the predictors created and used for the different forecast dates, ranging from 1 January to 1 June. The predictors are abbreviated, with *snowcov* and *sc* denoting the snow coverage in the catchment derived by the MODSNOW-tool, *precip* the station records of precipitation, *temp* the station records of temperature, and *Q* the discharge recorded at the river gauges. Catchment characteristics and the locations of the gauges are listed in Table 1. The data for all predictors are monthly values (mean for snow coverage, temperature and discharge, total for precipitation), with *jan* indicating January values, *feb* February values, *mar* March values, *apr* April values, *may* May values, and *jun* June values.

Multi-monthly values are mean values of the monthly values spanning over several months, whereas the range of the months included is indicated by the concatenation of the indicators of the months, e.g. *janapr* means multi-monthly means for the period January to April, or *febmar* indicates the mean of the months February and March. The predictor abbreviations are combined with the indicators for the months; *snowcov_apr* thus stands for the mean snow coverage of the catchment in April, or *precip_janmar* for the mean of the monthly total precipitation for the months January to March.

For the composites the predictors included are listed by their abbreviations, followed by the indicators for the months. For calculating the composites, the monthly values of the predictors denoted by the month indicators are multiplied. For example, *sc_temp_mar* thus means the product of the mean snow cover in March and the mean temperature in March, or *sc_temp_precip_janmay* denotes the product of the multi-monthly means January to May of snow coverage, temperature, and precipitation.

Predictors used for prediction on 1 January

Snowcover:
snowcov_dec, snowcov_nov, snowcov_oct, snowcov_octdec
Precipitation:
precip_dec, precip_nov, precip_oct, precip_novdec, precip_octdec
Temperature:
temp_dec, temp_nov, temp_oct, temp_novdec, temp_octdec
Composites snowcover × temperature:
sc_temp_octdec
Composites snowcover, × precipitation:
sc_precip_octdec
Composites temperature × precipitation:
temp_precip_dec, temp_precip_nov, temp_precip_oct, temp_precip_octdec
Composites snowcover, × temperature × precipitation:
sc_temp_precip_octdec
Antecedent discharge:
Q_dec, Q_nov, Q_oct, Q_novdec, Q_octdec

Predictors used for prediction on 1 February

Snowcover:
snowcov_jan, snowcov_dec, snowcov_nov, snowcov_oct, snowcov_octjan
Precipitation:
precip_jan, precip_dec, precip_nov, precip_oct, precip_decjan, precip_novjan, precip_octjan
Temperature:
temp_jan, temp_dec, temp_nov, temp_oct, temp_decjan, temp_novjan, temp_octjan, sc_temp_jan
Composites snowcover, × temperature:
sc_temp_jan
Composites snowcover × precipitation:
sc_precip_jan
Composites temperature × precipitation:
temp_precip_jan, temp_precip_dec, temp_precip_nov, temp_precip_oct, temp_precip_decjan, temp_precip_novjan, temp_precip_octjan
Composites snowcover × temperature × precipitation:
sc_temp_precip_octjan
Antecedent discharge:
Q_jan, Q_dec, Q_nov, Q_oct, Q_decjan, Q_novjan, Q_octjan

Predictors used for prediction on 1 March

Snowcover:
snowcov_feb, snowcov_jan, snowcov_janfeb, snowcov_dec, snowcov_nov, snowcov_oct, snowcov_octfeb
Precipitation:
precip_feb, precip_jan, precip_dec, precip_nov, precip_oct, precip_janfeb, precip_decfeb, precip_novfeb, precip_octfeb
Temperature:
temp_feb, temp_jan, temp_dec, temp_nov, temp_oct, temp_janfeb, temp_decfeb, temp_novfeb, temp_octfeb
Composites snowcover × temperature:
sc_temp_jan, sc_temp_feb, sc_temp_janfeb
Composites snowcover × precipitation:
sc_precip_jan, sc_precip_feb, sc_precip_janfeb
Composites temperature × precipitation:
temp_precip_jan, temp_precip_feb, temp_precip_dec, temp_precip_nov, temp_precip_oct, temp_precip_janfeb, temp_precip_novfeb, temp_precip_octfeb
Composites snowcover × temperature × precipitation:
sc_temp_precip_janfeb, sc_temp_precip_octfeb
Antecedent discharge:
Q_feb, Q_jan, Q_dec, Q_nov, Q_oct, Q_janfeb, Q_decfeb, Q_novfeb, Q_octfeb

Predictors used for prediction on 1 April

Snowcover:
snowcov_mar, snowcov_feb, snowcov_jan, snowcov_janmar, snowcov_febmar
Precipitation:
precip_mar, precip_feb, precip_jan, precip_dec, precip_nov, precip_oct, precip_febmar, precip_janmar, precip_decmar, precip_novmar, precip_octmar
Temperature:
temp_mar, temp_feb, temp_jan, temp_dec, temp_nov, temp_oct, temp_febmar, temp_janmar, temp_decmar, temp_novmar, temp_octmar
Composites snowcover × temperature:
sc_temp_mar, sc_temp_febmar, sc_temp_janmar
Composites snowcover × precipitation:
sc_precip_mar, sc_precip_febmar, sc_precip_janmar, sc_precip_mar_decmar, sc_precip_mar_novmar
Composites temperature × precipitation:
temp_precip_jan, temp_precip_feb, temp_precip_mar, temp_precip_febmar, temp_precip_janmar, temp_precip_decmar, temp_precip_novmar
Composites snowcover × temperature × precipitation:
sc_temp_precip_mar, sc_temp_precip_febmar, sc_temp_precip_janmar
Antecedent discharge:
Q_mar, Q_feb, Q_jan, Q_dec, Q_nov, Q_oct, Q_febmar, Q_janmar, Q_decmar, Q_novmar, Q_octmar

Predictors used for prediction on 1 May

Snowcover:
snowcov_apr, snowcov_mar, snowcov_feb, snowcov_janapr, snowcov_febapr, snowcov_marapr
Precipitation:
precip_apr, precip_mar, precip_feb, precip_jan, precip_marapr, precip_febapr, precip_janapr, precip_decapr, precip_novapr, precip_octapr
Temperature:
temp_apr, temp_mar, temp_feb, temp_jan, temp_marapr, temp_febapr, temp_janapr, temp_decapr, temp_novapr, temp_octapr
Composites snowcover × temperature:
sc_temp_mar, sc_temp_apr, sc_temp_marapr, sc_temp_febapr
Composites snowcover × precipitation:
sc_precip_mar, sc_precip_apr, sc_precip_marapr, sc_precip_febapr
Composites temperature × precipitation:
temp_precip_jan, temp_precip_feb, temp_precip_mar, temp_precip_apr, temp_precip_febapr, temp_precip_marapr, temp_precip_octapr
Composites snowcover × temperature × precipitation:
sc_temp_precip_mar, sc_temp_precip_apr, sc_temp_precip_marapr, sc_temp_precip_janapr
Antecedent discharge:
Q_apr, Q_mar, Q_feb, Q_jan, Q_marapr, Q_febapr, Q_janapr, Q_decapr, Q_novapr, Q_octapr

Predictors used for prediction on 1 June

Snowcover:
snowcov_apr, snowcov_mar, snowcov_feb, snowcov_janapr, snowcov_febapr, snowcov_marapr
Precipitation:
precip_may, precip_apr, precip_mar, precip_feb, precip_jan, precip_aprmay, precip_marmay, precip_febmay, precip_janmay, precip_octmay
Temperature:
temp_may, temp_apr, temp_mar, temp_feb, temp_jan, temp_aprmay, temp_marmay, temp_febmay, temp_janmay, temp_octmay
Composites snowcover × temperature:
sc_temp_mar, sc_temp_apr, sc_temp_marmay
Composites snowcover × precipitation:
sc_precip_mar, sc_precip_apr, sc_precip_marmay
Composites temperature × precipitation:
temp_precip_feb, temp_precip_mar, temp_precip_apr, temp_precip_may, temp_precip_marmay, temp_precip_octmay
Composites snowcover × temperature × precipitation:
sc_temp_precip_mar, sc_temp_precip_apr, sc_temp_precip_marmay, sc_temp_precip_janmay
Antecedent discharge:
Q_may, Q_apr, Q_mar, Q_feb, Q_jan, Q_aprmay, Q_marmay, Q_febmay, Q_janmay Q_octmay

A3 Autocorrelation of seasonal discharge time series

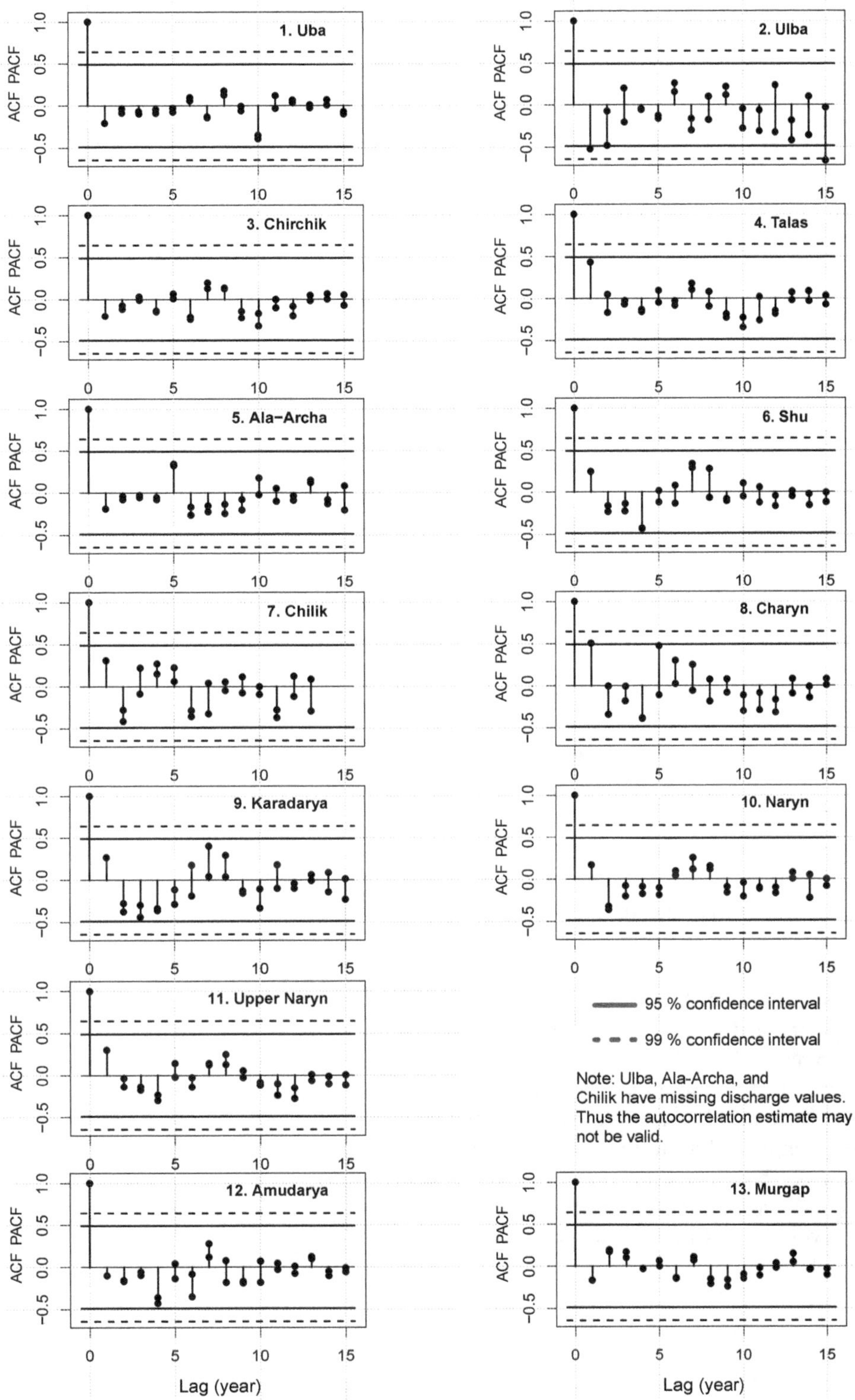

Figure A2. Auto-correlation (black) and partial auto-correlation (red) of the seasonal discharge tome series for all catchments and possible lags.

Table A1. Test for normal distributed residuals, for every catchment, prediction month, and the selected 20 models.

	January	February	March
Uba	1 1 1 1 1 1 1 1 1 1 1 1 1 1 1 1 1 1 1 1	1 0 1 0 1 0 1 1 0 1 1 0 1 0 0 1 0 1 1 0	1 0 0 0 0 1 1 1 0 1 1 1 0 0 0 0 1 0 1 1
Ulba	1 1 1 1 1 1 1 1 1 1 1 1 1 1 1 1 1 1 1 1	1 1 1 1 0 1 1 1 0 1 1 1 1 1 0 1 1 1 1 0	1 1 1 0 1 1 1 1 1 1 1 1 1 1 1 1 1 1 1 1
Chirchik	1 1 1 1 1 1 1 1 1 0 1 0 1 1 1 1 1 1 1 NA	1 1 1 1 1 1 1 1 1 1 1 1 1 1 1 1 1 1 1 1	1 1 1 1 1 1 1 1 1 1 1 1 1 1 1 1 1 1 1 1
Talas	1 1 1 1 1 1 1 1 1 1 0 0 0 1 1 1 1 1 1 1	1 1 1 1 1 1 1 1 1 1 1 0 1 1 1 1 1 1 1 1	1 1 1 1 1 1 1 1 1 1 1 1 1 1 1 1 1 1 1 1
Ala-Archa	1 1 1 1 1 0 1 1 1 1 1 1 0 0 1 1 1 1 1 1	1 0 1 1 0 1 1 1 1 1 1 1 1 1 1 1 1 0 1 1	1 1 1 0 1 1 1 1 1 1 1 1 1 1 0 0 1 1 1 1
Shu	1 1 1 1 0 1 1 1 1 1 1 1 1 NA NA NA NA NA NA NA	1 1 1 0 1 1 1 1 1 1 1 1 1 1 1 1 1 1 1 1	1 1 1 1 1 1 1 1 1 1 1 1 1 1 1 1 1 1 0 1
Chilik	1 1 1 1 1 1 1 1 1 1 1 1 1 1 1 1 1 1 1 1	1 1 1 1 1 1 1 1 1 1 1 1 1 1 1 1 1 1 1 1	1 0 1 1 1 1 1 1 0 1 1 0 1 1 1 1 1 1 1 1
Charyn	1 1 1 1 1 1 1 1 1 1 1 1 1 1 1 1 1 1 1 1	1 1 1 1 1 1 1 1 1 1 1 1 1 1 1 1 1 1 1 1	1 1 1 1 1 1 1 1 1 1 1 1 1 1 1 1 1 1 1 1
Karadarya	1 1 1 1 1 1 1 1 1 1 1 0 1 0 1 1 1 1 1 0	1 1 1 1 1 1 1 1 1 1 1 1 1 1 1 1 1 1 1 1	1 1 1 1 1 1 0 0 0 1 1 1 0 1 1 1 0 1 1 0
Naryn	1 1 1 1 1 1 1 0 1 1 1 1 1 1 1 1 1 1 1 1	1 1 1 1 1 1 1 1 1 1 1 1 1 1 1 1 1 1 1 1	1 1 1 1 1 1 1 1 1 1 1 1 1 1 1 1 1 1 1 1
Upper Naryn	1 1 1 1 1 1 1 1 1 1 1 1 1 1 1 1 1 1 1 1	1 1 1 1 1 0 1 1 1 0 0 0 1 1 1 1 1 1 1 1	0 0 1 0 1 1 1 1 1 1 1 1 1 1 1 1 1 1 1 0
Amudarya	1 1 1 1 1 1 1 1 1 1 1 1 1 NA NA NA NA NA NA NA	1 1 1 1 1 1 1 1 1 1 1 1 1 1 1 1 1 1 1 1	1 1 1 1 1 1 1 1 1 1 1 1 1 1 1 1 1 1 1 1
Murgap	1 1 1 1 1 1 1 1 1 1 1 1 1 1 1 1 1 1 0 1	1 1 1 1 1 1 1 1 1 1 1 1 1 0 1 1 1 1 1 1	1 1 1 1 1 1 1 1 1 1 1 1 0 1 1 1 1 1 1 1
	April	**May**	**June**
Uba	1 1 1 1 1 1 1 1 1 1 1 1 1 1 1 0 1 1 1 0	1 1 1 1 1 1 1 1 1 1 1 1 1 1 1 1 1 1 1 1	1 1 1 1 1 1 1 1 1 1 1 1 1 1 1 1 1 1 1 1
Ulba	1 1 1 1 1 1 1 1 1 1 1 1 1 1 1 1 1 1 1 1	1 1 1 1 1 1 1 1 1 1 1 0 0 0 0 1 0 0 0 0	1 1 1 1 1 1 1 1 1 1 1 1 1 1 1 1 1 1 1 1
Chirchik	1 1 1 1 1 1 1 1 1 1 1 1 1 1 1 1 0 1 0 0	0 0 0 0 0 1 1 1 1 0 0 0 0 0 0 0 0 0 0 1	0 0 1 1 1 0 0 1 1 1 0 1 1 1 1 1 1 1 1 1
Talas	1 1 1 1 1 1 1 1 1 1 1 1 1 1 1 1 1 1 1 1	1 1 1 1 1 1 1 1 1 1 1 1 1 1 1 1 1 1 1 1	1 0 1 0 1 1 1 1 0 1 1 1 1 1 1 1 1 0 0 1
Ala-Archa	1 1 1 1 1 1 1 1 1 1 1 1 1 1 1 1 1 1 1 1	1 1 1 0 1 1 0 1 0 1 1 1 1 1 1 1 0 1 1 1	1 1 1 1 1 1 1 1 1 1 1 1 1 1 1 1 1 1 1 1
Shu	1 1 1 1 0 0 0 1 1 0 1 1 1 1 1 1 0 0 1 1	1 1 1 1 1 1 1 1 1 1 1 1 1 1 0 1 1 1 1 1	1 1 1 1 1 1 1 1 1 1 1 1 1 1 1 1 1 1 1 1
Chilik	1 1 1 0 1 1 1 1 1 0 1 1 1 1 0 1 1 1 1 1	1 1 1 0 1 0 1 0 1 1 1 0 1 1 1 0 1 1 0 0	1 1 1 1 1 1 1 1 1 0 1 0 0 1 1 1 0 1 0 1
Charyn	1 1 1 1 1 1 1 1 1 1 1 1 1 1 1 1 1 1 1 1	1 1 1 1 1 1 1 1 1 1 1 1 1 1 1 1 1 1 1 1	1 1 1 1 1 1 1 1 1 1 1 1 1 1 1 1 1 1 1 1
Karadarya	1 1 1 1 1 1 1 1 1 1 1 1 1 1 1 1 0 1 1 1	1 1 1 1 1 1 1 1 1 1 1 1 1 1 1 1 1 1 1 1	1 1 1 0 1 1 1 0 0 1 1 0 0 1 1 1 1 1 0 0
Naryn	1 1 1 1 1 1 1 1 1 1 1 1 1 1 1 1 1 1 1 1	1 1 1 0 0 1 1 1 1 1 0 1 1 1 1 1 1 1 1 1	1 1 1 1 1 1 1 1 1 1 0 1 1 1 1 1 1 1 1 1
Upper Naryn	1 1 1 1 1 1 1 1 1 1 1 1 1 0 1 1 1 1 1 1	1 1 1 1 1 1 0 0 0 0 1 1 1 1 1 1 1 1 1 1	1 1 1 1 1 1 1 1 1 1 1 1 1 1 1 1 1 1 1 1
Amudarya	1 1 1 1 1 1 0 1 1 1 1 1 1 1 1 1 1 1 1 1	1 1 1 1 1 1 1 1 1 0 1 1 1 1 0 1 1 1 1 0	1 1 1 1 1 1 1 1 1 1 1 1 1 1 1 1 1 1 1 1
Murgap	1 1 1 1 1 1 1 1 1 0 1 1 1 0 0 1 1 1 0 1	1 1 1 1 1 1 1 1 1 1 0 1 1 1 1 0 1 1 1 1	1 1 1 1 1 0 1 0 0 1 1 1 1 1 1 0 1 1 0 1

1 = normal distributed. 0 = not normal distributed. NA = no valid model found.

A4 Formal test for MLR assumptions

The residuals of the models are tested for normality by the Shapiro–Wilk test for normality. Doing so, one has to bear in mind that this test is based on a sample size of maximal 16 values for each model only, so the test may not provide meaningful results. The table below shows the test result for every model, catchment, and forecast month. A "1" indicates normal distributed residuals, "0" not normal distributed residuals. "NA" indicates that no more models with significant predictors could be found. For every forecast month up to 20 indices are given according to the set of the best 20 models to be retained. The table shows that for most of the models (91 %) the test was positive, i.e. the residuals are normally distributed, even for this rather low and possibly not representative sample size.

Table A2. Test for autocorrelated (independent) residuals, for every catchment, prediction month, and the selected 20 models; lag = 1.

	January	February	March
Uba	0 0 0 0 0 0 0 0 0 0 1 0 0 1 0 0 0 0 1 0	1 0 0 0 1 0 0 0 0 0 0 0 1 0 1 1 1 0 0 0	0 0 0 0 0 0 0 0 0 0 0 0 0 0 0 0 0 0 0 0
Ulba	1 0 0 0 0 1 0 1 0 0 0 0 0 0 1 0 1 0 0 1	0 1 0 0 0 0 1 1 0 0 0 1 0 0 0 0 0 0 1 0	0 0 0 0 0 0 0 0 0 0 0 0 0 0 0 0 0 0 0 0
Chirchik	0 0 0 0 0 0 0 0 0 0 0 0 0 0 0 0 0 0 0 NA	0 0 0 0 0 0 0 0 0 0 0 0 0 0 0 0 0 0 0 0	0 0 0 0 0 0 0 0 0 0 0 0 0 0 0 0 0 0 0 0
Talas	0 0 0 0 0 0 0 0 0 0 0 0 0 0 0 0 0 0 0 0	0 0 0 0 0 0 0 0 0 0 0 0 0 0 0 0 0 0 0 0	0 0 0 0 0 0 0 0 0 0 0 0 0 0 0 0 0 0 0 0
Ala-Archa	0 0 0 0 0 0 0 0 0 0 0 0 0 0 0 0 0 0 0 0	0 0 0 0 0 0 0 0 0 0 0 0 0 0 0 0 1 1 0 1	0 0 0 0 0 0 0 0 0 0 0 0 0 0 0 0 0 0 0 0
Shu	0 0 0 0 0 0 0 0 0 0 0 0 0 NA NA NA NA NA NA NA	0 0 0 0 0 0 0 0 0 0 0 0 0 0 0 0 0 0 0 0	0 1 1 0 0 0 0 0 0 0 0 0 0 0 0 0 0 0 0 0
Chilik	0 0 0 0 0 0 0 0 0 0 0 0 0 0 0 0 0 0 0 0	0 0 0 0 0 0 0 0 0 0 0 0 0 0 0 0 0 0 0 0	0 0 0 0 0 0 0 0 0 0 0 0 0 0 0 0 0 0 0 0
Charyn	0 0 0 0 0 0 0 0 0 0 0 1 0 0 0 0 0 0 0 0	0 0 0 0 0 0 0 0 0 0 0 0 0 0 0 0 0 0 0 0	0 0 0 0 0 0 0 0 0 0 0 0 0 0 0 0 0 0 0 0
Karadarya	0 1 0 0 0 0 0 0 0 1 0 0 0 0 0 1 0 0 0 0	0 0 0 0 0 0 0 0 0 0 0 0 0 0 0 0 0 0 0 0	0 0 0 0 0 0 0 0 0 0 0 0 0 0 0 0 0 0 0 0
Naryn	0 0 0 0 0 0 0 0 0 0 0 0 0 0 0 0 0 0 0 0	0 0 0 0 0 0 0 0 0 0 0 0 0 0 0 0 0 0 0 0	0 0 0 0 0 0 0 0 0 0 0 0 0 0 0 0 0 0 0 0
Upper Naryn	0 0 0 0 0 0 0 0 0 0 0 0 0 0 0 0 0 0 0 0	0 0 0 0 0 0 0 0 0 0 0 0 0 0 0 0 0 0 0 0	0 0 0 0 0 1 0 0 0 0 0 0 0 0 0 0 0 0 0 0
Amudarya	0 0 0 0 0 0 0 0 0 0 0 0 0 NA NA NA NA NA NA NA	0 0 0 0 0 0 0 0 0 0 0 0 0 0 0 0 0 0 0 0	0 0 0 0 0 0 0 0 0 0 0 0 0 0 0 0 0 0 0 0
Murgap	0 0 0 0 0 0 0 0 0 0 0 0 0 0 0 0 0 0 0 0	1 0 0 0 1 0 0 0 0 0 0 0 0 0 0 0 0 0 0 0	0 0 0 0 0 0 0 0 0 0 0 0 0 0 0 0 0 0 0 0
	April	**May**	**June**
Uba	0 0 0 0 0 0 0 0 0 0 0 0 0 0 0 0 0 0 0 0	0 0 0 0 0 0 0 0 0 0 0 0 0 0 0 0 0 0 0 0	0 0 0 0 0 0 0 0 0 0 0 0 0 0 0 0 0 0 0 0
Ulba	0 0 0 0 0 0 0 0 0 0 0 0 0 0 0 0 0 0 0 0	0 0 0 0 0 0 0 0 0 0 0 0 0 0 0 0 0 0 0 0	1 1 1 1 0 1 0 1 1 1 0 0 1 0 0 0 0 0 0 0
Chirchik	0 0 0 0 0 0 0 0 0 0 0 0 0 0 0 0 0 0 0 0	0 0 0 0 0 0 0 0 0 0 0 0 0 0 0 0 0 0 0 0	0 0 0 0 0 0 0 0 0 0 0 0 0 0 0 0 0 0 0 0
Talas	0 0 0 0 0 0 0 0 0 0 0 0 0 0 0 0 0 0 0 0	0 0 0 0 0 0 0 0 0 0 0 0 0 0 0 0 0 0 0 0	0 0 0 0 0 0 0 0 0 0 0 0 0 0 0 0 0 0 0 0
Ala-Archa	0 0 0 0 0 0 0 0 0 0 0 0 0 0 0 0 0 0 0 0	0 0 0 0 0 0 0 0 0 0 0 0 0 0 0 0 0 0 0 0	0 0 0 0 0 0 0 0 0 0 0 0 0 0 0 0 0 0 0 0
Shu	0 0 0 0 0 0 0 0 0 0 0 0 0 0 0 0 0 0 0 0	0 0 0 0 0 0 0 0 0 0 0 0 0 0 0 0 0 0 0 0	0 0 0 0 0 1 1 0 0 0 0 0 0 0 1 0 0 0 0 0
Chilik	0 0 0 0 0 0 0 0 0 0 0 0 0 0 0 0 0 0 0 0	1 0 0 0 1 0 0 0 0 0 0 0 0 0 0 0 0 0 0 0	0 0 0 0 0 0 0 0 0 0 0 0 0 0 0 0 0 0 0 0
Charyn	0 0 0 0 0 0 0 0 0 0 0 0 0 0 0 0 0 0 0 0	0 0 0 0 0 0 0 0 0 0 0 0 0 0 0 0 0 0 0 0	0 0 0 0 0 0 0 0 0 0 0 0 0 0 0 0 0 1 0 0
Karadarya	0 0 0 0 0 0 0 0 0 0 0 0 0 0 0 0 0 0 0 0	0 0 0 0 0 0 0 0 0 0 0 0 0 0 0 0 0 0 0 0	0 0 0 0 0 0 0 0 0 0 0 0 0 0 0 0 0 0 0 0
Naryn	0 0 0 0 0 0 0 0 0 0 0 0 0 0 0 0 0 0 0 0	0 0 1 1 0 0 0 0 0 0 0 0 0 0 0 0 0 0 0 0	0 0 0 0 0 0 0 0 0 0 0 0 0 0 0 1 0 0 0 0
Upper Naryn	0 0 0 0 0 0 0 0 0 0 0 0 0 0 0 0 0 0 0 0	0 0 0 0 0 0 1 1 1 1 0 0 0 0 0 0 0 0 0 0	1 1 0 0 0 0 0 0 0 1 1 0 0 0 0 0 1 0 1 0
Amudarya	0 0 0 0 0 0 0 0 0 0 0 0 0 0 0 0 0 0 1 0	0 0 0 0 0 0 0 0 0 0 0 0 0 1 0 0 0 0 0 0	0 0 0 0 0 0 0 0 0 0 0 0 1 0 0 0 0 0 0 0
Murgap	0 0 0 0 0 0 0 1 0 0 0 0 0 0 0 0 0 0 0 0	1 0 0 0 0 0 0 0 0 0 0 0 0 0 0 0 0 0 0 0	0 0 0 0 0 0 0 0 0 0 0 0 0 0 0 0 0 0 0 0

1 = correlated. 0 = not correlated. NA = no valid model found.

Furthermore, testing was carried out to determine whether the residuals are independent applying a test for autocorrelation with lag 1 at significance level $p = 0.05$. In Table A2 a "0" indicates independence, a "1" dependence. It shows that 96 % of the models have independent residuals.

Table A3. Test for homoscedastic residuals, for every catchment, prediction month, and the selected 20 models.

	January	February	March
Uba	1 1 1 1 1 1 1 1 1 1 1 1 1 1 1 1 1 1 1 1	1 1 1 1 1 1 1 1 1 1 1 1 1 1 1 1 1 1 1 1	1 1 1 1 1 1 1 1 1 1 1 1 1 1 1 1 1 1 1 1
Ulba	1 1 1 1 1 1 1 1 1 1 1 1 1 1 1 1 1 1 1 1	1 1 1 1 1 1 1 1 1 1 1 1 1 1 1 1 1 1 1 1	1 1 1 1 1 1 1 1 1 1 1 1 1 1 1 1 1 1 1 1
Chirchik	1 1 1 1 1 1 1 1 1 1 1 1 1 1 1 1 1 1 1 NA	1 1 1 1 1 1 1 1 1 1 1 1 1 1 1 1 1 1 1 1	1 1 1 1 1 1 1 1 1 1 1 1 1 1 1 1 1 1 1 1
Talas	1 1 1 1 1 1 1 1 1 1 1 1 1 1 1 1 1 1 1 1	1 1 1 1 1 1 1 1 1 1 1 1 1 1 1 1 1 1 1 1	1 1 1 1 1 1 1 1 1 1 1 1 1 1 1 1 1 1 1 1
Ala-Archa	1 1 1 1 1 1 1 1 1 1 1 1 1 1 1 1 1 1 1 1	1 1 1 1 1 1 1 1 1 1 1 1 1 1 1 1 1 1 1 1	1 1 1 1 1 1 1 1 1 1 1 1 1 1 1 1 1 1 1 1
Shu	1 1 1 1 1 1 1 1 1 1 1 1 1 NA NA NA NA NA NA NA	1 1 1 1 1 1 1 1 1 1 1 1 1 1 1 1 1 1 1 1	1 1 1 1 1 1 1 1 1 1 1 1 1 1 1 1 1 1 1 1
Chilik	1 1 1 1 1 1 1 1 1 1 1 1 1 1 1 1 1 1 1 1	1 1 1 1 1 1 1 1 1 1 1 1 1 1 1 1 1 1 1 1	1 1 1 1 1 1 1 1 1 1 1 1 1 1 1 1 1 1 1 1
Charyn	1 1 1 1 1 1 1 1 1 1 1 1 1 1 1 1 1 1 1 1	1 1 1 1 1 1 1 1 1 1 1 1 1 1 1 1 1 1 1 1	1 1 1 1 1 1 1 1 1 1 1 1 1 1 1 1 1 1 1 1
Karadarya	1 1 1 1 1 1 1 1 1 1 1 1 1 1 1 1 1 1 1 1	1 1 1 1 1 1 1 1 1 1 1 1 1 1 1 1 1 1 1 1	1 1 1 1 1 1 1 1 1 1 1 1 1 1 1 1 1 1 1 1
Naryn	1 1 1 1 1 1 1 1 1 1 1 1 1 1 1 1 1 1 1 1	1 1 1 1 1 1 1 1 1 1 1 1 1 1 1 1 1 1 1 1	1 1 1 1 1 1 1 1 1 1 1 1 1 1 1 1 1 1 1 1
Upper Naryn	1 1 1 1 1 1 1 1 1 1 1 1 1 1 1 1 1 1 1 1	1 1 1 1 1 1 1 1 1 1 1 1 1 1 1 1 1 1 1 1	1 1 1 1 1 1 1 1 1 1 1 1 1 1 1 1 1 1 1 1
Amudarya	1 1 1 1 1 1 1 1 1 1 1 1 1 NA NA NA NA NA NA NA	1 1 1 1 1 1 1 1 1 1 1 1 1 1 1 1 1 1 1 1	1 1 1 1 1 0 1 1 1 1 1 1 1 1 1 1 1 1 1 1
	April	**May**	**June**
Uba	1 1 1 1 1 1 1 1 1 1 1 1 1 1 1 1 1 1 1 1	1 1 1 1 1 1 1 1 1 1 1 1 1 1 1 1 1 1 1 1	1 1 1 1 1 1 1 1 1 1 1 1 1 1 1 1 1 1 1 1
Ulba	1 1 1 1 1 1 1 1 1 1 1 1 1 1 1 1 1 1 1 1	1 1 1 1 1 1 1 1 1 1 1 1 1 1 1 1 1 1 1 1	1 1 1 1 1 1 1 1 1 1 1 1 1 1 1 1 1 1 1 1
Chirchik	1 1 1 1 1 1 1 1 1 1 1 1 1 1 1 1 1 1 1 1	1 1 1 1 1 1 1 1 1 1 1 1 1 1 1 1 1 1 1 1	1 1 1 1 1 1 1 1 1 1 1 1 1 1 1 1 1 1 1 1
Talas	1 0 1 1 1 1 1 1 1 1 1 1 1 1 0 1 1 1 1 1	1 1 1 1 1 1 1 1 1 1 1 1 1 1 1 1 1 1 1 1	1 1 1 1 1 1 1 1 1 1 1 1 1 1 1 1 1 1 1 1
Ala-Archa	1 1 1 1 1 1 1 1 1 1 1 1 1 1 1 1 1 1 1 1	1 1 1 1 1 1 1 1 1 1 1 1 1 1 1 1 1 1 1 1	1 1 1 1 1 1 1 1 1 1 1 1 1 1 1 1 1 1 1 1
Shu	1 1 1 1 1 1 1 1 1 1 1 1 1 1 1 1 1 1 1 1	1 1 1 1 1 1 1 1 1 1 1 1 1 1 1 1 1 1 1 1	1 1 1 1 1 1 1 1 1 1 1 1 1 1 1 1 1 1 1 1
Chilik	1 1 1 1 1 1 1 1 1 1 1 1 1 1 1 1 1 1 1 1	1 1 1 1 1 1 1 1 1 1 1 1 1 1 1 1 1 1 1 1	1 1 1 1 1 1 1 1 1 1 1 1 1 1 1 1 1 1 1 1
Charyn	1 1 1 1 1 1 1 1 1 1 1 1 1 1 1 1 1 1 1 1	1 1 1 1 1 1 1 1 1 1 1 1 1 1 1 1 1 1 1 1	1 1 1 1 1 1 1 1 1 1 1 1 1 1 1 1 1 1 1 1
Karadarya	1 1 1 1 1 1 1 1 1 1 1 1 1 1 1 1 1 1 1 1	1 1 1 1 1 1 1 1 1 1 1 1 1 1 1 1 1 1 1 1	1 1 1 1 1 1 1 1 1 1 1 1 1 1 1 1 1 1 1 1
Naryn	1 1 1 1 1 1 1 1 1 1 1 1 1 1 1 1 1 1 1 1	1 1 1 1 1 1 1 1 1 1 1 0 0 1 1 1 1 1 1 1	1 1 1 1 1 1 1 1 1 1 1 1 1 1 1 1 1 1 1 1
Upper Naryn	1 1 1 1 1 1 1 1 1 1 1 1 1 1 1 1 1 1 1 1	1 1 1 1 1 1 1 1 1 1 1 1 1 1 1 1 1 1 1 1	1 1 1 1 1 1 1 1 1 1 1 1 1 1 1 1 1 1 1 1
Amudarya	1 1 1 1 1 1 1 1 1 1 1 1 1 1 1 1 1 1 1 1	1 1 1 1 1 1 1 1 1 1 1 1 1 1 1 1 1 1 1 1	1 1 1 1 1 1 1 1 1 1 1 1 1 1 0 1 1 1 1 1

1 = homoscedasticity test (Breusch–Pagan test) passed. 0 = homoscedasticity test not passed. NA = no valid model found.

Finally, the Breusch–Pagan test for heteroscedasticity was applied to the residuals. This test shows that 99.5 % of the models have homoscedastic residuals. In Table A3 a “1” in- dicates homoscedastic residuals, a “0” heteroscedastic residuals according to the test.

Competing interests. The authors declare that they have no conflict of interest.

Special issue statement. This article is part of the special issue "Sub-seasonal to seasonal hydrological forecasting". It is not associated with a conference.

Acknowledgements. This work was undertaken within the framework of the CAWa project (http://www.cawa-project.net), funded by the German Federal Foreign Office as part of the German Water Initiative for Central Asia (grant number AA7090002).

Edited by: Ilias Pechlivanidis

References

Agaltseva, N. A., Borovikova, L. N., and Konovalov, V. G.: Automated system of runoff forecasting for the Amudarya River basin, in: Destructive Water: Water-Caused Natural Disasters, their Abatement and Control, Anaheim, California, 193–201, 1997.

Aizen, V. B., Aizen, E. M., and Melack, J. M.: Climate, snow cover, glaciers, and runoff in the Tien Shan, Central Asia, J. Am. Water Resour. Assoc., 31, 1113–1129, https://doi.org/10.1111/j.1752-1688.1995.tb03426.x, 1995.

Aizen, V. B., Aizen, E. M., and Melack, J. M.: Precipitation, melt and runoff in the northern Tien Shan, J. Hydrol., 186, 229–251, https://doi.org/10.1016/S0022-1694(96)03022-3, 1996.

Aizen, V. B., Aizen, E. M., and Kuzmichonok, V. A.: Glaciers and hydrological changes in the Tien Shan: simulation and prediction, Environ. Res. Lett., 2, 045019, https://doi.org/10.1088/1748-9326/2/4/045019, 2007.

Archer, D. R. and Fowler, H. J.: Using meteorological data to forecast seasonal runoff on the River Jhelum, Pakistan, J. Hydrol., 361, 10–23, https://doi.org/10.1016/j.jhydrol.2008.07.017, 2008.

Barlow, M. A. and Tippett, M. K.: Variability and Predictability of Central Asia River Flows: Antecedent Winter Precipitation and Large-Scale Teleconnections, J. Hydrometeorol., 9, 1334–1349, https://doi.org/10.1175/2008jhm976.1, 2008.

Bothe, O., Fraedrich, K., and Zhu, X.: Precipitation climate of Central Asia and the large-scale atmospheric circulation, Theor. Appl. Climatol., 108, 345–354, https://doi.org/10.1007/s00704-011-0537-2, 2012.

Conrad, C., Schonbrodt-Stitt, S., Low, F., Sorokin, D., and Paeth, H.: Cropping Intensity in the Aral Sea Basin and Its Dependency from the Runoff Formation 2000–2012, Remote Sensing, 8, 630, https://doi.org/10.3390/rs8080630, 2016.

Crochemore, L., Ramos, M. H., Pappenberger, F., and Perrin, C.: Seasonal streamflow forecasting by conditioning climatology with precipitation indices, Hydrol. Earth Syst. Sci., 21, 1573–1591, https://doi.org/10.5194/hess-21-1573-2017, 2017.

Delbart, N., Dunesme, S., Lavie, E., Madelin, M., and Goma, R.: Remote sensing of Andean mountain snow cover to forecast water discharge of Cuyo rivers, Journal of Alpine Research – Revue de géographie alpine, 103, 2–16, https://doi.org/10.4000/rga.2903, 2015.

Dixon, S. G. and Wilby, R. L.: Forecasting reservoir inflows using remotely sensed precipitation estimates: a pilot study for the River Naryn, Kyrgyzstan, Hydrolog. Sci. J., 61, 1–16, https://doi.org/10.1080/02626667.2015.1006227, 2015.

Duethmann, D., Peters, J., Blume, T., Vorogushyn, S., and Güntner, A.: The value of satellite-derived snow cover images for calibrating a hydrological model in snow-dominated catchments in Central Asia, Water Resour. Res., 50, 2002–2021, https://doi.org/10.1002/2013WR014382, 2014.

Duethmann, D., Bolch, T., Farinotti, D., Kriegel, D., Vorogushyn, S., Merz, B., Pieczonka, T., Jiang, T., Su, B., and Güntner, A.: Attribution of streamflow trends in snow and glacier melt-dominated catchments of the Tarim River, Central Asia, Water Resour. Res., 51, 4727–4750, https://doi.org/10.1002/2014WR016716, 2015.

Dukhovny, V. A. and de Schutter, J. L. G.: Water in Central Asia: Past, Present and Future, CRC Press/Balkema,Taylor & Francis Group, London, UK, 2011.

FAO: Irrigation in Central Asia in figures, AQUASTAT Survey-2012, http://www.fao.org/NR/WATER/AQUASTAT/countries_regions/asia_central/index.stm (last access: February 2017), 2013.

Feike, T., Mamitimin, Y., Li, L., and Doluschitz, R.: Development of agricultural land and water use and its driving forces along the Aksu and Tarim River, PR China, Environ. Earth Sci., 73, 517–531, https://doi.org/10.1007/s12665-014-3108-x, 2015.

Gafurov, A. and Bárdossy, A.: Cloud removal methodology from MODIS snow cover product, Hydrol. Earth Syst. Sci., 13, 1361–1373, https://doi.org/10.5194/hess-13-1361-2009, 2009.

Gafurov, A., Kriegel, D., Vorogushyn, S., and Merz, B.: Evaluation of remotely sensed snow cover product in Central Asia, Hydrol. Res., 44, 506–522, https://doi.org/10.2166/nh.2012.094, 2013.

Gafurov, A., Lüdtke, S., Unger-Shayesteh, K., Vorogushyn, S., Schöne, T., Schmidt, S., Kalashnikova, O., and Merz, B.: MODSNOW-Tool: an operational tool for daily snow cover monitoring using MODIS data, Environmental Earth Sciences, 75, 1-15, 10.1007/s12665-016-5869-x, 2016.

Gerlitz, L., Vorogushyn, S., Apel, H., Gafurov, A., Unger-Shayesteh, K., and Merz, B.: A statistically based seasonal precipitation forecast model with automatic predictor selection and its application to central and south Asia, Hydrol. Earth Syst. Sci., 20, 4605–4623, https://doi.org/10.5194/hess-20-4605-2016, 2016.

Grömping, U.: Relative importance for linear regression in R: The package relaimpo, J. Statist. Softw., 17, 27, https://doi.org/10.18637/jss.v017.i01, 2006.

Hagg, W., Mayer, C., Lambrecht, A., Kriegel, D., and Azizov, E.: Glacier changes in the Big Naryn basin, Central Tian Shan, Global Planet. Change, 110, 40–50, https://doi.org/10.1016/j.gloplacha.2012.07.010, 2013.

Hall, R. J., Jones, J. M., Hanna, E., Scaife, A. A., and Erdélyi, R.: Drivers and potential predictability of summer time North Atlantic polar front jet variability, Clim. Dynam., 48, 3869–3887, https://doi.org/10.1007/s00382-016-3307-0, 2017.

Kim, H. M., Webster, P. J., and Curry, J. A.: Seasonal prediction skill of ECMWF System 4 and NCEP CFSv2 retrospective forecast for the Northern Hemisphere Winter, Clim. Dynam., 39, 2957–2973, https://doi.org/10.1007/s00382-012-1364-6, 2012.

Pal, I., Lall, U., Robertson, A. W., Cane, M. A., and Bansal, R.: Predictability of Western Himalayan river flow: melt seasonal inflow into Bhakra Reservoir in northern India, Hydrol. Earth Syst. Sci., 17, 2131–2146, https://doi.org/10.5194/hess-17-2131-2013, 2013.

Pritchard, H. D.: Asia's glaciers are a regionally important buffer against drought, Nature, 545, 169–174, https://doi.org/10.1038/nature22062, 2017.

Renard, B., Kavetski, D., Kuczera, G., Thyer, M., and Franks, S. W.: Understanding predictive uncertainty in hydrologic modeling: The challenge of identifying input and structural errors, Water Resour. Res., 46, W05521, https://doi.org/10.1029/2009WR008328, 2010.

Rosenberg, E. A., Wood, A. W., and Steinemann, A. C.: Statistical applications of physically based hydrologic models to seasonal streamflow forecasts, Water Resour. Res., 47, W00h14, https://doi.org/10.1029/2010wr010101, 2011.

Schär, C., Vasilina, L., Pertziger, F., and Dirren, S.: Seasonal Runoff Forecasting Using Precipitation from Meteorological Data Assimilation Systems, J. Hydrometeorol., 5, 959–973, https://doi.org/10.1175/1525-7541(2004)005<0959:srfupf>2.0.co;2, 2004.

Schiemann, R., Luthi, D., Vidale, P. L., and Schar, C.: The precipitation climate of Central Asia – intercomparison of observational and numerical data sources in a remote semiarid region, Int. J. Climatol., 28, 295–314, https://doi.org/10.1002/joc.1532, 2008.

Schöne, T., Zech, C., Unger-Shayesteh, K., Rudenko, V., Thoss, H., Wetzel, H. U., Gafurov, A., Illigner, J., and Zubovich, A.: A new permanent multi-parameter monitoring network in Central Asian high mountains – from measurements to data bases, Geosci. Instrum. Meth. Data Syst., 2, 97–111, https://doi.org/10.5194/gi-2-97-2013, 2013.

Seibert, M., Merz, B., and Apel, H.: Seasonal forecasting of hydrological drought in the Limpopo Basin: a comparison of statistical methods, Hydrol. Earth Syst. Sci., 21, 1611–1629, https://doi.org/10.5194/hess-21-1611-2017, 2017.

Siebert, S., Burke, J., Faures, J. M., Frenken, K., Hoogeveen, J., Doll, P., and Portmann, F. T.: Groundwater use for irrigation – a global inventory, Hydrol. Earth Syst. Sci., 14, 1863–1880, https://doi.org/10.5194/hess-14-1863-2010, 2010.

Slater, L. J., Villarini, G., Bradley, A. A., and Vecchi, G. A.: A dynamical statistical framework for seasonal streamflow forecasting in an agricultural watershed, Clim. Dynam., https://doi.org/10.1007/s00382-017-3794-7, in press, 2017.

Sorg, A., Bolch, T., Stoffel, M., Solomina, O., and Beniston, M.: Climate change impacts on glaciers and runoff in Tien Shan (Central Asia), Nat. Clim. Change, 2, 725–731, https://doi.org/10.1038/Nclimate1592, 2012.

Unger-Shayesteh, K., Vorogushyn, S., Farinotti, D., Gafurov, A., Duethmann, D., Mandychev, A., and Merz, B.: What do we know about past changes in the water cycle of Central Asian headwaters? A review, Global Planet. Change, 110, 4–25, https://doi.org/10.1016/j.gloplacha.2013.02.004, 2013.

Viviroli, D., Durr, H. H., Messerli, B., Meybeck, M., and Weingartner, R.: Mountains of the world, water towers for humanity: Typology, mapping, and global significance, Water Resour. Res., 43, W07447, https://doi.org/10.1029/2006wr005653, 2007.

3

Hydrological modeling of the Peruvian–Ecuadorian Amazon Basin using GPM-IMERG satellite-based precipitation dataset

Ricardo Zubieta[1,2], Augusto Getirana[3,4], Jhan Carlo Espinoza[1,2], Waldo Lavado-Casimiro[5,2], and Luis Aragon[2]

[1]Subdirección de Ciencias de la Atmósfera e Hidrósfera (SCAH), Instituto Geofísico del Perú (IGP), Lima, Peru
[2]Programa de Doctorado en Recursos Hídricos, Universidad Nacional Agraria La Molina, Peru
[3]Hydrological Sciences Laboratory, NASA Goddard Space Flight Center, Greenbelt, MD, USA
[4]Earth System Science Interdisciplinary Center, College Park, MD, USA
[5]Servicio Nacional de Meteorología e Hidrología (SENAMHI), Lima, Peru

Correspondence to: Ricardo Zubieta (rzubieta@igp.gob.pe)

Abstract. In the last two decades, rainfall estimates provided by the Tropical Rainfall Measurement Mission (TRMM) have proven applicable in hydrological studies. The Global Precipitation Measurement (GPM) mission, which provides the new generation of rainfall estimates, is now considered a global successor to TRMM. The usefulness of GPM data in hydrological applications, however, has not yet been evaluated over the Andean and Amazonian regions. This study uses GPM data provided by the Integrated Multi-satellite Retrievals (IMERG) (product/final run) as input to a distributed hydrological model for the Amazon Basin of Peru and Ecuador for a 16-month period (from March 2014 to June 2015) when all datasets are available. TRMM products (TMPA V7 and TMPA RT datasets) and a gridded precipitation dataset processed from observed rainfall are used for comparison. The results indicate that precipitation data derived from GPM-IMERG correspond more closely to TMPA V7 than TMPA RT datasets, but both GPM-IMERG and TMPA V7 precipitation data tend to overestimate, compared to observed rainfall (by 11.1 and 15.7 %, respectively). In general, GPM-IMERG, TMPA V7 and TMPA RT correlate with observed rainfall, with a similar number of rain events correctly detected (~20 %). Statistical analysis of modeled streamflows indicates that GPM-IMERG is as useful as TMPA V7 or TMPA RT datasets in southern regions (Ucayali Basin). GPM-IMERG, TMPA V7 and TMPA RT do not properly simulate streamflows in northern regions (Marañón and Napo basins), probably because of the lack of adequate rainfall estimates in northern Peru and the Ecuadorian Amazon.

1 Introduction

Satellite-based precipitation data have been widely used for hydrometeorological applications, such as hydrological modeling, especially in data-sparse regions like the Amazon River basin (Collischonn et al., 2008; Getirana et al., 2011; Paiva et al., 2013; Zulkafli et al., 2014; Zubieta et al., 2015). Rainfall is extremely variable in both space and time, particularly over regions characterized by topographic contrast, such as the western Amazon Basin (Espinoza et al., 2009; Lavado et al., 2012). In this region, the Andes Mountains contribute to high spatiotemporal variability of rainfall (Laraque et al., 2007; Espinoza et al., 2015). To improve approximation and reduce uncertainty, detailed monitoring is needed using a high-density rain gauge network. Only a low-density rain gauge network is available in the Amazon Basin (AB), however, which limits understanding of hydrological processes and hydrological modeling over the region (Getirana et al., 2011; Paiva et al., 2013). Satellite-based datasets, uniformly distributed in both space and time, offer an alternative for modeling hydrological events. Their usefulness in Andean–Amazon basins and their applicability as input to hydrological models have been evaluated recently by comparing modeled and observed datasets. Results indicate that these datasets could be used for operational applications in some Andean–Amazon regions (Zulkafli et al., 2014; Zubieta et al., 2015). However, hydrological modeling using satellite-based precipitation data does not yield successful results in equatorial regions. This is mainly because of inadequate satellite estimates, because streamflows resulting from hydrological modeling using observed rainfall show acceptable performance in the Napo River basin in the equatorial region (Zubieta et al., 2015).

Hydrological modeling and forecasting are still poorly developed in the Andean and Amazonian regions. It is important to improve these tools, especially because of an intensification of extreme hydrological events in the Amazon Basin (Gloor et al., 2013), such as intense droughts in 2005 and 2010 (Marengo et al., 2008, 2011; Espinoza et al., 2011) and severe floods in 2009, 2012 and 2014 (Espinoza et al., 2012, 2013, 2014). Moreover, a high percentage of total annual precipitation can fall in just a few days, causing soil erosion and landslides (Zubieta et al., 2017)

In the last two decades, advances in satellite technology have improved rainfall estimation in much of the world (Huffman et al., 2007). The Tropical Rainfall Measuring Mission (TRMM) Multi-satellite Precipitation Analysis (TMPA) precipitation dataset (Huffman et al., 2007) has been important for research and for many hydrological applications in Amazon regions, and there is consensus among studies using TMPA in Amazon regions (Collischonn et al., 2008; Getirana et al., 2011; Paiva et al., 2013; Zulkafli et al., 2014; Zubieta et al., 2015). The TRMM mission ended on 8 April 2015, however, after the spacecraft depleted its fuel reserves (https://pmm.nasa.gov/trmm/mission-end). Despite TRMM's demise, this is not a substantive issue for some products, such as TMPA and TMPA-RT, which are expected to run in parallel with the new Global Precipitation Measurement (GPM) satellite until mid-2017 (Huffman et al., 2015). The GPM mission (Schwaller and Morris, 2011), launched in February 2014, comprises an international constellation of satellites that provide rainfall estimations with significant improvements in spatiotemporal resolution, compared to TMPA products. This is true of GPM products such as Integrated Multi-satellite Retrievals (IMERG) estimations. Recent studies highlight that the GPM-IMERG estimations can adequately substitute for TMPA estimations both hydrologically and statistically, despite limited data availability (Liu, 2016; Tang et al., 2016).

The aim of this paper is to evaluate the use of rainfall estimates from GPM-IMERG for obtaining streamflows over the Amazon Basin of Peru and Ecuador (ABPE) during a 16-month period (from March 2014 to June 2015) for which all datasets are available. It provides a comparative analysis of the GPM-IMERG, TMPA RT and TMPA V7 datasets and a ground-based precipitation dataset (PLU). PLU was developed by spatial interpolation using the Peruvian National Meteorology and Hydrology Service (SENAMHI) network. Each precipitation dataset was used as input for the MGB-IPH hydrological model (Collischonn et al., 2007), which was recently adapted to the ABPE (Zubieta et al., 2015).

The ABPE extends from the tropical Andes to the Peruvian Amazon, with elevations ranging up to 6300 m above sea level, a drainage area of 878 300 km^2 and a mean discharge of $\sim 35\,500\,\mathrm{m^3\,s^{-1}}$ at the Tabatinga station (Lavado et al., 2012). The ABPE is located in the northwestern AB (Fig. 1a), and its area corresponds to 14 % of the AB. It consists mainly of basins such as the Ucayali Basin (southern ABPE), Marañón Basin (west of the ABPE) and Napo Basin (northern ABPE) (Fig. 1b).

2 Datasets used

GPM is an international US–Japanese Earth science mission involving NASA and JAXA, respectively. The GPM mission improved and expanded on TRMM. GPM and TRMM provide precipitation data derived from different passive microwave (PMW) sources used in IMERG and TMPA, respectively (Huffman et al., 2015), including the Sounder for Atmospheric Profiling of Humidity in the Intertropics by Radiometry (SAPHIR), Advanced Technology Microwave Sounder (ATMS), Atmospheric Infrared Sounder (AIRS), Cross-Track Infrared Sounder (CRIS), and TRMM Combined Instrument (TCI) algorithms (2B31). They also include TRMM Microwave Image (TMI, data ended on 8 April 2015), GPM Microwave Imager (GMI), Advanced Microwave Scanning Radiometer for Earth Observing Systems (AMSR-E), Special Sensor Microwave Imager/Sounder (SSMIS), Microwave Humidity Sounder (MHS), Special Sensor Microwave Imager (SSM/I), Advanced Microwave Sounding Unit (AMSU), Operational Vertical Sounder (TOVS) and microwave-adjusted merged geo-infrared (IR). The precipitation datasets used in this study are as follows:

a. GPM (product IMERG-V03D) data at several levels of processing have been provided since March 2014 (GPM-IMERG data are available at http://pmm.nasa.gov/GPM, NASA, 2016). The input precipitation estimates are computed using raw satellite measurements, such as those from passive microwave sensors (TMI, AMSR-E, SSM/I, SSMIS, AMSU, MHS, SAPHIR, GMI, ATMS, TOVS, CRIS and AIRS), inter-calibrated to the GPM Combined Instrument (GCI, using GMI and Dual-frequency Precipitation Radar, DPR) and adjusted with monthly surface precipitation gauge analysis data (where available). All these datasets are used to obtain the best estimate of global precipitation maps. The temporal resolution of IMERG-V03D is *half-hourly*, and it has a 0.1° by 0.1° spatial resolution. Unlike other satellites, such as TRMM, GPM-IMERG can detect both light and heavy rain and snowfall.

b. TMPA 3B42 version 7 is obtained from the preprocessing of data provided by different satellite-based sensors

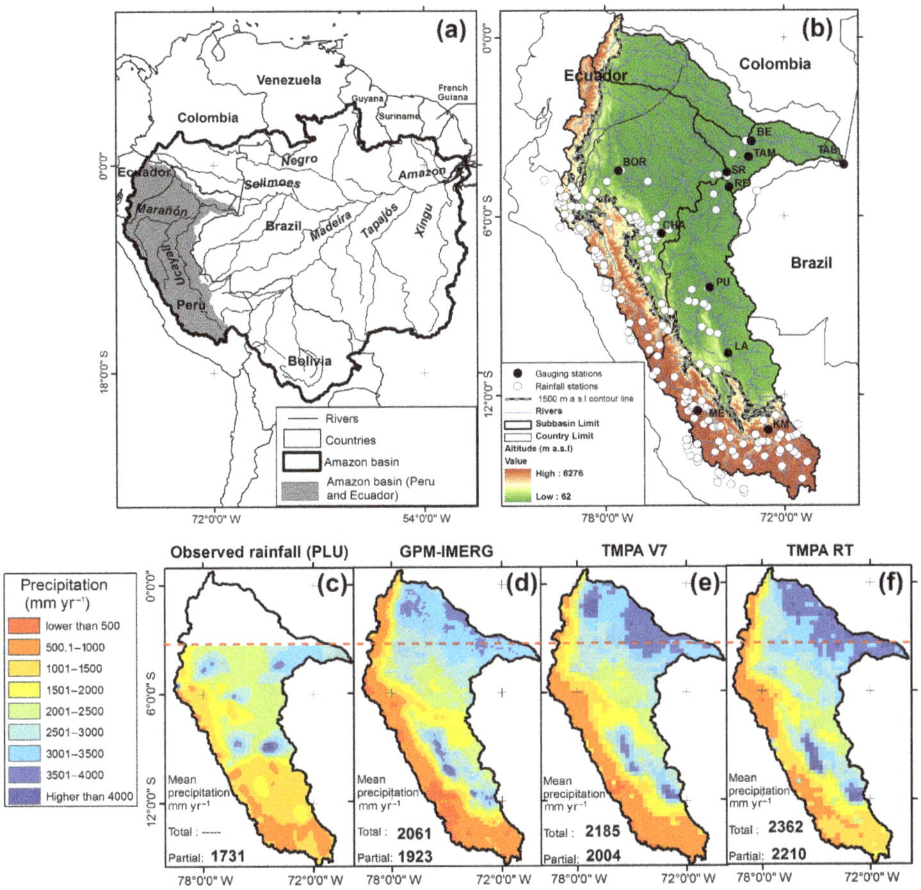

Figure 1. **(a)** Location of the Amazon Basin in South America. **(b)** The western Amazon Basin, gauging and rainfall stations used in this work; intermittent line represents main isohypse 1500 m a.s.l. Total annual precipitation estimated from **(c)** observed rainfall-PLU, **(d)** GPM-IMERG, **(e)** TMPA V7 and **(f)** TMPA RT over the Amazon Basin of Peru and Ecuador.

between 1998 and April 2015, in both real and near-real time (TMPA 3b42 data are available at https://pmm.nasa.gov/data-access/downloads/trmm, NASA, 2017). The 3B42 algorithm (every 3 h) combines precipitation estimates from TMI, AMSR, SSMIS, SSM/I, AMSU, MHS, TCI, MetOp-B and IR. After the preprocessing is complete, the 3-hourly multi-satellite estimations are summed for the month and combined with monthly rainfall obtained from the Global Precipitation Climatology Centre (GPCC), which uses ground-based precipitation. The last step is to scale each 3-hourly rainfall estimate for the month to sum to the monthly value (for each pixel separately, 0.25° by 0.25° spatial resolution).

c. TMPA RT (real time) precipitation data are related to TMPA V7, but do not include calibration measurements of rainy seasons, which are incorporated more than a month after the satellite data. (https://pmm.nasa.gov/data-access/downloads/trmm, NASA, 2017). As with TMPA V7, the final, gridded, subdaily temporal resolution of TMPA RT is usually every 3 hours, with a 0.25° by 0.25° spatial resolution.

d. To evaluate satellite-based datasets, a precipitation product was obtained using daily data series (PLU) from SENAMHI rainfall stations. We collected daily rainfall data for 202 rain stations during the selected period. Quality control based on the regional vector method (RVM) was used to select stations with the lowest probability of errors in their data series (Hiez, 1977; Brunet-Moret, 1979). Finally, 181 RVM-approved rainfall data series (distributed over 700 000 km^2) were selected, with data between March 2014 and June 2015 (Fig. 1b). The area with the highest data availability covers around 81 % of the ABPE (the 19 % without availability is mainly located in the northern region), where the largest distribution of rainfall stations is in the Andean regions, rather than Amazonian regions, of the Ucayali and Huallaga basins (the Huallaga is a subbasin of the Marañón Basin). For comparison, both regions with and without availability of rainfall data were considered for hydrological modeling. Rainfall observations were subsequently spatially interpolated to a resolution of $0.1° \times 0.1°$ by ordinary kriging, and a spherical semivariogram model was used to generate a gridded daily rainfall dataset. Data transformations and anisotropy were applied when necessary. This method has been used to interpolate environmental variables, such as rainfall in the Amazon and Andean regions (Guimberteau et al., 2012; Zubieta et al., 2017). To use each precipitation dataset as input to the hydrological model, subdaily data (for example, TMPA datasets with a temporal resolution of 3 h) were rescaled to a daily time step.

To evaluate model results, streamflow series from the SO-HYBAM Observatory (www.ore-hybam.org, HYBAM, 2016) and SENAMHI stations for the selected period were used; these were KM105 (KM), Mejorada (ME), Chazuta (CHA), Borja (BO), Bellavista (BE), Lagarto (LA), Pucallpa (PU), Requena (RE), San Regis (SR), Tamshiyacu (TAM) and Tabatinga (TAB) (Fig. 1b, Table 1). To describe climate characteristics, meteorological data from NCEP-DOE Reanalysis at surface level (Kanamitsu et al., 2002) were collected, including relative humidity, wind speed, solar radiation, air temperature and atmospheric pressure. Basin topography is derived from the Shuttle Radar Topography Mission (SRTM, version 2). Digital thematic maps correspond to vegetation and soil maps of Peru (http://ref.data.fao.org/es/web/guest/map?entryId=dfcc42e0-88fd-11da-a88f-000d939bc5d8, FAO, 2016) and a vegetation-type map of Ecuador (http://sociobosque.ambiente.gob.ec/files/Folletomapa-parte1.pdf, Socio Bosque, 2016). A soil map of Ecuador (SECS-Ecuador, http://www.gifex.com/detail/2011-10-31-14743/Mapa-General-de-Suelos-del-Ecuador-1986.html) and soil and land-use maps of Colombia (IGAC-Colombia, http://geoportal.igac.gov.co, IGAC-Colombia, 2016) were also considered. GPM-IMERG, TMPA V7, TMPA RT and PLU datasets were selected for the period corresponding to observed streamflows.

3 Methodology

The MGB-IPH model (Collischonn et al., 2007) has been used to simulate the hydrological behavior of the ABPE. It consists of modules for calculating soil water budget, evapotranspiration, flow propagation within a cell and flow routing through the drainage network. A HRU (hydrological response unit) (Kouwen et al., 1993) approach is used to perform soil water balance by mean spatial classification of all areas with a similar combination of soil and land cover. The benefit of using HRUs is the increased accuracy in streamflow simulations on smaller scales, as they make it possible to take better advantage of high spatial resolution databases for hydrological modeling applications. To create HRUs, the watershed is divided into regular elements (cells), which are interconnected by channels. A parameter set is calculated separately for each HRU of each pixel, considering only one layer of soil (Collischonn et al., 2007). The Muskingum–Cunge method is used for routing streamflows through the river network from runoff generated for different HRUs in the cells. Streamflows are adjusted for accuracy according to the stream-reach length and slope. A detailed description of the MGB-IPH model is provided in Collischonn et al. (2007).

The comparison of precipitation datasets was performed in two steps: first, an analysis of monthly averages and detected rain events at different precipitation thresholds (0.1, 1, 5, 10 and 20 mm day^{-1}) was conducted over the ABPE. The analysis was performed by computing the frequency bias index (FBI), probability of detection (POD), false alarm ratio (FAR) and equitable threat score (ETS) (see Table 2). These are calculated from a 2×2 contingency matrix composed of four parameters (a, b, c, d), where a is the number of observed rain events correctly detected, b is the number of observed rain events not detected, c is the number of rainfall events detected but not observed (false alarms), and d is the sum of cases in which neither observed nor detected rain events occurred. FBI allows analysis of overestimation or underestimation of rain events, POD provides information about sensitivity to non-detected and detected events, FAR is a function of false alarms, and ETS indicates the fraction of observed and/or detected rain events that were correctly detected. Comparison of rainfall estimates (GPM-IMERG, TMPA RT) to PLU has been also performed using the Heidke skill score (HSS). HSS is based on the number of correctly predicted data in which the category with the largest probability proves to be correct, as reflected in the formula: $\mathrm{HSS} = \frac{C-E}{N-E}$, where C is the number of correct predictions, E is the number of correct predictions expected by chance and N is the total number of predictions. $\mathrm{HSS} = 1$ refers to a perfect prediction, $\mathrm{HSS} = 0$ shows no skill and $\mathrm{HSS} < 0$, indicates that a prediction is worse than a random prediction.

Two performance coefficients were then used to evaluate the streamflow simulations: the Nash–Sutcliffe (NS) coefficient, and the difference between volumes calculated and observed (ΔV), shown in Eqs. (1) and (2):

$$\mathrm{NS} = 1 - \frac{\sum_{t=1}^{nt} (Q_{\mathrm{obs}}(t) + Q_{\mathrm{cal}}(t))^2}{\sum_{t=1}^{nt} \left(Q_{\mathrm{obs}}(t) - \overline{Q_{\mathrm{obs}}}\right)^2}, \quad (1)$$

$$\Delta V = \frac{\sum (Q_{\mathrm{obs}}(t)) - \sum (Q_{\mathrm{obs}}(t))}{\sum (Q_{\mathrm{obs}}(t))}, \quad (2)$$

with Q_{obs} observed and Q_{cal} modeled streamflows. The range of efficiency lies from $-\infty$ to 1. An efficiency of 1 ($E_{\mathrm{f}} = 1$) corresponds to a perfect fit of modeled streamflow and observed data, while an efficiency of less than zero indicates that the mean value of the time series (observed) would have been a better predictor than the model. A Taylor diagram was used to provide a graphic summary of how closely a pattern (or a set of simulated streamflows) matches observed streamflows. In this diagram, the similarity among three statistical patterns is quantified according to the amplitude of their coefficient of variation (CV %), correlation coef-

Table 1. Characteristics of streamflow gauging stations in the Amazon Basin of Peru and Ecuador: altitude, river, drainage area, annual mean streamflow (Q_{mean}), maximum streamflow (Q_{max}) and minimum streamflow (Q_{min}) in cubic meters per second.

N	Station	Altitude	River	Area (km^2)	Q_{mean} ($m^3\,s^{-1}$)	Q_{max} ($m^3\,s^{-1}$)	Q_{min} ($m^3\,s^{-1}$)
1	KM105 (KM)	2275	Ucayali	9635	98	446	30
2	Mejorada (ME)	2799	Ucayali	16 930	186	651	76
3	Chazuta (CHA)	226	Marañon	68 685	3430	8921	936
4	Borja (BOR)	163	Marañon	92 302	6123	13 250	1931
5	Bellavista (BE)	90	Napo	100 169	9338	13 110	4654
6	Lagarto (LA)	200	Ucayali	191 428	6194	30 460	1292
7	Pucallpa (PU)	141	Ucayali	260 418	10 833	21 830	3714
8	Requena (RE)	94	Ucayali	350 215	13 669	20 910	4088
9	San Regis (SR)	92	Marañon	359 883	20 119	26 610	9071
10	Tamshiyacu (TAM)	88	Amazon	682 970	37 380	53 840	15 000
11	Tabatinga (TAB)	62	Solimões	878 141	45 384	62 190	19 700

Table 2. Summary of rain event detection coefficients.

Coefficient	Name	Equation*	Range	Optimal score
FBI	Frequency bias index	$\mathrm{FBI} = (a+b)/(a+c)$	0–∞	1
POD	Probability of detection	$\mathrm{POD} = a/(a+c)$	0–1	1
FAR	False alarm ratio	$\mathrm{FAR} = c/(a+c)$	0–1	0
ETS	Equitable threat score	$\mathrm{ETS} = (a+\mathrm{He})/(a+b+c-\mathrm{He})$	−∞ to 1	1

* $\mathrm{He} = (a+b)\cdot(a+c)/N$ where N is the total number of estimates.

ficient and centered root-mean-square difference (RMSD %) (Taylor, 2001). This can be used to analyze the relative ability of hydrological models to simulate the spatial pattern of mean streamflow.

4 Results

4.1 Ground-based precipitation dataset (PLU)

To evaluate the ability of PLU to reproduce rainfall gradients in the Andes, the relationship between annual rainfall and altitude for 181 stations was compared. In this area, 100 rainfall stations are located above 2000 m a.s.l.; some record in excess of 1500 mm $year^{-1}$, while less than 1200 mm $year^{-1}$ is generally recorded above 3000 m a.s.l. At lower elevations, abundant rainfall is associated with warm, moist air and the release of a large quantity of water vapor over the first eastern slope of the Andes; as a result, the amount of rainfall decreases with altitude (Laraque et al., 2007; Espinoza et al., 2009). A group of 15 observed rainfall stations located above 2000 m a.s.l. show rainfall amount below 450 mm $year^{-1}$; this group cannot be adequately represented by PLU. Despite these differences, PLU and observed average rainfall show similar behavior at similar altitudes (Fig. S1 in the Supplement). Indeed, the observed average rainfall for 181 stations shows high correlation with PLU for the 2014–2015 period ($r = 0.77$ $p<0.01$; Fig. S2a). In contrast, observed average rainfall shows lower correlation with GPM-IMERG, TMPA V7 and TMPA RT (0.6, 0.56 and 0.61, respectively; Fig. S2b–d).

4.2 Comparison of GPM-IMERG and other rainfall datasets

Total annual rainfall over the ABPE during the selected period, using all four precipitation products, is shown in Fig. 1c–f. The satellite-based datasets (GPM-IMERG, TMPA V7 and TMPA RT) produce overestimates compared to observation (PLU) during this period (by 11.1, 15.7 and 27.7 %, respectively). As Fig. 1c–f show, the satellite-based products present similar spatial distributions. These products are comparable to PLU over (a) the Andean regions (for this paper, the Andean and Amazon regions are considered to be above and below 1500 m a.s.l., respectively; see Fig. 1b), with precipitation mainly between 500 and 1500 mm $year^{-1}$, and (b) the northern Amazon regions (3.0–6.0° S), with precipitation between 2000 and 3000 mm $year^{-1}$. There are some spatial differences over the southern Amazon regions. This can be attributed to greater uncertainty of the PLU dataset, however, because there are fewer rainfall stations in those regions, particularly the eastern Ucayali Basin (Fig. 1b).

A comparison of monthly rainfall over the Ucayali and Huallaga river basins (at the Requena and Chazuta stations) with satellite-based precipitation data during the selected period is shown in Figs. 2a and 3a. In these basins, spatial distribution of rainfall stations is greater in the Andes re-

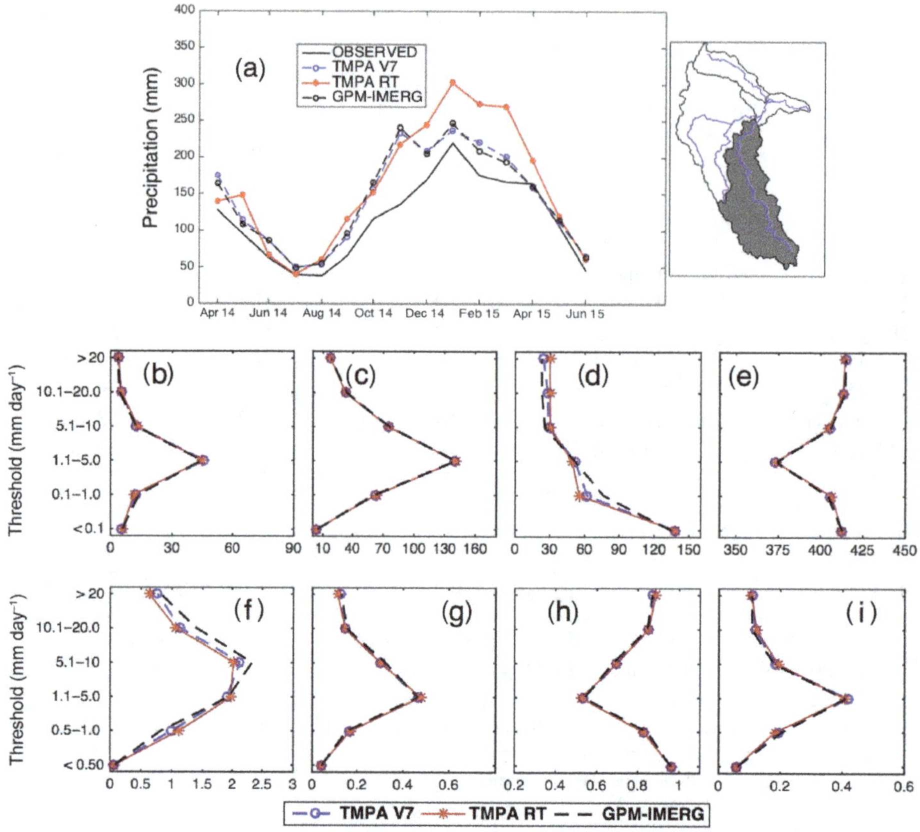

Figure 2. **(a)** Basin-average monthly rainfall for each precipitation dataset in the Ucayali Basin up to Requena station, **(b)** the number of observed rain events correctly detected, **(c)** the number of observed rain events not correctly detected, **(d)** the number of rain events detected but not observed (false alarms), **(e)** the sum of cases when neither observed nor detected rain events occurred, **(f)** coefficient frequency bias index (FBI), **(g)** probability of detection (POD), **(h)** false alarm ratio (FAR) and **(i)** equitable threat score (ETS).

gion than the Amazon region. The TMPA V7 and GPM-IMERG datasets are very similar to each other in the Ucayali and Huallaga river basins. A monthly rainfall analysis shows that TMPA V7 and GPM-IMERG tend to overestimate dry-season rainfall in the Ucayali Basin (April to September) by 10.6 %, compared to the PLU dataset (Fig. 2a). Both datasets tend to overestimate wet-season rainfall, by 23 %, compared to the PLU dataset. This overestimation is larger than that obtained by TMPA V7 or GPM-IMERG when TMPA RT is analyzed (44 %). The GPM-IMERG, TMPA V7 and TMPA RT datasets tend to underestimate dry- and wet-season rainfall in the Huallaga Basin by 30.7, 28.2 and 26.2 %, respectively, compared to PLU (Fig. 3a).

Building on the average number of total days of rain events (456), the number of rain events correctly detected ($\sim 20\,\%$) is similar for each satellite precipitation dataset, compared to the PLU dataset, over the Ucayali and Huallaga basins (Figs. 2b and 3b). The average number of events correctly and not correctly detected is also consistent – that is, all precipitation datasets are clearly better at identifying low- and moderate-precipitation events (1–5 mm day^{-1}) than the number of high- and very low-precipitation events (higher than 5 mm day^{-1} and lower than 1 mm day^{-1}, respectively; Figs. 2b–c and 3b–c). Average FBI values obtained for all datasets indicate a low ability to detect rain events greater than 5 mm day^{-1}, producing FBI values varying mainly between 1 and 2 in the Ucayali and Huallaga basins. This differs substantially from optimal conditions (~ 1; Figs. 2f and 3f). This variation is due to the high number of rain events that were not correctly detected ($\sim 80\,\%$; Figs. 2c and 3c). In general, the satellite-based datasets' limitation in representing rainfall may be due to their strong spatial variability in the Amazon–Andes region. The AB is distinguished by complex spatial distribution of rainfall because of the interactions between topography and large-scale humidity transport (Espinoza et al., 2015). High or extreme precipitation events can be variable in space and time, and the amount of rainfall recorded during extreme events in an Andean location may be normal in an Amazonian one.

Average POD values for all datasets indicate a moderate probability of detection (POD less than 0.55) of rain events greater than 5 mm day^{-1}; this probability decreases to ~ 0.2 for other events in the Ucayali and Huallaga basins (Figs. 2g and 3g). The average number of events correctly and not cor-

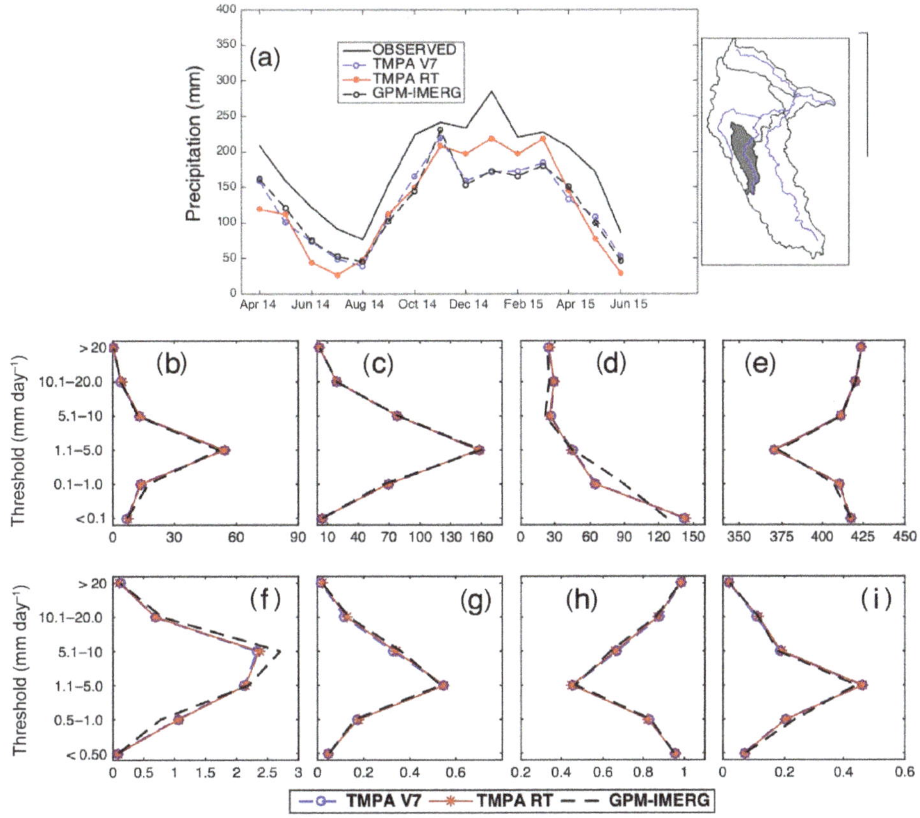

Figure 3. (a) Average monthly rainfall for each precipitation dataset in the Huallaga Basin up to the Chazuta station, **(b)** the number of observed rain events correctly detected, **(c)** the number of observed rain events not correctly detected, **(d)** the number of rain events detected but not observed (false alarms), **(e)** the sum of cases when neither observed nor detected rain events occurred, **(f)** coefficient frequency bias index (FBI), **(g)** probability of detection (POD), **(h)** false alarm ratio (FAR) and **(i)** equitable threat score (ETS).

rectly detected is also consistent – that is, all precipitation datasets are clearly better at identifying precipitation events of between 1 and 5 $\mathrm{mm\,day^{-1}}$. The low probability of detection is consistent with the fraction of rain events that were correctly detected (Figs. 2i and 3i). This is due to a high false alarm rate of between ~ 0.7 and ~ 0.9 for rain events higher than 5 $\mathrm{mm\,day^{-1}}$ and lower than 1 $\mathrm{mm\,day^{-1}}$ for all satellite precipitation datasets in both the Ucayali and Huallaga basins (Figs. 2h and 3h).

The limited ability to represent rainfall events of more than 5 $\mathrm{mm\,day^{-1}}$ using satellite precipitation datasets (GPM-IMERG, TMPA V7, TMPA RT), compared to PLU datasets (Figs. 2g and 3g), may be due to overestimation (in the Ucayali Basin) or high overestimation (in the Huallaga Basin), identified mainly during the wet season (Figs. 2a and 3a). Events exceeding 5 $\mathrm{mm\,day^{-1}}$ are more likely to occur during that period.

The HSS spatial distribution estimated from daily precipitation using each satellite dataset (GPM-IMERG, TMPA V7 and TMPA RT) and PLU was calculated using thresholds (0.1, 1, 5, 10 and 20 $\mathrm{mm\,day^{-1}}$) as a reference prediction (Fig. S3a–c). In general, for the daily scale, the HSS score varies between 0 and 0.4, indicating low skill. The mean HSS for GPM-IMERG shows a moderate HSS score of around 0.4 in the northern region (Fig. S3a). The lowest HSS values (lower than 0.2) for GPM-IMERG are mainly located in the Andean regions, where there are more rainfall stations than in the Amazonian regions. This could be due to strong spatial variability, which is characterized by rainfall decrease with altitude and by the leeward or windward position of the stations (Espinoza et al., 2009). Low scores are also observed in more scattered areas along the ABPE when TMPA V7 and TMPA RT are analyzed (lower than 0.15). Nevertheless, this relationship is slightly improved in the northern region of the Ucayali Basin (~ 0.2).

4.3 Streamflow simulation

To optimize the simulation of streamflows from precipitation datasets, different parameter sets were assigned to each basin in the ABPE during calibration. Analysis by subbasin is more reliable than assigning the same parameter set to the entire basin (Zubieta et al., 2015). Based on sensitivity analysis of the MGB-IPH model (Collischonn et al., 2007), six parameters were selected for calibration: Wm (mm), b (–), K_{int}

Table 3. Model parameters subjected to the process of automatic calibration for the Peruvian and Ecuadorian Amazon Basin.

Parameter	HRU	Hydrological process	First guess	Domain
Wm (mm)	Shrubs, agricultural areas/not deep soils	Water storage on the HRU	200	50–1200
	Shrubs, agricultural areas/deep soils		400	50–1200
	Forest/not deep soils		350	50–1200
	Forest/deep soils		600	50–1200
	Pasture/not deep soils		120	50–1200
	Pasture/deep soils		240	50–1200
K_{int} (mm d^{-1})	Shrubs, agricultural areas/not deep soils	Subsurface flow	80	50–150
	Shrubs, agricultural areas/deep soils		90	50–150
	Forest/not deep soils		100	50–150
	Forest/deep soils		120	50–150
	Pasture/not deep soils		70	50–150
	Pasture/deep soils		80	50–150
K_{bas} (mm d^{-1})	Shrubs, agricultural areas/not deep soils	Groundwater flow	30	10–100
	Shrubs, agricultural areas/deep soils		50	10–100
	Forest/not deep soils		70	10–100
	Forest/deep soils		80	10–100
	Pasture/not deep soils		55	10–100
	Pasture/deep soils		70	10–100
CS	All	Surface flow	15	0.35–40
CI (−)	All	Subsurface flow	120	1–200
b (−)	All	Variable infiltration curve	0.12	0.01–2

(mm d^{-1}), K_{bas} (mm d^{-1}), CS (–) and CI (–), where Wm represents water retained in the soil, which influences the evaporation process over time; K_{int} and K_{bas} control the amount of water in cases in which subsurface soil and groundwater, respectively, are saturated; and CS and CI allow for adjustment of retention time of flows (Collischonn, 2001). To determine optimal parameters, an automatic calibration process was used in order to reduce the domain extent; a previous manual adjustment of the values was performed (Table 3). To ensure impartiality, parameter sets were calibrated separately for each precipitation dataset. Different domains were considered initially for each parameter value, and a first value, determined by manual calibration, was defined as the relative centroid for each domain. The MOCOM-UA multi-criteria global optimization algorithm (Yapo et al., 1998) was then used to find optimal solutions for six parameters. This process results in an effective and efficient search on the Pareto optimum space (Boyle et al., 2000). To analyze the impacts on the calibrated parameters, average parameters were calculated for precipitation datasets and HRU (Table 4).

The results of the calibration process indicate that overestimation by TMPA RT compared to observed rainfall (PLU), GPM-IMERG and TMPA V7 (Fig. 2a) in several months is consistent with a mean increase in Wm (+53, +6 and +15 %, respectively), along with a predominantly mean decrease in K_{bas} (−18, −39 and −16 %, respectively) and K_{int} (−25, −15 and +2 %) to achieve water balance (Table 4). Meanwhile, the overestimation by PLU compared to GPM-IMERG, TMPA V7 and TMPA RT (Fig. 3a) is consistent with a mean increase in Wm (+33, +38 and +34 %, respectively), along with a mean decrease in K_{bas} (−30, −28 and −38 %, respectively) and K_{int} (−17, −16 and −17 %) to achieve water balance (Table 4).

Resulting simulated streamflows were compared to observations at 11 gauging stations (Fig. 1b, Table 5). The Ucayali and Huallaga basins (with greater availability of rainfall gauges) and the northern region of the ABPE (without rainfall gauge availability) were considered in the comparative analysis. In general, streamflows obtained from all satellite-based precipitation datasets show the same spatial pattern as those obtained by using PLU (Fig. 4a–b) and are similar to those obtained by Zubieta et al. (2015). This study shows that GPM-IMERG can also be a helpful alternative source of data (similar to TMPA V7 and TMPA RT) for rainfall–runoff simulation in areas where conventional rainfall data is lacking, such as the Andean–Amazon regions of the Ucayali Basin. The performance analysis over the equatorial regions does not agree well with observed streamflows (NS lower than 0.60), probably because of the lack of adequate rainfall estimates. Similar results are obtained using the TMPA V7 (Fig. 4c) and TMPA RT (Fig. 4d) satellite precipitation datasets in the hydrological modeling.

Figure 5a–f shows the ability of the MGB-IPH model to simulate observed streamflows using TMPA V7, TMPA RT, GPM-IMERG and PLU precipitation datasets. Simulated streamflows match observations at six stations: (a) Chazuta

Table 4. Values of the model mean parameters used in the Ucayali and Huallaga basins for each rainfall datasets for the 2014–2015 period.

		Ucayali Basin				Huallaga Basin			
Parameter	HRU	PLU	GPM-IMERG	TMPA V7	TMPA RT	PLU	GPM-IMERG	TMPA V7	TMPA RT
Wm (mm)	Shrubs, agricultural areas/not deep soils	268	351	294	373	100	60	65	60
	Shrubs, agricultural areas/deep soils	340	472	503	597	132	102	96	99
	Forest/not deep soils	300	408	273	344	130	101	99	96
	Forest/deep soils	422	453	445	435	250	203	180	209
	Pasture/not deep soils	144	350	261	321	101	60	66	59
	Pasture/deep soils	196	400	454	496	150	120	116	121
K_{int} ($\mathrm{mm\,d^{-1}}$)	Shrubs, agricultural areas/not deep soils	141	216	151	151	190	161	163	152
	Shrubs, agricultural areas/deep soils	180	236	156	163	220	189	195	198
	Forest/not deep soils	198	123	107	108	103	162	155	160
	Forest/deep soils	200	134	108	113	120	208	199	220
	Pasture/not deep soils	150	110	119	122	121	160	151	150
	Pasture/deep soils	180	113	126	128	132	193	201	190
K_{bas} ($\mathrm{mm\,d^{-1}}$)	Shrubs, agricultural areas/not deep soils	103	121	89	93	55	70	72	80
	Shrubs, agricultural areas/deep soils	113	123	100	103	61	90	94	100
	Forest/not deep soils	53	134	59	53	44	70	69	80
	Forest/deep soils	62	25	69	62	63	90	88	100
	Pasture/not deep soils	64	112	66	64	46	70	76	80
	Pasture/deep soils	74	113	71	71	63	90	66	100
CS	All	18	16	17	17	2.6	2.4	2.6	2.5
CI (−)	All	112	111	118	111	111	133	135	132
b (−)	All	0.13	0.17	0.15	0.12	0.12	0.15	0.14	0.14

Table 5. Summary of modeling results at 11 gauging stations in the Amazon Basin of Peru and Ecuador (to Tabatinga station in Brazil).

N	River	Station	Observed rainfall (PLU)		GPM-IMERG		TMPA V7		TMPA RT	
			NS	ΔV	NS	ΔV	NS	ΔV	NS	ΔV
1	Ucayali	KM105 (KM)	0.82	0.33	0.90	−0.25	0.80	−2.78	0.68	11.55
2	Ucayali	Mejorada (ME)	0.89	4.2	0.61	−18.5	0.61	−17.01	0.75	−6.49
3	Ucayali	Chazuta (CHA)	0.37	−18.27	−0.26	−31.96	−0.37	−33.51	−0.02	−29.55
4	Ucayali	Borja (BOR)	–	–	−3.94	−47.98	−3.09	−42.39	−3.91	−47.53
5	Ucayali	Bellavista (BE)	–	–	−2.17	−7.14	−18.24	−32.64	−20.93	−35.46
6	Marañon	Lagarto(LA)	0.74	−9.52	0.71	−0.13	0.80	−0.49	0.81	−0.18
7	Marañon	Pucallpa (PU)	0.48	−8.1	0.61	−17.2	0.60	−17.80	0.89	−8.3
8	Marañon	Requena (RE)	0.51	−10.6	−3.75	−23.59	−7.71	−33.28	−5.33	−23.32
9	Napo	San Regis (SR)	–	–	−5.40	−24.82	−5.68	−25.59	−4.90	−24.72
10	Amazon	Tamshiyacu (TAM)	–	–	−24.51	−32.22	−33.32	−37.57	−28.19	−33.19
11	Solimões	Tabatinga (TAB)	–	–	−3.85	−10.28	−12.88	−19.51	−5.21	−10.74

(CHA), (b) KM105 (KM), (c) Lagarto (LA), (d) Mejorada (ME), (e) Pucallpa (PU) and (e) Requena (RE). The location of each dataset on the plot quantifies how closely the modeled streamflows match observed streamflows in terms of CV, correlation coefficient and RMSD.

Figure 5a shows a Taylor diagram for the Chazuta station (Huallaga Basin), where modeled streamflows from the PLU dataset agree better with observed streamflows ($r = 0.84$, $p<0.01$, RMSD error (30 %) and CV of 29 %) than do those using data from satellite products (TMPA RT, TMPA V7 and GPM-IMERG). Analysis of the two smallest subbasins (in the Ucayali Basin) controlled at the KM (Fig. 5b) and ME (Fig. 5c) stations shows a correlation pattern of $r =\sim 0.9$ with RMSD of ∼ 40 % at KM and 24–40 % at ME (Fig. 5b–c). These results indicate that the streamflows from PLU and TMPA RT are more similar to the observed streamflow series mainly at ME, with RMSD lower than 30 %. The streamflow series at both KM and ME have a high CV (40–80 %), due to rainfall seasonality.

Analysis of the largest subbasins (in the Ucayali Basin) controlled at the LA, PU and RE stations shows greater similarity among them for the four streamflow series obtained from precipitation datasets (Fig. 5d–f). Their significant correlation patterns are between 0.8 and 0.9 ($r>0.9$ at the PU station), and RMSD is mainly between 20 and 25 % (PU and RE). It should be noted that streamflow data series have a lower CV in the larger subbasins, such as LA, with CV of 55 % (drainage area of 191 400 $\mathrm{km^2}$); PU, with CV ∼ of 42 %

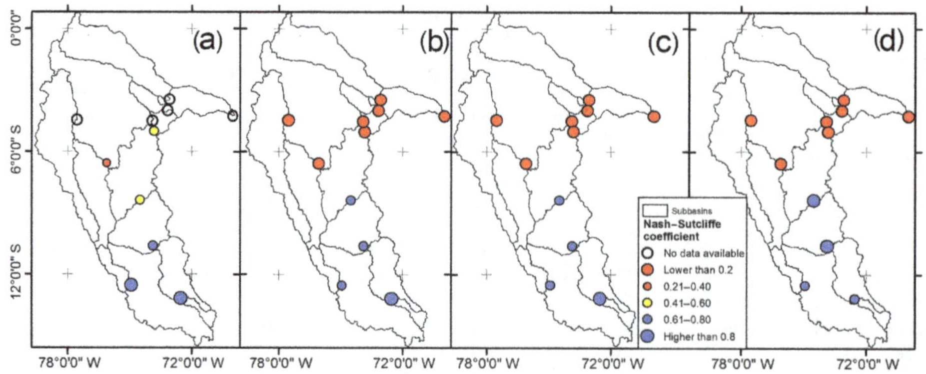

Figure 4. Nash–Sutcliffe efficiency coefficients map for simulations using **(a)** observed Rainfall (PLU), **(b)** GPM-IMERG, **(c)** TMPA V7 and **(d)** TMPA RT rainfall data.

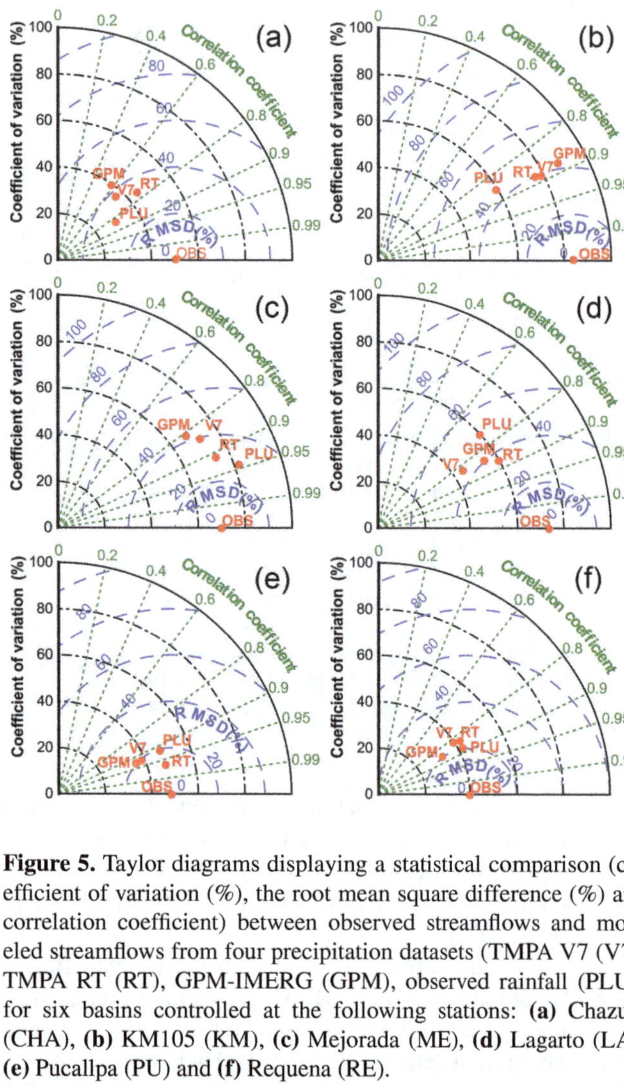

Figure 5. Taylor diagrams displaying a statistical comparison (coefficient of variation (%), the root mean square difference (%) and correlation coefficient) between observed streamflows and modeled streamflows from four precipitation datasets (TMPA V7 (V7), TMPA RT (RT), GPM-IMERG (GPM), observed rainfall (PLU)) for six basins controlled at the following stations: **(a)** Chazuta (CHA), **(b)** KM105 (KM), **(c)** Mejorada (ME), **(d)** Lagarto (LA), **(e)** Pucallpa (PU) and **(f)** Requena (RE).

(drainage area of 260 400 km^2); and RE, with CV of $\sim 40\,\%$ (drainage area of 350 200 km^2). This could be due to weaker seasonality of rainfall in the northern part of the basin. For simulations using satellite-based precipitation datasets, the correlation between simulated and observed streamflows is mainly between 0.6 and 0.9, and RMSD is relatively high (20–40 %), suggesting that a hydrological model using these datasets can represent seasonal streamflows.

The PLU dataset used as input to the hydrological model produced good results at the KM105 (NS $= 0.82$ and $\Delta V = 0.33\,\%$; Fig. 6a), Mejorada (NS $= 0.89$ and $\Delta V = 4.2\,\%$) and Lagarto (NS $= 0.74$ and $\Delta V = -9.52\,\%$) stations in the Ucayali Basin. This indicates its ability to represent extreme values (peak flow) with a low percentage of relative volume error ($\Delta V < 10\,\%$). However, the model's performance is low at the Pucallpa and Requena stations (NS < 0.51 and $\Delta V \sim 10\,\%$), where its predictions are not accurate. The low performance (NS < 0.60) is associated with drainage areas greater than the approximate threshold value of 200 000 km^2 in the Ucayali Basin. This could be due to greater uncertainty in the spatial distribution of rainfall in the Ucayali and Huallaga basins (northern region of the ABPE), because there are fewer rainfall stations in these regions. The Peruvian Andes are currently more instrumented than the Amazon regions (see Fig. 1b).

To analyze the usefulness of the GPM-IMERG datasets for hydrological modeling, hydrographs for the Ucayali Basin monitored at KM105 station (Fig. 6b) were analyzed, with streamflows from the PLU, TMPA V7 and TMPA RT datasets also considered (Fig. 6c–d). Visual analysis of the hydrographs shows that simulated streamflows using GPM-IMERG for the selected period agree fairly well with observed streamflows for the KM105 station. Although the Nash–Sutcliffe efficiency coefficient is generally acceptable (NS $= 0.90$ and $\Delta V = -0.25\,\%$), as shown in Fig. 6b, there is a slight overestimation of streamflow during the wet season, which could be due to overestimation of rainfall during that season. Other results indicate that the model's performance is minimally acceptable in comparison to observed streamflow at the Pucallpa (NS $= 0.61$, $\Delta V = -17.2\,\%$; Fig. 6g) and Mejorada stations (NS $= 0.61$, $\Delta V = -18.5\,\%$).

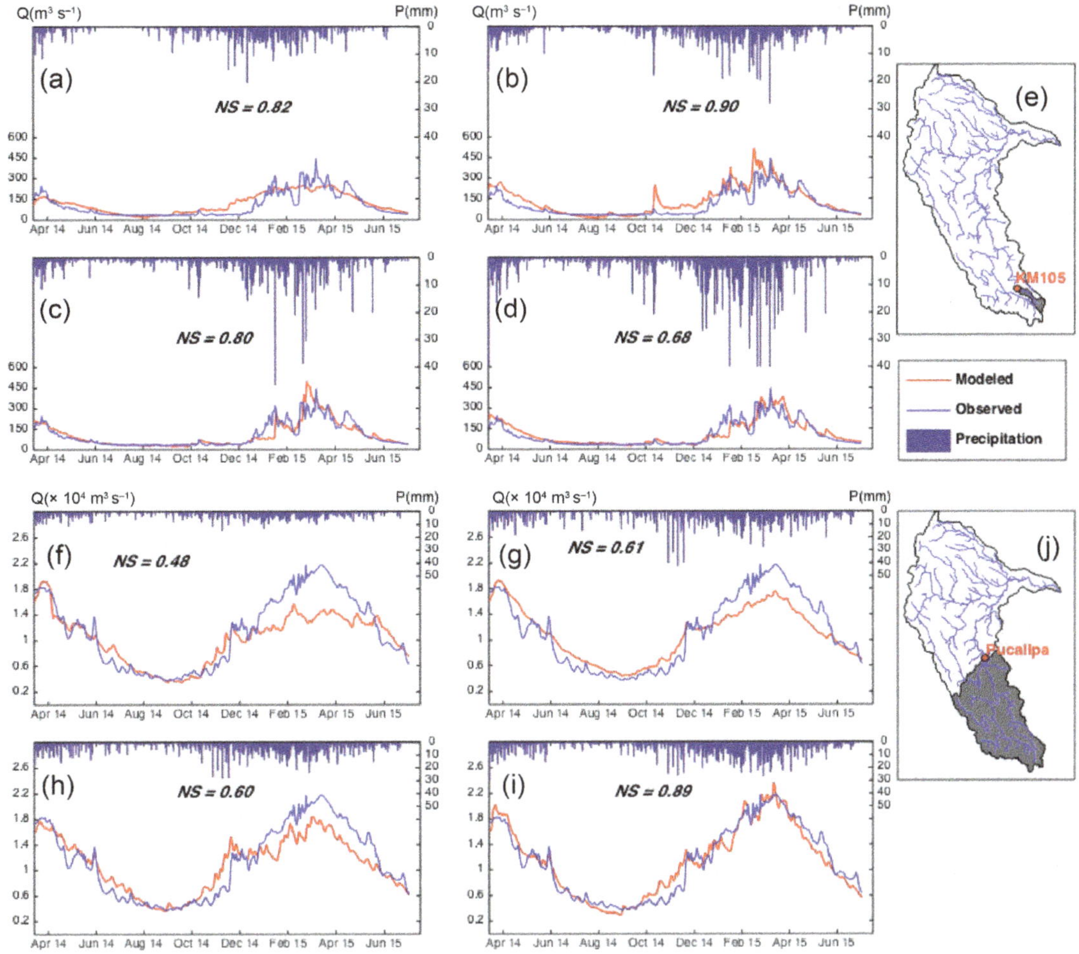

Figure 6. Observed and simulated streamflow hydrographs at KM105 station from 12 March 2014, to 30 June 2015, using the following precipitation datasets: **(a)** observed rainfall, **(b)** GPM-IMERG, **(c)** TMPA V7 and **(d)** TMPA RT. **(e)** Location of the drainage area controlled at the KM station. Observed and simulated streamflow hydrographs at the Pucallpa station from 12 March 2014, to 30 June 2015, using the following precipitation datasets: **(f)** observed rainfall, **(g)** GPM-IMERG, **(h)** TMPA V7 and **(i)** TMPA RT. **(j)** Location of the drainage area controlled at the Pucallpa station.

For the other stations, performance within the basin is less than zero.

Similar results were observed using TMPA V7 and TMPA RT, which reproduce the seasonal streamflow regime with similar performance at the KM105 (NS $= 0.80$ and $\Delta V = -2.78\,\%$, NS $= 0.68$ and $\Delta V = 11.5\,\%$, respectively; Fig. 6c–d) and Pucallpa (NS $= 0.60$ and $\Delta V = -17.8\,\%$, NS $= 0.89$ and $\Delta V = -8.3\,\%$, respectively) stations in the Ucayali Basin (Fig. 6h–i).

5 Concluding Remarks

Three satellite-based precipitation datasets (GPM-IMERG, TMPA V7 and TMPA RT) were evaluated against a rain-gauge-based dataset (PLU) obtained by spatial interpolation over the Amazon Basin of Peru and Ecuador. Each dataset was used as input for the MGB-IPH hydrological model to simulate streamflows for a 16-month period (from March 2014 to June 2015) in the Ucayali, Huallaga, Marañón, Napo, Amazon and Solimões river basins.

GPM-IMERG and TMPA V7 show high temporal and spatial similarity to PLU in the Ucayali Basin, but they tend to underestimate PLU in the Huallaga Basin during the wet season of the 2014–2015 period. TMPA RT tends to overestimate for the Ucayali Basin, compared to other precipitation datasets (PLU, TMPA V7 and GPM-IMERG), while it is more similar to other satellite-based precipitation datasets (TMPA V7 and GPM-IMERG) in the Huallaga Basin.

The GPM-IMERG dataset shows greater similarity to TMPA V7 than TMPA RT. This indicates that GPM-IMERG estimates are more similar to TMPA V7 both spatially and temporally when used as input for hydrological modeling over Andean and Amazonian basins. On average, rain event detection coefficients also suggest that GPM-IMERG, TMPA

V7 and TMPA RT are similar to PLU in the number of rain events correctly detected (∼ 20 %) for the Ucayali and Huallaga basins. Analysis of rain events from pixel value comparing PLU and estimated daily rainfall (GPM-IMERG, TMPA V7 and TMPA RT) suggests a low capacity for detection. This does not imply that they are not useful for hydrological modeling, because rain events not correctly detected for a region or a day could be correctly detected on another day or in nearby regions, compensating for the estimation of rainfall amount over large regions.

In general, the performance of the model when using the GPM-IMERG dataset indicates that these data are useful for estimating observed streamflows in Andean–Amazonian regions (Ucayali Basin, southern regions of the Peruvian and Ecuadorian Amazon Basin). These results are similar to those obtained from TMPA V7 estimates by Zubieta et al. (2015) for the 2003–2009 period. Streamflows obtained from the GPM-IMERG, TMPA V7 and TMPA RT datasets show the same spatial pattern as those obtained by using PLU (low and high performance in the northern and southern regions of the ABPE, respectively). The ability to represent seasonal streamflows in the southern region using these four precipitation datasets is validated with statistical evaluation.

It is important to note that the advantages of GPM-IMERG over TMPA-V7 for estimating streamflows, such as temporal resolution (30 min compared to 3 h, respectively), have not yet been fully analyzed. The use of subdaily rainfall data can be potentially useful for simulating discharge in the Andean rivers, where short convective rainfall episodes are more relevant for hydrological variability. In this study, precipitation and streamflows were analyzed at a daily time step. Further flash-flood modeling on smaller scales would reveal the effects of subdiurnal differences between datasets. Errors in streamflow simulations are mostly associated with input data uncertainty, including rainfall, limited representations of physical processes in models and parameters such as DEM and HRUs. Nevertheless, results show that it is possible to employ remote sensing data in large-scale hydrological models for streamflow simulations.

Competing interests. The authors declare that they have no conflict of interest.

Acknowledgements. The authors would like to thank SENAMHI for providing observed precipitation data series and the SO-HYBAM observatory for providing discharge data (www.ore-hybam.org). The authors also acknowledge GSFC/DAAC and PMM (NASA) for providing the TMPA (https://pmm.nasa.gov/data-access/downloads/trmm/) and GPM-IMERG (http://pmm.nasa.gov/GPM) datasets, respectively. Finally, Ricardo Zubieta and Jhan Carlo Espinoza wish to acknowledge the PNICP-Peru for providing funds through contract 397-PNICP-PIAP-2014.

Edited by: Niko Verhoest

References

Boyle, D. P., Gupta, H. V., and Sorooshian, S.: Toward improved calibration of hydrologic models: combining the strengths of manual and automatic methods, Water Resour. Res., 36, 3663–3674, 2000.

Brunet-Moret, Y.: Homogénéisation des précipitations. Cahiers ORSTOM, Série Hydrologie, 16, 3–4, 1979.

Collischonn, W.: Hydrologic simulation of large basins (in Portuguese), PhD Thesis, Inst. de Pesqui. Hidraul., Univ. Fed. do Rio Grande do Sul, Porto Alegre, Brazil, 2001.

Collischonn, W., Allasia, D. G., Silva, B. C., and Tucci, C. E. M.: The MGB-IPH model for large-scale rainfall-runoff modeling, J. Hydrol. Sci., 52, 878–895, https://doi.org/10.1623/hysj.52.5.878, 2007.

Collischonn, B., Collischonn, W., and Tucci, C .: Daily hydrological modeling in the Amazon basin using TRMM rainfall estimates, J. Hydrol., 360, 207–216, https://doi.org/10.1016/j.jhydrol.2008.07.032, 2008.

Espinoza, J. C, Ronchail, J., Guyot, J. L., Cochonneau, G., Filizola, N. P., Lavado, C., De Oliveira, E., Pombosa, R., and Vauchel, P.: Spatio-Temporal rainfall variability in the Amazon basin countries (Brazil, Peru, Bolivia, Colombia and Ecuador), J. Climatol., 29, 1574–1594, https://doi.org/10.1002/joc.1791, 2009.

Espinoza, J. C., Ronchail, J., Guyot, J. L., Junquas, C., Vauchel, P., Lavado, W.S., Drapeau, G., and Pombosa, R.: Climate variability and extremes drought in the upper Solimões River (Western Amazon Basin): Understanding the exceptional 2010 drought, Geophys, Res. Lett., 38, L13406, https://doi.org/10.1029/2011GL047862, 2011.

Espinoza, J. C., Ronchail, J. Guyot, J. L., Junquas, C., Drapeau, G., Martinez, J. M., Santini, W., Vauchel, P., Lavado, W., Ordoñez, J., and Espinoza, R.: From drought to flooding: understanding the abrupt 2010–11 hydrological annual cycle in the Amazonas River and tributaries, Environ. Res. Lett., 7, 024008, https://doi.org/10.1088/1748-9326/7/2/024008, 2012.

Espinoza, J. C., Ronchail, J., Frappart, F., Lavado, W., Santini, W., and Guyot, J. L.: The major floods in the Amazonas River and tributaries (Western Amazon basin) during the 1970–2012 period: A focus on the 2012 flood, J. Hydrometeorol., 14, 1000–1008, https://doi.org/10.1175/JHM-D-12-0100.1, 2013.

Espinoza, J. C., Marengo, J. A., Ronchail, J., Molina, J., Noriega, L., and Guyot, J. L.: The extreme 2014 flood in south-western Amazon basin: The role of tropical-subtropical south Atlantic SST gradient, Environ. Res. Lett., 9, 124007, https://doi.org/10.1088/1748-9326/9/12/124007, 2014.

Espinoza, J. C., Chavez, S., Ronchail, J., Junquas, C., Takahashi, K., and Lavado, W.: Rainfall hotspots over the southern tropical Andes: Spatial distribution, rainfall intensity and relations with large-scale atmospheric circulation, Water Resour. Res., 51, 3459–3475, https://doi.org/10.1002/2014WR016273, 2015.

Food and Agriculture Organization of the United Nations: http://www.fao.org, last access: June 2016.

Getirana, A. C. V., Espinoza, J. C., Ronchail, J., and Rotunno, O.: Assessment of different precipitation datasets and their impacts on the water balance of the Negro River basin, J. Hydrol., 404, 304–322, https://doi.org/10.1016/j.jhydrol.2011.04.037, 2011.

Gloor, M. R. J. W., Brienen, D., Galbraith, T. R., Feldpausch, J., Schöngart, W., Guyot, J. L., Espinoza, J. C., Lloyd, J., and Phillips, O. L.: Intensification of the Amazon hydrological cycle over the last two decades, Geophys. Res. Lett., 40, 1729–1733, https://doi.org/10.1002/grl.50377, 2013.

Guimberteau, M., Drapeau, G., Ronchail, J., Sultan, B., Polcher, J., Martinez, J.-M., Prigent, C., Guyot, J.-L., Cochonneau, G., Espinoza, J. C., Filizola, N., Fraizy, P., Lavado, W., De Oliveira, E., Pombosa, R., Noriega, L., and Vauchel, P.: Discharge simulation in the sub-basins of the Amazon using ORCHIDEE forced by new datasets, Hydrol. Earth Syst. Sci., 16, 911–935, https://doi.org/10.5194/hess-16-911-2012, 2012.

Hiez, G.: L'homogénéité des données pluviométriques, Cahier ORSTOM, série Hydrologie, 14, 129–172, 1977.

Huffman, G., Adler, R., Bolvin, D., Gu, G., Nelkin, E., Bowman, K., Hong, Y., Stocker, E., and Wolff, D.: The TRMM Multisatellite Precipitation Analysis (TCMA): quasi-global, multiyear, combined- sensor precipitation estimates at fine scales, J. Hydrometeorol., 8, 38–55, https://doi.org/10.1175/JHM560.1, 2007.

Huffman, G. J., Bolvin, D. T., and Nelkin, E. J.: Day 1 IMERG Final Run Release Notes; NASA Goddard Earth Sciences Data and Information Services Center: Greenbelt, MD, USA, 2015.

HYBAM: HYBAM dataset, available at: www.ore-hybam.org/, last access: June 2016.

IGAC-Colombia: http://geoportal.igac.gov.co/, last access: January, 2013.

Kanamitsu, M., Ebisuzaki, W., Woolen, J., Yang, S. K., Hnilo, J. J., Fiorino, M., and Potter, G. L.: NCEP-DOE AMIP-II Reanalysis (R-2), B. Am. Meteor. Soc., 83, 1631–1643, https://doi.org/10.1175/BAMS-83-11-1631, 2002.

Kouwen, N. and Mousavi, S. F.: WATFLOOD/SPL9: Hydrological model and flood forecasting system, in: Mathematical Models of Large Watershed Hydrology, edited by: Singh, V. P. and Frevert, D. K., Water Resources Publications, Highlands Ranch. Colorado, USA, 2002.

Kouwen, N., Soulis, E. D., Pietroniro, A., Donald, J., and Harrington, R. A.: Grouped response units for distributed hydrologic modelling, J. Water Resour. Plan.Manage., 119, 289–305, 1993.

Laraque, A., Ronchail, J., Cochonneau, G., Pombosa, R., and Guyot, J. L.: Heterogeneous distribution of rainfall and discharge regimes in the Ecuadorian Amazon basin, J. Hydrometeorol., 8, 1364–1381, https://doi.org/10.1175/2007JHM784.1, 2007.

Lavado, W., Labat, D., Ronchail, J., Espinoza, J. C., and Guyot, J. L.: Trends in rainfall and temperature in the Peruvian Amazon-Andes basin over the last 40 years (1965–2007), Hydrol. Process., 27, 2944–2957, https://doi.org/10.1002/hyp.9418, 2012.

Liu, Z.: Comparison of Integrated Multisatellite Retrievals for GPM (IMERG) and TRMM Multisatellite Precipitation Analysis (TMPA) monthly precipitation products: Initial results, J. Hydrometeor., 17, 777–790, https://doi.org/10.1175/JHM-D-15-0068.1, 2016.

Marengo, J. A., Nobre, C. A., Tomasella, J., Oyama, M.,. Sampaio, G., Camargo, H., and Alves, L.: The Drought of Amazonia in 2005, J. Climate, 21, 495–516, https://doi.org/10.1175/2007JCLI1600.1, 2008.

Marengo, J. A., Tomasella, J., Alves, L., Soares, W., and Rodriguez, D. A.: The Drought of 2010 in the Context of Historical Droughts in the Amazon Region, Geophys. Res. Lett., 38, 1–5, https://doi.org/10.1029/2011GL047436, 2011.

NASA: GPM-IMERG dataset, available at: http://pmm.nasa.gov/GPM/, last access: June 2016.

NASA: TRMM Data Downloads, available at: https://pmm.nasa.gov/data-access/downloads/trmm, last access: June 2017.

Paiva, R. C. D., Buarque, D. C., Collischonn, W., Bonnet, M. P., Frappart, F., Calmant, S., and Mendes, C. A. B.: Large scale hydrologic and hydrodynamic modeling of the Amazon River basin, Water Resour Res., 49, 1226–1243, https://doi.org/10.1002/wrcr.20067, 2013.

Schwaller, M. R. and Morris, K. R.: A Ground Validation Network for the Global Precipitation Measurement Mission, J. Atmos. Oceanic Technol., 28, 301–319, https://doi.org/10.1175/2010jtecha1403.1, 2011.

Socio Bosque: http://sociobosque.ambiente.gob.ec/, last access: January, 2012.

Tang, G., Zeng, Z., Long, D., Guo, X., Yong, B., Zhang, W., and Hong, Y.: Statistical and hydrological comparisons between TRMM and GPM level-3 products over a midlatitude basin: Is day-1 IMERG a good successor for TMPA 3B42V7?, J. Hydrometeor., 17, 121–137, https://doi.org/10.1175/jhm-d-15-0059.1, 2016.

Taylor, K. E.: Summarizing multiple aspects of model performance in a single diagram, J. Geophys. Res., 106, 7183–7192, https://doi.org/10.1029/2000JD900719, 2001.

Yapo, P. O., Gupta, H. V., and Sorooshian, S.: Multi-objective global optimization for hydrologic models, J. Hydrol., 204, 83–97, 1998.

Zubieta, R.,Geritana, A., Espinoza, J. C., and Lavado W.: Impacts of Satellite-based Precipitation Datasets on Rainfall-Runoff Modeling of the Western Amazon Basin of Peru and Ecuador, J. Hydrol., 528, 599–612, https://doi.org/10.1016/j.jhydrol.2015.06.064, 2015.

Zubieta, R., Saavedra, M., Silva, Y., and Giraldez, L.: Spatial analysis and temporal trends of daily precipitation concentration in the Mantaro River basin – Central Andes of Peru, Stochastic Environmental Research and Risk Assessment, 31, 1305–1318, https://doi.org/10.1007/s00477-016-1235-5, 2017.

Zulkafli, Z., Buytaert, W., Onof, C., Manf, B., Tarnavsky, E., Lavado, W., and Guyot, J. L.: A Comparative Performance Analysis of TRMM 3B42 (TMPA) Versions 6 and 7 for Hydrological Applications over Andean–Amazon River Basins, J. Hydrometeorol., 15, 581–592, https://doi.org/10.1175/JHM-D-13-094.1, 2014.

4

On the use of the GRACE normal equation of inter-satellite tracking data for estimation of soil moisture and groundwater in Australia

Natthachet Tangdamrongsub[1], **Shin-Chan Han**[1], **Mark Decker**[2], **In-Young Yeo**[1], **and Hyungjun Kim**[3]

[1]School of Engineering, University of Newcastle, Callaghan, New South Wales, Australia
[2]ARC Centre of Excellence for Climate System Science, University of New South Wales, Sydney, New South Wales, Australia
[3]Institute of Industrial Science, the University of Tokyo, Tokyo, Japan

Correspondence: Natthachet Tangdamrongsub (natthachet.tangdamrongsub@newcastle.edu.au)

Abstract. An accurate estimation of soil moisture and groundwater is essential for monitoring the availability of water supply in domestic and agricultural sectors. In order to improve the water storage estimates, previous studies assimilated terrestrial water storage variation (ΔTWS) derived from the Gravity Recovery and Climate Experiment (GRACE) into land surface models (LSMs). However, the GRACE-derived ΔTWS was generally computed from the high-level products (e.g. time-variable gravity fields, i.e. level 2, and land grid from the level 3 product). The gridded data products are subjected to several drawbacks such as signal attenuation and/or distortion caused by a posteriori filters and a lack of error covariance information. The post-processing of GRACE data might lead to the undesired alteration of the signal and its statistical property. This study uses the GRACE least-squares normal equation data to exploit the GRACE information rigorously and negate these limitations. Our approach combines GRACE's least-squares normal equation (obtained from ITSG-Grace2016 product) with the results from the Community Atmosphere Biosphere Land Exchange (CABLE) model to improve soil moisture and groundwater estimates. This study demonstrates, for the first time, an importance of using the GRACE raw data. The GRACE-combined (GC) approach is developed for optimal least-squares combination and the approach is applied to estimate the soil moisture and groundwater over 10 Australian river basins. The results are validated against the satellite soil moisture observation and the in situ groundwater data. Comparing to CABLE, we demonstrate the GC approach delivers evident improvement of water storage estimates, consistently from all basins, yielding better agreement on seasonal and inter-annual timescales. Significant improvement is found in groundwater storage while marginal improvement is observed in surface soil moisture estimates.

1 Introduction

The changes in terrestrial water storage (ΔTWS) derived from the Gravity Recovery And Climate Experiment (GRACE) data products have been used in the last decade to study global water resources, including groundwater depletion in India and the Middle East (Rodell et al., 2009; Voss et al., 2013), water storage accumulation in Canada (Lambert et al., 2013), and flood-influenced water storage fluctuation in Cambodia (Tangdamrongsub et al., 2016). The gravity data obtained from GRACE satellites are commonly processed and released in three different product levels (L) that increase in the amount of processing, L1B – satellite tracking data (e.g. Wu et al., 2006), L2 – global gravitational Stokes coefficients (e.g. Bettadpur, 2012), and L3 – global grids (e.g. Landerer and Swenson, 2012). The original (L1B) GRACE information is inevitably altered due to data processing and successive post-processing filterings, because the error covariance information is not propagated through each post-processing step.

The GRACE-derived ΔTWS has been computed widely from the higher-level products (e.g. L2 and L3) on which various ad hoc post-processing filters were applied (e.g. Gaussian smoothing filter, Jekeli, 1981; destripe filter, Swenson and Wahr, 2006). ΔTWS obtained from these filters lacks proper error covariance information and is attenuated and distorted. To overcome the signal attenuation in GRACE high-level products, empirical approaches have been developed, including the application of scale factors computed from land surface models (LSMs; Landerer and Swenson, 2012) to the GRACE L3 products. GRACE uncertainty in the high-level product is usually unknown or assumed. For example, Zaitchik et al. (2008) derived empirically a global average uncertainty that is variable depending on choices of post-processing filters (Sakumura et al., 2014). Furthermore, GRACE error and sensitivity is dependent on latitudes due to the orbit convergence toward poles (Wahr et al., 2006) and any post-processing filters will alter the GRACE data and their error information. Rigorous statistical error information is of equal importance to derivation of ΔTWS for data assimilation and model calibration (Tangdamrongsub et al., 2017; Schumacher et al., 2016, 2018). ΔTWS and its uncertainty estimates should be formulated directly from L1B data considering the complete statistical information.

The GRACE information is not fully exploited in many studies. For example, groundwater storage variation (ΔGWS) is often computed by subtracting the soil moisture variation (ΔSM) component simulated by the land surface model from GRACE-derived ΔTWS data (Rodell et al., 2009; Famiglietti et al., 2011), assuming the model ΔSM is error-free. This may result in the inaccurate ΔGWS and the associated error estimate as the uncertainties in observation and of the land surface model outputs are neglected in the combination (or regression) of two noisy data (e.g. Long et al., 2016). In data assimilation, the GRACE uncertainty is often derived empirically, not necessarily reflecting the actual GRACE error characteristics (e.g. Zaitchik et al., 2008; Tangdamrongsub et al., 2015; Tian et al., 2017). For example, Girotto et al. (2016) used the L3 product and showed that it was necessary to adjust GRACE observation and its uncertainty in order to make their water storage estimates more accurate. Similarly, Tian et al. (2017) reported the need for the application of a scale factor to GRACE uncertainty (from mascon, mass concentration, product) in their GRACE assimilation process. It is apparent that the use of post-processed GRACE products often requires data tuning, leading possibly to an integration of the altered gravity information into the data assimilation system. Some recent studies began to employ the full variance–covariance information in the data assimilation scheme to enhance the quality of the estimates (Eicker et al., 2014; Schumacher et al., 2016; Tangdamrongsub et al., 2017; Khaki et al., 2017a, b).

This study aims to use the GRACE information of ΔTWS measurement directly from the least-squares normal equation data. The approach optimally combines GRACE's normal equations with the model simulation results from the Community Atmosphere Biosphere Land Exchange (CABLE, Decker, 2015) to improve ΔSM and ΔGWS estimates. The proposed approach presents three main advantages. Firstly,

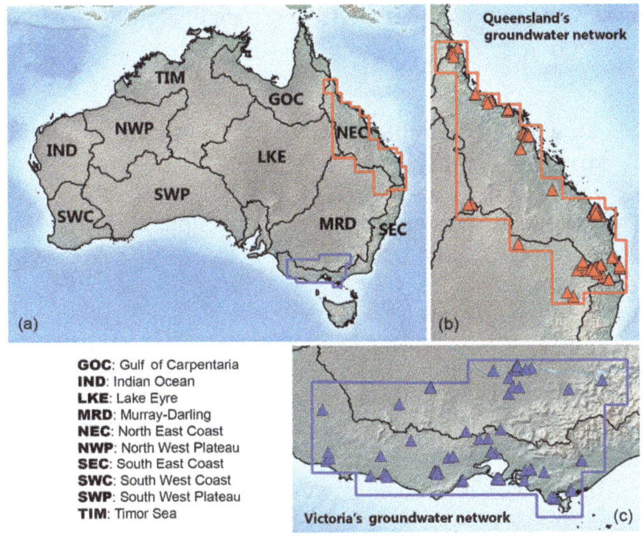

Figure 1. (a) Geographical location of 10 Australian river basins. Red and blue polygons indicate the boundaries of the groundwater networks in Queensland **(b)** and Victoria **(c)**, respectively. Triangles (in **b** and **c**) represent the selected bore locations used in this study.

one can exploit the full GRACE signal and error information by using the normal equation data sets. Secondly, the approach is developed for optimal least-squares combination (e.g. Ramillien et al., 2004), which maximizes the model and observation strength while simultaneously suppressing their weaknesses. Finally, the method bypasses empirical, multiple-step post-processing filters.

The main objective of this study is to present the GRACE-combined (GC) approach to improve the model-estimated ΔSM and ΔGWS on regional scales. We demonstrate our approach applied to 10 Australian river basins (Fig. 1a). One advantage of the study area is that the state vector can be simply defined by ΔSM and ΔGWS as other hydrological components (e.g. snow, glacier) are negligible. We validate the top layer of ΔSM estimates against the satellite soil moisture observation (the Advanced Microwave Scanning Radiometer aboard the Earth Observing System (EOS) – AMSR-E, Njoku et al., 2003) over all 10 basins and the ΔGWS estimates against the in situ groundwater data available over Queensland and Victoria (Fig. 1b and c).

This paper is outlined as follows: firstly, the derivation of the GC approach is presented in Sect. 2 while the description of GRACE data processing, including the use of the GRACE normal equation, is given in Sect. 3. Secondly, the CABLE modelling is outlined in Sect. 4. This includes the derivation of model uncertainty based on the quality of precipitation data and the model parameter inputs. The processing of validation data is also described in Sect. 4. Thirdly, Sect. 5 presents the result of ΔSM and ΔGWS estimates and comparison to in situ data. The long-term trends in the Australian mass variation over the last 13 years are also investigated in this section.

2 A method of combining GRACE L1B data with land surface model outputs

The statistical information of ΔTWS computed from a land surface model can be written as

$$\tilde{\boldsymbol{h}} = \boldsymbol{h} + \boldsymbol{\epsilon};\ \boldsymbol{\epsilon} \sim N(\mathbf{0}, \mathbf{C}), \tag{1}$$

where $\boldsymbol{h}$ is the "truth" (unknown) model state vector while $\tilde{\boldsymbol{h}}$ is the calculated state vector characterized with the model error $\boldsymbol{\epsilon}$. The model error is assumed to have zero mean and covariance $\mathbf{C}$.

The term $\boldsymbol{h}$ is used to represent a vector including the global ΔTWS grid, and terms with a subscript "R" (e.g. $\boldsymbol{h}_{\mathrm{R}}$, $\mathbf{C}_{\mathrm{R}}$) are used to represent only a regional set of ΔTWS (for example, in Australia). As such, the observation equation over a region can be rewritten as

$$\tilde{\boldsymbol{h}}_{\mathrm{R}} = \boldsymbol{h}_{\mathrm{R}} + \boldsymbol{\epsilon};\ \boldsymbol{\epsilon} \sim N(\mathbf{0}, \mathbf{C}_{\mathrm{R}}). \tag{2}$$

As soil moisture and groundwater are the major components of ΔTWS in Australia (surface water storage being insignificant), the vector $\boldsymbol{h}_{\mathrm{R}}$ can be defined as

$$\boldsymbol{h}_{\mathrm{R}} = \begin{bmatrix} \Delta \boldsymbol{SM}_{\mathrm{top}} & \Delta \boldsymbol{SM}_{rz} & \Delta \boldsymbol{GWS} \end{bmatrix}^T, \tag{3}$$

where $\Delta \boldsymbol{SM}_{\mathrm{top}}$, $\Delta \boldsymbol{SM}_{rz}$, and $\Delta \boldsymbol{GWS}$ represent the vectors of top (surface) soil moisture, root zone soil moisture, and groundwater storage variations, respectively.

A least-squares normal equation of GRACE can be written as

$$\mathbf{N}\boldsymbol{x} = \boldsymbol{c}, \tag{4}$$

where $\mathbf{N}$ is a normal matrix, $\boldsymbol{x}$ contains the spherical harmonic coefficients (SHC) of the geopotential, and $\boldsymbol{c}$ is the normal vector. In this study, $\mathbf{N}$ and $\boldsymbol{c}$ can be obtained from the ITSG-Grace2016 products (Mayer-Gürr et al., 2016; https://www.tugraz.at/institute/ifg/downloads/gravity-field-models/itsg-grace2016, see more details in Sect. 3.1). Equation (4) can be written in terms of $\boldsymbol{h}$ as follows (see Appendix A for the derivation):

$$\left(\mathbf{H}^T\mathbf{Y}^T\mathbf{N}\mathbf{Y}\mathbf{H}\right)\hat{\boldsymbol{h}} = \mathbf{H}^T\mathbf{Y}^T\boldsymbol{c}, \tag{5}$$

where $\mathbf{Y}$ converts ΔTWS to geopotential coefficients considering the load Love numbers (e.g. Wahr et al., 1998) and $\mathbf{H}$ is the operational matrix converting $\Delta \boldsymbol{SM}_{\mathrm{top}}$, $\Delta \boldsymbol{SM}_{rz}$, and $\Delta \boldsymbol{GWS}$ to ΔTWS. Equation (5) is based on the assumption that the GRACE orbital perturbation is a result of ΔTWS variation on the surface. If M is the number of model grid cells, N_{max} is the maximum degree of the geopotential coefficients, and $L = (N_{\mathrm{max}} + 1)^2 - 4$ is the number of geopotential coefficients from GRACE; the dimension of $\mathbf{Y}$, $\mathbf{H}$, and $\boldsymbol{h}$ are $L \times M$, $M \times 3M$, and $3M \times 1$, respectively. Note that Eq. (5) is defined with the global grid of $\boldsymbol{h}$. For a regional application, Eq. (5) can be modified as

$$\left[\mathbf{H}_{\mathrm{R}}^T\mathbf{Y}_{\mathrm{R}}^T | \mathbf{H}_{\mathrm{o}}^T\mathbf{Y}_{\mathrm{o}}^T\right]\mathbf{N}\begin{bmatrix} \mathbf{Y}_{\mathrm{R}}\mathbf{H}_{\mathrm{R}} \\ \mathbf{Y}_{\mathrm{o}}\mathbf{H}_{\mathrm{o}} \end{bmatrix}\begin{bmatrix} \hat{\boldsymbol{h}}_{\mathrm{R}} \\ \hat{\boldsymbol{h}}_{\mathrm{o}} \end{bmatrix} = \left[\mathbf{H}_{\mathrm{R}}^T\mathbf{Y}_{\mathrm{R}}^T | \mathbf{H}_{\mathrm{o}}^T\mathbf{Y}_{\mathrm{o}}^T\right]\boldsymbol{c}, \tag{6}$$

where the subscript "R" indicates the grid ΔTWS only in a region of interest and "o" for the rest of the globe. The number of the model grid cells associated with R is J, and the number of grid cells outside cells is $M - J$. As such, the dimensions of $\mathbf{Y}_{\mathrm{R}}$, $\mathbf{H}_{\mathrm{R}}$, $\hat{\boldsymbol{h}}_{\mathrm{R}}$, $\mathbf{Y}_{\mathrm{o}}$, $\mathbf{H}_{\mathrm{o}}$, $\hat{\boldsymbol{h}}_{\mathrm{o}}$ are $L \times J$, $J \times 3J$, $3J \times 1$, $L \times (M - J)$, $(M - J) \times 3(M - J)$, $3(M - J) \times 1$, respectively. The dimension of $\mathbf{N}$ and $\boldsymbol{c}$ remain unchanged, since they are essentially from the normal equations of the original GRACE L1B data (to be discussed in the following section).

From Eq. (6), the normal equations associated with ΔTWS in the region of interest can then be written as

$$\mathbf{H}_{\mathrm{R}}^T\mathbf{Y}_{\mathrm{R}}^T\mathbf{N}\mathbf{Y}_{\mathrm{R}}\mathbf{H}_{\mathrm{R}}\hat{\boldsymbol{h}}_{\mathrm{R}} = \mathbf{H}_{\mathrm{R}}^T\mathbf{Y}_{\mathrm{R}}^T\boldsymbol{c} - \mathbf{H}_{\mathrm{R}}^T\mathbf{Y}_{\mathrm{R}}^T\mathbf{N}\mathbf{Y}_{\mathrm{o}}\mathbf{H}_{\mathrm{o}}\hat{\boldsymbol{h}}_{\mathrm{o}} \tag{7}$$

or

$$\mathbf{N}_{\mathrm{R}}\hat{\boldsymbol{h}}_{\mathrm{R}} = \boldsymbol{c}_{\mathrm{R}}, \tag{8}$$

where $\mathbf{N}_{\mathrm{R}} = \mathbf{H}_{\mathrm{R}}^T\mathbf{Y}_{\mathrm{R}}^T\mathbf{N}\mathbf{Y}_{\mathrm{R}}\mathbf{H}_{\mathrm{R}}$ and $\boldsymbol{c}_{\mathrm{R}} = \mathbf{H}_{\mathrm{R}}^T\mathbf{Y}_{\mathrm{R}}^T\boldsymbol{c} - \mathbf{H}_{\mathrm{R}}^T\mathbf{Y}_{\mathrm{R}}^T\mathbf{N}\mathbf{Y}_{\mathrm{o}}\mathbf{H}_{\mathrm{o}}\hat{\boldsymbol{h}}_{\mathrm{o}}$. As seen, Eq. (7) is the regional representation of Eq. (5) where only the grid cells inside the study region are used, while the contribution from the grid cells outside the region needs to be removed or corrected. Combining the normal equation of Eqs. (2) and (8), the optimal combined solution of $\hat{h}_{\mathrm{R}}$ can be resolved as follows:

$$\hat{\boldsymbol{h}}_{\mathrm{R}} = \left(\mathbf{C}_{\mathrm{R}}^{-1} + \mathbf{N}_{\mathrm{R}}\right)^{-1}\left(\mathbf{C}_{\mathrm{R}}^{-1}\tilde{\boldsymbol{h}}_{\mathrm{R}} + \boldsymbol{c}_{\mathrm{R}}.\right). \tag{9}$$

The computation of model covariance matrix $\mathbf{C}_{\mathrm{R}}$ will be discussed in Sect. 4.2. The a posteriori covariance of $\hat{\boldsymbol{h}}_{\mathrm{R}}$ can be estimated as follows:

$$\hat{\mathbf{\Sigma}} = \left(\mathbf{C}_{\mathrm{R}}^{-1} + \mathbf{N}_{\mathrm{R}}\right)^{-1}, \tag{10}$$

and the uncertainty estimate of $\hat{h}_{\mathrm{R}}$ is simply calculated as

$$\boldsymbol{\sigma}_{\hat{h}} = \sqrt{\mathrm{diag}(\hat{\mathbf{\Sigma}})}, \tag{11}$$

where diag( ) represents the diagonal element of the given matrix.

3 GRACE data

3.1 GRACE least-squares normal equations

In this study, the least-squares normal equations are obtained from the ITSG-Grace2016 products between January 2003 and March 2016. All L1B data including K-band range (KBR) inter-satellite tracking data, attitude, accelerometer, GPS-based kinematic orbit data, and AOD1B corrections are reduced in terms of the normal equations. These data products are usually used to compute the Earth's geopotential field to the maximum harmonic degree and order of 90, or at a spatial resolution of ~ 220 km. The products contain the information of the normal matrix $\mathbf{N}$ and the vector $\boldsymbol{c}$ (as shown in Eq. 4) as well as the a priori time-varying gravity field coefficients predicted with the GOCO05s solution (Mayer-Gürr et al., 2015). Note that the solution of the ITSG-Grace2016 normal equation is the anomalous geopotential coefficient vector ($\Delta\boldsymbol{x}$), which is referenced to the a priori time-varying gravity field ($\boldsymbol{x}_0$), through:

$$\mathbf{N}\Delta\boldsymbol{x} = \boldsymbol{d}, \tag{12}$$

where $\boldsymbol{d}$ and $\boldsymbol{x}_0$ are given. To obtain a complete gravity field variation between the study period ($\boldsymbol{x}$ term in Eq. 4), the a priori time-varying gravity field, $\boldsymbol{x}_0$, is firstly restored to Eq. (12), and the mean gravity field ($\overline{\boldsymbol{x}}_0$) computed from all $\boldsymbol{x}_0$ between January 2003 and March 2016 is then removed as follows:

$$\mathbf{N}(\Delta\boldsymbol{x} + \boldsymbol{x}_0 - \overline{\boldsymbol{x}}_0) = \boldsymbol{d} + \mathbf{N}(\boldsymbol{x}_0 - \overline{\boldsymbol{x}}_0), \tag{13}$$

$$\mathbf{N}\boldsymbol{x} = \boldsymbol{d} + \mathbf{N}(\boldsymbol{x}_0 - \overline{\boldsymbol{x}}_0). \tag{14}$$

Therefore, in Sect. 2 (e.g. Eq. 5), the matrix $\mathbf{N}$ remains unchanged while the vector $\boldsymbol{c}$ can be simply replaced by $\boldsymbol{c} = \boldsymbol{d} + \mathbf{N}(\boldsymbol{x}_0 - \overline{\boldsymbol{x}}_0)$.

3.2 GRACE-derived ΔTWS products

Three monthly GRACE-derived ΔTWS products are also used: the ITSG-Grace2016 DDK5 solution (ITSG-DDK5 for short, http://icgem.gfz-potsdam.de/series/99_non-iso/ITSG-Grace2016), the CNES/GRGS Release 3 (RL3) (GRGS for short, Lemoine et al., 2015; http://grgs.obs-mip.fr/grace/variable-models-grace-lageos/grace-solutions-release-03), and the JPL RL05M mascon-CRI version 2 product (mascon for short, Watkins et al., 2015; Wiese et al., 2016; http://grace.jpl.nasa.gov/data/get-data/jpl_global_mascons).
The ITSG-DDK5 product is the post-processed version of the ITSG L2 solution where the non-isotropic filter DDK5 (Kusche et al., 2009) is applied. The DDK5 solution is empirically selected here to be a good balance between the over-smoothed (e.g. DDK1) and noisy (e.g. DDK8) solutions. The GRGS solution provides ΔTWS at $1^\circ \times 1^\circ$ globally, derived from the Earth's geopotential coefficients up to the maximum degree and order 80, and no filter nor scale factor is applied (L2 data product). Mascon provides ΔTWS at equal-area 3° spherical cap grid globally. In contrast to the ITSG-DDK5 and GRGS solutions, the mascon uses a gain factor derived from the land surface model to restore mitigated signals and reduce leakage errors (L3 data products) (Watkins et al., 2015; Wiese et al., 2016). Additionally, mascon provides the ΔTWS uncertainty together with the solution. The uncertainty is computed based on several geophysical models (see Watkins et al., 2015, and Wiese et al., 2016, for more details). The uncertainty information is not available in the ITSG-DDK5 or GRGS product.

The GRACE data are obtained between January 2003 and March 2016. After retrieval, the long-term mean value between January 2003 and March 2016 is computed and subtracted from the monthly products. To be consistent with CABLE grid spacing (see Sect. 4), the ΔTWS is computed using 0.5° spatial resolution. The coarse-scale datasets (e.g. mascon, GRGS) are resampled to $0.5^\circ \times 0.5^\circ$ using the nearest grid values.

In this study, the independent GRACE solutions are used for two main reasons:

1. To obtain the ΔTWS values outside Australia. As shown in Eq. (7), the $\hat{\boldsymbol{h}}_o$ vector needs to be known, which can be from the GRACE-derived ΔTWS solution. We use the GRGS solutions as the GRGS solution is not subject to the filter choice and it provides ΔTWS at a spatial resolution comparable to the normal equation data.

2. To compare with the ΔTWS estimates from our approaches. All solutions are used to compare and validate our ΔTWS estimates.

4 Hydrology model and validation data

4.1 Model setup

The extensive description of the CABLE model is given in Decker (2015) and Ukkola et al. (2016). This section describes the model setup and specific changes applied to this study. CABLE is a public available land surface model and can be used to estimate soil moisture and groundwater in terms of volumetric water content every 3 h at a $0.5^\circ \times 0.5^\circ$ spatial resolution. The soil moisture and groundwater storage can be simply computed by multiplying the estimates with thicknesses of various layers. For soil moisture, the thickness of six soil layers is 0.022, 0.058, 0.154, 0.409, 1.085, and 2.872 m, from top to bottom, respectively. The thickness of the groundwater layer is modelled to be 20 m uniformly. Recalling Eq. (3), $\Delta \boldsymbol{SM}_{\text{top}}$ is defined as the soil moisture storage variation at the top 0.022 m thick layer, while $\Delta \boldsymbol{SM}_{rz}$ is the variation accumulated over the second to the bottom soil layers (depth between 0.022 and 4.6 m).

CABLE is initially forced with the data from the Global Soil Wetness Project Phase 3 (GSWP3), which is currently available until December 2010 (http://hydro.iis.u-tokyo.ac.jp/GSWP3, https://doi.org/10.20783/dias.501). We replace GSWP3 forcing data with GLDAS data (Rodell et al., 2004)

Table 1. Precipitation data from seven different products used in this study, the Global Soil Wetness Project Phase 3 (GSWP3), the Global Land Data Assimilation System (GLDAS), the Tropical Rainfall Measuring Mission (TRMM), the Modern-Era Retrospective Analysis for Research and Applications (MERRA), the European Centre for Medium-Range Weather Forecasts (ECMWF), Princeton's Global Meteorological Forcing Dataset (Princeton), and the Precipitation Estimation from Remotely Sensed Information using Artificial Neural Networks (PERSIANN). The temporal resolution of all products is 3 h. Most products are available to present while GSWP3, MERRA, and Princeton terminate earlier.

Product	Availability	Spatial resolution	References
GSWP3	Jan 1901–Dec 2010	$0.5° \times 0.5°$	http://dx.doi.org/10.20783/DIAS.501
GLDAS (NOAH025SUBP 3H)	Mar 2000–present	$0.25° \times 0.25°$	Rodell et al. (2004)
TRMM (3B42)	Jan 1998–present	$0.25° \times 0.25°$	Huffman et al. (2007)
MERRA (MSTMNXMLD.5.2.0)	Jan 1980–Feb 2016	$0.5° \times 0.67°$	Rienecker et al. (2011)
ECMWF (ERA-Interim)	Jan 1979–present	$0.75° \times 0.75°$	Dee et al. (2011)
Princeton (V2 0.5°)	Jan 1987–Dec 2012	$0.5° \times 0.5°$	Sheffield et al. (2005)
PERSIANN (3 h)	Mar 2002–present	$0.25° \times 0.25°$	Sorooshian et al. (2000)

Table 2. Model parameters that are sensitive to SM and GWS estimates. The following parameters were perturbed using the additive noise with the boundary conditions given in the last column. The further parameter description can be found in Decker (2015) and Ukkola et al. (2016).

Parameter	Name	Spatial variability	Perturbed range
$f_{\mathrm{clay}}, f_{\mathrm{sand}}, f_{\mathrm{silt}}$	Fraction of clay, sand, and silt	Yes	0–1
f_{sat}	Fraction of grid cell that is saturated	No	810–990
q_{sub}	Maximum rate of subsurface drainage assuming a fully saturated soil column	No	0.009–0.01
f_{p}	Tuneable parameter controlling drainage speed	No	1.9–2.2

to compute the water storage changes to 2016. The forcing data used in CABLE are precipitation, air temperature, snowfall rate, wind speed, humidity, surface pressure, and shortwave and long-wave downward radiations. To investigate the impact of different forcing data, the offline sensitivity study is conducted by comparing the water storage estimates computed using

1. all eight forcing data components of GSWP3 and
2. GSWP3 data with the replacement of one component obtained from GLDAS forcing data.

It is found that the water storage estimate is most sensitive to the replacement of precipitation data, as expected, and relatively less sensitive to the change in other forcing components. We use the GLDAS forcing data in this study and also further test seven different precipitation data products (see more details in Sect. 4.2). The forcing data are up- or down-sampled to a $0.5° \times 0.5°$ spatial grid to reconcile with the CABLE spatial resolution.

4.2 Model uncertainty

In this study, the CABLE uncertainty is derived from 210 ensemble estimates associated with different forcing data and model parameters. The seven different precipitation products (see Table 1) are used to run the model independently. Most products are available to present day while GSWP3, Princeton, and MERRA are only available until December 2010, December 2012, and February 2016, respectively. For each precipitation forcing, 30 ensembles are generated by perturbing the model parameters within ±10 % of the nominal values. The perturbed size of 10 % is similar to Dumedah and Walker (2014). Based on the CABLE structure, the $\Delta \boldsymbol{SM}$ and $\Delta \boldsymbol{GWS}$ estimates are most sensitive to the model parameters listed in Table 2. For example, the fractions of clay, sand, and silt ($f_{\mathrm{clay}}, f_{\mathrm{sand}}, f_{\mathrm{silt}}$) are used to compute soil parameters including field capacity, hydraulic conductivity, and soil saturation which mainly affect soil moisture storage. Similarly, the drainage parameters (e.g. $q_{\mathrm{sub}}, f_{\mathrm{p}}$) control the amount of subsurface runoff, which has a direct impact on root zone soil moisture and groundwater storages.

From ensemble generations, total $K = 210$ sets of the ensemble water storage estimates ($\boldsymbol{h}_{\mathrm{e}}$) are obtained:

$$\mathcal{H}_{\mathrm{R}} = \left[\ \boldsymbol{h}_{\mathrm{e}}|_{k=1}\quad \boldsymbol{h}_{\mathrm{e}}|_{k=2}\quad \boldsymbol{h}_{\mathrm{e}}|_{k=3}\quad \cdots\quad \boldsymbol{h}_{\mathrm{e}}|_{k=K}\right], \tag{15}$$

and the mean value of $\mathcal{H}_{\mathrm{R}}$ is computed as follows:

$$\widetilde{\boldsymbol{h}}_{\mathrm{R}} = \frac{1}{K}\sum_{k=1}^{K} \boldsymbol{h}_{\mathrm{e}}|_{k}. \tag{16}$$

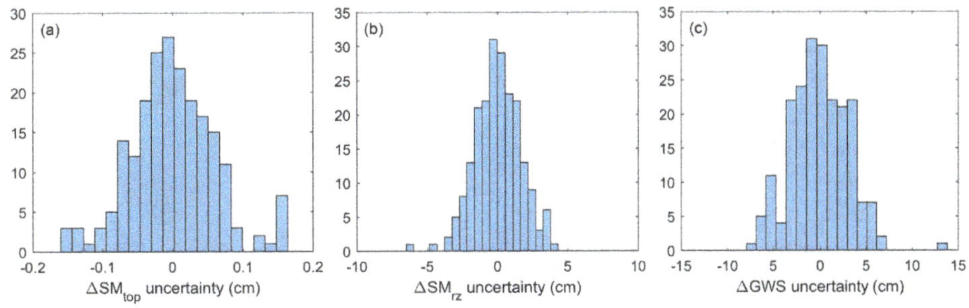

Figure 2. Histograms of the model errors computed from 210 ensemble members ($\mathcal{H}'_R$) without the mean. The basin averaged values (from all 10 Australian basins) of January 2003, for example, are shown.

Note that due to the absence of GSWP3, Princeton, and MERRA data, the number of ensembles reduces to $K = 180$ after December 2010, $K = 150$ after December 2012, and $K = 120$ after February 2016, respectively. The GC approach assumes that model errors are normally distributed with zero mean. Any violation of this assumption will yield a bias in the combined solutions. Therefore, the mean value is removed from each ensemble member, $\mathcal{H}'_R = \mathcal{H}_R - \widetilde{\boldsymbol{h}}_R$, and the error covariance matrix of the model is empirically computed as

$$\mathbf{C}_R = \mathcal{H}'_R(\mathcal{H}'_R)^T/(K-1). \tag{17}$$

The $\widetilde{\boldsymbol{h}}_R$ (Eq. 16) and $\mathbf{C}_R$ (Eq. 17) terms can be directly used in Eq. (9). The distribution of model errors is demonstrated in Fig. 2. The figure illustrates the histogram of model errors ($\mathcal{H}'_R$) computed using 210 ensemble members of the model-estimated ΔSM and ΔGWS in January 2003. The histogram indicates that the model error may be approximately described by a normal distribution as introduced in Eq. (1).

Furthermore, in practice, the sampling error caused by finite sample size might lead to spurious correlations in the model covariance matrix (Hamill et al., 2001). The effect can be reduced by applying an exponential decay with a particular spatial correlation length to $\mathbf{C}_R$. In this study, the correlation length is determined based on the empirical covariance of model-estimated ΔTWS. The covariance function of ΔTWS is firstly assumed isotropic, and it is computed empirically based on the method given in Tscherning and Rapp (1974). The distance where the maximum value of the variance decreases to half is defined as the correlation length. The obtained values vary month to month, and the mean value of 250 km is used in this study.

It is emphasized that the model omission error caused by imperfect modelling of hydrological processes within the LSM is not taken into account in the above description. The omission error may increase the model covariance and introduce a bias as well. We account for the omission error by increasing 20 % of the model covariance. (i.e. multiplying $\mathbf{C}_R$ by 1.2). We determine such omission error based on trial and error such that it increases the model error (due to the omission error) but does not exceed the model error value reported by Dumedah and Walker (2014). We acknowledge that this is only a simple practical way of accounting for the omission error into the total model error.

4.3 Validation data

4.3.1 Satellite soil moisture observation

The satellite-observed surface soil moisture data are obtained from the Advanced Microwave Scanning Radiometer aboard the Earth Observing System using the Land Parameter Retrieval Model (Njoku et al., 2003). The observation is used to validate our estimates of top soil moisture changes (ΔSM_{top}). The AMSR-E product provides volumetric water content in the top layer derived from a passive microwave data (from NASA EOS Aqua satellite) and forward radiative transfer model. In this study, the level 3 product, available daily between June 2002 and June 2011 at 0.25° × 0.25° spatial resolution, is used (Owe et al., 2008). The measurements from ascending and descending overpasses are averaged for each frequency band (C and X). Then, the monthly mean value is computed by averaging the daily data within a month. To obtain the variation in the surface soil moisture, the long-term mean between June 2002 and June 2011 is removed from the monthly data. Regarding the different depth measured in CABLE and AMSR-E, the cumulative density function (CDF)-matching technique (Reichle and Koster, 2004) is used to reduce the bias between the top soil moisture model and the observation. The CDF is built using the 2003–2004 data, and it is used for the entire period. There are no satellite-observed or ground-measured root zone soil moisture data for meaningful comparison with our results, particularly on a continental scale. Validation of ΔSM_{rz} on regional and continental scales is currently unachievable due to a complete lack of observations on this spatial scale.

4.3.2 In situ groundwater

The in situ groundwater levels from bore measurements are obtained from two different ground observation networks (see Fig. 1). The data in Queensland are obtained from the Department of Natural Resources and Mines (DNRM) while the data in Victoria are from the Department of Environment and Primary Industries (DWPI). More than 10 000 measurements are available from each network, but the data gap and outliers are present. Therefore, the bore measurement is firstly filtered by removing the sites that present no data or data gap longer than 30 months during the study period.

To obtain the monthly mean value, the hourly or daily data are averaged in a particular month. The outliers are detected and fixed using the Hampel filter (Pearson, 2005) where the remaining data gaps are filled using the cubic spline interpolation. To obtain the groundwater level variation, the long-term mean groundwater level computed between the study period is removed from the monthly values. The groundwater level variation (ΔL) is then converted to ΔGWS using $\Delta \mathrm{GWS} = S_y \cdot \Delta L$, where S_y is specific yield. Based on Chen et al. (2016), $S_y = 0.1$ is used for the Victoria network. Specific yields of Queensland's network have been found ranging from 0.045 (Rassam et al., 2013) to 0.06 (Welsh, 2008), and an averaged $S_y = 0.05$ is used in this study. Finally, the mean value computed from all data (in each network) is used to represent the in situ data of the network.

5 Results

5.1 Model-only performance

We study the model ΔTWS changes under different meteorological forcing and land parameterization. A total of 210 estimates of monthly TWS (sum of $\mathrm{SM_{top}}$, SM_{rz}, and GWS) are obtained between January 2003 and March 2016 from the ensemble run based on seven different precipitation inputs. Then, the averaged values of the TWS estimates are computed from the 30 precipitation-associated ensemble members. This results in seven sets of monthly mean TWS estimates from seven different precipitation data. For each set, the monthly ΔTWS is computed by removing the long-term mean computed between January 2003 and March 2016.

The precipitation-based ΔTWS values are then compared with the GRACE mascon solution (see Sect. 3.2) over 10 different Australian basins. The comparison is carried out between January 2003 and March 2016. Due to the availability of the data, the periods used are shorter in cases of GSWP3, Princeton, and MERRA precipitation (see Table 1). The metric used to evaluate a goodness of fit between the CABLE run and GRACE mascon estimates is the Nash–Sutcliffe (NS) coefficient (see Eq. B1) (Fig. 3).

Figure 3 demonstrates CABLE ΔTWS varies noticeably by precipitation as well as locations. The area-weighted average values (see Eq. B2) computed from Princeton, GSWP3, and TRMM yield the model ΔTWS, reasonably agreeing with GRACE by giving an NS coefficient greater than 0.45, while MERRA, PERSIANN, and GLDAS show NS $= \sim 0.3$.

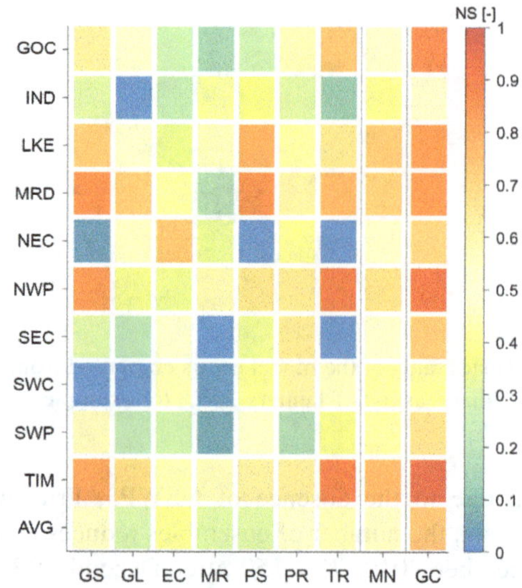

Figure 3. NS coefficients between the model and GRACE mascon ΔTWS over 10 Australian basins (in ordinate). The NS values were computed based on CABLE ΔTWS computed with seven different precipitation data (in abscissa), including GSWP3 (GS), GLDAS (GL), ECMWF (EC), MERRA (MR), PERSIANN (PR), TRMM (TR). The NS value of the mean ΔTWS estimates (the average of seven variants) is also shown (MN). The area-weighted average NS value over all basins is also shown (AVG). The NS value of ΔTWS from the GRACE-combined (GC) approach is shown in the last column. The full name of the basins can be found in Fig. 1.

The lower agreement is mainly due to the quality of rainfall estimates over Australia. The NS of ECMWF is around 0.4.

All model ensembles are consistent with the GRACE data over the Timor Sea and inner parts of Australia (e.g. Lake Eyre, LKE; Murray–Darling, MRD; North West Plateau, NWP) where the NS value can reach as high as 0.9 (see e.g. TRMM over TIM, Timor Sea). On the contrary, the lower agreement is found mostly over the coastal basins. Very small or even negative NS values indicate the misfit between the CABLE and GRACE mascon solutions, and they are observed over the Indian Ocean (see GLDAS), North East Coast (NEC; see GSWP3, PERSIANN, TRMM), South East Coast (see MERRA, TRMM), South West Coast (SWC; see GSWP3, GLDAS, MERRA), and South West Plateau (SWP; see MERRA).

By averaging all ΔTWS estimates from seven different precipitation datasets, the mean-ensemble estimate (MN) delivers the best agreement with GRACE as seen by the highest average NS value (MN of AVG $= 0.55$) among all ensembles. Particularly, NS values are greater than 0.4 in all basins and no negative NS values are presented in MN. On average, it can be clearly seen that using the mean value (MN) is a viable option to increase the overall performance of the ΔTWS estimates. Therefore, only the CABLE MN result will be used in further analyses. The comparison with the GRGS GRACE solution was also evaluated (not shown here) and the overall results are similar to Fig. 3.

5.2 Impact of GRACE on storage estimates

5.2.1 Contribution of GRACE

This section investigates the impact of the GC approach on the estimates of various water storage components. The ΔTWS estimate obtained from the GC approach is demonstrated in Sect. 5.1, by comparing with the independent GRACE mascon solution. Figure 3 shows the GC result yields the highest NS values in all basins, outperforming all other CABLE runs. In the average (AVG), the NS value increases by $\sim 35\,\%$ (0.55 to 0.74) from the MN case. Similar behaviour is also seen when compared with the GRGS GRACE solution (not shown); the average NS value increases from 0.50 to 0.74. This is not surprising as the GC approach uses the fundamental GRACE tracking data as GRACE mascon and GRGS solutions do. Improvement of the NS coefficient indicates merely the successfulness of integrating GRACE data and the model estimates.

Figures 4 and 5 show the GC results of ΔTWS as well as ΔSM_{top}, ΔSM_{rz}, and ΔGWS in different basins. The monthly time series and the deseasonalized time series are shown. In general, GRACE tends to increase ΔTWS when the model ΔTWS (MN) is predicted to be underestimated (see e.g. LKE, MRD, NWP, SWP, TIM between 2011 and 2012) and decrease ΔTWS when it is determined to be overestimated (see all basins between 2008 and 2010). A clear example is seen over the Gulf of Carpentaria (GOC, Fig. 4d), where CABLE overestimates ΔTWS and produces phase delay between 2008 and 2010. The overestimated amplitude and phase delay seen in CABLE ΔGWS during this above period (Fig. 4c) is caused by an overestimation of soil and groundwater storage. The positively biased soil and groundwater storage causes a phase delay by increasing the amount of time required for the subsurface drainage (baseflow) to reduce to soil and groundwater stores. The overestimation of water storage is the result of overestimated precipitation or underestimated evapotranspiration. The amplitude and phase of the water storage estimate are adjusted toward GRACE observation in the GC approach.

The impact of GRACE varies across the individual storage as well as across the geographical location (climate regime). In general, the major contributors to ΔTWS are ΔSM_{rz} and ΔGWS. Due to a small store size (only ~ 2 cm thick), ΔSM_{top} contributes only $\sim 2\,\%$ to ΔTWS. As such, ΔSM_{rz} and ΔGWS have greater variations, which commonly lead to greater uncertainty compared to ΔSM_{top}, and, therefore, the stores anticipate greater shares from the GRACE update. This behaviour is seen over all basins where the differences between the CABLE-simulated and GC ΔSM_{rz}, and ΔGWS estimates are greater (compared to ΔSM_{top}).

Furthermore, the impact of GRACE on ΔSM_{rz} and ΔGWS is different across the continent. For example, over central and southern Australia (see e.g. LKE, MRD, NWP, SWP), the dry climate is responsible for a small amount of groundwater recharge and most of the infiltration is stored in soil compartments. In this climate condition, ΔSM_{rz} amplitude is significantly larger than ΔGWS and it plays a greater role in ΔTWS, and, consequently, the GRACE contribution is mostly seen in the ΔSM_{rz} component. Different behaviour is seen over northern Australia (GOC, NEC, TIM) where ΔGWS amplitudes are greater ($\sim 40\,\%$ of ΔTWS) compared to other basins (only $\sim 17\,\%$ of ΔTWS). This is due to the sufficient amount of rainfall over the wet climate region, replenishing groundwater recharges and resulting in greater variability in ΔGWS. Therefore, compared to the dry climate basin, GRACE contributes to ΔGWS over these basins by a larger amount.

5.2.2 Impact on long-term trend estimates

The spatial patterns of the long-term trends of water storage changes over January 2003 and March 2016 are analysed before and after applying the GC approach (Fig. 6). For comparison, the long-term trends of ΔTWS derived from the ITSG-DDK5, mascon, and GRGS solutions are shown in Fig. 7. From Fig. 6b, GRACE effectively changes the long-term trend estimates in most basins in a way the spatial pattern of the ΔTWS trend of the GC solution consistent to the GRACE solutions while satisfying the model processes and keeping the spatial resolution. The trend of ΔSM_{top} is insignificant (Fig. 6c) and the GC approach does not change (Fig. 6d). The largest adjustment is seen in ΔSM_{rz} and ΔGWS components, to be consistent with the GRACE data in most basins (Fig. 6 and h).

GRACE shows significant changes in the ΔTWS trend estimates particularly over the northern and western parts of the continent (Fig. 7). The model estimates around the Gulf of Carpentaria basin show a strong negative trend that is inconsistent from the GRACE data. It is found that underestimated precipitation after 2012 is likely the cause of such an incompatible negative trend (see Fig. 4d). Applying the GC approach clearly improves the trend (Fig. 6a vs. Fig. 6b). The other example is seen over the western part of the continent (see rectangular area in Fig. 6a and b) where the averaged long-term trend of ΔTWS was predicted to be $-0.4\,\mathrm{cm\,yr^{-1}}$ but changed to be $-1.2\,\mathrm{cm\,yr^{-1}}$ (see also Sect. 5.4) by the GC approach. The precipitation over Western Australia is understood to be overestimated after 2012, evidently seen by the fact that the model ΔTWS is always greater than the GC solution (see e.g. Figs. 4h, 5d and p). The GC approach reveals that the water loss over Western Australia is at least 2 times greater than what has been predicted by the CABLE model.

In addition, the shortage of water storage in the south-eastern part of the continent from the Millennium Drought

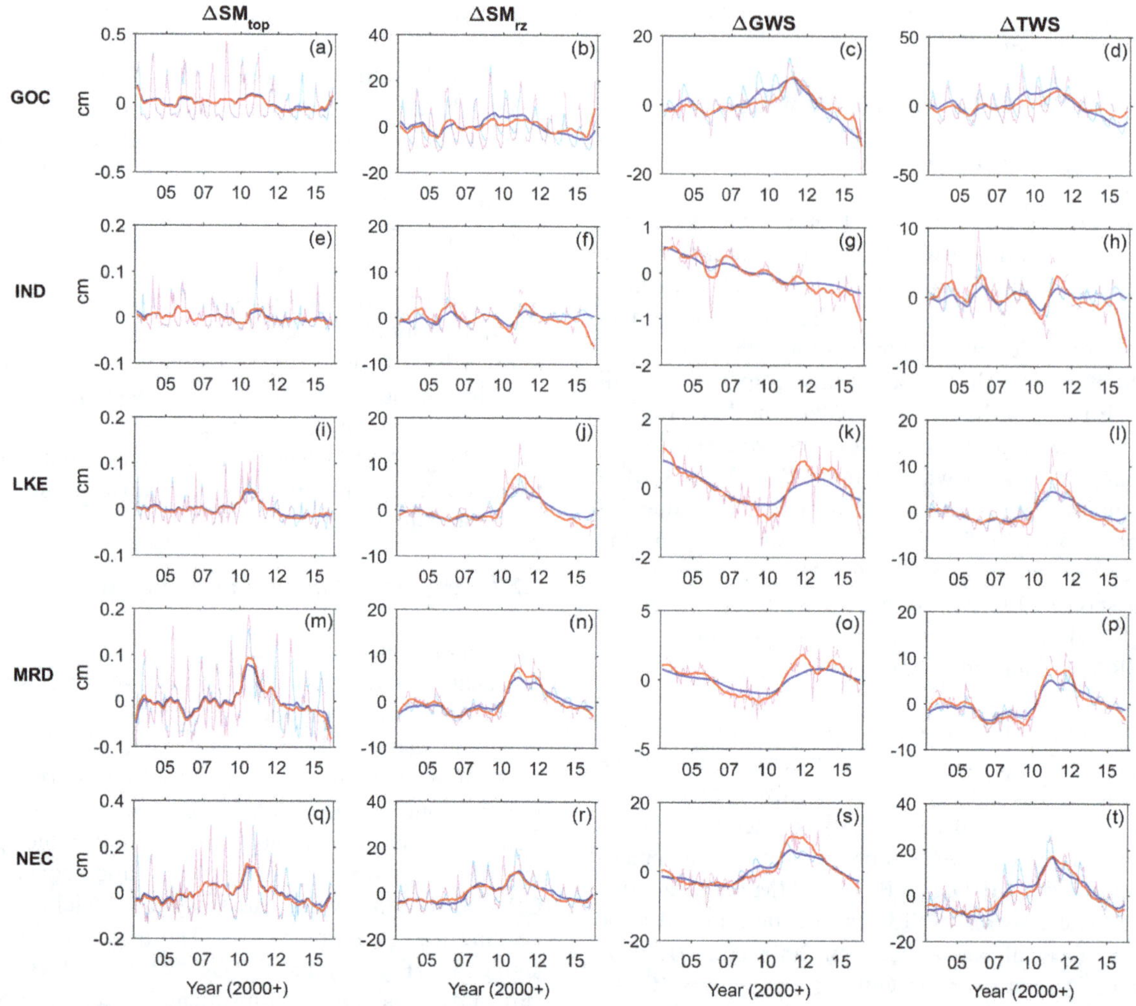

Figure 4. The monthly time series of ΔSM_{top}, ΔSM_{rz}, ΔGWS, and ΔTWS estimated from model (blue) and GC (red) solutions over the Gulf of Carpentaria (GOC), Indian Ocean (IND), Lake Eyre (LKE), Murray–Darling (MRD), and North East Coast (NEC). The deseasonalized time series is also shown.

(McGrath et al., 2012) has been recovered (seen as a positive water storage trend in Fig. 6) after the rainfall between 2009 and 2012, while the western part is still drying out (seen as negative trends). The trend estimates in terms of mass change are discussed in more detail in Sect. 5.4.

5.2.3 Reduction of uncertainty

Influenced by climate pattern, the uncertainty in water storage estimates significantly varies across Australia. The uncertainty in the model estimate is computed from the variability induced by different precipitation and model parameters while the uncertainty in the GC solution is computed using Eq. (11). As expected, larger uncertainties are observed in ΔSM_{rz} and ΔGWS than in ΔSM_{top} (an order of magnitude smaller) since ΔSM_{top} is smaller than others (Fig. 8). Over the wet basins, larger amplitude of the water storage leads to larger uncertainty, seen over the Gulf of Carpentaria, North East Coast, South East Coast, and Timor Sea where the CABLE-simulated ΔTWS uncertainty is approximately 28 % larger than other basins. The smaller uncertainty is found over the dry regions (e.g. LKE, SWP). In most basins, the uncertainty in ΔSM_{rz} is larger than the ΔGWS, except the wet basins (e.g. GOC, NEC, TIM) where the greater groundwater recharge leads to a larger uncertainty in ΔGWS.

Figure 8 demonstrates how much the formal error of each of the storage components is reduced by the GC approach. Overall, the estimated CABLE uncertainties averaged over all basins (AVG) are 0.2, 4.0, 4.0, and 5.7 cm for ΔSM_{top}, ΔSM_{rz}, ΔGWS, and ΔTWS, respectively. With the GC approach, the uncertainties in ΔSM_{top}, ΔSM_{rz}, ΔGWS, and ΔTWS decrease by approximately 26, 35, 39, and 37 %, respectively.

It is worth mentioning that the model uncertainty is mainly influenced by the meteorological forcing data. The uncertainty in precipitation derived from seven different precipi-

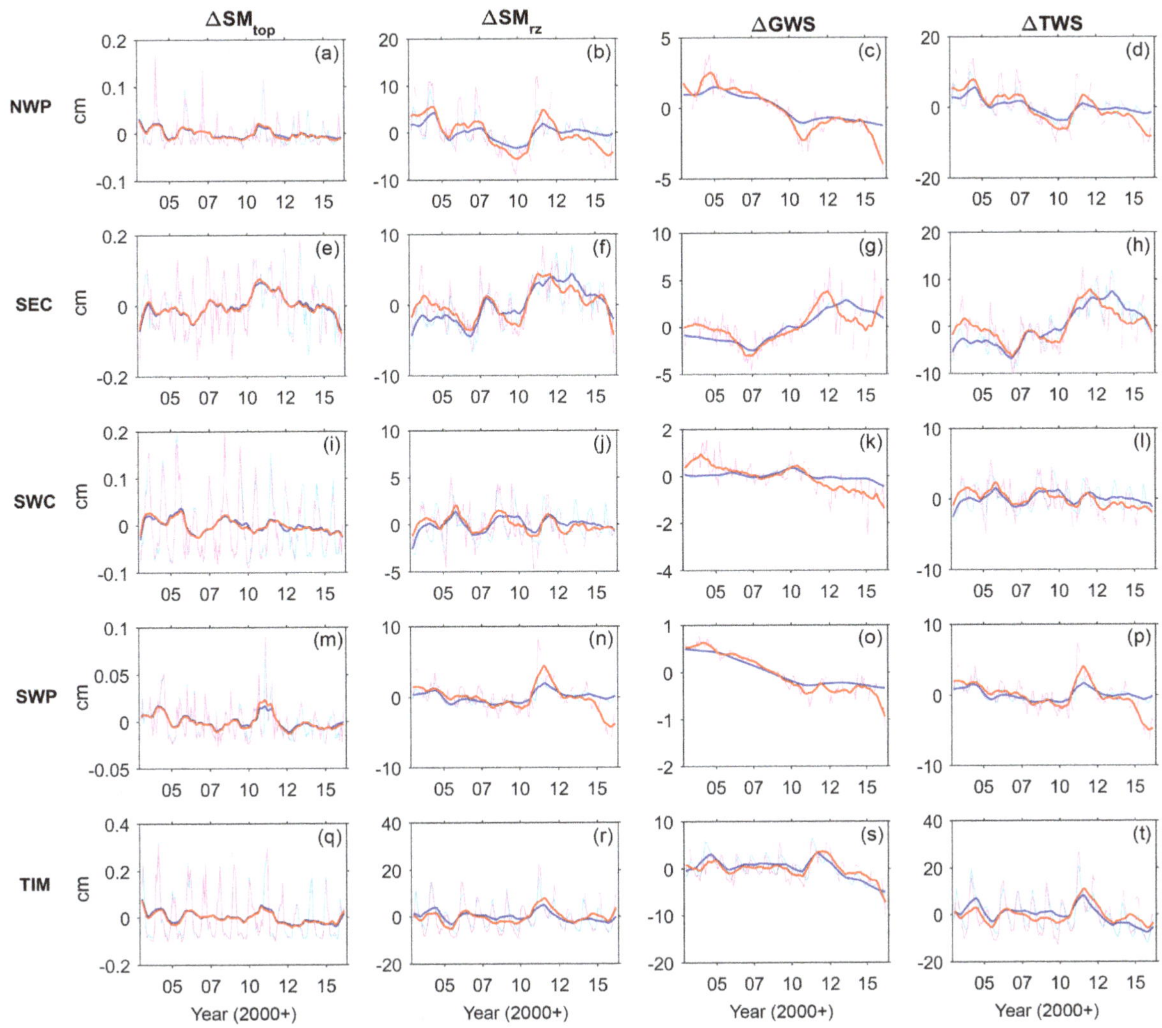

Figure 5. Similar to Fig. 3, but estimated over the North West Plateau (NWP), South East Coast (SEC), South West Coast (SWC), South West Plateau (SWP), and Timor Sea (TIM).

tation products is shown in Fig. 8e. The spatial pattern of the precipitation uncertainty is correlated with the uncertainty in water storage estimates. The larger water storage uncertainty is deduced from the larger precipitation uncertainty. The quality of precipitation forcing data is found to be an important factor to determine the accuracy of water storage computation.

5.3 Comparison with independent data

5.3.1 Soil moisture

The ΔSM_{top} estimates are compared with the AMSR-E-derived soil moisture. The processing of AMSR-E data is described in Sect. 4.3.1. The performance is assessed using Nash–Sutcliffe coefficients, given in Table 3. In general, CABLE (MN) shows a good performance in the top soil moisture simulation showing an NS value of > 0.4 for most of the basins. The top soil moisture estimate shows slightly better agreement with the C-band measurement of the AMSR-E product. This is likely caused by the greater emitting depth of the C-band measurement (~ 1 cm), which is closer to the depth of the top soil layer (~ 2 cm) used in this study (Njoku et al., 2003).

The GC approach leads to a small bit of improvement of the top soil estimate consistently from C- and X-band measurements and from all basins. No degradation of the NS value is observed in the GC solutions. The largest improvement is seen over LKE and NEC, where NS increases by 10–15 %. For other regions, the change in the NS coefficient may be incremental.

5.3.2 Groundwater

The ΔGWS estimates from the model and the GC method are compared with the in situ data obtained from two different ground networks in Queensland and Victoria. For each network, all ΔGWS data inside the groundwater network

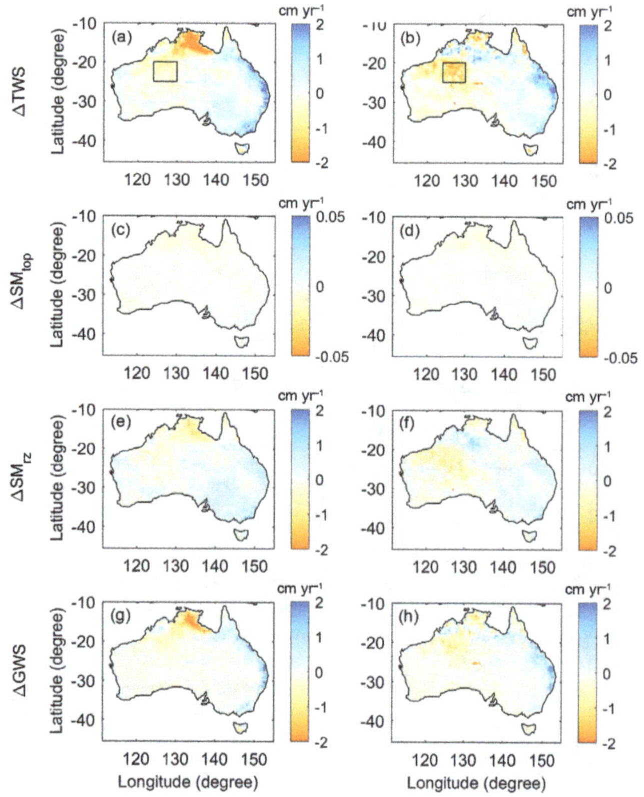

Figure 6. Long-term trends of ΔTWS **(a, b)**, ΔSM_{top} **(c, d)**, ΔSM_{rz} **(e, f)**, and ΔGWS **(g, h)** estimated from the model-only (left) and the GC solutions (right). The eastern part of North West Plateau basin is shown as a rectangle polygon in **(a)** and **(b)**.

boundary (see polygons in Fig. 1) are used to compute the average ΔGWS time series. From the comparison given in Fig. 9, it is found that the GC solutions of ΔGWS follow the overall inter-annual pattern of CABLE but with a considerably larger amplitude. This results in a better agreement with the in situ ΔGWS data seen from both networks. The NS coefficient of ΔGWS between the estimates and the in situ data are given in Table 4. The CABLE ΔGWS performs significantly better in Queensland (NS = ~0.5) than Victoria (NS = ~0.3). Significant improvement is found from the GC solutions in both networks, where the NS value increases from 0.5 to 0.6 (~22 %) in Queensland and from 0.3 to 0.6 (~85 %) in Victoria. Even greater improvement is seen when the inter-annual patterns are compared. The NS value increases from 0.5 to 0.7 (~32 %) and 0.4 to 0.8 (~93 %) in Queensland and Victoria, respectively.

The comparison of the long-term trend of ΔGWS is also evaluated. The estimated trends in Queensland and Victoria are given in Table 4. Beneficially from the GC approach, the ΔGWS trend is improved by approximately 20 % (from 0.4 to 0.6, compared to 1.6 cm yr^{-1}) in Queensland. The increase in ΔGWS is mainly influenced by the large amount of rainfall during the 2009–2012 La Niña episodes (see Fig. 9a). In

Table 3. NS coefficients between top soil moisture estimates and the satellite soil moisture observations from AMSR-E products over 10 different Australian basins. The area-weighted average value (AVG) is also shown.

	C band		X band	
	CABLE	GC	CABLE	GC
GOC	0.67	0.68	0.58	0.60
IND	0.53	0.54	0.41	0.41
LKE	0.48	0.53	0.36	0.42
MRD	0.77	0.80	0.75	0.78
NEC	0.34	0.39	0.14	0.19
NWP	0.33	0.36	0.38	0.42
SEC	0.68	0.68	0.69	0.71
SWC	0.85	0.85	0.89	0.89
SWP	0.55	0.56	0.46	0.48
TIM	0.44	0.45	0.16	0.16
AVG	0.53	0.56	0.47	0.50

Victoria, a significant improvement of the ΔGWS trend by about 76 % (from 0.1 to −0.2, compared to −0.3 cm yr^{-1}) is observed. Similar improvement of long-term trend estimates is seen in deseasonalized time series (improves by ~15 % in Queensland and by ~74 % in Victoria). The decrease in ΔGWS in Victoria is mainly due to the highly demanded groundwater consumption by agriculture and domestic activities (Van Dijk et al., 2007; Chen et al., 2016). As the groundwater consumption is not parameterized in CABLE, the decrease in the ΔGWS estimate cannot be properly captured in the model simulation. Applying GC approach effectively reduces the model deficiency and improves the quality of the groundwater estimations.

5.4 Assessment of mass variation in the past 13 years

Australia experiences significant climate variability; for example, the Millennium Drought starting from late 1990 (Van Dijk et al., 2013) and extremely wet conditions during several La Niña episodes (Trenberth, 2012; Han, 2017). These periods are referred as "Big Dry" and "Big Wet" (Ummenhofer et al., 2011; Xie et al., 2016). To understand the total water storage (mass) variation influenced by these two distinct climate variabilities, the water storage change obtained from the GC approach during Big Dry and Big Wet is separately investigated over 10 basins. The time window between January 2003 and December 2009 is defined as the Big Dry period while between January 2010 and December 2012 it is defined as the Big Wet period following Xie et al. (2016). In each period, the long-term trends of GC estimates of ΔTWS, ΔSM_{top}, ΔSM_{rz}, and ΔGWS are firstly calculated. Then, the total water storage variation (in m) is simply obtained by multiplying the long-term trend (in m yr^{-1}) with the number of years in the specific period – 7 years for Big Dry and 3 years for Big Wet. To obtain the mass variation,

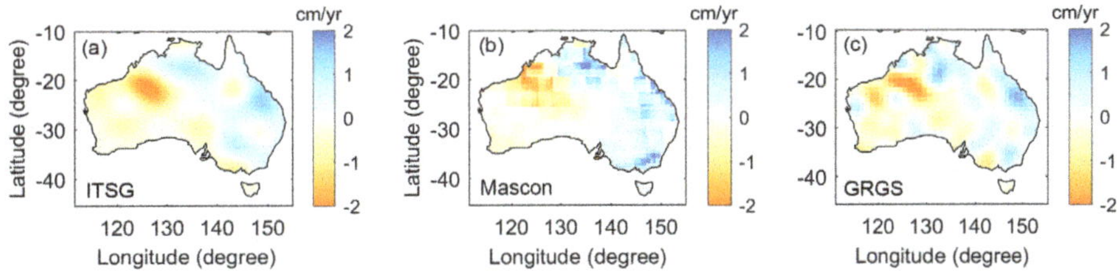

Figure 7. Long-term trends of GRACE-derived ΔTWS from ITSG-DDK5 **(a)**, mascon **(b)**, and GRGS solution **(c)**.

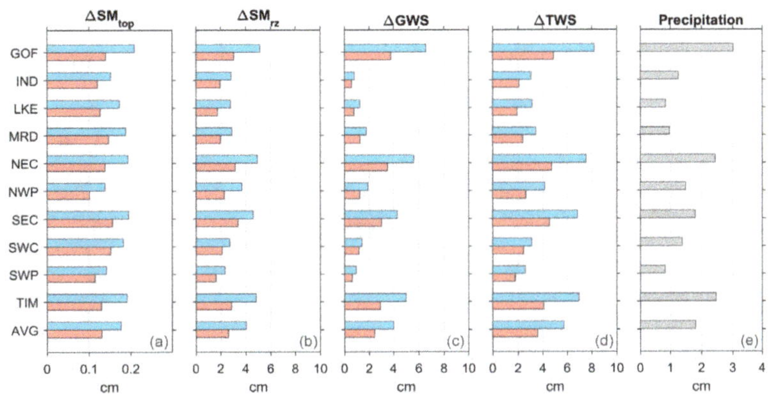

Figure 8. Uncertainties in ΔSM_{top}, ΔSM_{rz}, ΔGWS, and ΔTWS estimated from the model (blue) and the GC solutions (red) in 10 different Australian basins. The uncertainty in the precipitation is shown in **(e)**. The area-weighted average value (AVG) is also shown.

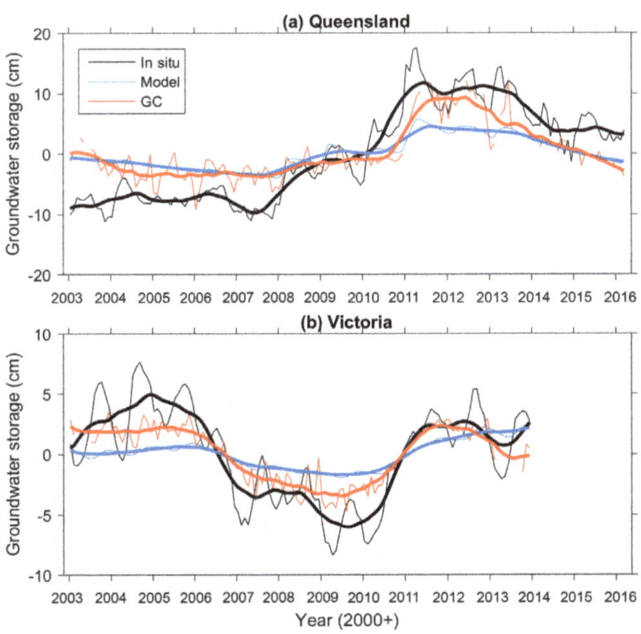

Figure 9. The monthly time series of ΔGWS estimated from the model, GC solutions, and measured from the in situ groundwater network in Queensland **(a)** and Victoria **(b)**. Deseasonalized time series are shown in thick lines.

the water storage variation is multiplied by the area of the basin and the density of water ($1000\,\mathrm{kg\,m^{-3}}$). The estimated mass variations during Big Dry and Big Wet are displayed in Fig. 10. The long-term mass variation in the entire period (January 2003–March 2016) is also shown.

During Big Dry (2003–2009), a significant loss of total storage (40–60 Gt over 7 years) is observed over the LKE, MRD, NWP, and SWP basins. The largest groundwater loss of > 20 Gt is found from LKE and MRD. No significant change is observed over the tropical climate regions (e.g. GOC, NEC). The mass loss mostly occurs in the root zone and groundwater compartments where the sum of ΔSM_{rz} and ΔGWS explains more than 90 % of the ΔTWS value. The mass loss is also observed in ΔSM_{top} but is > 10 times smaller than ΔSM_{rz} and ΔGWS.

During Big Wet (2010–2012), the basins such as LKE, MRD, and TIM exhibit the significant total storage gain of > 100 Gt. The gain is particularly larger in ΔSM_{rz} over the basins that experienced the significant loss during Big Dry. For example, over LKE and MRD, the gain of ΔSM_{rz} is approximately 2–3 times greater than ΔGWS. It implies that most of the infiltration (from the 2009–2012 La Niña rainfall) is stored as soil moisture through the long drought period and that the groundwater recharge is secondary to the ΔSM_{rz} increase.

The opposite behaviour is observed over the basins (such as NEC and GOC) that experienced mass gain during Big

Table 4. NS coefficient and long-term trend of ΔGWS estimated from the model-only and GC solutions in Queensland and Victoria groundwater network. The long-term trend of the in situ data is also shown.

	Queensland			Victoria		
	In situ	CABLE	GC	In situ	CABLE	GC
Original time series						
NS (–)	–	0.49	0.60	–	0.34	0.63
Trend (cm yr^{-1})	1.60 ± 0.05	0.39 ± 0.02	0.63 ± 0.05	-0.27 ± 0.05	0.10 ± 0.02	-0.18 ± 0.03
Deseasonalized time series						
NS (–)	–	0.50	0.66	–	0.43	0.83
Trend (cm yr^{-1})	1.60 ± 0.05	0.39 ± 0.02	0.57 ± 0.04	-0.25 ± 0.05	0.10 ± 0.02	-0.16 ± 0.03

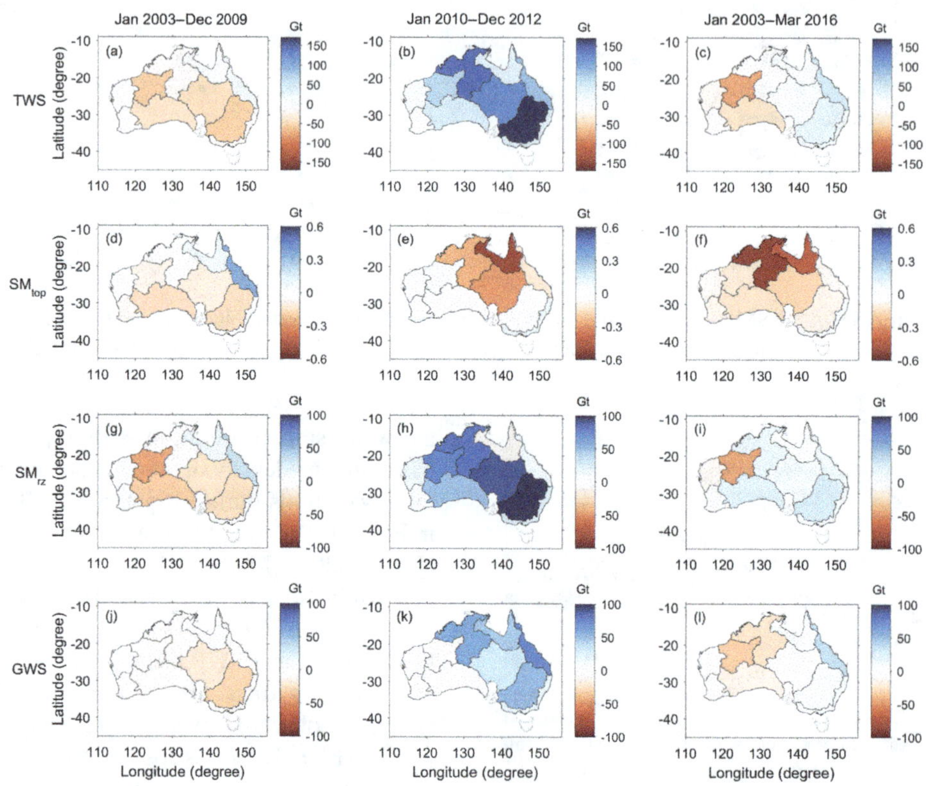

Figure 10. Mass changes (Gt, Gigatonne) of ΔTWS, ΔSM_{top}, ΔSM_{rz}, and ΔGWS estimated from GC solutions over 10 Australian basins in three different periods: Big Dry (January 2003–December 2009), Big Wet (January 2010–December 2012), and the entire period (January 2003–March 2016).

Dry. The water storage gain is greater in ΔGWS compared to ΔSM_{rz}. In NEC, ΔGWS gain is ~ 8 times larger than ΔSM_{rz} during Big Wet. The soil compartment may be saturated during Big Dry and additional infiltration from the Big Wet precipitation leads to an increased groundwater recharge. The ΔSM_{rz} loss observed over GOC is simply caused by the timing selection of the Big Wet period, which ends earlier ($\sim$2011) in GOC than in other basins. The ΔSM_{rz} gain becomes $\sim$26 Gt if the Big Wet period is defined as 2008–2011. During the post-Big-Wet period (2012 and afterwards), the decreasing trend of water storage is observed from all basins (see Figs. 4 and 5). This is mainly caused by the decrease in precipitation after 2012 and by gradual water loss through evapotranspiration (Fasullo et al., 2013).

The overall water storage change in the last 13 years demonstrates that the severe water loss from most basins during Big Dry (the Millennium Drought) is balanced with the gain during Big Wet (the La Niña). The negative ΔTWS estimated during Big Dry becomes positive in LKE, MRD, and

SEC and less negative in TIM, and the greatest gain is observed from NEC by ~50 Gt during the 13-year period (see Fig. 10c). However, the water mass loss is still detected over the western basins (e.g. IND, NWP, SWP, SWC), and their magnitudes are even larger than the mass loss during Big Dry. For example, the greatest ΔTWS loss of ~79 Gt is observed over NWP, which is ~25 Gt greater than the loss during Big Dry (see Fig. 10a and c). The basin is less affected by the La Niña, and the rainfall during Big Wet is clearly inadequate to support the water storage recovery in the basin. Rainfall deficiency also reduces the groundwater recharge, resulting in even more decreasing of ΔGWS, compared to the Millennium Drought period (see Fig. 10j and l). The continual decrease in water storage over western basins is likely caused by the interaction of complex climate patterns such as the El Niño–Southern Oscillation, Indian Ocean Dipole, and Southern Annular Mode cycles (Australian Bureau of Meteorology, 2012; Xie et al., 2016).

5.5 Comparison of the GC approach with alternatives

The simplest approach to estimate ΔGWS is to subtract the model soil moisture component from GRACE ΔTWS data, without considering uncertainty in the model output, as used in Rodell et al. (2009) and Famiglietti et al. (2011). This method is called Approach 1 (App 1). In Approach 2 (App 2) as in Tangdamrongsub et al. (2017), by accounting for the uncertainty in model outputs and GRACE data, the water storage states are updated through a Kalman filter:

$$\hat{\boldsymbol{h}}_{\mathrm{R}} = \tilde{\boldsymbol{h}}_{\mathrm{R}} + \mathbf{H}\mathbf{C}_{\mathrm{R}}^{T}\left(\mathbf{H}\mathbf{R}\mathbf{H}^{T} + \mathbf{C}_{\mathrm{R}}\right)^{-1}\left(\boldsymbol{b} - \mathbf{H}\tilde{\boldsymbol{h}}_{\mathrm{R}}\right), \tag{18}$$

where $\tilde{\boldsymbol{h}}_{\mathrm{R}}$, $\mathbf{H}$, $\mathbf{C}_{\mathrm{R}}$ are described in Sect. 2; $\boldsymbol{b}$ is an observation vector containing GRACE-derived ΔTWS; and $\mathbf{R}$ is an error variance–covariance matrix of the observation. The GRACE-derived ΔTWS and its error information is obtained from the mascon solution. The matrix $\mathbf{R}$ is a (diagonal) error variance matrix since no covariance information is given in the mascon product. Note that the model uncertainty remains the same as in GC approach (Sect. 4.2). The different results from App 1 and App 2 are mainly attributed to the different estimates of the uncertainty.

The ΔGWS estimates from App 1, App 2, and GC in Queensland and Victoria are shown in Fig. 11. It is clearly seen that ΔGWS values from App 1 are overestimated while the one from App 2 fits the ground data significantly better. This behaviour was also seen in Tangdamrongsub et al. (2017), where the water storage estimates tend to be overestimated when error components such as spatial correlation error were neglected as in App 1. ΔGWS from App 2 shows clear improvements in terms of NS coefficients in both networks. Considering the deseasonalized ΔGWS estimates, in Queensland, the trend increases from 0.39 ± 0.03 to 0.42 ± 0.03 cm yr^{-1} (improves by 1.5 %) and the NS value increases from 0.46 to 0.53. In Victoria, the trend decreases

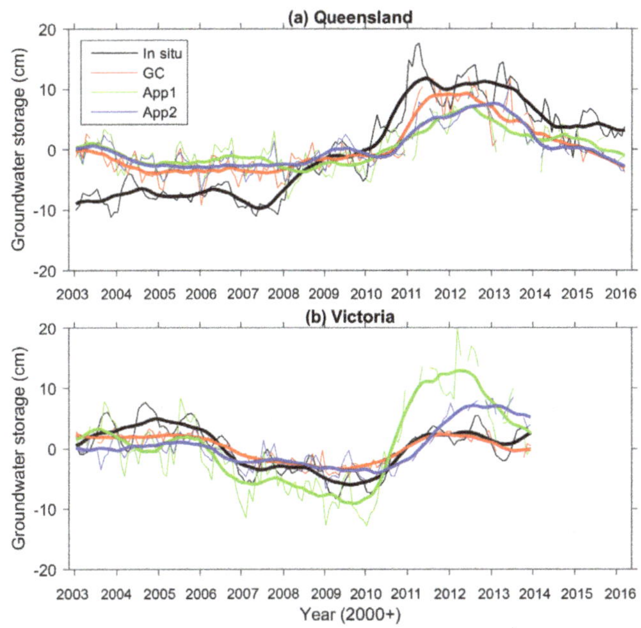

Figure 11. ΔGWS estimated from Approach 1 (App 1) and Approach 2 (App 2) in Queensland **(a)** and Victoria **(b)**. The in situ groundwater network data and the GC solutions are also shown. Deseasonalized time series are shown in thick lines.

from 0.73 ± 0.10 to 0.46 ± 0.05 cm yr^{-1} (improves by 27 %) and the NS value increases from −0.89 to 0.30. Although App 2 is not yet as good as the GC solution based on the most comprehensive error propagation, this simple test demonstrates the importance of considering the uncertainty. The reason for App 2 being less accurate than GC is likely the too-simplified error information implemented in App 2.

6 Conclusion

This study presents an approach to combine the raw GRACE observation with model simulation to improve water storage estimates over Australia. Distinct from other methods, we exploit the fundamental GRACE satellite tracking data and the full data error variance–covariance information to avoid alteration of signal and measurement error information present in higher-level data products.

We compare groundwater storage estimates from the GC approach and two other approaches, subject to inclusion of GRACE uncertainty in the ΔGWS calculation. Validating three results of ΔGWS against the in situ groundwater data, we find that the GC approach delivers the most accurate groundwater estimate, followed by the approach based on incomplete information of GRACE's data error. The poorest estimate of groundwater storage is seen when the GRACE uncertainty is completely ignored. This confirms the critical value of using the complete GRACE signal and error information at the raw data level.

The analysis of water storage change between 2003 and 2016 reveals that half of the continent (5 out of 10 basins) has still not fully recovered from the Millennium Drought. The TWS decrease in Western Australia has been most characteristic, and the GC approach finds that the water loss mainly occurs in the groundwater layer. Rainfall inadequacy is attributed to the continual dry condition, leading to a greater decrease in groundwater recharge and storage over Western Australia.

The land surface model we used is deficient in anthropogenic groundwater consumption. The model calibration will never help, and the groundwater consumption must be brought in by external sources. On the contrary, statistical approaches similar to our GC approach may be useful to fill in the missing component and lead to a more comprehensive water storage inventory.

However, it is difficult to constrain different water storage components by only using total storage observation such as GRACE. In addition, it is challenging to improve surface soil moisture varying rapidly in time, using a monthly mean GRACE observation. Tian et al. (2017) utilized the satellite soil moisture observation from Soil Moisture and Ocean Salinity (SMOS, Kerr et al., 2001) in addition to GRACE data for their data assimilation and showed a clear improvement in the top soil moisture estimate. The GC approach with complementary observations at higher temporal resolution should be considered, particularly to enhance the surface soil moisture computation.

Furthermore, the GC approach can be simply extended for GRACE data assimilation. Assimilating the raw GRACE data into land surface models like CABLE enables the model state and parameter to be adjusted with the realistic error information, allowing more reliable storage computation.

Appendix A: Least-squares normal equation of GRACE

A linearized GRACE satellite-tracking observation equation is formulated as

$$\boldsymbol{y} = \mathbf{A}\boldsymbol{x} + \boldsymbol{e};\ \boldsymbol{e} \sim N(\mathbf{0}, \boldsymbol{\Sigma}), \tag{A1}$$

where $\boldsymbol{y}$ is the observation vector containing various kinds of L1B data including the inter-satellite ranging data; $\mathbf{A}$ is the design (partial derivative) matrix relating the data and the Earth gravity field variations; $\boldsymbol{x}$ contains the Stokes coefficients of time-varying geopotential fields (e.g. Wahr et al., 1998); and $\boldsymbol{e}$ is the L1B data noise, which has zero mean and covariance $\boldsymbol{\Sigma}$. Equation (A1) can be modified explicitly in terms of soil moisture and groundwater storage variations as

$$\boldsymbol{y} = \mathbf{A}\mathbf{S}\overline{\mathbf{Y}}\mathbf{H}\boldsymbol{h} + \boldsymbol{e};\ \boldsymbol{e} \sim N(\mathbf{0}, \boldsymbol{\Sigma}), \tag{A2}$$

where $\mathbf{S}$ contains a factor used to convert ΔTWS to geopotential coefficients considering the load Love numbers (e.g. Wahr et al., 1998) and $\overline{\mathbf{Y}}$ converts the gridded data into the corresponding spherical harmonic coefficients. For convenience, the term $\mathbf{Y} = \mathbf{S}\overline{\mathbf{Y}}$ is used in the further derivation. A least-squares solution of Eq. (A2) is given as

$$\left(\mathbf{H}^T\mathbf{Y}^T\mathbf{A}^T\boldsymbol{\Sigma}^{-1}\mathbf{A}\mathbf{Y}\mathbf{H}\right)\hat{\boldsymbol{h}} = \mathbf{H}^T\mathbf{Y}^T\mathbf{A}^T\boldsymbol{\Sigma}^{-1}\boldsymbol{y}. \tag{A3}$$

It can be simplified as

$$\mathbf{H}^T\mathbf{Y}^T\mathbf{N}\mathbf{Y}\mathbf{H}\hat{\boldsymbol{h}} = \mathbf{H}^T\mathbf{Y}^T\boldsymbol{c}, \tag{A4}$$

where $\mathbf{N} = \mathbf{A}^T\boldsymbol{\Sigma}^{-1}\mathbf{A}$ and $\boldsymbol{c} = \mathbf{A}^T\boldsymbol{\Sigma}^{-1}\boldsymbol{y}$. Equation (A4) is identical to Eq. (5).

Appendix B: Nash–Sutcliffe coefficient and area-weighted average

The Nash–Sutcliffe coefficient (NS) is computed as follows:

$$\mathrm{NS} = 1 - \frac{\sum_{i=1}^{N}\left(y_i - \hat{x}_i\right)^2}{\sum_{i=1}^{N}\left(y_i - \overline{y}\right)^2}, \tag{B1}$$

where $\boldsymbol{y}$ is an observation vector, $\overline{\boldsymbol{y}}$ is the mean of the observation, $\hat{\boldsymbol{x}}$ is a vector containing the simulated result, i is the index of observation, and N is the number of observations.

Area-weighted average ($\overline{Z}$) is compute as follows:

$$\overline{Z} = \frac{\sum_{j=1}^{M} w_j \overline{z}_j}{\sum_{j=1}^{M} w_j}, \tag{B2}$$

where w is the area size, $\overline{z}$ is the mean value inside the considered area, j is the area index, and M is the number of considered areas.

Competing interests. The authors declare that they have no conflict of interest.

Acknowledgements. This work is funded by The University of Newcastle to support NASA's GRACE and GRACE Follow-On projects as an international science team member to the missions. Mark Decker was supported by the ARC Centre of Excellence for Climate Systems Science. Hyungjun Kim was supported by Japan Society for the Promotion of Science KAKENHI (16H06291). We thank Torsten Mayer-Gürr for GRACE data products in the form of the least-squares normal equations. We also thank three anonymous reviewers for helping us improve the manuscript.

Edited by: Louise Slater

References

Australian Bureau of Meteorology: Record-breaking La Niña events: An analysis of the La Niña life cycle and the impacts and significance of the 2010–11 and 2011–12 La Niña events in Australia, National Climate Centre, Bureau of Meteorology, http://www.bom.gov.au/climate/enso/history/La-Nina-2010-12.pdf (last access: 5 January 2017), 2012.

Bettadpur, S.: CSR Level-2 Processing Standards Document for Product Release 05, GRACE 327-742, Center for Space Research, The University of Texas, Austin, 2012.

Chen, J. L., Wilson, C. R., Tapley, B. D., Scanlon, B., and Güntner, A.: Long-term groundwater storage change in Victoria, Australia from satellite gravity and in situ observations, Global Planet. Change, 139, 56–65, https://doi.org/10.1016/j.gloplacha.2016.01.002, 2016.

Decker, M.: Development and evaluation of a new soil moisture and runoff parameterization for the CABLE LSM including subgrid-scale processes, J. Adv. Model. Earth Syst., 7, 1788–1809, https://doi.org/10.1002/2015MS000507, 2015.

Dee, D. P., Uppala, S. M., Simmons, A. J., Berrisford, P., Poli, P., Kobayashi, S., Andrae, U., Balmaseda, M. A., Balsamo, G., Bauer, P., Bechtold, P., Beljaars, A. C. M., van de Berg, L., Bidlot, J., Bormann, N., Delsol, C., Dragani, R., Fuentes, M., Geer, A. J., Haiberger, L., Healy, S. B., Hersbach, H., Hólm, E. V., Isaksen, L., Kållberg, P., Köhler, M., Matricardi, M., McNally, A. P., Monge-Sanz, B. M., Morcrette, J. J., Park, B. K., Peubey, C., de Rosnay, P., Tavolato, C., Thépaut, J. N., and Vitart, F.: The ERA-Interim reanalysis: configuration and performance of the data assimilation system, Q. J. Roy. Meteorol. Soc., 137, 553–597, https://doi.org/10.1002/qj.828, 2011.

Dumedah, G. and Walker, J. P.: Intercomparison of the JULES and CALBE land surface models through assimilation of remote sensed soil moisture in southeast Australia, Adv. Water Resour., 74, 231–244, https://doi.org/10.1016/j.advwatres.2014.09.011, 2014.

Eicker, A., Schumacher, M., Kusche, J., Döll, P., and Müller Schmied, H.: Calibration data assimilation approach for integrating GRACE data into the WaterGAP Global Hydrology Model (WGHM) using an Ensemble Kalman Filter: First Results, Surv. Geophys., 35, 1285–1309, https://doi.org/10.1007/s10712-014-9309-8, 2014.

Famiglietti, J. S., Lo, M., Ho, S. L., Bethune, J., Anderson, K. J., Syed, T. H., Swenson, S. C., de Linage, C. R., and Rodell, M.: Satellites measure recent rates of groundwater depletion in California's Central Valley, Geophys. Res. Lett., 38, L03403, https://doi.org/10.1029/2010GL046442, 2011.

Fasullo, J. T., Boening, C., Landerer, F. W., and Nerem, R. S.: Australia's unique influence on global sea level in 2010–2011, Geophys. Res. Lett., 40, 4368–4373, https://doi.org/10.1002/grl.50834, 2013.

Girotto, M., De Lannoy, G. J. M., Reichle, R. H., and Rodell, M.: Assimilation of gridded terrestrial water storage observations from GRACE into a land surface model, Water Resour. Res., 52, 4164–4183, https://doi.org/10.1002/2015WR018417, 2016.

Hamill, T. M., Whitaker, J. S., and Snyder, C.: Distance-Dependent Filtering of Background Error Covariance Estimates in an Ensemble Kalman Filter, Mon. Weather Rev., 129, 2776–2790, 2001.

Han, S.-C.: Elastic deformation of the Australian continent induced by seasonal water cycles and the 2010–11 La Niña determined using GPS and GRACE, Geophys. Res. Lett., 44, 2763–2772, https://doi.org/10.1002/2017GL072999, 2017.

Huffman, G. J., Adler, R. F., Bolvin, D. T., Gu, G., Nelkin, E. J., Bowman, K. P., Hong, Y., Stocker, E. F., and Wolf, D. B.: The TRMM multisatellite precipitation analysis (TMPA): Quasi-global, multiyear, combined-sensor precipitation estimates at fine scales, J. Hydrometeorol., 8, 38–55, https://doi.org/10.1175/JHM560.1, 2007.

Jekeli, C.: Alternative methods to smooth the Earth's gravity field, Rep. 327, Dept. of Geod. Sci. and Surv., Ohio State Univ., Columbus, 1981.

Kerr, Y. H., Waldteufel, P., Wigneron, J.-P., Martinuzzi, J.-M., Font, J., and Berger, M.: Soil moisture retrieval from space: The soil moisture and ocean salinity (SMOS) mission, IEEE T. Geosci. Remote, 39, 1729–1735, 2001.

Khaki, M., Hoteit, I., Kuhn, M., Awange, J., Forootan, E., van Dijk, A., Schumacher, M., and Pattiaratchi, C.: Assessing sequential data assimilation techniques for integrating GRACE data into a hydrological model, Adv. Water Resour., 107, 301–316, https://doi.org/10.1016/j.advwatres.2017.07.001, 2017a.

Khaki, M., Schumacher, M., Forootan, E., Kuhn, M., Awange, J., and van Dijk, A.: Accounting for spatial correlation errors in the assimilation of GRACE into hydrological models through localization. Adv. Water Resour., 108, 99–112, https://doi.org/10.1016/j.advwatres.2017.07.024, 2017b.

Kusche, J., Schmidt, R., Petrovic, S., and Rietbroek, R.: Decorrelated GRACE time-variable gravity solutions by GFZ, and their validation using a hydrological model, J. Geodesy, 83, 903–913, https://doi.org/10.1007/s00190-009-0308-3, 2009.

Lambert, A., Huang, J., van der Kamp, G., Henton, J., Mazzotti, S., James, T. S., Courtier, N., and Barr, A. G.: Measuring water accumulation rates using GRACE data in areas experiencing glacial isostatic adjustment: The Nelson River basin, Geophys. Res. Lett., 40, 6118–6122, https://doi.org/10.1002/2013GL057973, 2013.

Landerer, F. W. and Swenson, S. C.: Accuracy of scaled GRACE terrestrial water storage estimates, Water Resour. Res., 48, W04531, https://doi.org/10.1029/2011WR011453, 2012.

Lemoine, J. M., Bourgogne, S., Bruinsma, S., Gégout, P., Reinquin, F., and Biancale, R.: GRACE RL03-v2 monthly time series of solutions from CNES/GRGS, EGU2015-14461, EGU General Assembly 2015, Vienna, Austria, 2015.

Long, D., Chen, X., Scanlon, B. R., Wada, Y., Hong, Y., Singh, V. P., Chen, Y., Wang, C., Han, Z., and Yang, W.: Have GRACE satellites overestimated groundwater depletion in the Northwest India Aquifer?, Sci. Rep., 6, 24398, https://doi.org/10.1038/srep24398, 2016.

Mayer-Gürr, T., Behzadpour, S., Ellmer, M., Kvas, A., Klinger, B., and Zehentner, N.: ITSG-Grace2016 – Monthly and Daily Gravity Field Solutions from GRACE, GFZ Data Services, https://doi.org/10.5880/icgem.2016.007, 2016.

Mayer-Gürr, T., Pail, R., Gruber, T., Fecher, T., Rexer, M., Schuh, W.-D., Kusche, J., Brockmann, J.-M., Rieser, D., Zehentner, N., Kvas, A., Klinger, B., Baur, O., Höck, E., Krauss, S., and Jäggi, A.: The combined satellite gravity field model GOCO05s, EGU 2015, Vienna, 2015.

McGrath, G. S., Sadler, R.,Fleming, K., Tregoning, P., Hinz, C., and Veneklaas, E. J.: Tropical cyclones and the ecohydrology of Australia's recent continental-scale drought, Geophys. Res. Lett., 39, L03404, https://doi.org/10.1029/2011GL050263, 2012.

Njoku, E. G., Jackson, T. L., Lakshmi, V., Chan, T., and Nghiem, S. V.: Soil Moisture Retrieval from AMSR-E, IEEE T. Geosci. Remote, 41, 215–229, 2003.

Owe, M., de Jeu, R., and Holmes, T.: Multisensor historical climatology of satellite-derived global land surface moisture, J. Geophys. Res., 113, F01002, https://doi.org/10.1029/2007JF000769, 2008.

Pearson, E. K.: Mining imperfect data: Dealing with contamination and incomplete records, ProSanos Corporation, Harrisburg, Pennsylvania, https://doi.org/10.1137/1.9780898717884, 2005.

Ramillien, G., Cazenave, A., and Brunau, O., Global time variations of hydrological signals from GRACE satellite gravimetry, Geophys. J. Int., 158, 813–826, 2004.

Rassam, D. W., Peeters, L., Pickett, T., Jolly, I., and Holz, L.: Accounting for surfaceegroundwater interactions and their uncertainty in river and groundwater models: A case study in the Namoi River, Australia, Environ. Model. Softw., 50, 108–119, https://doi.org/10.1016/j.envsoft.2013.09.004, 2013.

Reichle, R. H. and Koster, R. D.: Bias reduction in short records of satellite soil moisture, Geophys. Res. Lett., 31, L19501, https://doi.org/10.1029/2004GL020938, 2004.

Rienecker, M. M., Suarez, M. J., Gelaro, R., Todling, R., Bacmeister, J., Liu, E., Bosilovich, M. G., Schubert, S. D., Takacs, L., Kim, G.-K., Bloom, S., Chen, J., Collins, D., Conaty, A., da Silva, A., Gu, W., Joiner, J., Koster, R. D., Lucchesi, R., Molod, A., Owens, T., Pawson, S., Pegion, P., Redder, C. R., Reichle, R., Robertson, F. R., Ruddick, A. G., Sienkiewicz, M., and Woollen, J.: MERRA – NASA's Modern-Era Retrospective Analysis for Research and Applications, J. Climate, 24, 3624–3648, https://doi.org/10.1175/JCLI-D-11-00015.1, 2011.

Rodell, M., Houser, P. R., Jambor, U., Gottschalck, J., Mitchell, K., Meng, C. J., Arsenault, K., Cosgrove, B., Radakovich, J., Bosilovich, M., Entin, J. K., Walker, J. P., Lohmann, D., and Toll, D.: The global land data assimilation system, B. Am. Meteorol. Soc., 85, 381–394, 2004.

Rodell, M., Velicogna, I., and Famiglietti, J. S.: Satellite-based estimates of groundwater depletion in India, Nature, 460, 999–1002, https://doi.org/10.1038/nature08238, 2009.

Sakumura, C., Bettadpur, S., and Bruinsma, S.: Ensemble prediction and intercomparison analysis of GRACE time-variable gravity field models, Geophys. Res. Lett., 41, 1389–1397, https://doi.org/10.1002/2013GL058632, 2014.

Schumacher, M., Kusche, J., and Döll, P.: A Systematic Impact Assessment of GRACE Error Correlation on Data Assimilation in Hydrological Models, J. Geodesy, 90, 537–559, https://doi.org/10.1007/s00190-016-0892-y, 2016.

Schumacher, M., Forootan, E., van Dijk, A., Schmied, H. M., Crosbie, R., Kusche, J., and Döll, P.: Improving drought simulations within the Murray-Darling Basin by combined calibration/assimilation of GRACE data into the WaterGAP Global Hydrology Model, Remote Sens. Environ., 204, 212–228, https://doi.org/10.1016/j.rse.2017.10.029, 2018.

Sheffield, J., Goteti, G., and Wood, E. F.: Development of a 50-yr high-resolution global dataset of meteorological forcings for land surface modeling, J. Climate, 19, 3088–3111, 2005.

Sorooshian, S., Hsu, K., Gao, X., Gupta, H. V., Imam, B., and Braithwaite, D.: Evaluation of PERSIANN System Satellite-Based Estimates of Tropical Rainfall, B. Am. Meteorol. Soc., 81, 2035–2046, 2000.

Swenson, S. C. and Wahr, J.: Post-processing removal of correlated errors in GRACE data, Geophys. Res. Lett., 33, L08402, https://doi.org/10.1029/2005GL025285, 2006.

Tangdamrongsub, N., Ditmar, P. G., Steele-Dunne, S. C., Gunter, B. C., and Sutanudjaja, E. H.: Assessing total water storage and identifying flood events over Tonlé Sap basin in Cambodia using GRACE and MODIS satellite observations combined with hydrological models, Remote Sens. Environ., 181, 162–173, https://doi.org/10.1016/j.rse.2016.03.030, 2016.

Tangdamrongsub, N., Steele-Dunne, S. C., Gunter, B. C., Ditmar, P. G., and Weerts, A. H.: Data assimilation of GRACE terrestrial water storage estimates into a regional hydrological model of the Rhine River basin, Hydrol. Earth Syst. Sci., 19, 2079–2100, https://doi.org/10.5194/hess-19-2079-2015, 2015.

Tangdamrongsub, N., Steele-Dunne, S. C., Gunter, B. C., Ditmar, P. G., Sutanudjaja, E. H., Xie, T, Wang, Z.: Improving estimates of water resources in a semi-arid region by assimilating GRACE data into the PCR-GLOBWB hydrological model, Hydrol. Earth Syst. Sci., 21, 2053–2074, https://doi.org/10.5194/hess-21-2053-2017, 2017.

Tian, S., Tregoning, P., Renzullo, L. J., van Dijk, A. I. J. M., Walker, J. P., Pauwels, V. R. N., and Allgeyer, S.: Improved water balance component estimates through joint assimilation of GRACE water storage and SMOS soil moisture retrievals, Water Resour. Res., 53, 1–21, https://doi.org/10.1002/2016WR019641, 2017.

Trenberth, K. E.: Framing the way to relate climate extremes to climate change, Climatic Change, 115, 283–290, https://doi.org/10.1007/s10584-012-0441-5, 2012.

Tscherning, C. C. and Rapp, R. H.: Closed covariance expressions for gravity anomalies, geoid undulations, and deflections of the vertical implied by anomaly degree variance models, Rep. 208, Dep. of Geod. Sci. and Surv., Ohio State Univ., Columbus, 1974.

Ukkola, A. M., Pitman, A. J., Decker, M., De Kauwe, M. G., Abramowitz, G., Kala, J., and Wang, Y.-P.: Modelling evapotranspiration during precipitation deficits: identifying critical processes in a land surface model, Hydrol. Earth Syst. Sci., 20, 2403–2419, https://doi.org/10.5194/hess-20-2403-2016, 2016.

Ummenhofer, C. C., Gupta, A., Briggs, P. R., England, M. H., McIntosh, P. C., Meyers, G. A., Pook, M. J., Raupach, M. R., and Risbey, J. S.: Indian and Pacific Ocean influences on Southeast Australian drought and soil moisture, J. Climate, 24, 1313–1336, https://doi.org/10.1175/2010JCLI3475.1, 2011.

Van Dijk, A., Podger, G., and Kirby, M.: Integrated water resources management in the Murray-Darling Basin, in: Increasing demands on decreasing supplies, Reducing the Vulnerability of Societies to Water Related Risks at the Basin Scale, edited by: Schumann, A. and Pahlow, M., IAHS Publ., Bochum, Germany, 24–30, 2007.

Van Dijk, A., Beck, H. E., Crosbie, R. S., De Jeu, E. A. M., Liu, Y. Y., Podger, G. M., Timbal, B., and Viney, N. R.: The Millennium Drought in southeast Australia (2001–2009): Natural and human causes and implications for water resources, ecosystems, economy, and society, Water Resour. Res., 49, 1040–1057, https://doi.org/10.1002/wrcr.20123, 2013.

Voss, K. A., Famiglietti, J. S., Lo, M., de Linage, C., Rodell, M., and Swenson, S. C.: Groundwater depletion in the Middle East from GRACE with implications for transboundary water management in the Tigris-Euphrates-Western Iran region, Water Resour. Res., 49, 904–914, https://doi.org/10.1002/wrcr.20078, 2013.

Wahr, J., Molenaar, M., and Bryan, F.: Time variability of the Earth's gravity field: Hydrological and oceanic effects and their possible detection using GRACE, J. Geophys. Res., 103, 30205–30229, 1998.

Wahr, J., Swenson, S., and Velicogna, I.: Accuracy of GRACE mass estimates, Geophys. Res. Lett., 33, L06401, https://doi.org/10.1029/2005GL025305, 2006.

Watkins, M. M., Wiese, D. N., Yuan, D.-N., Boening, C., and Landerer, F. W.: Improved methods for observing Earth's time variable mass distribution with GRACE using spherical cap mascons, J. Geophys. Res.-Solid, 120, 2648–2671, https://doi.org/10.1002/2014JB011547, 2015.

Welsh, W. D.: Water balance modelling in Bowen, Queensland, and the ten iterative steps in model development and evaluation, Environ. Model. Softw., 23, 195–205, 2008.

Wiese, D. N., Landerer, F. W., and Watkins, M. M.: Quantifying and reducing leakage errors in the JPL RL05M GRACE mascon solution, Water Resour. Res., 52, 7490–7502, https://doi.org/10.1002/2016WR019344, 2016.

Wu, S. C., Kruizinga, G., and Bertiger, W.: Algorithm theoretical basis document for GRACE Level-1B data processing V1.2, JPL D-27672, Jet Propul. Lab., Pasadena, California, 2006.

Xie, Z., Huete, A., Restrepo-Coupea, N., Maa, X., Devadasa, R., and Caprarellib, G.: Spatial partitioning and temporal evolution of Australia's total water storage under extreme hydroclimatic impacts, Remote Sens. Environ., 183, 43–52, https://doi.org/10.1016/j.rse.2016.05.017, 2016.

Zaitchik, B. F., Rodell, M., and Reichle, E. H.: Assimilation of GRACE terrestrial water storage data into a land surface model: Results for the Mississippi basin, J. Hydrometeorol., 9, 535–548, 2008.

5

Large-scale hydrological model river storage and discharge correction using a satellite altimetry-based discharge product

Charlotte Marie Emery[1,a], **Adrien Paris**[1,2,3,b], **Sylvain Biancamaria**[1], **Aaron Boone**[4], **Stéphane Calmant**[1], **Pierre-André Garambois**[5], **and Joecila Santos da Silva**[6]

[1]LEGOS, Université de Toulouse, CNES, CNRS, IRD, UPS, Toulouse, France
[2]GET, Université de Toulouse, UPS, CNRS, IRD, Toulouse, France
[3]LMI OCE IRD/UNB, Campus Darcy Ribeiro, Brasilia, Brazil
[4]Meteo France CNRS, CNRM UMR 3589, Toulouse, France
[5]ICUBE – UMR 7357, Fluid Mechanics Team, INSA, Strasbourg, France
[6]CESTU, Universidade do Estado do Amazonas, Manaus, Brazil
[a]now at: JPL, Pasadena, CA, USA
[b]now at: CLS, Ramonville-Saint-Agne, France

Correspondence: Charlotte Marie Emery (charlotte.emery@jpl.nasa.gov)

Abstract. Land surface models (LSMs) are widely used to study the continental part of the water cycle. However, even though their accuracy is increasing, inherent model uncertainties can not be avoided. In the meantime, remotely sensed observations of the continental water cycle variables such as soil moisture, lakes and river elevations are more frequent and accurate. Therefore, those two different types of information can be combined, using data assimilation techniques to reduce a model's uncertainties in its state variables or/and in its input parameters. The objective of this study is to present a data assimilation platform that assimilates into the large-scale ISBA-CTRIP LSM a punctual river discharge product, derived from ENVISAT nadir altimeter water elevation measurements and rating curves, over the whole Amazon basin. To deal with the scale difference between the model and the observation, the study also presents an initial development for a localization treatment that allows one to limit the impact of observations to areas close to the observation and in the same hydrological network.

This assimi-lation platform is based on the ensemble Kalman filter and can correct either the CTRIP river water storage or the dis-charge. Root mean square error (RMSE) compared to gauge discharges is globally reduced until 21 % and at Óbidos, near the outlet, RMSE is reduced by up to 52 % compared to ENVISAT-based discharge. Finally, it is shown that localization improves results along the main tributaries.

1 Introduction

The continental part of the water cycle is commonly studied, at large scale, with hydrological modelling. These models are generally issued from the coupling of a land surface model (LSM) with a river routing model (RRM). The LSM determines the water and energy budget at the surface by spreading precipitations between the soil and the canopy. Meanwhile, the RRM transfers water mass through the basin to the outlet and gives an estimate of river discharge.

RRMs are mainly based on kinematic (e.g. Oki and Sud, 1998) or sometimes also diffusive wave models (e.g. Yamazaki et al., 2011). Several RRMs have been developed since the 1990s and they differ mainly in their modelling of the flow velocity and the inclusion or not of groundwater and floodplain dynamics (Oki and Sud, 1998; Coe, 1998; Hagemann and Dümenil, 1998; Ducharne et al., 2003; Ngo-Duc et al., 2007; Paiva et al., 2011; Yamazaki et al., 2011; Decharme et al., 2012). The modelling of the river flow velocity is addressed in several ways in the literature. Coe (1998) and Oki and Sud (1998) considered a uniform and constant flow velocity over the entire basin in SWAM and TRIP RRMs, respectively. Other studies rather use a spatially distributed, but still constant in time, flow velocity based on topography and river channel characteristics (Vörösmarty et al., 1989; Hagemann and Dümenil, 1998; Ducharne et al., 2003). However, most recent models rely on a time-varying and spatially distributed flow velocity estimation based on the Manning formula (Manning, 1891), e.g. Arora et al. (1999), Ngo-Duc et al. (2007), Lucas-Picher et al. (2010) and Decharme et al. (2012). The first RRMs only modelled water flowing in the river channel (Vörösmarty et al., 1989; Coe, 1998; Oki and Sud, 1998). Subsequent RRM developments included the modelling of the groundwater inflow to the river (Hagemann and Dümenil, 1998; Arora et al., 1999; Ducharne et al., 2003; Decharme et al., 2008; Lucas-Picher et al., 2010) as well as the floodplains–river dynamics (Decharme et al., 2008; Paiva et al., 2011; Yamazaki et al., 2011). The routing network is derived from a digital elevation model (DEM): some of them remain defined on a regular mesh grid (Oki and Sud, 1998; Decharme et al., 2012), while others use an irregular discretization by sub-catchments, such as MGB-IPH (Paiva et al., 2011) and CaMa Flood (Yamazaki et al., 2011), or by river reaches, such as RAPID (David et al., 2011). More information on some global LSMs and their associated RRMs (with the corresponding references) could be found, for example, in Schellekens et al. (2017).

However, even if hydrological models become more and more accurate, inherent model uncertainties are unavoidable. They originate from several sources: simplification and lack of knowledge in the real physics, numerization and discretization-induced errors and uncertainties in the input parameters and forcing. All these uncertainties impact a model's outputs. In the worst case, all those uncertainties could accumulate and result in the collapse of the model. The model gives therefore an approximate view of the system's real state.

Observations of the system can be used to calibrate and/or validate the model and reduce its errors. These observations can be obtained from in situ or remote techniques. In situ techniques mainly focus on measuring river water elevations at a gauge station. Another important variable of interest in river hydrology is the river discharge, which is sparsely measured compared to water elevation. Based on river discharges and elevations measured at the same time and at the same location, it is possible to build a rating curve that represents the elevation–discharge relationship. This rating curve is then applied to water elevation to set continuous discharge time series. Institutions delivering in situ data provide mainly discharge. Even though in situ measures are generally quite accurate with a high time sampling (i.e. sub-daily), their main limitation is their local and spatially sparse sampling over the river network. Furthermore, nowadays, remotely sensed data from satellite missions are more and more available and provide useful observations of rivers. The most straightforward and used instrument to measure river water elevations is the nadir altimeter.

Altimeters were initially developed to measure ocean topography with satellite missions GEOS-3 (1975–1978) and SEASAT launched in 1978 (MacArthur, 1980). Nadir altimetry consists in estimating water surface elevation at the vertical (or at the nadir) of the satellite. It therefore produces punctual water elevation observations along the satellite ground track. These missions were followed by a long series of other missions: GEOSAT (1985–1990), TOPEX-Poseidon (1992–2006), JASON-1 (2001–2013), JASON-2 (2008–now), JASON-3 (2016–now), GFO (1998–2008), ERS-1 (1991–2000), ERS-2 (1995–2003), ENVISAT (2002–2012), SARAL (2013–now) and Sentinel-3 (2016–now). It is with TOPEX-Poseidon that the use of nadir altimetry to monitor lakes (Birkett, 1995; Hwang et al., 2005; Cretaux et al., 2009), floodplains (Birkett et al., 2002) and rivers (Birkett, 1998; Kouraev et al., 2004; Silva et al., 2010) developed widely. However, the main limitations of nadir altimetry are their punctual measurements (at the location where the satellite track crosses a river stream) and their temporal sampling (from 10 to 35 days, depending on the mission). Besides, in contrast to ocean surfaces, the signal over continental surfaces is impacted by vegetation and topography surrounding the river. Therefore, the purpose of this study is to combine model outputs and altimetry-based products using data assimilation (DA) techniques, in order to get more precise discharge estimates within the Amazon basin.

DA aims to improve model skills to forecast/simulate the physical system evolution. To do so, DA techniques focus on either correcting the model's input parameters (parameter estimation) or the model's outputs (state estimation). State estimation (SE) consists in using observations to directly correct the model output state. It is based on the assumption that the model (and the observations) are known to be imperfect. So, SE aims at correcting model outputs, whose errors result from all sources of uncertainties previously described.

SE has been widely used in oceanography and meteorology (Evensen and Leeuwen, 1996; Houtekamer and Mitchell, 2001; Gustafsson et al., 2012; Tanajura et al., 2015). However, DA of remotely sensed observations to correct hydrological model states is more recent. Moreover, it is more developed for LSMs than RRMs, as shown for example by the global-scale Land Data Assimilation System of the NASA Goddard Earth Observing System (Reichle et al., 2014). This platform assimilates simultaneously the SMOS soil moisture product, MODIS snow cover extent fraction and integrated GRACE terrestrial water storage variations into an ensemble Kalman filter (EnKF) to correct the states of several LSMs. Other studies assimilate similar kinds of observations, along with in situ data, into smaller-scale hydrological models (Trudel et al., 2014; López López et al., 2016). As for RRMs, to the authors' knowledge, there are few studies where remotely sensed and/or in situ data are assimilated into global-scale RRMs. However, in the liter-

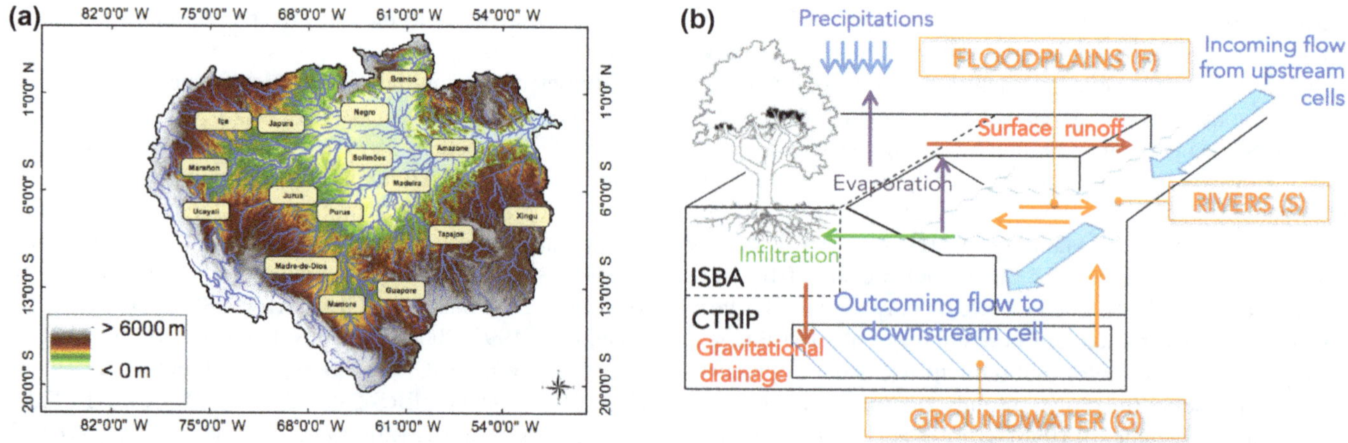

Figure 1. **(a)** The Amazon basin main tributaries and rivers with the underlying topography from SRTM. **(b)** Schematic representation of the ISBA-CTRIP system for a given grid cell. ISBA surface runoff ($Q_{\mathrm{ISBA,sur}}$) flows into the river/surface reservoir S; ISBA gravitational drainage ($Q_{\mathrm{ISBA,sub}}$) feeds the groundwater reservoir G. The surface water is transferred from one cell to another following the TRIP river routing network.

ature, there are several studies that used assimilation techniques at smaller and local scales with finer spatial resolution than global RRMs, using mostly in situ data (Schumann and Domeneghetti, 2016). For example, Clark et al. (2008) applied the EnKF to correct soil, aquifer and surface water storage in a small river in New Zealand (the Wairu). More particularly, they used gauge discharge data from four gauges to correct water storages in the 380 sub-catchments dividing the study zone. Paiva et al. (2013a) also used an EnKF over the Amazon basin for three different experiments, which assimilate, first, ENVISAT water surface anomalies from 212 virtual stations, and then discharge data from 109 gauges and finally remotely sensed discharges from 287 stations obtained from Getirana and Peters-Lidard (2013). This study aimed at correcting discharge estimated by the MGB-IPH hydrological model over more than 5000 sub-catchments comprising the Amazon River basin. Moreover, in two different studies, Michailovsky et al. (2013) and Michailovsky and Bauer-Gottwein (2014) assimilated for the Brahmaputra River in Asia and the Zambezi River in Africa, respectively, using an extended Kalman filter, ENVISAT water surface elevation measurements from six and nine virtual stations in Michailovsky et al. (2013) and Michailovsky and Bauer-Gottwein (2014), respectively, to correct simulated water volumes in 18 and 37 sub-catchments, respectively.

The objective of the present study is to investigate the contribution of remotely sensed data, and in particular measurements derived from nadir altimeters that provide local information, to improve a large-scale RRM via DA. The scale difference between the observations and the model leads us to also study the need to use localization methods within our DA framework. We used an ensemble Kalman filter, to which we added a simple localization module, to assimilate discharges derived from ENVISAT water surface elevation measurements. These observations are used to correct the state of the large-scale Total Runoff Integrated Pathways (TRIP, Oki and Sud, 1998) RRM version included in the Surfaces Externalisées land surface modelling platform (SurfEx, Masson et al., 2013) and developed at the Centre National de Recherches en Météorologie (CNRM, France). This particular version is denoted by the CTRIP acronym hereinafter. CTRIP is coupled with the Interactions-Soil-Biosphere-Atmosphere (ISBA, Boone et al., 1999) LSM at a resolution of $0.5° \times 0.5°$.

In Sect. 2, we present the study domain along with the ISBA-CTRIP model version and remotely sensed product used in this study. Section 3 provides first a general presentation of the ensemble Kalman filter (EnKF) DA method. Then we introduce the special features associated with the study and the description of the assimilation strategy. Then, in Sect. 4, we present results for a series of DA experiments testing the ensemble generation strategy and the correction of different state variables. Finally, Sect. 5 discusses these results and some perspectives. The last section gives the conclusions and some perspectives of the study.

2 Study domain, model and data used

2.1 Study domain: the Amazon basin

The study is focused on the Amazon River basin (see Fig. 1a). It is the world's largest river in terms of averaged discharge ($2 \times 10^5\ \mathrm{m^3\,s^{-1}}$) and drainage area ($6.15 \times 10^6\ \mathrm{km^2}$). The discharge at its mouth represents 30 % of total freshwater inflow to the Atlantic Ocean (Wisser et al., 2010) and its catchment area covers about 40 % of South America. The river source is located in the Peruvian Andes and flows through the Brazilian rainforest while receiving water from several important tributaries: first, the Ucayali, the Japurá River, the Purus River and, at Manaus, the Negro River (14 % of the total discharge). At this point, the river has reached 56 % of its total discharge. From Manaus to its mouth, it receives water from the Madeira River (17 % of the total discharge), the Tapajós River and the Xingu River (11 % of the total discharge all together) (Molinier et al., 1993).

The Amazon basin's geology can be divided into three major morpho-structural units: the western Andean Cordillera, the central Amazon trough and the shields at the eastern part of the basin (Guiana shield to the north and the Brazilian shield to the south). The northern and southern regions of the basin are under a tropical climate with a dry and a wet season, but the maximum rainfall season for the two parts occurs at different periods during the year (Meade et al., 1991). This implies that annual peak discharge in southern tributaries occurs a few months earlier than in northern tributaries. Meanwhile, the central basin is under an equatorial climate zone, implying high surface temperatures, air humidity and, especially, precipitation. Thus, a vast floodplain along the mainstream is filled every year, leading to the damping of discharge extremes.

2.2 ISBA-CTRIP model

2.2.1 Model presentation

The ISBA model (Noilhan and Planton, 1989) is a relatively standard land surface model (LSM) defined over a regular mesh grid at global scale. The model's equations are solved for each grid cell separately from the others. All grid cells are only correlated through the spatial patterns of atmospheric (especially precipitation) and radiative inputs, vegetation cover and soil composition. By taking into account the heterogeneity in precipitation, topography and vegetation within each grid cell (Decharme and Douville, 2006) and based on the force-restore method (Blackadar, 1976), ISBA gives a diagnosis of the water and energy budgets in each grid cell. Especially the ISBA-3L configuration (Boone et al., 1999), used in Alkama et al. (2010) and Decharme et al. (2012), has been chosen for the present study. In this version, the soil is divided into three layers: the superficial layer, the root zone and the sub-root zone. Precipitation can either fall directly on the soil surface or be intercepted by the canopy. The soil water content varies with canopy dripping, surface infiltration, soil evaporation, plant evapotranspiration, surface runoff and deep drainage (for more details, see Alkama et al., 2010; Decharme et al., 2012). Then, ISBA gives a diagnostic of each water budget component, in particular the surface runoff ($Q_{\mathrm{ISBA,sur}}$) and gravitational drainage ($Q_{\mathrm{ISBA,sub}}$) which are the main inputs for CTRIP.

The CTRIP RRM is also defined over a regular mesh grid. In this study, it is run at the same resolution as ISBA ($0.5° \times 0.5°$). CTRIP is dedicated to the lateral transfer of water from one cell to the other, up to the continent–ocean interface following a river network (Oki and Sud, 1998). The CTRIP version (Decharme et al., 2010) used in this study is coupled with the ISBA LSM and was subsequently developed by Decharme et al. (2010, 2012). It consists of a system of three reservoirs (see Fig. 1b): the surface reservoir S (kg) modelling the river, the groundwater reservoir G (kg) and the floodplain reservoir F (kg).

Only the surface reservoir S sends water from cell to cell based on the TRIP routing network. A cell can receive water from several upstream cells, but sends water into a unique downstream cell based on a space and time-varying flow velocity $v(t)$ estimated with the Manning formula (Manning, 1891). For any given cell, TRIP inputs are the TRIP outflow from upstream cells $Q^{\mathrm{S}}_{\mathrm{in,TRIP}}(t)$ and the ISBA surface runoff for that cell $Q_{\mathrm{ISBA,sur}}$. Moreover, S receives water from the groundwater reservoir G, $Q^{\mathrm{G}}_{\mathrm{out}}(t)$, and can exchange water mass with the floodplain F, $Q^{\mathrm{F}}_{\mathrm{out}}(t) - Q^{\mathrm{F}}_{\mathrm{in}}(t)$.

G receives inflows from ISBA gravitational drainage $Q_{\mathrm{ISBA,sub}}$ and outflows to the river reservoir S. This outflow represents more a delayed contribution of the gravitational drainage to the river than a real groundwater dynamic.

The floodplain scheme activates when the water height in the river, h_{S}, exceeds a given critical bankful height H_{c}. Then, part of the precipitation is intercepted by the floodplain ($P_{\mathrm{F}}(t)$) and the water in the floodplain can either evaporate ($E_{\mathrm{F}}(t)$) or infiltrate into the soil ($I_{\mathrm{F}}(t)$). A detailed description of the floodplain scheme is given in Decharme et al. (2008, 2010, 2012).

2.2.2 Model implementation over the Amazon

ISBA-CTRIP is run in offline mode. This implies that external atmospheric data are needed to force the model. Here, the atmospheric data from the Global Soil Wetness Projet 3 (GSWP3, http://hydro.iis.u-tokyo.ac.jp/GSWP3) are used. The project consists of three global-scale experiments with the objective of investigating long-term changes in the energy–water–carbon cycle components and their interactions. The 3-hourly resolution atmospheric boundary conditions used in the present study were generated by dynamically downscaling the global 2°-resolution 20th Century Reanalysis (Compo et al., 2011). This reanalysis assimilates several atmospheric observations into the Climate Forecast System (CFS) operational model from NCEP (National Centers for Environmental Prediction).

For ISBA-CTRIP, the Amazon basin is composed of a total number of 2028 cells. A sensitivity analysis (SA) of the ISBA-CTRIP has been conducted by Emery et al. (2016). In this analysis, the basin was divided into nine hydrogeomorphological zones which are shown in Fig. 2. These zones were designed to take into account different components: (1) a hydrological component (the main course is separated from the tributaries which have their own zones); and (2) a geological component (the three major morpho-

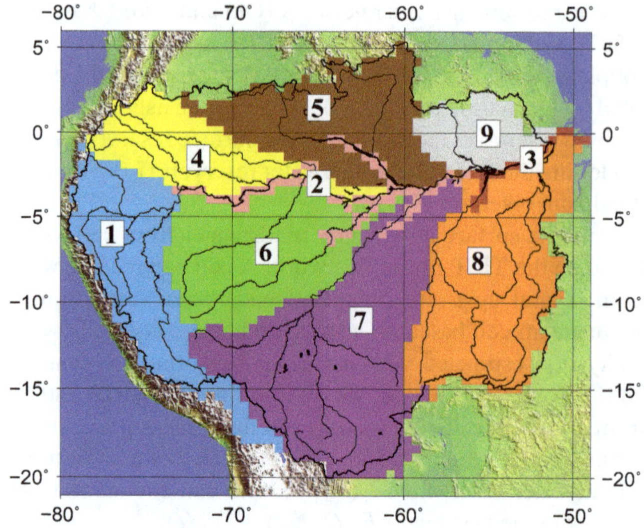

Figure 2. Map of hydro-geomorphological zones defined over the Amazon basin.

structural units are distinguishable). The nine zones are the following: (1) the upstream Andean part of the basin until the city of Iquitos, Peru; (2) the mainstream from Iquitos to Óbidos; (3) the mainstream from Óbidos to the river mouth; (4) left-bank tributaries from the Napo River to the Japurá River; (5) left-bank tributaries from the Japurá River to Óbidos, including the Negro River and its drainage area; (6) right-bank tributaries from Iquitos to the Purus River confluence at Anamã; (7) right-bank tributaries from Anamã to Óbidos, including the Madeira River; (8) right-bank tributaries exiting in zone 3, including the Tapajós River and the Xingu River; and (9) left-bank tributaries exiting in zone 3. This subdivision will be used within the DA platform.

2.3 Observations

2.3.1 Altimetry-based discharge product

The altimetry-based discharge product used in this study is derived from water surface elevations measured by the ENVISAT Radar Altimeter-2 altimeter instrument at Virtual Station (VS). VS is computed where the altimeter track crosses the river. The ENVISAT mission operated from September 2002 to October 2010 on its nominal orbit, which has a 35-day repeat period and an 80 km inter-track distance at the Equator. The water surface elevations measured over the Amazon basin were initially generated by Silva et al. (2010). The final product was referenced to the EGM2008 geoid (Palvis et al., 2012) and the vertical precision ranged from 12 to 30–40 cm for most of the stations (and can reach several metres for the worst stations).

Turning water surface elevation measures into an equivalent discharge requires the use of elevation–discharge rating curves. The rating curves used in this study have been built and validated by Paris et al. (2016), using water surface elevations from ENVISAT (Silva et al., 2010, 2012, 2014) and discharges simulated by hydrological–hydrodynamic model MGB-IPH (Model de Grandes Bacias-Instituto de Pesquisas Hidráulicas, Collischon et al., 2007). The model's original version, developed over the Amazon River basin, consists of a large-scale distributed hydrological model coupled with a hydrodynamic module that uses a simple storage scheme for floodplains (Paiva et al., 2011). The entire basin is divided into 5765 elementary catchments with an area varying between 100 and 5000 km^2. A surface scheme is applied for each mini-basin to estimate the main flows and a routing network is used to direct the flows from one elementary catchment to another, down to the outlet. Two approaches are used to estimate the river discharge: 1 – the Muskingum–Cunge method (MC) for basin heads and small tributaries; 2 – the Saint-Venant equations (HD for hydrodynamic) for the main stem and main tributaries. The DEM used is SRTM (Farr et al., 2007), and parameters such as the river width and depth are determined using geomorphological relationships calibrated over the sub-basins (Paiva et al., 2013a). Moreover, the model version used to determine the rating curves is the version developed by Paiva et al. (2013b), where gauge discharges are assimilated into the model via an ensemble Kalman filter (EnKF, Evensen, 2003), over the period between 1998 and 2010. The assimilated discharges allow one to correct the simulated discharges over both gauged and ungauged elementary catchments. With a better estimation of discharges, Paiva et al. (2013b) also provide an estimation of discharge uncertainty (modelled as a white noise) for each elementary catchment.

The MGB-IPH discharges were used by Paris et al. (2016) as a baseline to estimate the altimetric rating curves such that

$$\forall j, \forall i, \forall t_{\text{alti}}, \; Q_{\text{mgb},j(i)}\left(t_{\text{alti},i}\right) = a_i \times \left(H_{\text{alti},i} - z_{0,i}\right)^{b_i}, \quad (1)$$

where

- $H_{\text{alti},i}$ is the altimetric water surface elevation at the ith virtual station which is available at time $t_{\text{alti},i}$,
- $Q_{\text{mgb},j(i)}(t_{\text{alti},i})$ is the discharge estimated by the MGB-IPH model, at time $t_{\text{alti},i}$, in the jth mini-basin which contains the ith virtual station, and
- a_i, $z_{0,i}$ and b_i are the rating curve parameters to be determined. Those parameters are constant in time but vary from one virtual station to the other.

To calculate those parameters at each virtual station, a global optimization algorithm, the Shuffled Complex Evolution Metropolis developed by Vrugt et al. (2003), was used. It allowed determination of rating curves for 767 ENVISAT virtual stations. More details about the rating curve computation can be found in Paris et al. (2016). Once rating curve parameters are determined, altimetric water surface elevations are easily converted into equivalent "altimetric discharges".

Moreover, the altimetric discharges are provided with an estimation of their uncertainty including the normalized deviation from the MGB discharge.

The quality assurance of the discharge product has been made by constraining the rating curve coefficients within a physical range of values (Paris et al., 2016). Paris et al. (2016) also conducted a sensitivity analysis that showed a small sensitivity of the coefficient estimation to their first guess value. The quality check was done by comparing the satellite-derived discharge to the modelled discharge over a validation time period distinct from the calibration period used to derive rating curves. Discharge was also compared to some in situ gauges. Satellite-derived discharge is of course heavily correlated with the model accuracy. Overall, a comparison to 51 gauge measurements led to a mean Nash–Sutcliffe coefficient of around 0.8 and a normalized root mean square error of around 10 % over the validation period (Table 8 in Paris et al., 2016).

Altimetric discharges have then to be compared to ISBA-CTRIP discharges. However, while the virtual stations are irregularly distributed over the entire basin, the model is defined over a coarse regular mesh grid of $0.5^\circ \times 0.5^\circ$. A preliminary treatment of the virtual stations is applied to associate each ENVISAT virtual station with an ISBA-CTRIP cell with respect to their localization and the drainage network. The following algorithm has been used.

- The CTRIP river network is compared to a realistic river system (produced with GoogleEarth) to properly associate ISBA-CTRIP cells with a given tributary in the basin.
- Then, each virtual station is coupled with the closest ISBA-CTRIP cell along the same tributary. It may be the cell containing the virtual station or an adjacent cell according to the river network.

This algorithm allowed association of most of the virtual stations with a unique CTRIP cell. However, some particular cases have been treated. First, some virtual stations were located on tributaries too small to be represented on the CTRIP river network. In this case, the virtual station was not included in the study. Then, there were several very close virtual stations associated with the same ISBA-CTRIP cell. In this second case, the virtual stations with the lowest deviation from the MGB discharge were conserved. Finally, over the 767 ENVISAT virtual stations initially available, 368 ENVISAT virtual stations were kept and associated with an ISBA-CTRIP cell. Among them, 19 % (69 virtual stations out of 368) have been associated with an adjacent cell.

2.3.2 In situ discharge product

At a national or basin scale, water agencies can share discharge time series, such as the Agencia Nacional de Agua (ANA, hidroweb.ana.gov.br) in Brazil for the Amazon River basin. For the present study, we retrieved discharge time series from 145 ANA in situ stations over the entire basin. These gauge discharges have then been used to evaluate the performances of the DA (but they have not been assimilated into ISBA-CTRIP).

3 Method

The purpose of the SE DA is to correct model outputs using observations while taking into consideration uncertainties in both the model and the observations. In this work, as observed data correspond to discharge estimates, we chose to correct model output variables such as discharge or river storage. Indeed, following the results from the ISBA-CTRIP sensitivity analysis (SA; Emery et al., 2016), discharge is mainly sensitive to water inflow. Figure 3 presents the general DA method in the present study. The figure reads from top to bottom and from left to right. Three types of state variables will be considered: the river initial storage, the river final storage (which are both the main ISBA-CTRIP state variables) and the river discharge that will all be compared to the observed discharge. All three can be corrected through assimilation with specific treatment that will be detailed in the following sections. The DA will use several operators (in Fig. 3, $\mathcal{M}_{[k-1,k]}$, $\mathcal{Z}_k$ and $\mathcal{H}_k$) that link state variables with each other.

3.1 Data assimilation variables

The DA technique implemented in the present study is a sequential EnKF (Evensen, 2003). Here we shortly give the mathematical formalism used in the rest of the paper and a brief description of the EnKF method.

First of all, the the term "assimilation window" used hereafter corresponds to the period during which a complete assimilation cycle is conducted. It is delineated by two consecutive observation times and will be denoted by $[k-1, k]$. From now on, the kth assimilation cycle will be the cycle starting at time $k-1$ and assimilating the available observation(s) at time k.

3.1.1 Control variables

The vector $\boldsymbol{x}_k \in \mathbb{R}^{n_x}$ is called the control vector. It includes the n_x uncertain variables to be estimated during the kth DA cycle (within the time interval $[k-1, k]$). As stated before, control variables are prognostic or diagnostic variables of the ISBA-CTRIP model. Prognostic variables are the physical unknown in the differential equations' system that describes the model's behaviour. Diagnostic variables are also physical variables, but they are estimated from the prognostic variables. The choice of the control variables determines the observation operator $\mathcal{H}_k$ that maps the control variables into the observation space:

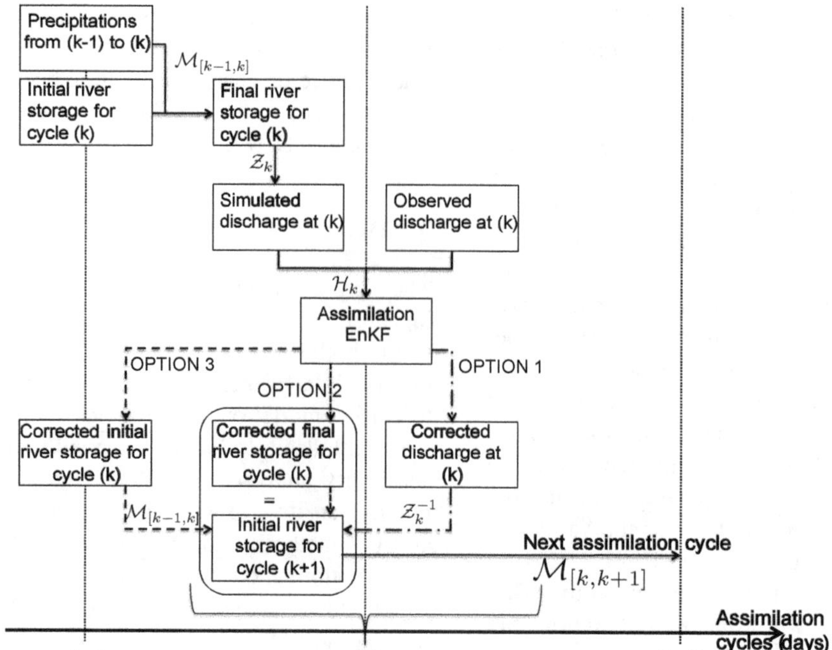

Figure 3. General framework of the DA method at a kth assimilation cycle. The figure reads from top to bottom and from left to right. The three main variables involved are the river initial storage, the river final storage and the river discharge. $\mathcal{M}_{[k-1,k]}$ is the model operator that maps the initial river storage into final river storage, $\mathcal{Z}_k$ is the diagnostic operator and $\mathcal{H}_k$ is the observation operator that maps simulated discharge into observed discharge for assimilation.

$$\boldsymbol{y}_k = \mathcal{H}_k(\boldsymbol{x}_k), \tag{2}$$

where $\boldsymbol{y}_k$ are the control variables equivalent in the observation space, also called model observations. They are then compared to the measured observations $\boldsymbol{y}_k^{\mathrm{o}}$ (described in Sect. 3.1.2) during the DA step.

Unlike hydrodynamic models, which directly solve Saint-Venant equations and for which discharge is a model state variable (or prognostic variable), the hydrological model ISBA-CTRIP solve differential equations describing the time evolution of water stock in the river (S), the groundwater (G) and the floodplain (F). Then, water elevation and river discharge are diagnostic variables derived from these prognostic variables. In CTRIP, river discharge $Q_{\mathrm{out}}^{\mathrm{S}}$ is computed as follows:

$$Q_{\mathrm{out}}^{\mathrm{S}} = L^{-1} v S \left[\mathrm{kg\,s^{-1}}\right], \tag{3}$$

with L (m) river section length, v the flow velocity (estimated from the Manning formula) and S the surface water mass.

Therefore, three types of variables can be considered as control variables in the data assimilation scheme: the discharge $Q_{\mathrm{out}}^{\mathrm{S}}$ (denoted Q in the remaining of the study to simplify notation, which is a diagnostic variable), the river final water stock S_{end} (a prognostic variable) or the river initial water stock 5also a prognostic variable). Definition and complexity of the observation operator $\mathcal{H}_k$, that maps the control space into the observation space, depends on the nature of the control variable. These three options are presented below.

- *Option 1: correcting ISBA-CTRIP discharges*: for this option, the control variables, gathered into the vector $\boldsymbol{x}_k$, are the ISBA-CTRIP discharges $Q_{i,k}$, $i = 1 \ldots n_x = 2028$ (number of TRIP cells in the Amazon basin) estimated for all 2028 cells in the TRIP Amazon basin, at the assimilation cycle k.

  The observation operator $\mathcal{H}_k$ resumes to a selection operator $\mathcal{S}_k$ which selects the observed TRIP cells at the current assimilation cycle:

  $$\mathcal{H}_k = \mathcal{S}_k. \tag{4}$$

  This operator is linear. The difficulty with this operator is that, once the assimilation analysis is produced, it is necessary to convert the analysis discharge $Q_{i,k}^{\mathrm{a}}$, $i = 1 \ldots n_x$ (i.e. the corrected discharge obtained after assimilation) into the equivalent final water stock $S_{\mathrm{end},i,k}^{\mathrm{a}}$. Indeed, as already stated, in ISBA-CTRIP, discharge is not a prognostic variable. Correction from the assimilation step needs to be propagated to the model prognostic variables, here the river final stock. Moreover, the analysis of the final water stock $S_{\mathrm{end},i,k}^{\mathrm{a}}$ will be used as an initial condition for the model run until the next assimilation cycle: $\forall i = 1 \ldots n_x$, $\forall k$, $S_{\mathrm{end},i,k}^{\mathrm{a}} = S_{\mathrm{ini},i,k+1}^{\mathrm{b}}$. However, the exact relationship linking discharge to the final river stock is unknown.

A possible solution consists in inverting Eq. (3). Assuming that the discharge estimated by ISBA-CTRIP $Q^{\mathrm{a}}_{i,k}$ is the instantaneous flow at the final time of the integration window,

$$Q^{\mathrm{a}}_{k,i,[\mathrm{kg\,s^{-1}}]} = L^{-1} v S^{\mathrm{a}}_{\mathrm{end},k,i,[\mathrm{kg}]} \Longleftrightarrow Q^{\mathrm{a}}_{k,[\mathrm{m^3\,s^{-1}}]} = \rho^{-1} L^{-1} v S^{\mathrm{a}}_{\mathrm{end},k,i,[\mathrm{kg}]}.$$

We obtain that (for more details on this approximation, see Appendix A)

$$S^{\mathrm{a}}_{\mathrm{end},k,i,[\mathrm{kg}]} \approx \rho L W^{2/5} s^{-3/10} n^{3/5} \left(Q^{\mathrm{a}}_{k,i,[\mathrm{kg\,s^{-1}}]} \right)^{3/5}, \quad (5)$$

with ρ ($\mathrm{m^3\,kg^{-1}}$) the water density, L (m) the river section length, W (m) the river width, s (–) the riverbed slope and n (–) the Manning coefficient in the riverbed. Then, for experiments with discharges as control variables, the formula in Eq. (5) will be used to convert corrected discharges into river stock and then propagate the model to the next observation time.

- *Option 2: correcting ISBA-CTRIP final water stock*: for this option, the control variables, gathered into the vector $\boldsymbol{x}_k$, are the ISBA-CTRIP final water stock $S_{\mathrm{end},i,k}$, $i = 1 \ldots n_x$ estimated for all 2028 cells in the TRIP Amazon basin, at the assimilation cycle k.

  The computational cost for this option is the same as for the first option but, now, the observation operator is defined as

$$\mathcal{H}_k = \mathcal{S}_k \cdot \mathcal{Z}_k, \quad (6)$$

  where $\mathcal{Z}_k$ is the operator (implicitly defined within TRIP) that turns the river final stock $S_{\mathrm{end},i,k}$ into equivalent discharge $Q_{i,k}$. Even though $\mathcal{H}_k$ is not linear any more, this option presents the advantage of correcting the river final stock $S^{\mathrm{a}}_{k,\mathrm{end}}$ that can be directly used for the next assimilation cycle and no additional uncertainties are introduced. However, the corresponding analysis discharge $Q^{\mathrm{a}}_{i,k}$ is now unknown as the explicit expression of $\mathcal{Z}_k$ is also unknown. A potential formula to determine $Q^{\mathrm{a}}_{i,k}$ can be deduced from Eq. (5). Such a formula will be necessary to make comparative statistics to ENVISAT and in situ discharges and be able to evaluate the assimilation performances.

- *Option 3: correcting ISBA-CTRIP initial water stock*: for this final option, the control variables, gathered into the vector $\boldsymbol{x}_k$, are the ISBA-CTRIP initial water stock $S_{\mathrm{ini},i,k}$, $i = 1 \ldots n_x$ estimated for all 2028 cells in the TRIP Amazon basin, at the assimilation cycle k.

  The discharge observations are used to correct the surface water stock at the time prior to the observation time or, in other words, at the initial time of the integrating window. Therefore, the observation operator is written as the composition of the model operator $\mathcal{M}_{[k-1,k]}$ with the observation operator defined in Eq. (6):

$$\mathcal{H}_k = \mathcal{S}_k \cdot \mathcal{Z}_k \cdot \mathcal{M}_{[k-1,k]}. \quad (7)$$

  This operator is highly non-linear as it contains the full model operator. However, it is the only option where no uncertainties are added from the use of an external formula to compute corrected discharge at the observation time. Uncertainties in the stock–discharge relationship are only due to the model uncertainties. Indeed, once the analysis initial water stock $S^{\mathrm{a}}_{\mathrm{ini},k,i}$ is determined, the control variable update must be propagated through the next assimilation time by re-integrating the ISBA-CTRIP model over the assimilation window. The initial water stock $S^{\mathrm{a}}_{\mathrm{end},k,i}$ and the analysis discharge $Q^{\mathrm{a}}_{k,i}$ are automatically determined during this run.

3.1.2 Observation variables

In the framework of the state estimation, the observation variables, at a given day within the Amazon basin, are the discharge estimates derived from ENVISAT water surface elevations at the virtual stations associated with an ISBA-CTRIP cell. The ENVISAT repeatability is 35 days, and therefore a given virtual station will provide an observation every 35 days at best. During the data assimilation experiments, all virtual stations will be used simultaneously. Because of the ENVISAT orbit, the number of available observations at a given day will vary between 0 and 15, and these observations will be assimilated daily via the EnKF. Then, the observation vector $\boldsymbol{y}^{\mathrm{o}}_k$ at the assimilation cycle k (equivalently, at the simulation day k) is composed of the $n_{y,k}$ discharge measures available at day k:

$$\boldsymbol{y}^{\mathrm{o}}_k = \left[y^{\mathrm{o}}_{k,1}, \quad y^{\mathrm{o}}_{k,2}, \quad \ldots, \quad y^{\mathrm{o}}_{k,n_{y,k}} \right], \quad (8)$$

where $y^{\mathrm{o}}_{k,j}$, $j = 1 \ldots n_{y,k}$ is the jth observation among the $n_{y,k}$ at cycle k.

Measurement errors $\boldsymbol{\epsilon}^{\mathrm{m}}_k$ come from errors in ENVISAT water surface elevations, errors in MGB discharges and uncertainties in the rating curve parameters used to turn water surface elevation into discharge. Sorooshian and Dracup (1980), Clark et al. (2008) and Paris et al. (2016) noticed that the concavity of the elevation–discharge relationship implies that the higher a water elevation, the more uncertain the corresponding discharge. Therefore, the observation error standard deviation $\sigma^{\mathrm{o}}_{k,j}$, associated with the jth observation at cycle k, is defined with respect to the instantaneous discharge measure $y^{\mathrm{o}}_{k,j}$, i.e.:

$$\sigma^{\mathrm{o}}_{k,j} = \eta^{\mathrm{o}}_j \times y^{\mathrm{o}}_{k,j}, \quad j = 1 \ldots n_{y,k}, \quad (9)$$

where $\eta^{\mathrm{o}}_j \in [0, 1]$ is a constant depending on the virtual station, such that η^{o}_j models the relative error. The observation error standard deviation $\sigma^{\mathrm{o}}_{k,j}$ is then a fraction of the instantaneous discharge. For each virtual station, the value of η^{o}_j depends on, first, the deviation from the MGB discharge, noted σ_j^{mgb} [%] and determined by Paris et al. (2016). As MGB discharges were used to determine ENVISAT discharges from ENVISAT water elevations, σ_j^{mgb} represents the deviation between the two discharges data. Second, to take into account that MGB discharge is not perfect (in other words, to take into account some deviation from the real discharge), 0.05 is added to σ_j^{mgb} and the sum is rounding up to the nearest whole number, giving

$$\eta_j^{\mathrm{mgb}} = E\left(\sigma_j^{\mathrm{mgb}} + 0.05\right),$$

where the function E is the ceiling function. Finally, η^{o}_j is equal to the first multiple of 5, above η_j^{mgb}. At the end, η^{o}_j varies from 0.10 to 0.35 over the entire basin and is constant in time. Besides, the representativeness error ϵ^{r}_k induced when a virtual station is associated with cells of the coarse TRIP mesh grid is neglected here.

Moreover, for a given assimilation cycle and also between different cycles, the observations are considered uncorrelated in space and time. The observation error covariance matrix at cycle k $\mathbf{R}_k$ is then a diagonal definite positive square matrix.

3.2 The ensemble Kalman filter (EnKF) for ISBA-CTRIP state estimation

3.2.1 The EnKF sequential estimation

In the EnKF framework, the model and observation operator are not linear. Therefore, the main idea is to use stochastic ensembles to represent the control variables PDFs along with the error models (Evensen, 1994, 2003). First, the background control variables $\boldsymbol{x}^{\mathrm{b}}_k$ are stochastically represented by an ensemble of n_{e} members:

$$\mathbf{X}^{\mathrm{b}}_{\mathrm{e},k} = \left[\boldsymbol{x}_k^{\mathrm{b},[1]} \quad \boldsymbol{x}_k^{\mathrm{b},[2]} \quad \ldots \quad \boldsymbol{x}_k^{\mathrm{b},[n_{\mathrm{e}}]}\right]. \tag{10}$$

Each member is estimated separately through the EnKF prediction step. For each control variable case (see Sect. 3.1.1), each member of the control ensemble $\mathbf{X}^{\mathrm{b}}_{\mathrm{e},k}$ is estimated by integrating the model operator from the corresponding analysis member at the previous assimilation cycle, while adding external uncertainties (see Sect. 3.2.3):

$$\forall l = 1 \ldots n_{\mathrm{e}}, \boldsymbol{x}_k^{\mathrm{b},[l]} = \mathcal{M}_{[k-1,k]}\left(\boldsymbol{x}_{k-1}^{\mathrm{a},[l]}\right). \tag{11}$$

Then, the background control ensemble must be compared to the observations. Depending on the control variables nature, the model operator is already included (option 3) or not (option 1 and 2) within the observation operator. Besides, following Burgers et al. (1998), an additional noise $\boldsymbol{\epsilon}^{\mathrm{o}}_k$ is added to the observation vector $\boldsymbol{y}^{\mathrm{o}}_k$ to avoid ensemble collapse. The observation ensemble thus obtained is noted:

$$\mathbf{Y}^{\mathrm{o}}_{\mathrm{e},k} = \left[\boldsymbol{y}_k^{\mathrm{o},[1]} \quad \boldsymbol{y}_k^{\mathrm{o},[2]} \quad \ldots \quad \boldsymbol{y}_k^{\mathrm{o},[n_{\mathrm{e}}]}\right]. \tag{12}$$

Finally, the EnKF analysis step is applied to each member of the ensemble such that

$$\forall l = 1 \ldots n_{\mathrm{e}}, \boldsymbol{x}_k^{\mathrm{a},[l]} = \boldsymbol{x}_k^{\mathrm{b},[l]} + \mathbf{K}_{k,\mathrm{e}}\left(\boldsymbol{y}_k^{\mathrm{o},[l]} - \mathcal{H}_k\left(\boldsymbol{x}_k^{\mathrm{b},[l]}\right)\right), \tag{13}$$

where the direct non-linear observation operator is applied to convert the control variables into equivalent model observations.

The particularity of the EnKF is that the Kalman gain ($\mathbf{K}_{\mathrm{e},k}$) is stochastically estimated from the different control and model observation ensemble, as follows:

$$\mathbf{K}_{\mathrm{e},k} = \left[\mathbf{PH}^T\right]_{\mathrm{e},k}\left(\left[\mathbf{HPH}^T\right]_{\mathrm{e},k} + \mathbf{R}_k\right)^{-1}. \tag{14}$$

3.2.2 Localization of the error covariance matrices

In the framework of state estimation, the sampling error can introduce artificial correlations into the background/analysis error covariance matrices, and generate spurious correlations between two distant grid cells in the mesh (Anderson, 2007). The ensemble size n_{e} is limited by computational constraints. Therefore, before the EnKF analysis step, a numerical processing of the matrices $[\mathbf{PH}^T]_{\mathrm{e},k}$ and $[\mathbf{HPH}^T]_{\mathrm{e},k}$ is necessary to suppress spurious correlations that can potentially degrade the analysis. Localization methods are designed to reduce these problems.

There exist two types of localization techniques (Greybush et al., 2011; Sakov and Bertino, 2011). The first one is called B-localization. It is based on explicitly modifying the background error covariance matrix $\mathbf{P}^{\mathrm{b}}_{\mathrm{e},k}$. It consists in multiplying the matrix $\mathbf{P}^{\mathrm{b}}_{\mathrm{e},k}$ by a correlation matrix generated from a radial function, namely a function of the two/three spatial dimensions which monotonously decrease with the distance between control variables (Hamill et al., 2001; Houtekamer and Mitchell, 2001, 2005). This modified matrix replaces $\mathbf{P}^{\mathrm{b}}_{\mathrm{e},k}$ in the calculation of the Kalman gain matrix $\mathbf{K}_{\mathrm{e},k}$. The other common localization technique is called R-localization or local analysis. This one consists in performing the analysis step in characteristic sub-spaces of the overall problem space.

However, all these localization techniques described above have been developed for atmospheric modelling where problems are in two or three dimensions. The use of localization in hydrology is more limited. Several studies exist to improve subsurface flow modelling (Devegowda et al., 2010; Delijani et al., 2014), but these approaches have a dimensionality close to atmospheric approaches, as they take place in a continuous medium in two or three dimensions. Other studies using localization allow estimation of better model parameters, still continuously defined in two or three dimensions (Sun et al., 2009; Rasmussen et al., 2015).

The localization method used with the CTRIP river routing model is of the B-localization type. However, it can not be simply defined on a two-dimensional radial function. Indeed, the river flow is along several one-dimensional flow directions, modelled by the routing network. The localization technique must consider the routing network to decorrelate adjacent cells on the mesh grid but located in two different sub-catchments. Nevertheless, along a same flow direction, the correlation between two distinct cells depends on the distance between the two cells. Then, for each assimilation cycle, the localization consists in a localization mask delimiting an influence area for each observation. These influence areas gather a limited number of neighbouring downstream and upstream cells around the observed cell with respect to the river routing network. We chose a fixed localization scale for simplicity and as a first step in the feasibility study of the development of a localization method for a hydrology application.

To determine the number of cells defining the influence area, the basin subdivision into nine hydro-geomorphological zones is used with a mean flow velocity for each zone. The influence area, for a given observed cell, is given by the criteria below. For an influence area of size p cells, the area is composed of the following.

- The observed cell.
- The p downstream cells according to the river routing network.
- all the cells upstream the observed one covering p upstream levels. However, the going up stops when the hydro-geomorphological zone is different from the one of the observed cell.

The number of cells within the influence area depends on the mean flow velocities (averaged over a year of simulation) in the zone in which the considered cell is situated. Those mean velocities are calculated from the free run simulation, namely the ISBA-CTRIP simulation realized without any assimilation step. The ISBA-CTRIP resolution is $0.5° \times 0.5°$, or approximately 50 km × 50 km. Given the river meanderings, the minimal covered distance through a cell is of 50 km. Furthermore, by comparing the free run discharge to in situ and ENVISAT discharges, it seems that the free run underestimate discharge (and so the flow velocity). Consequently, to fix the number p of cells defining the influence area in each hydro-geomorphological zone, the following steps have been performed:

1. the mean velocity for the cells into a given zone is converted into an equivalent distance in km,
2. the maximal distance within the zone is kept and rounded to the closest higher multiple of 50,
3. the number p determining the size of the influence area is the number of cells covered by the maximal rounded distance, knowing that 50 km = 1 cell.

The number of cells into the influence area is presented, for each zone, in the Table 2.

The final localization mask is presented into a matrix of size $n_x \times n_{y,k}$ (with n_x, the number of control variables, and $n_{y,k}$, the number of observation variables, at the assimilation cycle k) containing only 0 and 1. The localization mask $\mathbf{S}$ has the same dimension as the matrix $[\mathbf{PH}^T]_{e,k}$. So, the localization matrix is term-to-term multiplied (sign "×" in Eq. 15) to the error covariance matrix $[\mathbf{PH}^T]_{e,k}$ such as in Moore (1973) and Biancamaria et al. (2011):

$$\boldsymbol{x}^{a} = \boldsymbol{x}^{b} + \left[\mathbf{S} \times \left(\mathbf{P}^{b}\mathbf{H}^{T}\right)\right]\left\{\mathbf{H}\left[\mathbf{S} \times \left(\mathbf{P}^{b}\mathbf{H}^{T}\right)\right] + \mathbf{R}\right\}^{-1} \left(\boldsymbol{y}^{o} - \mathbf{H}\boldsymbol{x}^{b}\right). \quad (15)$$

To then extend the localization to the error covariance matrix $[\mathbf{HPH}^T]_{e,k}$, the lines in $[\mathbf{PH}^T]_{e,k}$ corresponding to the observed cells are extracted to form the second localization matrix. This second matrix is also term-to-term multiplied to $[\mathbf{HPH}^T]_{e,k}$. This localization step is applied in each assimilation cycle with respect to the activated ENVISAT virtual stations at the current assimilation cycle.

3.2.3 Generating the ensembles

The background error covariance matrices $[\mathbf{PH}^T]_{e,k}$ et $[\mathbf{HPH}^T]_{e,k}$ are estimated from the control variable ensemble using the definition suggested by Evensen (2004), Moradkhani et al. (2005) and Durand et al. (2008). To get a large ensemble, while maintaining a reasonable computational time, the ensemble size n_e has been set to a hundred members. Details on how they are exactly calculated are given in Appendix B. These matrices have a $n_x \times n_{y,k}$ and $n_{y,k} \times n_{y,k}$ size, respectively. The elements in the error covariance matrices, depend only on the definition of the background ensemble stored in the control matrix $\mathbf{X}^{b}_{e,k}$ and the parameter uncertainties taken into consideration to generate $\mathcal{H}(\mathbf{X}^{b}_{e,k})$. In the framework of state estimation, we choose to consider uncertainty into the initial condition and uncertainty into the precipitation forcing (Nijssen and Lettenmaier, 2004; Andreadis and Lettenmaier, 2006; Clark et al., 2008; Paiva et al., 2013b).

- *Perturbation of the initial condition*: the vector containing initial surface reservoir storage for each $n_x = 2028$ CTRIP cell at the assimilation cycle k is called $\boldsymbol{c}_k$. To ease the notations, we will omit, as much as possible, the assimilation cycle k subscript, knowing that, for all randomly perturbed variables and constants, a new ensemble is generated at each cycle.

  We used the Amazon basin division into $n_s = 9$ hydro-geomorphological zones (see Fig. 2). Initial conditions are perturbed by applying a multiplying factor over each zone $\eta_s^{c,[l]}$ such that

$$\forall s = 1 \ldots n_\mathrm{s},\ \forall l = 1 \ldots n_\mathrm{e},\ \boldsymbol{c}_{\mathrm{s},k}^{[l]} = \eta_i^{\mathrm{c},[l]} \cdot \boldsymbol{c}_{\mathrm{s},k}, \tag{16}$$

where $\boldsymbol{c}_{\mathrm{s},k}$ is the reduction of the initial condition $\boldsymbol{c}_k$ to the only zone s. For the perturbation to vary from one member to another, the value $\eta_\mathrm{s}^{\mathrm{c},[l]}$ is the realization of a Gaussian distribution, different for each member $[l] = 1 \ldots n_\mathrm{e}$ and for each hydrogeomorphological zone s. The Gaussian distributions used have a mean value of 1 and a standard deviation of $\sigma_{\eta_{\mathrm{s},k}^\mathrm{c}}$ whose values are detailed in Table 1.

The $\eta_i^{\mathrm{c},[l]}$ values depend on the assimilation cycle k and on the hydrogeomorphological zone in which the ith cell is.

- Firstly, a more important perturbation is applied to cells situated on the river mainstream (zones 2 and 3), as we assume that the uncertainties are more important in those zones. Indeed, discharges in these zones are the highest of the entire basin. Besides, several cells are confluence cells and are subject to backwater effects. As ISBA-CTRIP does not model the backwater effects, the water stock uncertainties in these cells are increased.
- Secondly, at the first assimilation cycle, the initial condition before perturbation $\boldsymbol{c}_1$ is identical for every member. At this particular cycle, the ensemble variance after perturbation depends only on the perturbation method presented in Eq. (16) while, for the other assimilation cycles, the successive previous assimilation cycles introduced an additional variability between members, before the perturbation step in Eq. (16). Therefore, the initial condition variance is more important at the second assimilation cycle and after. Then, to generate a larger variability at the first assimilation cycle, the standard deviation $\sigma_{\eta_{\mathrm{s},k}^\mathrm{c}}$ of the variable $\eta_i^{\mathrm{c},[l]}$ is more important at the first cycle than for the others.

- *Perturbation of precipitations*: another source of uncertainties for the generation of the ensemble $\mathcal{H}_k(\mathbf{X}_\mathrm{e}^\mathrm{b})$ lies in the precipitation fields. Atmospheric forcing comes from the GSWP3 product (see Sect. 2.2.2). Precipitation forcing $\mathbf{F}$ has been perturbed using the procedure presented by Clark et al. (2008). The ensemble of perturbed precipitation fields $\widetilde{\mathbf{F}}_\mathrm{e}$ is defined such that

$$\begin{aligned}\widetilde{\mathbf{F}}_\mathrm{e} &= \left\{\widetilde{\mathbf{F}}^{[1]},\ \widetilde{\mathbf{F}}^{[2]},\ \ldots\ \widetilde{\mathbf{F}}^{[n_\mathrm{e}]}\right\}\\ &= \left\{\boldsymbol{\varphi}_p^{[1]} \cdot \mathbf{F},\ \boldsymbol{\varphi}_p^{[2]} \cdot \mathbf{F},\ \ldots\ \boldsymbol{\varphi}_p^{[n_\mathrm{e}]} \cdot \mathbf{F}\right\},\end{aligned} \tag{17}$$

where

- $\mathbf{F}$ is the two-dimensional field of precipitation forcing before perturbation (with a time step of 3 h),
- $\widetilde{\mathbf{F}}^{[l]}$, for $l = 1 \ldots n_\mathrm{e}$, is the lth perturbed precipitation field,
- $\boldsymbol{\varphi}_p^{[l]}$, for $l = 1 \ldots n_\mathrm{e}$, is the lth multiplying uniformly distributed field of $\mathbf{F}$ to generate $\widetilde{\mathbf{F}}^{[l]}$. More details on how the fields $\boldsymbol{\varphi}_p^{[l]}$ have been generated are given in Appendix C.

Table 1. Constant values used to generate the background control ensemble $\mathbf{X}_{\mathrm{e},k}^\mathrm{b}$, the observation ensemble $\mathbf{Y}_{\mathrm{e},k}^\mathrm{o}$ and the model observation ensemble $\mathcal{H}(\mathbf{X}_{\mathrm{e},k}^\mathrm{b})$. k is the assimilation index and s is the basin zone index.

Variable	Description	Value	Eq.
n_e	Ens size	101	–
n_x	Control space size	2028	–
n_y	Obs space size	$\in [0, 15]$	–
η^o	Obs error	0.2	(9)
$\sigma_{\eta_{\mathrm{s},k}^x}$	Error on initial condition	$s = 2.3, k = 1 : 0.25$ $s = 2.3, k > 1 : 0.05$ $s = 1.4 : 9, k = 1 : 0.10$ $s = 1.4 : 9, k > 1 : 0.02$	(16)

Table 2. Size of the influence area for the localization process.

Zone	Real max dist (km)	Rounded max dist (km)	Influence area (number of cells)
1	84.16	100	2
2	174.24	200	4
3	233.80	250	5
4	93.29	100	2
5	82.03	100	2
6	80.03	100	2
7	69.57	100	2
8	87.86	100	2
9	67.02	100	2

3.3 Assimilation diagnostics

During the assimilation experiment, it is necessary to quantify the assimilation performances. The quality of the assimilation will be evaluated in a given cell i by estimating the root mean square error (RMSE) between the simulated discharge Q_i^* and the observed discharge $Q_i^\dagger$, giving

$$\mathrm{RMSE}_i^{*,\dagger} = \sqrt{\frac{1}{K^\dagger}\sum_{k=1}^{K^\dagger}\left(Q_{i,k}^* - Q_{i,k}^\dagger\right)^2}\ \left[\mathrm{m^3\,s^{-1}}\right]. \tag{18}$$

$K^\dagger$ represents the total number of available observed discharges $Q_i^\dagger$ at the studied cell for the study period. The "*" superscript can be either the superscript "f" for the free run

Table 3. Presentation of the different state estimation experiments. The "SE" acronym stands for "state estimation", indexes "1", "2" or "3" are to differentiate the control variables ("1": initial river storage, "2": final river storage and "3": discharge) and the suffixes "direct", "diag" and "local" indicate the localization scheme ("direct": without localization, "diag": diagonal error covariance matrices and "local": with localization).

Exp. name	Control variable	Localization scheme
SE1-direct	initial storage	No – $[\mathbf{PH}^T]_{e,k}$ and $[\mathbf{HPH}^T]_{e,k}$ defined in Eqs. (B1)–(B2)
SE1-diag	initial storage	No – diagonal $[\mathbf{PH}^T]_{e,k}$ and $[\mathbf{HPH}^T]_{e,k}$
SE1-local	initial storage	Yes – see Sect. 3.2.2
SE2-local	final storage	Yes – see Sect. 3.2.2
SE3-local	discharge	Yes – see Sect. 3.2.2

(without assimilation) or the superscript "a" for the analysis run (after assimilation). The "†" superscript can be either "o" for the observed ENVISAT discharge or "situ" for the gauge discharge.

Based on this definition, the assimilation performance will be estimated at each cell with the normalized RMSE (RMSEn) defined by

$$\mathrm{RMSEn}_i^{*,\dagger} = 100 \times \frac{\mathrm{RMSE}_i^{*,\dagger}}{\overline{Q_{i,\bullet}^{\dagger}}}, \quad [-], \tag{19}$$

where $\overline{Q_{i,\bullet}^{\dagger}}$ corresponds to the mean of $Q_{i,k}^{\dagger}$ averaged over the available time steps k.

Also, to evaluate the global performance of the assimilation over the entire basin, a global RMSEn ($\mathrm{RMSEn}_{\mathrm{global}}$) will be determined by

$$\mathrm{RMSEn}_{\mathrm{global}}^{*,\dagger} = 100 \times \frac{1}{\left(\sum_{i=1}^{I^\dagger} w_i\right)} \sum_{i=1}^{I^\dagger} w_i \cdot \mathrm{RMSEn}_i^{\dagger}, \quad [-], \tag{20}$$

where $I^\dagger$ is the total number of stations and w_i a weighting constant depending on the maximal discharge at the ith station ($\max_k(Q_{i,.}^{\dagger})$) and the maximal discharge over the basin ($\max_{i,k}(Q_{.,.}^{\dagger})$) such that

$$w_i = \frac{\log_{10}\left[\max_k\left(Q_{i,.}^{\dagger}\right)\right]}{\log_{10}\left[\max_{i,k}\left(Q_{.,.}^{\dagger}\right)\right]}. \tag{21}$$

With this weighting, cells along the mainstream and the largest tributaries (with the highest discharges) have more importance in the global statistics than cells located in basin heads.

Besides, the analysis run is available as an ensemble. The statistics will then be estimated for each member of the ensemble and the mean (see Eq. 22) of the ensemble will be presented, such as

$$\overline{\mathrm{RMSEn}_i^{\mathrm{a},\dagger}} = \frac{1}{n_\mathrm{e}} \sum_{l=1}^{n_\mathrm{e}} \mathrm{RMSEn}_i^{\mathrm{a},[l],\dagger}, \tag{22}$$

where $\mathrm{RMSEn}_i^{\mathrm{a},[l],\dagger}$ is the normalized root mean square deviation of the lth member of the analysis discharge ensemble.

3.4 Assimilation strategy

The state estimation experiments have the objective of testing the different control variables described in Sect. 3.1.1. Also, another objective is to test, validate and criticize the localization mask introduced in Sect. 3.2.2. In the following, experiment names using the localization will have the "-local" suffix, ones without any localization will have the "-direct" suffix and ones with no correlation between cells will have the "-diag" suffix. The objective of this study is to determine the best strategy for assimilating ENVISAT discharges into the ISBA-CTRIP model using the EnKF to correct the model state variables. The different experiments are presented in Table 3. After analysing these five elementary simulations over a single year, a last experiment will be run over the entire ENVISAT observing period (from September 2002 to June 2010), based on the best configurations.

For all the DA experiments, the observation errors are those described in Sect. 3.1.2, and the model errors are those presented in Sect. 3.2.3. Moreover, each experiment in Table 3 lasts 365 assimilation cycles of 1 day (so 1 year of assimilation) from 1 January to 31 December 2009. We chose this period as it overlaps with another altimetry mission (namely JASON-2), and future works may include comparison of the two datasets' contribution. The numerous ensemble ISBA-CTRIP simulations were realized with the CALMIP high-performance computation platform (https://www.calmip.univ-toulouse.fr/spip/) with the EOS supercomputer.

In the SE1-direct experiment, ENVISAT discharges are assimilated to correct the initial surface reservoir storage in TRIP (and inherently TRIP simulated discharges). For this first experiment, a classical EnKF, without any localization treatment of the error covariance matrices $[\mathbf{PH}^T]_{e,k}$ and $[\mathbf{HPH}^T]_{e,k}$, is conducted. This first experiment will be compared to the two next experiments, SE1-diag and SE1-local, which will highlight the contributions and/or limitations of the chosen localization approach. Finally, the last two experiments, SE2-local and SE3-local, will test the other possible control variables and the reliability of the operator $\mathcal{Z}_k$. More particularly, the SE2-local experiment is based on control vector option 2 (see Sect. 3.1.1) that assimilates discharges to correct the final rive storage, and SE3-local is based on control vector option 1 (see Sect. 3.1.1) that assimilates discharges to directly correct the ISBA-CTRIP discharges.

4 Results

4.1 Free run performances

The current section briefly presents the model performance without assimilation called the free run. As all in situ and ENVISAT VS have been associated with a unique ISBA-CTRIP cell, it is possible to compare observed discharge at these stations to corresponding ISBA-CTRIP simulated discharge. To begin with, a sample of 12 in situ stations, spread over the entire basin (over the mainstream and the main tributaries), is selected. The location and the name of these stations are represented in Fig. 4. Figure 5 compares ISBA-CTRIP free run discharges to in situ and ENVISAT discharges at the 12 stations over 1 year of simulation (year 2009). From this comparison the following observations can be drawn.

- Over the majority of cells where there are both an in situ station and a virtual station, the two discharge time series are similar (but not identical; see Fig. 5, panels 1–3, 5–6, 8–9, and 12). These results are due to the fact that gauge discharges were directly assimilated into the MGB-IPH hydrological model to correct the MGB-IPH estimated discharges (Paiva et al., 2013a). Then, those same estimated discharges were used to calculate parameters of the rating curves between ENVISAT water elevations and MGB discharges (Paris et al., 2016). Even though these rating curves have been derived from a model that assimilated in situ data, ENVISAT-derived discharges depend essentially on the remotely sensed water surface elevation variations (Paris et al., 2016). Therefore, ENVISAT discharges remain independent enough of in situ data.

- A strong difference between the in situ and ENVISAT discharges could indicate either that the rating curve parameters were not correctly estimated or that in situ/ENVISAT/MGB-IPH discharges have strong errors. As an example, see Fig. 5, panel 11, at Itaituba, where the gauge discharge is discontinuous and is even equal to 0 for some dates. Another example is the gauge discharge at Manicoré, in Fig. 5, panel 10, (Paris et al., 2016).

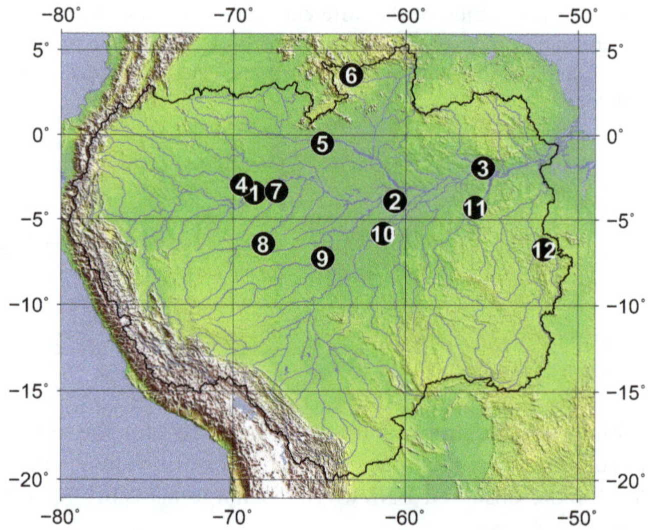

Figure 4. Map of the 12 in situ stations used to evaluate assimilation performance: (1) São Paulo de Olivenca (Solimões), (2) Manacapuru (Solimões), (3) Óbidos (Amazonas), (4) Ipiranga (Putumayo/Icá), (5) Serrinha (Negro), (6) Uaicás (Branco), (7) Porto Seguro (Jutaí), (8) Santos Dumont (Juruá), (9) Lábrea (Purus), (10) Manicoré (Madeira), (11) Itaituba (Tapajós), and (12) Boa Sorte (Xingu).

- Finally, in most cases, the free run discharge is quite different from the observed discharge. At downstream mainstream stations (at Manacapuru and Óbidos in Fig. 5, panels 2 and 3), the ISBA-CTRIP model is not able to reproduce flooding occurring between June and August. Therefore, in the free run, the discharge peak occurs earlier in the year and the discharge variations in this period are faster than the observed discharge variations. Similarly, at most of the right-bank tributary stations, the free run discharge peak is higher than the observed discharge peak (see Fig. 5, panels 7–12). However, the seasonal cycle is well reproduced for all these stations. These results illustrate the necessity of conducting the DA experiments.

Then, Fig. 6 displays the global performances of the free run. For each ENVISAT virtual station (see Fig. 6a) and each in situ station (see Fig. 6b), the RMSEn (defined in Eq. 19) between the simulated and observed discharges is calculated and its value is indicated by a colour at the location of the station over the basin. The results are similar between ENVISAT and gauge discharges, confirming good concordance between the two discharge datasets. RMSEn shows important deviations in basin heads on most of the tributaries as well as at the confluence between right-bank tributaries and the mainstream. Apart from the confluence and basin heads, the largest tributaries, such as the Negro and the Madeira, are well represented. Concerning global statistics (see Eq. 20), $\mathrm{RMSE}^{\mathrm{f,o}}_{\mathrm{global}}$ is equal to 71.12 % compared to ENVISAT dis-

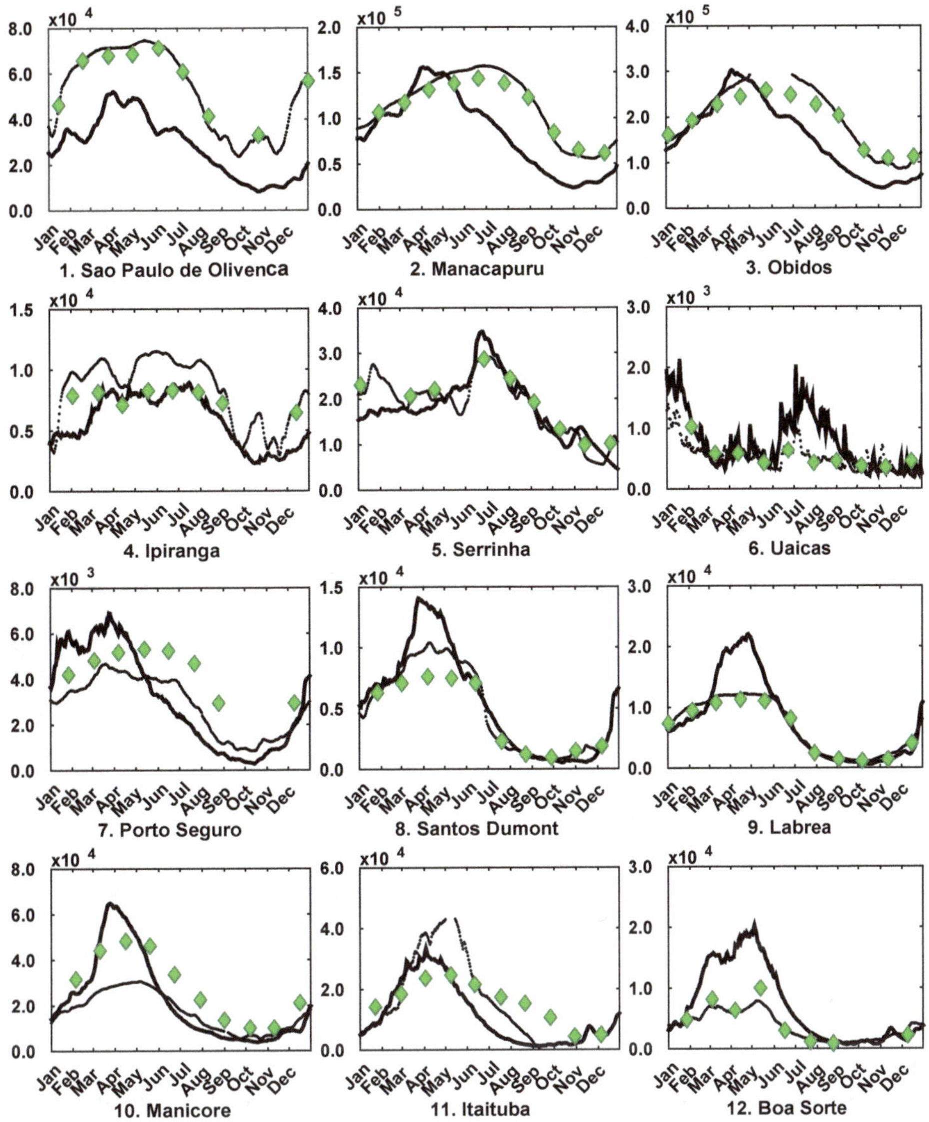

Figure 5. Comparison between the ISBA-CTRIP free run (blue line), ENVISAT-derived observed discharges (green markers) and ANA gauge discharges (black dots) over the year 2009. For each panel, the x-axis represents time (in days) and the y-axis represents discharge (in $m^3\,s^{-1}$).

charges and $\text{RMSE}^{\text{f,simu}}_{\text{global}}$ is equal to 68.96 % compared to gauge discharges. These deviations are likely due to atmospheric forcing, parametrization and modelling errors, especially floodplain parametrization. The DA experiments will focus on correcting the model outputs which result from those uncertainties.

4.2 Evaluation of the localization method

The first series of experiments assimilates ENVISAT discharges to correct the ISBA-CTRIP initial river stock (see the three first rows in Table 3). They differ on the definition of the background error covariance matrices $[\mathbf{PH}^T]_{e,k}$ and $[\mathbf{HPH}^T]_{e,k}$. The SE1-direct experiment uses the complete stochastic matrix defined in Eqs. (B1) and (B2). In the SE1-diag experiment, these matrices are processed such that covariance between two different CTRIP cells is set to 0 if the two variables are situated in two different CTRIP cells. Lastly, SE1-local is based on the localized version of the matrices presented in Sect. 3.2.2. So, Table 4 displays the global RMSEn (see the definition in Eq. 20) for the three experiments compared to the free run global statistics. From Table 4, we can see that the RMSE between the free run discharge and both the ENVISAT and gauge discharges is reduced for all experiments, showing that the data assimilation platform is working correctly. The SE1-diag experiment gives the worst results when compared to both the ENVISAT discharge and the gauge discharge. Then, compared

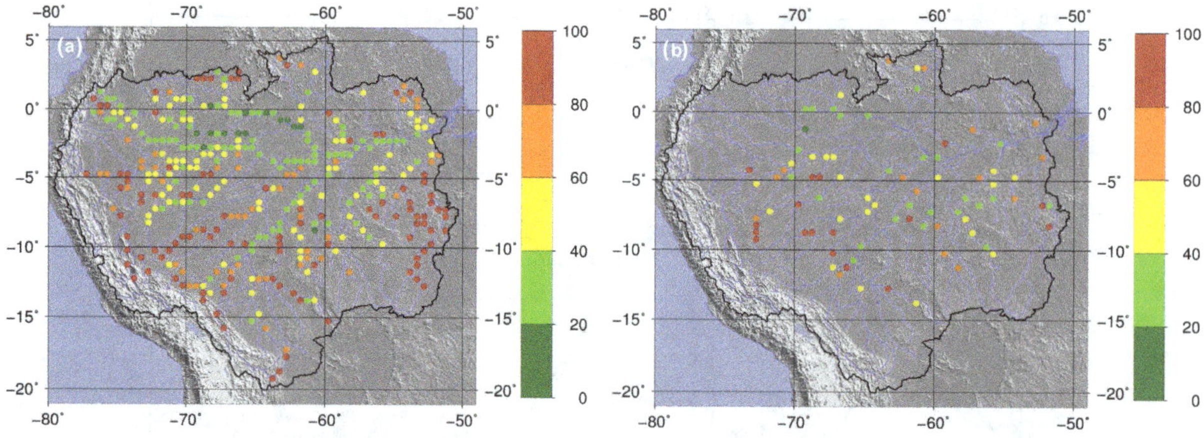

Figure 6. RMSEn for the free run simulation compared to the ENVISAT discharges **(a)** and the gauge discharges **(b)**.

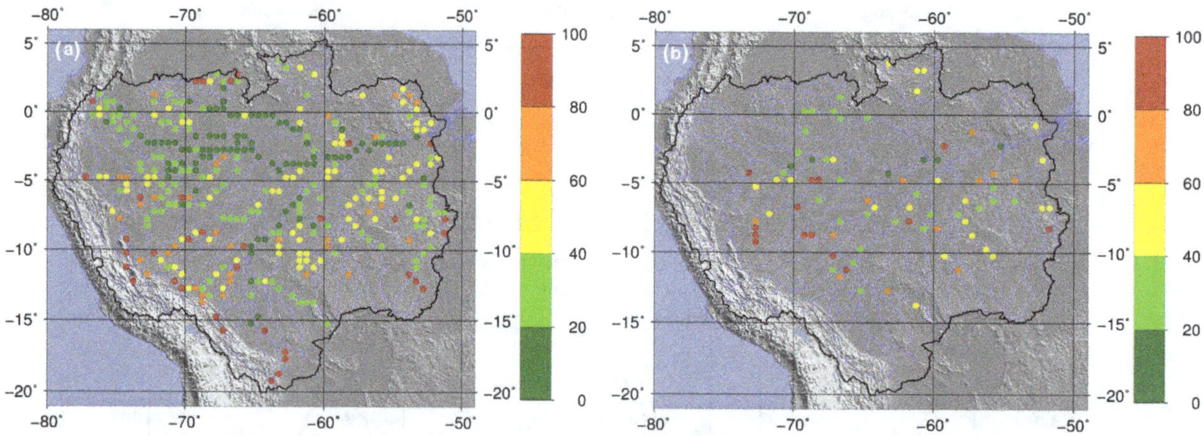

Figure 7. Analysis RMSEn for the SE1-direct experiment with respect to **(a)** the ENVISAT discharge and **(b)** the gauge discharge.

to ENVISAT discharges, SE1-local gives the best results by reducing the global RMSEn by more than 56 % (49 % for SE1-direct), while SE1-direct presents slightly better global statistics than SE1-local when compared to gauge discharges (RMSEn is reduced by 16.5 % for SE1-direct and by 15.25 % for SE1-local). Overall, the global statistics are more reduced when compared to ENVISAT discharges than to gauge discharges. This is due to the fact that gauge discharges are not directly assimilated, unlike ENVISAT discharges. The next subsections present and analyse in more detail results from each experiment.

4.2.1 SE1-direct results

Figure 7 displays the mean analysis RMSEn ($\overline{\mathrm{RMSEn}_i^{a,\dagger}}$ defined in Eq. 22) for each ENVISAT virtual station (Fig. 7a) and for each in situ station (Fig. 7b). First of all, results between the ENVISAT RMSEn and the in situ RMSEn are similar, due to the similarity between ENVISAT and gauge discharge time series at most stations. According to Fig. 7a, the assimilation worked quite well along the mainstream and the main left-bank tributaries, namely the Negro River, the Japurá and the Icá, with several stations where RMSEn is below 20 %. The assimilation performances are more moderate over right-bank tributaries, where RMSEn is mostly between 20 and 60 %. Over the entire basin, RMSEn remains high in all basin heads, along small tributaries and also at most confluences; see for example the Jutaí–Solimões confluence (RMSEn above 60 %), the Purus–Solimões/Madeira–Solimões/Tapajós–Amazon confluences (RMSEn above 40 %) or the Xingu–Amazon confluence (RMSEn above 80 %).

Figure 8 compares the mean analysis discharge in red line at the 12 stations previously introduced in Sect. 4.1. For most stations, we can see that the mean analysis discharges recovers a seasonal cycle closer to the observations than the free run. It is especially true for stations along the mainstream, namely Sao Paulo de Olivenca, Manacapuru and Óbidos (Fig. 8, panels 1–3). Also, for stations along right-bank tributaries (Fig. 8, panels 7–12), the analysis seasonal discharge peak is lowered compared to the free run seasonal discharge

Table 4. Global statistics for experiments with different localization schemes.

Statistics	Free run ($*=f$)	SE1-direct ($*=a$)	SE1-diag ($*=a$)	SE1-local ($*=a$)	Units
$\text{RMSE}^{*,o}_{\text{global}}$	8.3795×10^3	3.2110×10^3	4.8626×10^3	2.4377×10^3	$m^3\,s^{-1}$
$\text{RMSEn}^{*,o}_{\text{global}}$	71.12	36.30	44.30	31.16	%
$\text{RMSE}^{*,\text{situ}}_{\text{global}}$	7.1478×10^3	4.2489×10^3	5.4300×10^3	4.1542×10^3	$m^3\,s^{-1}$
$\text{RMSEn}^{*,\text{situ}}_{\text{global}}$	68.96	57.54	63.12	58.44	%

peak and fits better the observations. This shows the good functioning of the assimilation platform.

Nevertheless, mean analysis discharge for all displayed stations presents a chaotic behaviour with numerous local minima and maxima. We can assume that this behaviour is present for all CTRIP cells in the basin. Moreover, for a given cell, most of these sudden variations are asynchronous with ENVISAT observation dates for this cell. For example, at Serrinha in the left panel in Fig. 9, an ENVISAT observation is available on the 4th day of the 35-day repeat period when big off-peaks appear on the 25th day of the same 35-day repeat period. The right panel in Fig. 9 displays the Serrinha station (red circle) with all ENVISAT observations available during the 25th day (yellow circles). On inspection of the contribution of all these observations to the analysis control variable at Serrinha (not shown here), we find that it is observation number 4 that has the highest impact on the analysis (and not observation number 5, as could be expected). This observation 4, located on a very small Negro tributary, has a low discharge value and is responsible for the low corrected discharge at Serrinha after the assimilation step.

These abrupt variations are completely artificial and directly result from the assimilation processing. Indeed, for days with unrealistic peaks/off-peaks, there are multiple ENVISAT observations available on the basin, which impact many cells all over the basin, even if they are located on other sub-catchments or tributaries. This is due to the construction of the error covariance matrices $[\mathbf{PH}^T]_{e,k}$ and $[\mathbf{HPH}^T]_{e,k}$. As these matrices are generated from the ensemble with a limited number of members, some spurious elements may appear in the matrices and link two cells that are very distant in the basin or even situated on different sub-basin. This first experiment highlights the necessity to treat the error covariance matrices to limit such spurious elements.

4.2.2 SE1-diag results

In the SE1-diag experiment, the error covariance matrices are forced to be diagonal. The objective of such processing on the error covariance matrices is to limit the impact of a given observation only to the observed cell. According to Table 4, the assimilation experiment allowed one to reduce the global ENVISAT, an in situ RMSE, when compared to the free run. However, among all three experiments, it is the one which gives the worst global performances. In this experiment, the chaotic behaviour of the mean analysis discharge is not present any more (not shown here). Nevertheless, the mean analysis discharge remains close to the free run discharge, except for regular peaks/off-peaks at an observation time when it is closer to the observed discharge. Therefore, the information brought by only one local observation is not enough. With the localization (see next section), whose results are presented in the following section, the information of several neighbouring VS is used and should constrain the analysis discharge more.

4.2.3 SE1-local results

SE1-local uses the localization treatment presented in Sect. 3.2.2. Figure 10 displays the RMSEn evolution from the SE1-direct to SE1-local experiments for both ENVISAT and gauge discharge. Green colours indicate that the SE1-local experiment reduced the RMSEn compared to the SE1-direct experiment, while yellow to red colours indicate that the SE1-local experiment increased them. The RMSEn is mostly improved over the entire basin and more particularly along major right-bank tributaries. However, the RMSEn is generally degraded along the mainstream. These maps show the good performances of the localization method over tributaries. Now, it appears that, compared to gauge discharges (see Fig. 10b), the SE1-direct experiment gives better results, especially along the mainstream. Indeed, Table 5 details the local $\text{RMSEn}_i^{a,T}$ at ENVISAT/in situ stations located along the mainstream and confirms that SE1-direct gives better results. As the global RMSEn is defined as the mean of the RMSEn weighted by the maximum discharge at the station (see Eq. 20), this explains why the SE1-direct experiment gives a better global RMSEn compared to gauge discharge.

Then, Fig. 11 displays the mean analysis discharge for SE1-local compared to the free run discharge, with corresponding ENVISAT discharge and gauge discharge, at the 12 in situ stations already used in Figs. 5 and 8. Except for stations along the mainstream (Fig. 11, panels 1–3) and also at Boa Sorte (Fig. 11, panel 12), the analysis discharge shows less sharp variations. From these results, we can say that the

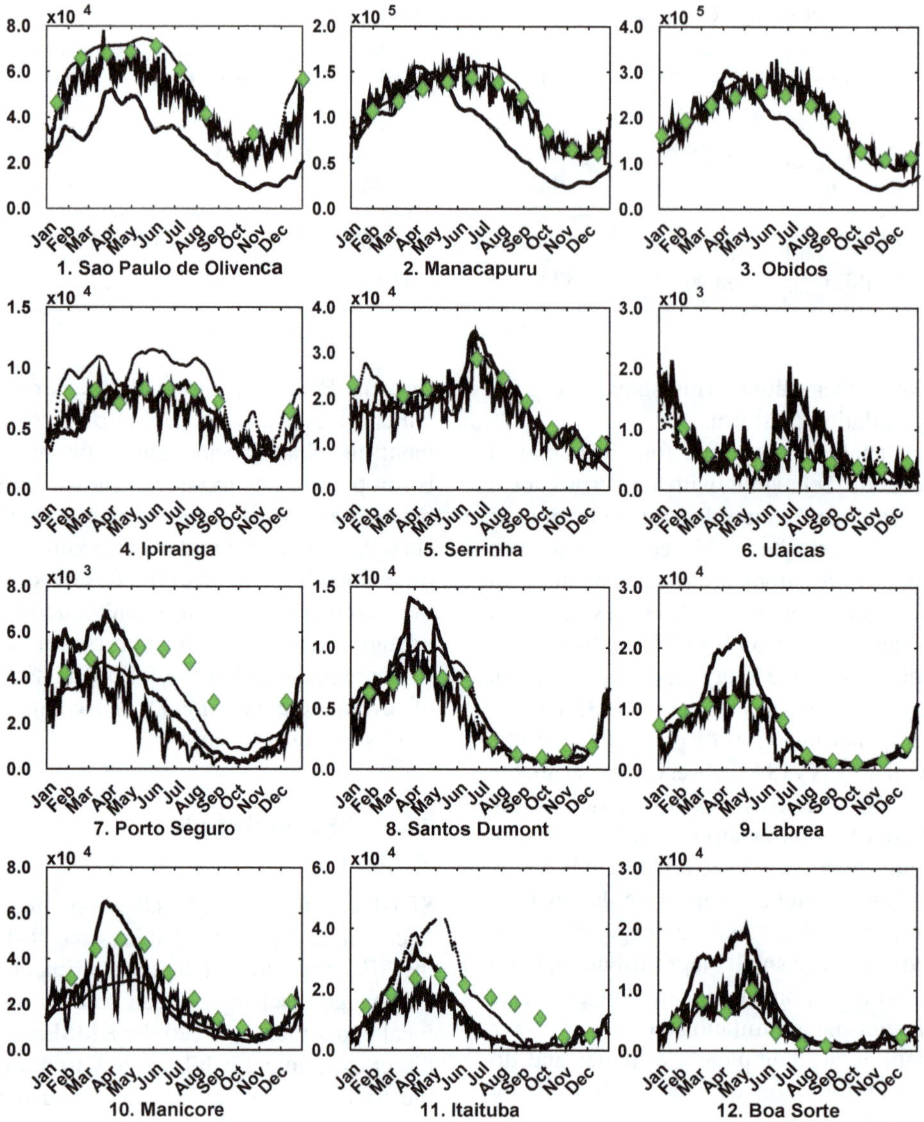

Figure 8. SE1-direct ensemble mean analysis discharge (red line) compared to the free run discharge (blue line), the ENVISAT observed discharges (green markers) and the measured gauge discharges (black dots) over the year 2009. For each panel, the x-axis represents time (in days) and the y-axis represents discharge (in $m^3\ s^{-1}$).

localization scheme is necessary and improves the assimilation.

4.2.4 Discussions on the localization scheme

The localization mask has been built to avoid the effect of spurious correlations between distant cells or ones situated on different sub-basins. The current localization scheme meets this constraint. Indeed, results for the SE1-local experiment are globally improved compared to the previous experiments.

Nevertheless, along the mainstream, the initial experiment without localization gives better results. We can interpret that by the fact that discharge along the mainstream integrates hydrological processes from all the upstream basins. So when, in the SE1-local experiment, we limit the impact of the observation to only close cells, we suppress part of the information brought by distant cells to mainstream cells.

Therefore, the current localization mask should be improved. The main difficulty here is to determine the size of the influence area for each observation. Currently, this size is predetermined and is constant in time according to averaged flow velocity. A potential development is to consider an influence area size that can vary in time, according to the hydrological season (high-flow/low-flow season). For example, during high-flow season, the flow velocity is higher so is the size of the influence area. Thus, the error covariance matrices would depend on the river time and space dynamic

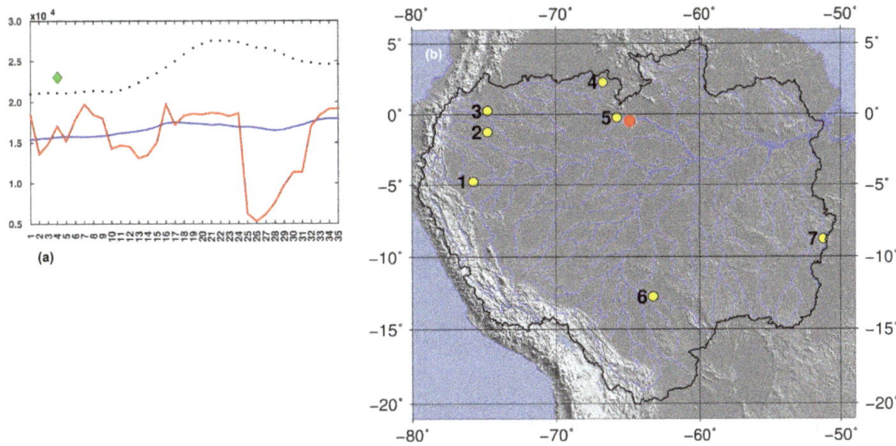

Figure 9. **(a)** Same as Fig. 8, panel 5 but only over the 35 first days of simulation. **(b)** Location of all active ENVISAT VS on the 25th day of the assimilation (yellow circles) compared to the location of the Serrinha stations (red circle).

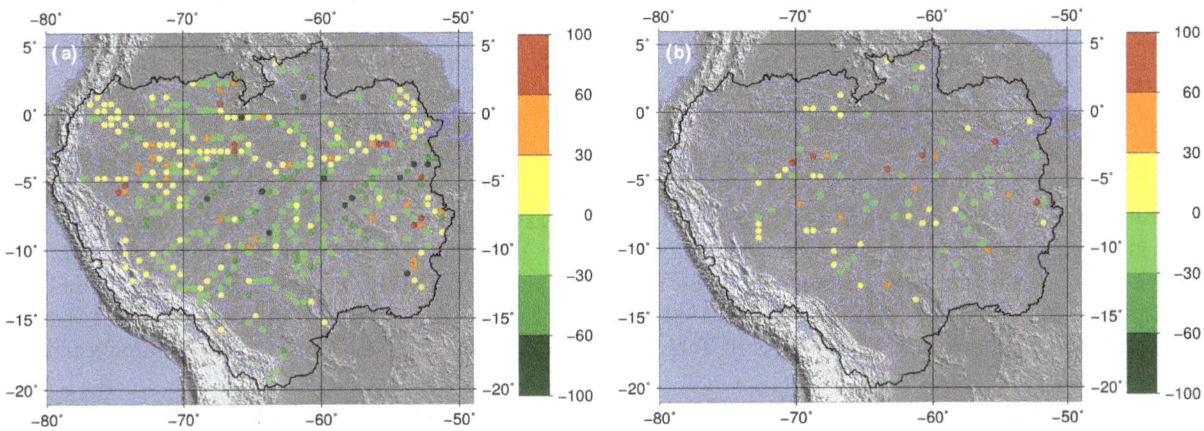

Figure 10. Analysis RMSEn difference between SE1-direct and the SE1-local experiment with respect to **(a)** the ENVISAT discharge and **(b)** gauge discharge. Negative RMSEn differences (green colours) mean that the results of the SE1-local experiment are better than the SE1-direct results at the given CTRIP cell. Positive RMSEn differences (yellow, orange and red colours) mean that the results of the SE1-direct experiment are better that the SE1-local results at the given CTRIP.

Table 5. Local $\overline{\mathrm{RMSEn}_i^{\mathrm{a,situ}}}$ for the SE1-direct and SE1-local experiments at in situ stations along the mainstream (from the most upstream to the most downstream).

Station	$\overline{\mathrm{RMSEn}_i^{\mathrm{a,situ}}}$	
	SE1-direct	SE1-local
Tamishiyacu	100.97	92.54
Tabatinga	22.24	38.81
São Paulo de Olivenca	21.77	34.86
Itapéua	17.20	28.58
Manacapuru	18.15	30.38
Jatuarana	19.72	31.01
Óbidos	16.60	28.11

(as if there were defined from a well-sampled and significant ensemble).

4.3 Impact of the chosen control variables

In the second series of experiments, all of them uses the localization scheme (see Sect. 3.2.2) to correct different types of state variables. After assimilating ENVISAT discharge to correct river initial storage (SE1-local experiment), we are testing in a second experiment the assimilation of ENVISAT discharge to correct river final storage (SE2-local) and, in a last experiment, the assimilation of ENVISAT discharge to directly correct river discharge (SE3-direct). These two other experiments need to use an empirical relationship (see Eq. 5) linking simulated river final storage to simulated discharge. For the SE2-local experiment, the formula is used to convert analysis final rive storage to discharge. Indeed, experiment statistics are based on discharge and, when correcting the final river storage, we do not have an equivalent discharge. For the SE3-local experiment, the formula is used during the assimilation steps to convert the analysis discharge into an equivalent river storage to propagate in time the corrected discharge.

Table 6 displays the global RMSEn for the three experiments compared to the free run global statistics. For all experiments, the assimilation enables to improve the RMSEn compared to the free run. Also, compared to both ENVISAT and gauge discharge, SE3-local experiment (discharge is the control variable) gives the best results, followed by SE1-local experiment (initial river storage is the control variable). Finally, it is SE2-local experiment (final river storage is the control variable) that gives the worst results, even if it is still improving the RMSEn compared to the free run.

Figures 12–13 display, for each ENVISAT (Figs. 12a and 13a) and for each in situ stations (Figs. 12b and 13b), mean RMSEn difference (in percent) between SE1-local experiment and SE2-local (Fig. 12) and between SE1-local experiment and SE3-local experiment (Fig. 13). Figure 12 shows a slight increase of the RMSEn in SE2-local experiment globally over the Amazon basin except for some basin heads. Also, the upstream part of the mainstream is more degraded (RMSEn increased of more than 60 %). These degraded results imply that the assimilation of discharges may not be adapted to correct the final river stock. However, we need to keep in mind that the analysis discharges are determined from the analysis final river stock using Eq. (5). The bad SE2-local experiment results can either be due to bad assimilation results or to an unadapted formula to convert the final river storage into discharge. On the other hand, SE3-local experiment gives better general results. As in Table 6 and Fig. 13, the SE3-local experiment shows a global improvement of the RMSEn compared to the SE1-local experiment (apart from a few cells upstream the Amazon mainstream). Indeed, even if Eq. (5) is still used to convert the analysis discharges back into river stock, it is used within the assimilation experiment (and not afterwards as for the SE2-local experiment). Therefore, the formula uncertainties are accounted for within the EnKF. Also, as the observed discharges are directly used to correct the simulated discharge, it appears logical that the assimilation gives better results as the link between the observed and the simulated variables is immediate.

5 Discussions

From the different approaches tested in this paper, it appears that there is no one specific configuration that gives the best results for all rivers, when compared to both ENVISAT and gauge discharges. In contrast, the most effective configuration depends on the size and location of the rivers. Along the river mainstream (the Solimões and the Amazon in Fig. 1a), the SE1-direct experiment clearly gives the best results (see the three first rows in Table 7). When the contribution of observations on tributaries is suppressed with the localization, the assimilation is less effective along the mainstream cells (see panels 1 to 3 in Fig. 8 for SE1-direct and compare to the same panels in Fig. 11 for SE1-local). This could be due to the fact that discharge along the mainstream is the result of hydrological processes from the entire drainage area. So, using all available observations helps the EnKF to correct the most efficiently discharge on the mainstream. However, it is different for cells along tributaries. As presented in Table 7, the localization method improves assimilation results for most cells along tributaries compared to the SE1-direct experiment. Along these cells, the localization allows the impact of observation from different sub-basins to be suppressed, especially the ones that are not connected to these cells. Finally, the comparison between the two experiments with localization (i.e. SE1-diag and SE1-local) shows that the local experiment (SE1-local) performs better than SE1-diag. This result was expected, as the localization mask is more realistic in SE1-local, because there is more than one cell impacted by the correction from one observation, in contrast to SE1-diag.

Nevertheless, among all experiments (see Table 3), the one producing the best results globally is SE3-local, where the localization method is used to directly correct the discharge. Therefore, the SE3-local configuration is used for an 8-year experiment, from 25 September 2002 (first date with an ENVISAT observation of the study domain) to 24 September 2010 (last date with an ENVISAT observation). At the basin scale, RMSEn between model outputs and gauge discharges is reduced by 27.11 % (it decreases from 96.71 to 70.49 %) and RMSEn between model outputs and ENVISAT discharges is reduced by 63.28 % (it decreases from 75.10 to 27.58 %). $\mathrm{RMSEn}^{\mathrm{f,situ}}_{\mathrm{global}}$ is high, because a large fraction of in situ stations (25 out of 108) are situated along very small tributaries or at basin heads, where the local $\mathrm{RMSEn}^{\mathrm{f,situ}}$ is largely over 100 %. These very high $\mathrm{RMSEn}^{f,\mathrm{situ}}$ have a huge impact on the global $\mathrm{RMSEn}^{\mathrm{f,situ}}_{\mathrm{global}}$ (despite the weighting used to calculate $\mathrm{RMSEn}^{\mathrm{f,situ}}_{\mathrm{global}}$). If the statistics are computed using only cells with a $\mathrm{RMSEn}^{\mathrm{f,situ}}$ below 100 %, we find that the global $\mathrm{RMSEn}^{\mathrm{f,situ}}_{\mathrm{global}}$ is reduced by 14.66 % (it goes from 49.80 to 42.50 %) and the $\mathrm{RMSEn}^{\mathrm{f,o}}_{\mathrm{global}}$ is reduced by 50.21 % (it goes from 51.74 to 25.76 %). This shows the limitation of this assimilation scheme, as ISBA-CTRIP resolution (roughly 50 km by 50 km) does not simulate basin heads well (rivers are too small to be correctly represented in a coarse grid).

Figure 14 displays, for the 12 in situ stations (see Fig. 4 for their locations) already used in Figs. 5 and 8, the mean analysis discharge over the whole experiment time period (red line), which is compared to the free run discharge at the station (blue line), the ENVISAT discharge (green markers) and the gauge discharge (black markers). Overall, analysis discharge is quite close to observed discharges (ENVISAT and in situ).

However, despite the use of the localization, the analysis discharge keeps presenting a quite chaotic behaviour: more

Table 6. Global statistics for experiments with different types of control variables.

Statistics	Free run ($*=f$)	SE1-local ($*=a$)	SE2-local ($*=a$)	SE3-local ($*=a$)	Units
$RMSE^{*,o}_{global}$	8.3795×10^3	2.4377×10^3	4.3069×10^3	2.2298×10^3	$m^3 s^{-1}$
$RMSEn^{*,o}_{global}$	71.12	31.16	36.37	24.73	%
$RMSE^{*,situ}_{global}$	7.1478×10^3	4.2489×10^3	5.4300×10^3	4.1542×10^3	$m^3 s^{-1}$
$RMSEn^{*,situ}_{global}$	68.96	58.44	61.69	54.46	%

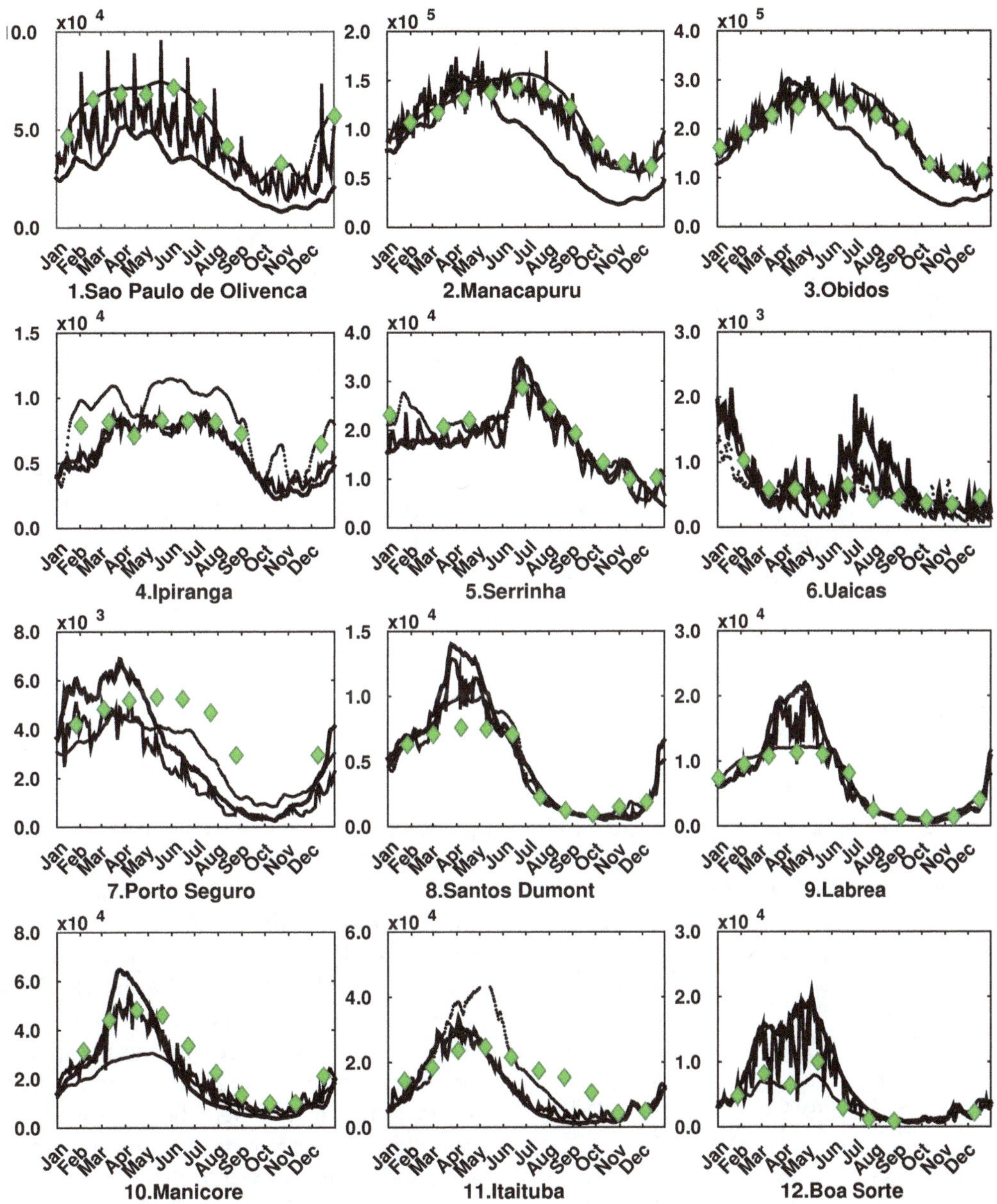

Figure 11. SE1-local ensemble mean analysis discharge (red line) compared to the free run discharge (blue line), the ENVISAT observed discharges (green markers) and the measured gauge discharges (black dots) over the year 2009. For each panel, the x-axis represents time (in days) and the y-axis represents discharge (in $m^3 s^{-1}$).

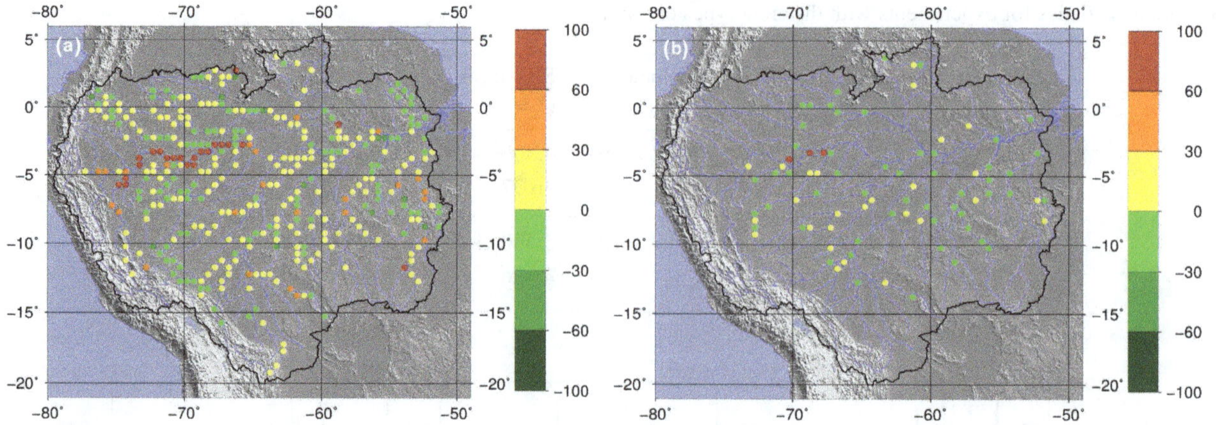

Figure 12. Analysis RMSEn differences between SE1-local and SE2-local experiments with respect to **(a)** the ENVISAT discharge and **(b)** gauge discharge.

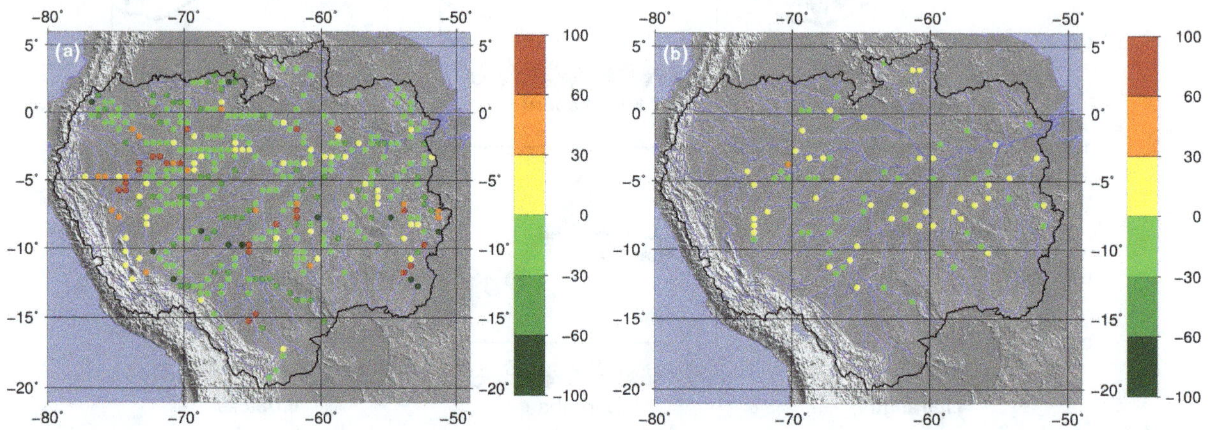

Figure 13. Analysis RMSEn differences between SE1-local and SE3-local experiments with respect to **(a)** the ENVISAT discharge and **(b)** gauge discharge.

Table 7. Statistics between analysis and in situ stations for the different assimilation experiments. Italic values indicate the best result among the three experiments for all tested gages.

Statistics	$\overline{\text{RMSEn}_i^{*,\text{situ}}}$			
	Free run	SE1-direct	SE1-local	SE3-local
1. São Paulo de Olivenca	49.67	*21.77*	34.86	40.98
2. Manacapuru	36.01	*18.15*	30.38	30.76
3. Óbidos	34.65	*16.60*	28.11	28.33
4. Ipiranga	34.63	35.43	*32.75*	33.54
5. Serrinha	*20.65*	26.82	23.99	28.24
6. Uaicás	79.53	51.28	51.33	*48.92*
7. Porto Seguro	44.16	46.32	39.81	*38.93*
8. Santos Dumont	28.37	35.54	*27.55*	33.63
9. Lábrea	50.62	40.39	40.33	*39.86*
10. Manicoré	72.36	*35.04*	54.38	61.96
11. Itaituba	*43.33*	66.43	47.50	46.23
12. Boa Sorte	149.38	*58.99*	112.58	82.75

particularly at Sao Paulo de Olivenca (Fig. 14, panel 1), Manacapuru (Fig. 14, panel 2) and during high flow seasons along right-bank tributaries (Fig. 14, panels 7 to 12). This shows the limit of assimilating 35-day repeat period ENVISAT observations. If no data are missing at a given VS, it means that there will be, at most, 11 available observations during 1 year. Moreover, in a state estimation context, only the model output state is corrected and not the model parametrization or, in our set-up, forcings. Therefore, if the model is not constrained by direct or neighbouring observations, it naturally goes back to free run discharge. The performance of the assimilation, with respect to the daily in situ data, is therefore often limited by the low ENVISAT observation frequency. In future works, it will be interesting to study the assimilation of similar data with a higher frequency, such as the JASON-2 altimeter data (which has a 10-day repeat period but a coarser spatial sampling).

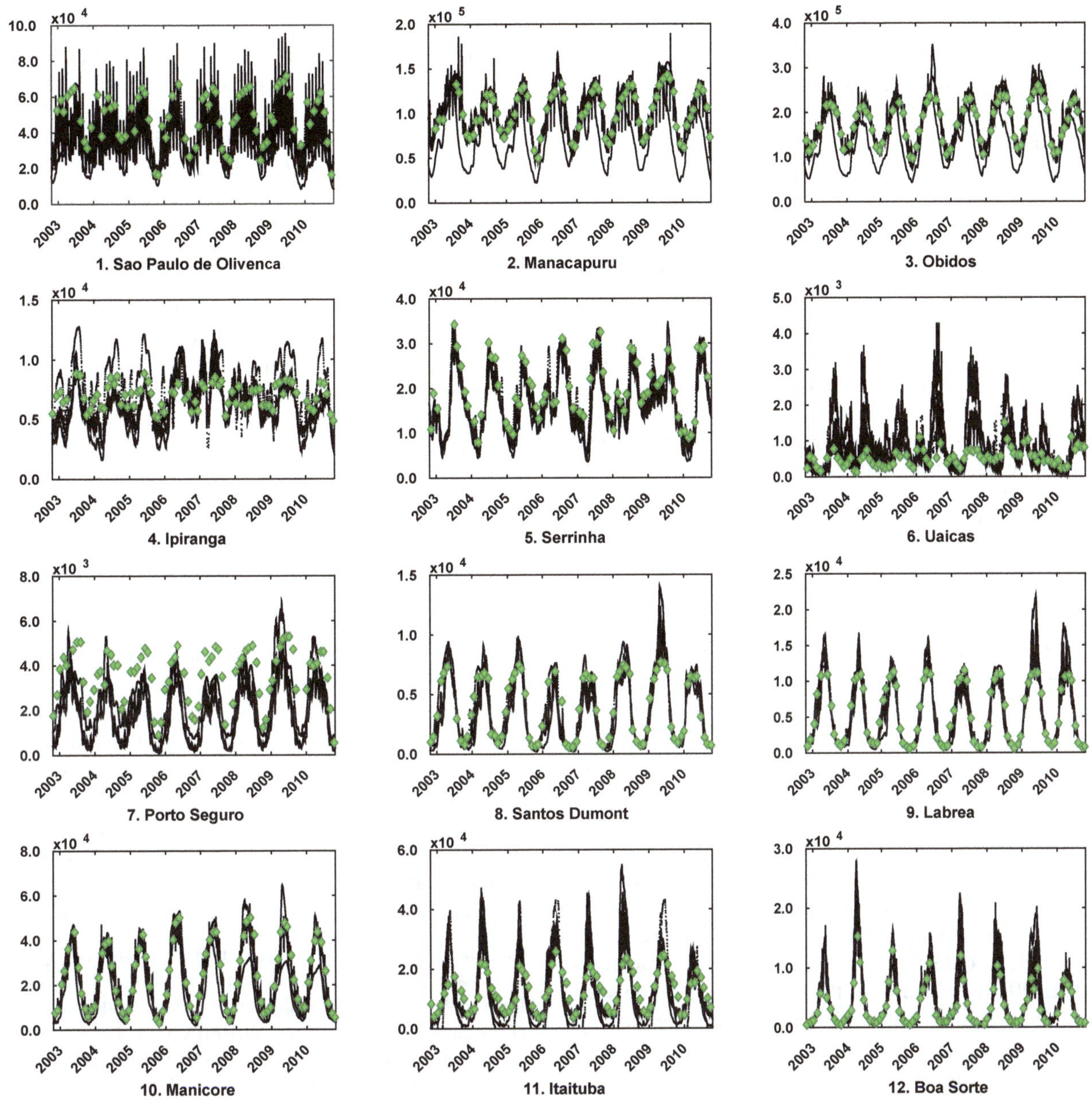

Figure 14. SE3-local ensemble mean analysis discharge (red line) compared to the free run discharge (blue line), the ENVISAT observed discharges (green markers) and the measured gauge discharges (black dots) from 25 September 2002 and for 8 years. In each panel, the x-axis represents time (in days) and the y-axis represents discharge (in $m^3 s^{-1}$).

6 Conclusions and perspectives

This study presents, over the Amazon basin, the assimilation of a satellite-derived discharge product into a large-scale hydrological model to correct its state variables. The remotely sensed discharge data are derived from the ENVISAT nadir altimeter and are assimilated into the ISBA-CTRIP model using an ensemble Kalman filter. Five experiments were carried out over the year 2009. For all experiments, the assimilations were able to reduce the modelling errors compared to both observed and gauge discharges.

The first experiments tested different definition of the background error covariance matrices, where the influence of a given observation is either reduced to the only observed cell (SE1-diag), or limited to a few close cells on the hydrological network (SE1-local), or not limited and can potentially impact the entire basin (SE1-direct). Results showed that the complete stochastic matrices gave the best results along the mainstream and the localization treatment appeared necessary along the tributaries. The need for the localization is explained by the spurious elements in the error covariance matrix due to the limited ensemble size and the methodology used to generate it.

The last tests compared the corrections of different state variables: the river initial storage (SE1-local), or the river final storage (SE2-local), or the river discharge (SE3-local). The main difficulty with these different types of variables is, on the one hand, the relationship between the control and the observed variables (gathered in the observation operator) and, on the other hand, the reciprocal relationship to generate inputs for the next DA cycle. Results showed that correcting river discharge gives the best global results over the entire basin, as the link between the observed and corrected variables is the most straightforward. Therefore, the final experiment (SE3-local-long) uses the SE3-local configuration over the whole ENVISAT observation period (from September 2002 to 2010) and confirms the possibility of using such low-resolution remotely sensed data in a large-scale model.

These experiments offer several perspectives. First, the localization treatment could be improved by combining the three tested approaches according to the cell's position on the river: discharge correction for cells along the mainstream should be impacted by all upriver observations, while correction for cells on tributaries should be impacted only by close observations along the same sub-catchment. Moreover, the size of the area of influence for a given observations could also vary in time according to the season (high flow/low flow). With ulterior developments of the localization method, new challenges may appear such as the risk of imbalance, already studied in the field atmospheric DA (e.g. Greybush et al., 2011). The analysis of imbalance may need to be considered in future works.

A main limitation of assimilating ENVISAT data is their low repeat period (one observation every 35 days, at best). Indeed, corrected discharges often present strong sudden variations between unobserved and observed dates, as the model goes back to its free run when it is not constrained by an observation. However, there are other satellite altimetry missions with different repeat periods, for example JASON-2 (10-day repeat period from June 2008 to October 2016), JASON-3 (10-day repeat period, launched in January 2016), Sentinel-3A (27-day repeat period, launched in February 2016) or Sentinel-3B (27-day repeat period, which should be launched in 2018). Also, the incoming SWOT (Surface Water and Ocean Topography, launch scheduled for 2021) wide-swath altimetry mission will provide a remotely sensed discharge product. SWOT will have a 21-day repeat period, with an almost global spatial coverage thanks to its two 50 km swaths. All these data could be combined with ENVISAT data (during the overlapping period) within the assimilation scheme to have a denser network of observation over the study domain, to get a better estimate of discharge (similar to a reanalysis) over a multi-decadal time frame (Tourian et al., 2017).

To improve these DA results, several aspects could be investigated. For example, one could study whether a more realistic ensemble method generation could be helpful. In the present study, only the model initial condition and the precipitation forcing are perturbed to generate the background forecast ensemble. More uncertainties in this ensemble could be added by also perturbing CTRIP parameters and/or ISBA outputs. Another DA aspect to look into is the potential use of a smoothing data assimilation algorithm, such as the ensemble Kalman smoother (Evensen and Leeuwen, 2000). A smoother could help to have less "variability" in the corrected discharge. Finally, the assimilation scheme presented in this study could be applied to other river basins in the world, as ISBA-CTRIP is a global LSM. However, more work is needed to apply the DA platform at a global scale.

Code and data availability. The CTRIP code is open source and is available as a part of the surface modelling platform called SURFEX, which can be downloaded at http://www.cnrm-game-meteo.fr/surfex/. SURFEX is updated approximately every 3 to 6 months and the CTRIP version presented in this paper is from SURFEX version 7.3. If more frequent updates are needed, please follow the procedure to obtain a SVN or Git account in order to access real-time modifications of the code (see the instructions at the previous link). The ISBA-CTRIP model is coupled to the DA codes via the OpenPalm coupler available at http://www.cerfacs.fr/globc/PALM_WEB/. To get the DA routines coupled to ISBA-CTRIP with OpenPalm, please directly contact C. Emery (charlotte.emery@jpl.nasa.gov) or S. Biancamaria (sylvain.biancamaria@legos.obs-mip.fr). To obtain the GSWP3 forcings, please refer to the following url: http://search.diasjp.net/en/dataset/GSWP3_EXP1_Forcing (https://doi.org/10.20783/DIAS.501). The ENVISAT-based discharge data are available by contacting A. Paris (aparis@cls.fr). Finally, the in situ discharges used to validate the results are available on the ANA website (http://hidroweb.ana.gov.br/default.asp).

Appendix A: Equations to compute river storage from discharge using the Manning formula

This Appendix provides more details and the approximation used to derive Eq. (5). This equation allows conversion of simulated discharges $Q^{\mathrm{a}}_{i,k}$ to equivalent final river storage $S^{\mathrm{a}}_{\mathrm{end},k,i}$ using the Manning formula. We chose to invert Eq. (3). Assuming that the discharge estimated by ISBA-CTRIP $Q^{\mathrm{a}}_{i,k}$ is the instantaneous flow at the final time of the integration window,

$$Q^{\mathrm{a}}_{k,i,[\mathrm{kg\,s^{-1}}]} = L^{-1} v S^{\mathrm{a}}_{\mathrm{end},k,i,[\mathrm{kg}]} \Longleftrightarrow Q^{\mathrm{a}}_{k,[\mathrm{m^3\,s^{-1}}]}$$

$$= \rho^{-1} L^{-1} v S^{\mathrm{a}}_{\mathrm{end},k,i,[\mathrm{kg}]}.$$

To ease the notations, we will skip the units in the following equations knowing that discharges are expressed in $\mathrm{m^3\,s^{-1}}$ and water stock in kg. Then, $\forall k$, $\forall i$:

$$Q^{\mathrm{a}}_{k,i} = L^{-1} \rho^{-1} s^{1/2} n^{-1} \left(\frac{W h_{\mathrm{S}}}{W + 2h - S} \right)^{2/3} S^{\mathrm{a}}_{\mathrm{end},k,i}.$$

We suppose : $W \gg h_{\mathrm{S}}$

$$Q^{\mathrm{a}}_{k,i} \approx L^{-1} \rho^{-1} s^{1/2} n^{-1} h^{2/3} S^{\mathrm{a}}_{\mathrm{end},k,i}.$$

Yet$S = \rho L W h_{\mathrm{S}}$, so

$$Q^{\mathrm{a}}_{k,i} \approx L^{-5/3} \rho^{-5/3} W^{2/3} s^{1/2} n^{-1} \left(S^{\mathrm{a}}_{\mathrm{end},k,i} \right)^{5/3}.$$

Finally giving Eq. (5):

$$S^{\mathrm{a}}_{\mathrm{end},k,i} \approx \rho L W^{2/5} s^{-3/10} n^{3/5} \left(Q^{\mathrm{a}}_{k,i} \right)^{3/5}, \qquad \text{(A1)}$$

with ρ ($\mathrm{m^3\,kg^{-1}}$) the water density, L (m) the river section length, W (m) the river width, s (–) the riverbed slope and n ($\mathrm{s\,m^{-1/3}}$) the Manning coefficient in the riverbed. Then, for experiments with discharges as control variables, the formula in Eq. (5) will be used to turn back corrected discharges into river stock and propagate the model up to the next observation time.

Appendix B: Definition of error covariance matrices

The background error covariance matrices $[\mathbf{PH}^T]_{\mathrm{e},k}$ and $[\mathbf{HPH}^T]_{\mathrm{e},k}$ are estimated from the definition suggested by Evensen (2004), Moradkhani et al. (2005) and Durand et al. (2008) such that

$$\left[\mathbf{PH}^T\right]_{\mathrm{e},k} = (n_{\mathrm{e}} - 1)^{-1} \left(\mathbf{X}^{\mathrm{b}}_{\mathrm{e},k} - \overline{\mathbf{X}^{\mathrm{b}}_{\bullet,k}} \cdot \mathbf{1}^T_{n_{\mathrm{e}}} \right)$$

$$\left(\mathcal{H}\left(\mathbf{X}^{\mathrm{b}}_{\mathrm{e},k}\right) - \overline{\mathcal{H}\left(\mathbf{X}^{\mathrm{b}}_{\bullet,k}\right)} \cdot \mathbf{1}^T_{n_{\mathrm{e}}} \right)^T, \qquad \text{(B1)}$$

and

$$\left[\mathbf{HPH}^T\right]_{\mathrm{e},k} = (n_{\mathrm{e}} - 1)^{-1} \left(\mathcal{H}\left(\mathbf{X}^{\mathrm{b}}_{\mathrm{e},k}\right) - \overline{\mathcal{H}\left(\mathbf{X}^{\mathrm{b}}_{\bullet,k}\right)} \cdot \mathbf{1}^T_{n_{\mathrm{e}}} \right)$$

$$\left(\mathcal{H}\left(\mathbf{X}^{b}_{\mathrm{e},k}\right) - \overline{\mathcal{H}\left(\mathbf{X}^{\mathrm{b}}_{\bullet,k}\right)} \cdot \mathbf{1}^T_{n_{\mathrm{e}}} \right)^T, \qquad \text{(B2)}$$

with $\mathbf{X}^{\mathrm{b}}_{\mathrm{e},k}$ the control matrix containing the n_{e} control vector $\boldsymbol{x}^{\mathrm{b},[l]}_k$, $l = 1 \ldots n_{\mathrm{e}}$ from the background ensemble such that

$$\mathbf{X}^{\mathrm{b}}_{\mathrm{e},k} = \left[\boldsymbol{x}^{\mathrm{b},[1]}_k \ldots \boldsymbol{x}^{\mathrm{b},[N_{\mathrm{e}}]}_k \right]$$

and $\mathcal{H}(\mathbf{X}^{\mathrm{b}}_{\mathrm{e},k})$ the same control matrix expressed in the observation space such that

$$\mathcal{H}\left(\mathbf{X}^{\mathrm{b}}_{\mathrm{e},k}\right) = \left[\mathcal{H}\left(\boldsymbol{x}^{\mathrm{b},[1]}_k\right) \ldots \mathcal{H}\left(\boldsymbol{x}^{\mathrm{b},[n_{\mathrm{e}}]}_k\right) \right].$$

$\overline{\mathbf{X}^{\mathrm{b}}_{\bullet,k}}$ and $\overline{\mathcal{H}(\mathbf{X}^{\mathrm{b}}_{\bullet,k})}$ are the ensemble sample expectations of the control matrix $\mathbf{X}^{\mathrm{b}}_{\mathrm{e},k}$ and its mapping on the observation space $\mathcal{H}(\mathbf{X}^{\mathrm{b}}_{\mathrm{e},k})$ respectively such that

$$\overline{\mathbf{X}^{\mathrm{b}}_{\bullet,k}} = \frac{1}{n_e} \sum_{l=1}^{n_{\mathrm{e}}} \boldsymbol{x}^{\mathrm{b},[l]}_k \quad \overline{\mathcal{H}\left(\mathbf{X}^{\mathrm{b}}_{\bullet,k}\right)} = \frac{1}{n_e} \sum_{l=1}^{n_e} \mathcal{H}(\boldsymbol{x}^{b,[l]}_k).$$

The vectors' dimensions are n_x and $n_{y,k}$, respectively, and $\mathbf{1}_{n_{\mathrm{e}}}$ is a vector of size n_{e} containing only 1 s.

Appendix C: Perturbations of precipitations

The ensemble of perturbed precipitation fields $\widetilde{\mathbf{F}}_e$ is defined such that

$$\widetilde{\mathbf{F}}_e = \left\{\widetilde{\mathbf{F}}^{[1]}, \quad \widetilde{\mathbf{F}}^{[2]}, \quad \ldots \quad \widetilde{\mathbf{F}}^{[n_e]}\right\} = \left\{\boldsymbol{\varphi}_p^{[1]} \cdot \mathbf{F}, \quad \boldsymbol{\varphi}_p^{[2]} \cdot \mathbf{F}, \quad \ldots \quad \boldsymbol{\varphi}_p^{[n_e]} . \mathbf{F}\right\}, \tag{C1}$$

where

- $\mathbf{F}$ is the two-dimensional field of precipitation forcing before perturbation (with a time step of 3 h),
- $\widetilde{\mathbf{F}}^{[l]}$, for $l = 1 \ldots n_e$, is the lth perturbed precipitation field,
- $\boldsymbol{\varphi}_p^{[l]}$, for $l = 1 \ldots n_e$, is the lth multiplying uniformly distributed field of $\mathbf{F}$ to generate $\widetilde{\mathbf{F}}^{[l]}$.

The precipitation field $\mathbf{F}$ is then perturbed by applying a random multiplying field such that

$$\boldsymbol{\varphi}_p^{[l]} = \left(1 - \eta^{\mathrm{F},[l]}\right) + 2\mathbf{U}_{\mathrm{F}}^{[l]} \cdot \eta^{\mathrm{F},[l]}, \tag{C2}$$

where

- $\mathbf{U}_{\mathrm{F}}^{[l]}$ is random field following a uniform law between 0 and 1,
- $\eta^{\mathrm{F},[l]}$ is a scalar representing the relative error of the precipitations.

Therefore, $\boldsymbol{\varphi}_p^{[l]}$ is a random field following a uniform law between $(1 - \eta^{\mathrm{F},[l]})$ and $(1 + \eta^{\mathrm{F},[l]})$.

The precipitation relative error η^{F} quantifies the uncertainties in the precipitation intensity. The variable η^{F} is different for each member of the ensemble and follows a Gaussian law with expectation $\overline{\eta^{\mathrm{F}}} = 30\,\%$ and standard deviation $\sigma_{\eta^{\mathrm{F}}} = 0.1\,\%$ (Clark et al., 2008; Paiva et al., 2013b).

The fields $\mathbf{U}_{\mathrm{F}}^{[l]}$, for $l = 1 \ldots n_e$, allow one to introduce a time and space correlation in the precipitation error and are generated with the algorithm presented in Evensen (2003). This algorithm generates two-dimensional Gaussian random fields $\mathbf{S}^{[l]}$ with a zero mean and a space-correlation length of e^{-1}. These Gaussian random fields are turned into uniform random fields by applying the complementary error function erfc():

$$\mathbf{U}_{\mathrm{F},k'}^{[l]} = \frac{1}{2}\mathrm{erfc}\left(\frac{\mathbf{S}_{k'}^{[l]}}{\sqrt{2}}\right), \tag{C3}$$

where k' is the atmospheric forcing proper time step, equal to 3 h in ISBA-CTRIP, and shorter than the ISBA-CTRIP output time step, equal to 24 h. The space PDF of $\mathbf{U}_{\mathrm{F}}^{[l]}$ decreases of e^{-1} when the distance is equal to the space-correlation length τ_x (here, the letter x exceptionally denotes the spacial dimension). For the simulations, τ_x is fixed to 1.0° (Clark et al., 2008; Paiva et al., 2013b) and is invariant from one member to another and from one assimilation cycle to another.

For the time correlation, the parameter $\vartheta^{[l]}$

$$\vartheta^{[l]} = 1 - \frac{\Delta k'}{\tau_k^{[l]}} \tag{C4}$$

Table C1. Constant values used to perturb the precipitation fields. k is the assimilation index.

Variable	Description	Value	Eq.
τ_x (°)	Precip spatial corr.	1.0	
$\overline{\eta^{\mathrm{F}}}$	Precip relative error mean	0.3	(C2)
$\sigma_{\eta^{\mathrm{F}}}$	Precip relative error mean	0.1	(C2)
$\overline{\tau_k}$ (s)	Precip temp corr mean	43 200	(C4)
σ_{τ_k} (s)	Precip temp corr std	10 800	(C4)

determines the time correlation length between the different fields $\mathbf{S}_{k'}^{[l]}$. It is concretely generated by combining the random field from the previous time step $\mathbf{S}_{k'-1}^{[l]}$ and an auxiliary random field $\mathbf{W}_{k'}^{[l]}$ with the same properties such that

$$\mathbf{S}_{k'}^{[l]} = \vartheta^{[l]}\mathbf{S}_{k'-1}^{[l]} + \sqrt{1 - (\vartheta^{[l]})^2}\mathbf{W}_{k'}^{[l]}, \tag{C5}$$

with $\Delta k' = 3$ h the forcing time step and $\tau_k^{[l]}$ the time constant characterizing $\vartheta^{[l]}$. $\vartheta^{[l]} = 0$ generates a white noise (which means a perturbation uncorrelated in time), while $\vartheta^{[l]} = 1$ makes the perturbation constant in time. The variable τ_k takes a different value for each member as it follows a Gaussian law with an expectation equal to $\overline{\tau_k} = 12$ h (or 43 200 s) and a standard deviation of $\sigma_{\tau_k} = 3$ h (or 10 800 s). These values are chosen so that the time correlation has effects during an assimilation window of 1 day. All variables used to generate the ensembles with their value are summarized in Table C1.

Competing interests. The authors declare that they have no conflict of interest.

Acknowledgements. This work has been performed using HPC resources from CALMIP (grant 2016-P1408). The GSWP3 team is acknowledged for letting the authors use their different forcing fields. This work was supported by the CNES, through a grant from the Terre-Océan-Surfaces Continentales-Atmosphère (TOSCA) committee affiliated with the project entitled "Towards an improved understanding of the global hydrological cycle using SWOT measurements". The European Space Agency (ESA) is also thanked for providing to the scientific community observations from the RA2 altimeter embarked on ENVISAT. Charlotte Marie Emery was supported by a CNES/région Midi-Pyrénées PhD grant. This work was done as a private venture and not in the author's capacity as an employee of the Jet Propulsion Laboratory, California Institute of Technology.

Edited by: Albrecht Weerts

References

Alkama, R., Decharme, B., Douville, H., Becker, M., Cazenave, A., Sheffield, J., Voldoire, A., Tyteca, S., and Moigne, P. L.: Global Evaluation of the ISBA-TRIP Continental Hydrological System. Part I: Comparison to GRACE Terrestrial Water Storage Estimates and In Situ River Discharges, J. Hydrometeorol., 11, 583–600, 2010.

Anderson, J. L.: Exploring the need for localization in ensemble data assimilation using a hierarchical ensemble filter, Physica D, 230, 99–111, 2007.

Andreadis, K. M. and Lettenmaier, D. P.: Assimilating remotely sensed snow observations into a macroscale hydrology model, Adv. Water Resour., 29, 872–886, 2006.

Arora, V., Chiew, F., and Grayson, R.: A river flow routing scheme for general circulation models, J. Geophys. Res., 104, 14347–14357, 1999.

Biancamaria, S., Durant, M., Andreadis, K. M., Bates, P. D., Boone, A., Mognard, N. M., Rodriguez, E., Alsdorf, D. E., Lettenmaier, D. P., and Clark, E. A.: Assimilation of virtual wide swath altimetry to improve Arctic river modeling, Remote Sens. Environ., 115, 373–381, 2011.

Birkett, C. M.: The contribution of TOPEX/Poseidon to the global monitoring of climatically sensitive lakes, J. Geophys. Res., 100, 25179–25204, 1995.

Birkett, C. M.: Contribution of the TOPEX NASA altimeter to the global monitoring of large rivers and wetlands, Water Resour. Res., 34, 1223–1239, 1998.

Birkett, C. M., Mertes, L. A. K., Dunne, T., Costa, M. H., and Jasinski, M. J.: Surface water dynamics in the Amazon basin: Application of satellite radar altimetry, J. Geophys. Res., 107, 8059, https://doi.org/10.1029/2001JD000609, 2002.

Blackadar, A.-F.: Modeling the nocturnal boundary layer, in: Third Symposium on Atmospheric Turbulence, Diffusion and Air Quality, American Meteorological Society, Boston, 46–49, 1976.

Boone, A., Calvet, J.-C., and Noilhan, J.: Inclusion of a Third Soil Layer in a Land Surface Scheme Using the Force-Restore Method, J. Hydrometeorol., 38, 1611–1630, 1999.

Burgers, G., Leeuwen, P. J. V., and Evensen, G.: Analysis Scheme in the Ensemble Kalman Filter, Mon. Weather Rev., 126, 1719–1724, 1998.

Clark, M. P., Rupp, D. E., Woods, R. A., Zheng, X., Ibbitt, R. P., Slater, A. G., Schmidt, J., and Uddstrom, M. J.: Hydrological data assimilation with the ensemble Kalman filter: Use of streamflow observations to update states in a distributed hydrological model, Adv. Water Resour., 31, 1309–1324, 2008.

Coe, M.: A linked global model of terrestrial hydrologic processes: Simulation of modern rivers, lakes, and wetlands, J. Geophys. Res., 103, 8885–8899, 1998.

Collischon, W., Allasia, D., da Silva, B. C., and Tucci, C. E. M.: The MGB-IPH model for large-scale rainfall-runoff modelling, Hydrolog. Sci. J., 52, 878–895, 2007.

Compo, G., Whitaker, J., Sardeshmukh, P., Matsui, N., Allan, R., Yin Jr., X., B. G., Vose, R., Rutledge, G., Bessemoulin, P., Brönnimann, S., Brunet, M., Crouthamel, R., Grant, A., Groisman, P., Jones, P., Kruk, M., Kruger, A., Marshall, G., Maugeri, M., Mok, H., Nordli, O., Ross, T., Trigo, R., Wang, X., Woodruff, S., and Worley, S.: The Twentieth Century Reanalysis Project, Q. J. Roy. Meteorol. Soc., 137, 1–28, 2011.

Cretaux, J.-F., Calmant, S., Romanoski, V., Shabunin, A., Lyard, F., Berge-Nguyen, M., Cazenave, A., Hernandez, F., and Perosanz, F.: An absolute calibration site for radar altimeters in the continental domain: Lake Issykkul in Central Asia, J. Geodesy, 83, 723–735, 2009.

David, C. H., Maidment, D. R., Niu, G.-Y., Yang, Z.-L., Habets, F., and Eijkhout, V.: River Network Routing on the NHDPlus Dataset, J. Hydrometeorol., 12, 913–934, https://doi.org/10.1175/2011JHM1345.1, 2011.

Decharme, B. and Douville, H.: Introduction of a subgrid hydrology in the ISBA land surface model, Clim. Dynam., 26, 65–78, 2006.

Decharme, B., Douville, H., Prigent, C., Papa, F., and Aires, F.: A new river flooding scheme for global climate applications: Off-line evaluation over South America, J. Geophys. Res., 113, D11110, https://doi.org/10.1029/2007JD009376, 2008.

Decharme, B., Alkama, R., Douville, H., Becker, M., and Cazenave, A.: Global Evaluation of the ISBA-TRIP Continental Hydrological System. Part II: Uncertainties in River Routing Simulation Related to Flow Velocity and Groundwater Storage, J. Hydrometeorol., 11, 601–617, 2010.

Decharme, B., Alkama, R., Papa, F., Faroux, S., Douville, H., and Prigent, C.: Global off-line evaluation of the ISBA-TRIP flood model, Clim. Dynam., 38, 1389, https://doi.org/10.1007/s00382-011-1054-9, 2012.

Delijani, E. B., Pishvaie, M. R., and Boozarjomehry, R. B.: Subsurface characterization with localized ensemble Kalman filter employing adaptative thresholding, Adv. Water Resour., 69, 181–196, 2014.

Devegowda, D., Arroyo-Negrete, E., and Datta-Gupta, A.: Flow relevant covariance localization during dynamic data assimilation using EnKF, Adv. Water Resour., 33, 129–145, 2010.

Ducharne, A., Golaz, C., Leblois, E., Laval, K., Polcher, J., Ledoux, E., and de Marsily, G.: Development of a high resolution runoff routing model, calibration and application to assess runoff from the LMD GCM, J. Hydrol., 280, 207–228, 2003.

Durand, M., Andreadis, K., Alsdorf, D., Lettenmaier, D., Moller, D., and Wilson, M.: Estimation of bathymetric depth and slope from data assimilation of swath altimetry into a hydrodynamic model, Geophys. Res. Lett., 35, L20401, https://doi.org/10.1029/2008GL034150, 2008.

Emery, C. M., Biancamaria, S., Boone, A., Garambois, P.-A., Ricci, S., Rochoux, M. C., and Decharme, B.: Temporal variance-based sensitivity analysis of the river routing component of the large scale hydrological model ISBA-TRIP: Application on the Amazon Basin, J. Hydrometeorol., 17, 3007–3027, 2016.

Evensen, G.: Sequential data assimilation with a nonlinear quasi-geostropic model using Monte Carlo methods to forecast error statistics, J. Geophys. Res., 99, 10143–10162, 1994.

Evensen, G.: The Ensemble Kalman Filter: theoretical formulation and practical implementation, Ocean Dynam., 53, 343–367, 2003.

Evensen, G.: Sampling strategies and square root analysis schemes for the EnKF, Ocean Dynam., 54, 539–560, 2004.

Evensen, G. and Leeuwen, P. J. V.: Assimilation of Geosat altimeter data for the Agulhas current using the ensemble Kalman filter with a quasi-geostrophic model, Mon. Weather Rev., 124, 85–96, 1996.

Evensen, G. and Leeuwen, P. J. V.: An Ensemble Kalman Smoother for Nonlinear Dynamics, Mon. Weather Rev., 128, 1852–1867, 2000.

Farr, T. G., Rosen, P. A., Caro, E., Crippen, R., Duren, R., Hensley, S., Kobrick, M., Paller, M., Rodriguez, E., Roth, L., Seal, D., Shaffer, S., Shimada, J., Werner, J. U. M., Oskin, M., Burbank, D., and Alsdorf, D.: The Shuttle Radar Topography Mission, Rev. Geophys., 45, RG2004, https://doi.org/10.1029/2005RG000183, 2007.

Getirana, A. C. V. and Peters-Lidard, C.: Estimating water discharge from large radar altimetry datasets, Hydrol. Earth Syst. Sci., 17, 923–933, https://doi.org/10.5194/hess-17-923-2013, 2013.

Greybush, S. J., Kalnay, E., Miyoshi, T., Ide, K., and Hunt, B. R.: Balance and Ensemble Kalman Filter Localization Techniques, Mon. Weather Rev., 139, 511–522, 2011.

Gustafsson, N., Huang, X. Y., Yang, X. H., Mogensen, K., Lindskog, M., Vignes, O., Wilhelmsson, T., and Thorsteinsson, S.: Four-dimensional data assimilation for a limited area model, Tellus A, 64, 14985, https://doi.org/10.3402/tellusa.v64i0.14985, 2012.

Hagemann, S. and Dümenil, L.: A parameterization of the lateral waterflow for the global scale, Clim. Dynam., 14, 17–31, 1998.

Hamill, T. M., Whitaker, J. S., and Snyder, C.: Distance-Dependent Filtering of Background Error Covariance Estimates in an Ensemble Kalman Filter, Mon. Weather Rev., 129, 2776–2790, 2001.

Houtekamer, P. L. and Mitchell, H. L.: A sequential ensemble kalman filter for atmospheric data assimilation, Mon. Weather Rev., 129, 124–137, 2001.

Houtekamer, P. L. and Mitchell, H. L.: Ensemble Kalman filtering, Q. J. Roy. Meteorol. Soc., 131, 3269–3289, 2005.

Hwang, C., Peng, M.-F., Ning, J., Luo, J., and Sui, C.-H.: Lake level variations in China from TOPEX/Poseidon altimetry: Data quality assessment and links to precipitation and ENSO, Geophys. J. Int., 161, 1–11, 2005.

Kouraev, A. V., Zakharovab, E. A., Samainc, O., Mognard, N. M., and Cazenave, A.: Ob' River discharge from TOPEX/Poseidon satellite altimetry (1992–2002), Remote Sens. Environ., 93, 238–245, 2004.

López López, P., Wanders, N., Schellekens, J., Renzullo, L. J., Sutanudjaja, E. H., and Bierkens, M. F. P.: Improved large-scale hydrological modelling through the assimilation of streamflow and downscaled satellite soil moisture observations, Hydrol. Earth Syst. Sci., 20, 3059–3076, https://doi.org/10.5194/hess-20-3059-2016, 2016.

Lucas-Picher, P., Arora, V., Caya, D., and Laprisse, R.: Implementation of a large-scale variable velocity river flow routing algorithm in the Canadian Regional Climate Model (CRCM), Atmos.-Ocean, 41, 139–153, 2010.

MacArthur, J. L.: Altimeter designs – Seasat-1 and future missions, Mar. Geodesy, 3, 39–61, 1980.

Manning, R.: On the flow of water in open channels and pipes, Inst. Civ. Eng. Ireland, 20, 161–207, 1891.

Masson, V., Le Moigne, P., Martin, E., Faroux, S., Alias, A., Alkama, R., Belamari, S., Barbu, A., Boone, A., Bouyssel, F., Brousseau, P., Brun, E., Calvet, J.-C., Carrer, D., Decharme, B., Delire, C., Donier, S., Essaouini, K., Gibelin, A.-L., Giordani, H., Habets, F., Jidane, M., Kerdraon, G., Kourzeneva, E., Lafaysse, M., Lafont, S., Lebeaupin Brossier, C., Lemonsu, A., Mahfouf, J.-F., Marguinaud, P., Mokhtari, M., Morin, S., Pigeon, G., Salgado, R., Seity, Y., Taillefer, F., Tanguy, G., Tulet, P., Vincendon, B., Vionnet, V., and Voldoire, A.: The SURFEXv7.2 land and ocean surface platform for coupled or offline simulation of earth surface variables and fluxes, Geosci. Model Dev., 6, 929–960, https://doi.org/10.5194/gmd-6-929-2013, 2013.

Meade, R., Rayol, J., Conceicão, S. D., and Natividade, J.: Backwater Effects in the Amazon River Basin of Brazil, Environ. Geol. Water Sci., 18, 105–114, 1991.

Michailovsky, C. I. and Bauer-Gottwein, P.: Operational reservoir inflow forecasting with radar altimetry: the Zambezi case study, Hydrol. Earth Syst. Sci., 18, 997–1007, https://doi.org/10.5194/hess-18-997-2014, 2014.

Michailovsky, C. I., Milzow, C., and Bauer-Gottwein, P.: Assimilation of radar altimetry to a routing model of the Brahmaputra river, Water Ressour. Res., 49, 4807–4816, 2013.

Molinier, M., Guyot, J.-L., Orstom, B., Guimarães, V., de Oliveira, E., and Dnaee, B.: Hydrologie du bassin de l'Amazone, in: Grands Bassins Fluviaux Périatlantiques, edited by: Boulègue, J. and Olivry, J.-C., 3PEGI-INSA-CNRS-ORSTOM, Paris, 335–345, 1993.

Moore, J. B.: Discrete-time fixed-lag smoothing algorithms, Automatica, 9, 163–173, 1973.

Moradkhani, H., Hsu, K.-L., Gupta, H., and Sorooshian, S.: Uncertainty assessment of hydrologic model states and parameters: Sequential data assimilation using particle filter, Water Resour. Res., 41, W05012, https://doi.org/10.1029/2004WR003604, 2005.

Ngo-Duc, T., Oki, T., and Kanae, S.: A variable streamflow velocity method for global river routing model: model description and preliminary results, Hydrol. Earth Syst. Sci. Discuss., 4, 4389–4414, https://doi.org/10.5194/hessd-4-4389-2007, 2007.

Nijssen, B. and Lettenmaier, D. P.: Effect of precipitation sempling error on simulated hydrological fluxes and states: Anticipating the Global Precipitation Measurement satellites, J. Geophys. Res., 109, D02103, https://doi.org/10.1029/2003JD00349, 2004.

Noilhan, J. and Planton, S.: A simple parameterization of land surface processes for meteorological models, Mon. Weather Rev., 117, 536–549, 1989.

Oki, T. and Sud, Y. C.: Design of Total Integrating Pathways (TRIP) – A Global River Channel Network, Earth Interact., 2, 1–36, 1998.

Paiva, R. C. D., Collischonn, W., and Tucci, C. E. M.: Large scale hydrologic and hydrodynamique modelling using limited data and a GIS based approach, J. Hydrol., 406, 170–181, 2011.

Paiva, R. C. D., Buarque, D. C., Collischonn, W., Bonnet, M.-P., Frappart, F., Calmant, S., and Mendes, C. A. B.: Large scale hydrological and hydrodynamic modeling of the Amazon River basin, Water Resour. Res., 49, 1226–1243, 2013a.

Paiva, R. C. D., Collischonn, W., and Buarque, D. C.: Validation of a full hydrodynamic model for large-scale hydrological modelling in the Amazon, Hydrol. Process., 27, 333–346, 2013b.

Palvis, N. K., Holmes, S. A., Kengon, S. C., and Factor, J. K.: The development and evaluation of the Earth Gravitational Model 2008 (EGM2008), J. Geophys. Res., 117, B04406, https://doi.org/10.1029/2011JB008916, 2012.

Paris, A., Paiva, R. C. D., Silva, J. S. D., Moreira, D., Calmant, S., Garambois, P.-A., Collischonn, W., Bonnet, M.-P., and Seyler, F.: Stage-discharge rating curves based on satellite altimetry and modeled discharge in the Amazon basin, Water Resour. Res., 52, 3787–3814, https://doi.org/10.1002/2014WR016618, 2016.

Rasmussen, J., Madsen, H., Jensen, K. H., and Refsgaard, J. C.: Data assimilation in integrated hydrological modeling using ensemble Kalman filtering: evaluating the effect of ensemble size and localization on filter performance, Hydrol. Earth Syst. Sci., 19, 2999–3013, https://doi.org/10.5194/hess-19-2999-2015, 2015.

Reichle, R. H., DeLannoy, G. J. M., Draper, B. A. F. C. S., and Liu, Q.: Connecting Satellite Observations with Water Cycle Variables Through Land Data Assimilation: Examples Using the NASA GEOS-5 LDAS, Surv. Geophys., 35, 577–606, https://doi.org/10.1007/s10712-013-9220-8, 2014.

Sakov, P. and Bertino, L.: Relation between two common localisation methods for the EnKF, Comput. Geosci., 15, 225–237, 2011.

Schellekens, J., Dutra, E., Martínez-de la Torre, A., Balsamo, G., van Dijk, A., Sperna Weiland, F., Minvielle, M., Calvet, J.-C., Decharme, B., Eisner, S., Fink, G., Flörke, M., Peßenteiner, S., van Beek, R., Polcher, J., Beck, H., Orth, R., Calton, B., Burke, S., Dorigo, W., and Weedon, G. P.: A global water resources ensemble of hydrological models: the eartH2Observe Tier-1 dataset, Earth Syst. Sci. Data, 9, 389–413, https://doi.org/10.5194/essd-9-389-2017, 2017.

Schumann, G. J.-P. and Domeneghetti, A.: Exploiting the proliferation of current and future satellite observations of rivers, Hydrol. Process., 30, 2891–2896, 2016.

Silva, J. S. D., Calmant, S., Seyler, F., Filho, O. C. R., Cochonneau, G., and Mansur, W. J.: Water levels in the Amazon basin derived from the ERS 2 and ENVISAT radar altimetry missions, Remote Sens. Environ., 114, 2160–2181, 2010.

Silva, J. S. D., Calmant, S., Seyler, F., Lee, H., and Shum, C. K.: Mapping of extreme stage variations using ENVISAT altimetry in the Amazon basin rivers, Int. Water Technol. J., 2, 14–25, 2012.

Silva, J. S. D., Calmant, S., Seyler, F., Moreira, D. M., Oliveira, D., and Monteiro, A.: Radar altimetry aids managing gauge networks, Water Resour. Manage., 28, 587–603, 2014.

Sorooshian, S. and Dracup, J. A.: Stochastic parameter estimation procedures for hydrologic rainfall-runoff models: correlated and heteroscedastic error cases, Water Resour. Res., 16, 430–442, 1980.

Sun, A. Y., Morris, A. P., and Mohanty, S.: Sequential updating of multimodal hydrogeologic parameter fields using localisation and clustering techniques, Water Resour. Res., 45, W07424, https://doi.org/10.1029/2008WR007443, 2009.

Tanajura, C. A. S., Lima, L. N., and Belyav, K. P.: Assimilation of satellite surface-height anomalies data into a Hybrid Coordinate Ocean Model (HYCOM) over the Atlantic Ocean, Mar. Phys., 55, 667–678, https://doi.org/10.1134/S0001437015050161, 2015.

Tourian, M. J., Schwatke, C., and Sneeuw, N.: River discharge estimation at daily resolution from satellite altimetry over an entire river basin, J. Hydrol., 546, 230–247, 2017.

Trudel, M., Leconte, R., and Paniconi, C.: Analysis of the hydrological response of a distributed physically-based model using post-assimilation (EnKF) diagnostics of streamflow and in situ soil moisture observations, J. Hydrol., 514, 192–201, https://doi.org/10.1016/j.jhydrol.2014.03.072, 2014.

Vörösmarty, C., Moore, B., Grace, A., Gildea, M., Melillo, J., Peterson, B., Rastetter, E., and Steudler, P.: Continental scale models of water balance and fluvial transport: An application to South America, Global Biogeochem. Cy., 3, 241–265, 1989.

Vrugt, J. A., Gupta, H. V., Bouten, W., and Sorooshian, S.: A Shuffled Complex Evolution Metropolis algorithm for optimization and uncertainty assessment of hydrologic model parameters, Water Resour. Res., 39, 1201, https://doi.org/10.1029/2002WR001642, 2003.

Wisser, D., Fekete, B. M., Vörösmarty, C. J., and Schumann, A. H.: Reconstructing 20th century global hydrography: a contribution to the Global Terrestrial Network-Hydrology (GTN-H), Hydrol. Earth Syst. Sci., 14, 1–24, https://doi.org/10.5194/hess-14-1-2010, 2010.

Yamazaki, D., Kanae, S., Kim, H., and Oki, T.: A physically based descrpition of floodplain inundation dynamics in a global river routing model, Water Resour. Res., 47, W04501, https://doi.org/10.1029/2010WR009726, 2011.

6

Using rainfall thresholds and ensemble precipitation forecasts to issue and improve urban inundation alerts

Tsun-Hua Yang[1], **Gong-Do Hwang**[1,2], **Chin-Cheng Tsai**[1], **and Jui-Yi Ho**[1]

[1]Taiwan Typhoon and Flood Research Institute (TTFRI), National Applied Research Laboratories (NARLabs), Taipei, 10093, Taiwan

[2]Department of Atmospheric Science, National Taiwan University, Taipei, 10617, Taiwan

Correspondence to: Tsun-Hua Yang (tshyang@narlabs.org.tw)

Abstract. Urban inundation forecasting with extended lead times is useful in saving lives and property. This study proposes the integration of rainfall thresholds and ensemble precipitation forecasts to provide probabilistic urban inundation forecasts. Utilization of ensemble precipitation forecasts can extend forecast lead times to 72 h, predicting peak flows and to allow response agencies to take necessary preparatory measures. However, ensemble precipitation forecasting is time- and resource-intensive. Using rainfall thresholds to estimate urban areas' inundation risk can decrease this complexity and save computation time. This study evaluated the performance of this system using 352 townships in Taiwan and seven typhoons during the period 2013–2015. The levels of forecast probability needed to issue inundation alerts were addressed because ensemble forecasts are probability based. This study applied six levels of forecast probability and evaluated their performance using five measures. The results showed that this forecasting system performed better before a typhoon made landfall. Geography had a strong impact at the start of the numerical weather modeling, resulting in the underestimation of rainfall forecasts. Regardless of this finding, the inundation forecast performance was highly contingent on the rainfall forecast skill. This study then tested a hybrid approach of on-site observations and rainfall forecasts to decrease the influence of numerical weather predictions and improve the forecast performance. The results of this combined system showed that forecasts with a 24 h lead time improved significantly. These findings and the hybrid approach can be applied to other hydrometeorological early warning systems to improve hazard-related forecasts.

1 Introduction

Flooding is one of the most destructive disasters in the world and results in enormous losses of life and property annually (Gruntfest and Handmer, 2001; Barredo, 2009; Hallegatte et al., 2013; Sampson et al., 2015). Global flood risk is likely to increase under climate change; as a result, numerous adaption strategies should be considered (Hirabayashi et al., 2013). Establishing an early flood warning system to reduce disaster losses is the most cost-effective solution of all of the structural and non-structural measures studied (Alfieri et al., 2012; Hallegatte, 2012). Several flood warning systems have been developed and implemented in response to floods (Pappenberger et al., 2005; Thielen et al., 2009; López-Trujillo, 2010; De Kleermaeker et al., 2012; Doong et al., 2012).

Various approaches are used to simulate flooding based on the available rainfall data. Complex models such as the one- or two-dimensional Saint-Venant equations better describe flow behaviors and provide detailed spatial information as part of their flood forecasts (e.g., Nguyen et al., 2015; Huthoff et al., 2015). However, the high computation costs and substantial data requirements involved in solving these detailed models limit the application of these models during an emergency response or real-time forecast. Therefore, a variety of alternatives, such as simplified equation-based systems, data-driven models, and rainfall threshold-based approaches, have been developed to improve the computing efficiency of the models.

- Simplified equation-based systems (e.g., Cirbus and Podhoranyi, 2013; Liu et al., 2015; Shao et al., 2015) use simplified equations, such as Manning's equation, to describe water spreading, thereby improving the calculation efficiency of the forecasting models. However, the data required, including digital elevation models (DEMs) and surface roughness, are sometimes difficult to collect. As a result, data preparedness is still a practical concern for the abovementioned models.

- Data-driven models are based on computational intelligence or machines. Flood forecasting is just one of the applications of these models (e.g., Chang et al., 2010; Lin et al., 2013). As indicated by the name, the quality and quantity of data used in the model have a considerable impact on the performance of data-driven models. To collect accurate flood inundation data is a challenge in itself. In addition, the performance of data-driven models deteriorates as forecast time increases (e.g., Lin and Jhong, 2015; Badrzadeh et al., 2015). Data-driven models also cannot provide forecasts with longer lead times.

- A rainfall threshold approach is commonly applied to evaluate landslide risk (e.g., Crosta and Frattini, 2003; Guzzetti et al., 2007; Posner and Georgakakos, 2015). Meteorological organizations generally issue flood forecasts/warnings if a critical value – namely, a rainfall threshold – is exceeded by the observed or predicted rainfall (Martina et al., 2006). Several operational meteorological agencies throughout the world issue warnings based on Flash Flood Guidance (FFG) values (Gourley et al., 2014). The US National Weather Service (NWS) developed FFG values for flash flooding (Carpenter et al., 1999). Based on these values, floods are predicted, and flood warnings are issued Georgakakos (2005, 2006) studied operational flash flood warning systems based on FFG and provided analytical results. These studies found that an FFG threshold is likely to produce a high probability of detection in regions where flash floods are frequent. The European Flood Awareness System (EFAS) uses numerical weather predictions and the Enhanced Runoff Index based on Climatology, which is based on simulated climatology, an FFG-related concept, to provide flash flood warnings (Raynaud et al., 2015). In countries such as Kenya and Haiti that do not have enough well-trained operators and resources to set up an efficient flood warning system, the approach is a viable alternative that allows for the mitigation of flood damage (Georgakakos et al., 2013; Shamir et al., 2013; Hoedjes et al., 2014). The rainfall threshold approach has proven successful in identifying a number of flash floods across Europe (Alfieri et al., 2014). Although it should not be considered a substitute for complex hydro-meteorological models because of its simplicity, using a rainfall threshold approach to develop a flood warning system can be an immediately useful tool for a variety of decision makers interested in early warnings and flash floods (Martina et al., 2006). Only a few studies (Jang, 2015; Wu et al., 2015) have applied rainfall thresholds to evaluate urban inundation risk. The present study represents the first of its kind to use the rainfall threshold approach and quantitative precipitation forecasts (QPFs) to evaluate inundation risk in Taiwan. By directly comparing QPFs with critical rainfall thresholds, this study aims to propose an early warning system that provides forecasts, allows for the possibility of issuing urban inundation warnings and gives response agencies enough lead time to implement emergency preparedness plans.

A flood warning system that uses QPFs as the rainfall input could increase the forecasting horizon from a few hours to a few days (Pappenberger et al., 2005; Shi et al., 2015). Georgakakos (2005) concluded that the dominant source of uncertainty in applying a rainfall thresholds approach to evaluate flood risk is precipitation. The uncertainty in forecasted rainfall values is generally higher than that for observed rainfall data. Nevertheless, to extend the forecast lead time, operational and research flood forecasting systems around the world are increasingly moving toward using QPFs to provide early warnings (Cloke and Pappenberger, 2009). Martina et al. (2006) discussed the possibility of providing flood warnings at given river reaches by directly comparing the QPF to a critical rainfall threshold value. Regardless of the forecasts' uncertainty, considering which probabilistic forecast levels should be used to issue inundation alerts or take actions is a challenging topic. Higher levels of probabilistic forecasts usually give the practitioner more confidence in the results. Dale et al. (2014) proposed a risk-based decision-support framework that could be easily applied in an operational flood forecast and early warning context. Other studies have also discussed the selection of appropriate probabilistic forecasts in terms of the economic and practical consequences of taking action (Coughlan de Perez et al., 2015, 2016). Therefore, the present study evaluates the system's performance in terms of different levels of forecast probability. In addition, this study proposes a data assimilation technique that uses real-time observations to decrease the uncertainty from rainfall forecasts and increases the 24 h forecast accuracy.

2 System development

The proposed inundation early warning system integrates ensemble precipitation forecasts, rainfall thresholds, and a real-time data assimilation technique to assess the possibility of issuing inundation alerts. Figure 1 shows the system's operational process during a typhoon event. The forecast results are intended to be provided to practitioners through a web

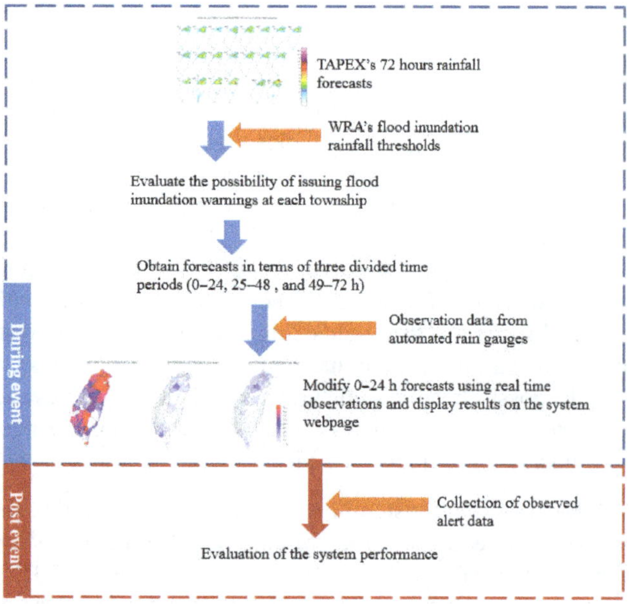

Figure 1. The operational flow chart for the proposed urban inundation early warning system.

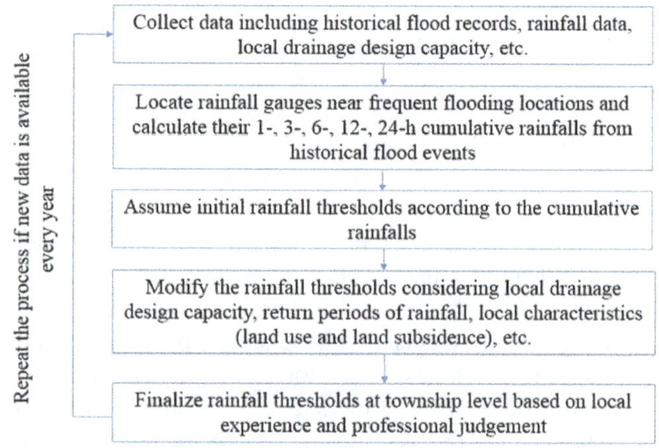

Figure 2. The identification process of rainfall thresholds (modified from Wu and Wang, 2009).

page. Due to a limitation in the computing resources and data retrieval tools available, the system generates a forecast every 6 h and updates the results on the web page. The details of each component in the system are as follows.

2.1 Ensemble precipitation forecasts for system input

This study used rainfall forecasts from a precipitation ensemble forecast experiment, namely, the Taiwan cooperative precipitation ensemble forecast experiment (TAPEX). TAPEX is a collective effort among academic institutes and government agencies such as National Taiwan University (NTU), National Central University (NCU), National Taiwan Normal University (NTNU), Chinese Culture University (CCU), the Central Weather Bureau (CWB), the National Center for High-Performance Computing (NCHC), the Taiwan Typhoon and Flood Research Institute (TTFRI), and the National Science and Technology Center for Disaster Reduction (NCDR). The experiment began in 2010 and was the first attempt to design a high-resolution numerical ensemble weather model in Taiwan. The experiment collects worldwide observation data, including temperature, wind, surface pressure, and relative humidity, from satellites, atmospheric sounding devices, buoys, aviation routine weather reports, ships, and other available sources (e.g., Hsiao et al., 2012, 2013). TAPEX uses the outputs from the Global Forecast System (GFS) produced by the National Centers for Environment Prediction (NCEP), along with observation data, as the initial and boundary conditions for its forecasts. Various model physics schemes and data assimilation strategies are used to perturb the numerical weather models and create differentiated ensemble members. To date, 20 ensemble members and 4 different regional models (AWR-WRF, HWRF, MM5 and CreSS) have been established for precipitation forecasting. The experiment aims to provide 24, 48, and 72 h typhoon rainfall forecasts and generates four runs per day at a 5 km spatial resolution. TAPEX's rainfall forecasts can extend the inundation forecast lead time to 72 h, which exceeds the average rainfall-runoff concentration time and the lag between observed peak participation and flooding in Taiwan. This lead time is thus considered sufficient for decision-making processes to be implemented prior to inundation.

2.2 Rainfall threshold for urban inundation alerts

Coughlan de Perez et al. (2016) defined the danger level of flooding as the 95th percentile of a flood model's forecasts at a lead time of 0 h. The present study considered rainfall thresholds as danger levels related to the likelihood of urban inundation. In Taiwan, the Water Resources Agency (WRA) has developed rainfall thresholds for all townships (Wu and Wang, 2009). Figure 2 shows the WRA's identification process for rainfall thresholds at the township level. It starts by collecting historical flood records that show when and where a flood occurred. The initial rainfall thresholds can then be estimated by determining the cumulative rainfall amounts at nearby rain gauges. The finalized rainfall thresholds for different townships are based on further investigations of local drainage capacity, local characteristics (e.g., land subsidence), and the professional judgement of local experts. The WRA reviews the rainfall thresholds every year once the new records are available. Inundation alerts are issued when observed rainfall meets or exceeds a given rainfall threshold. Local governments and civil agencies take necessary measures such as evacuating residents and deploying dewatering pumps based on the alerts. Given the historical record, the WRA assumes that inundations are directly related to accumulated rainfall and use a regression analysis to identify a two-level alarm for five duration periods. The five duration periods are 1, 3, 6, 12, and 24 h; a total of 10 rainfall thresholds are used to issue urban inundation alerts. The two levels of alarms are defined as follows:

- First-level alert: If the rain continues, the roads and villages subject to a high risk of flooding in the alerted townships may flood.
- Second-level alert: If the rain continues, the roads and villages subject to a high risk of flooding in the alerted townships will flood in the next 3 h.

The WRA has associated different rain gauges with different townships and issues warnings by comparing the observations with the associated rain gauges. The rainfall thresholds for the first and second alerts are different. There is a 3 h lead time before flooding if the accumulated rainfall reaches the second-level alert. The first-level alert is at an immediate risk of flooding. The WRA identified the rainfall thresholds of the second-level alerts for the purpose of precaution so that the responding authorities have time to take action. An inundation alert is issued if any of the rainfall thresholds is met by the observed rainfall. Wu (2013) compared the alerts to collected inundation records in 2012 and 2013 and concluded that the forecast accuracy rate is above 60 %. As the only rainfall threshold approach used to issue inundation alerts in Taiwan, it has proven its applicability in predicting flood inundation. This study used the rainfall thresholds of the second-level alerts to develop an early flood warning system.

2.3 Inundation risk evaluation and a data assimilation technique to modify the forecasts

In practice, the WRA issues inundation alerts when the cumulative rainfall exceeds the rainfall threshold at time T (Fig. 3). However, WRA compares real-time precipitation observations to the rainfall thresholds, and thus the lead time is usually not long enough to allow communities to implement emergency preparedness measures. This study proposes a practical early warning system that compares cumulative projected rainfall instead of observed rainfall to provide probabilistic urban inundation forecasts. The system uses TAPEX's forecasted rainfall to extend the model's lead time to 72 h. Figure 4 shows the forecast length during a real-time operation. TAPEX uses available observations at $t-6$ as its model's initial conditions, and its numerical weather model computation process took 6 h to produce rainfall forecasts from t to $t+72$ h; 352 Taiwanese townships were used in this study to evaluate the proposed system's performance. Equations (1) and (2) were used to calculate the probability of inundation in any given township; the forecasts were displayed over three distinct time periods (1–24, 25–48, and 49–72 h). A rolling window approach was applied to estimate the probability of issuing an inundation alert: each hour

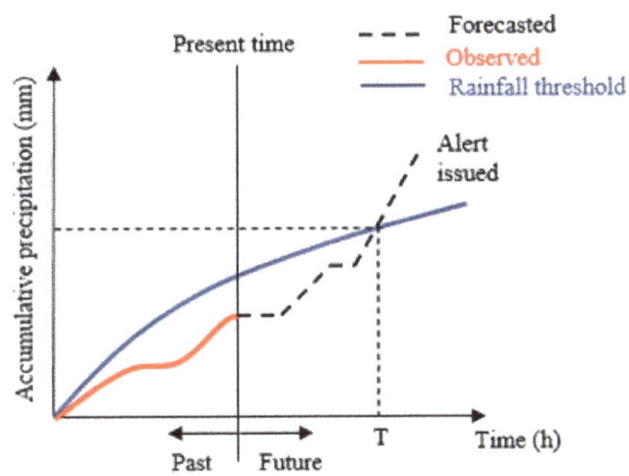

Figure 3. WRA issues an inundation alert when observed rainfalls meet or exceed any given rainfall thresholds (modified from Martina et al., 2006).

of the forecasting period was considered an evaluation end point, and the cumulative rainfall was calculated for the different durations.

$$f_i = \begin{cases} 1 & \text{if } \mathrm{PF}_{i,\mathrm{accu}} \geq \mathrm{PT}_{\mathrm{dur}} \\ 0 & \text{if } \mathrm{PF}_{i,\mathrm{accu}} < \mathrm{PT}_{\mathrm{dur}} \end{cases} \quad \mathrm{dur} = \{1, 3, 6, 12, 24\,\mathrm{h}\}, \qquad (1)$$

where $\mathrm{PF}_{i,\mathrm{accu}}$ is the cumulative forecasted rainfall of the ith ensemble member in TAPEX. $\mathrm{PT}_{\mathrm{dur}}$ represents cumulative rainfall thresholds for the different durations (dur) (1, 3, 6, 12, and 24 h). An inundation occurred ($f_i = 1$) if the cumulative forecasted rainfall exceeded any of these thresholds.

$$\mathrm{Pr} = \frac{1}{N}\sum_{i=1}^{N} f_i \times 100, \quad N = 1, 2, \ldots, 20 \qquad (2)$$

There are 20 ensemble members ($N = 20$) in TAPEX. Equation (2) sums the f_i values to obtain a probability (Pr), which represents the inundation risk for any given township. Each township's inundation risk can be obtained by repeating the above steps and comparing the results to TAPEX's 72 h rainfall forecasts. Three separate time periods (1–24, 25–48, and 49–72 h) illustrate the township's future inundation risk.

The accuracy of the rainfall forecasts has a considerable impact on the flood inundation forecasts. McBride and Ebert (2000) revealed that most numerical global models over-predicted and slightly under-predicted the rainfall frequency of various thresholds in Australia in summer and winter, respectively. These authors used a bias score (bias) to address the over- or under-estimation issue. A prediction is underestimated if the bias is less than 1.0. McBride and Elbert (2000) found that the biases of most numerical models were less than 1.0 for rainfall thresholds greater than 20 mm day^{-1}. The TAPEX under-predicted the rainfall frequencies during a rainfall event greater than 100 mm day^{-1} according to the forecast results in 2016. The

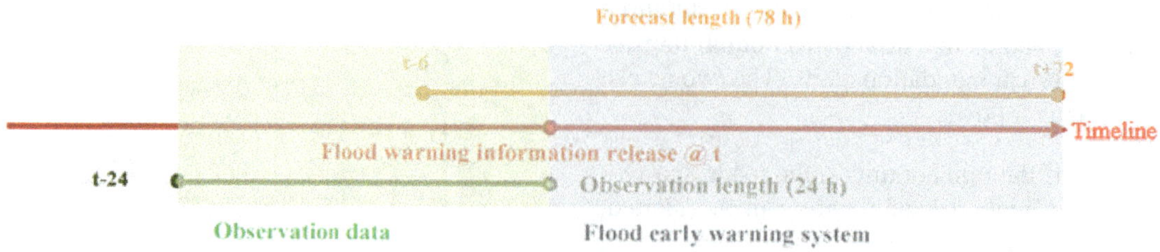

Figure 4. A combination of real-time rainfall observations and forecasts to improve 1 to 24 h inundation forecasts.

biases were 0.49 and 0.12 for rainfall thresholds of 100 and 350 mm day^{-1}, respectively. The complexity of the earth–atmosphere system and associated physical interactions adds uncertainty to the ensemble rainfall forecasts. However, this is beyond the scope of this study. The purpose of this study is to provide flood warnings by adopting the existing uncertainties in numerical weather forecasts and to improve the forecasts by using a data assimilation technique that uses real-time rainfall observations. This study used the technique to modify the 24 h urban inundation forecast performance. Figure 4 illustrates the combination of observations and forecasts used in the forecasting process. This study utilized five rainfall thresholds to represent different rainfall durations. However, these five thresholds could not be applied to evaluate the inundation risk at every hour within the first 24 h forecast. For example, only one rainfall threshold covers the 1 h period, which can be considered time t in Fig. 3; however, there is a lack of forecasts for $t-1$ and the preceding hours. When $t=t+2$, only rainfall thresholds for 1 and 3 h can be adopted. This shortcoming results in the underestimation of inundation forecasts. Given the above assumption, all five duration periods are applicable after the 25th hour. This study proposes a data assimilation technique using observed rainfall data to address the absence of rainfall forecasts. It applies available observation data from $t-24$ to $t-1$ prior to issuing inundation forecasts at t (Fig. 4). Figure 2 combines the observation data (red line) and forecasts (dash line) with all rainfall thresholds (solid blue line). Alerts are issued if the combination exceeds the rainfall threshold at any given duration. In other words, the inundation forecast is improved within the first 24 h.

3 Study area and data

3.1 Study area

Taiwan has an area of approximately 36 000 km^2, and approximately 70 % of the island is covered by mountains. A mountain range runs through the center of the island from north to south and forms a ridge dividing the east- and west-bound rivers. The rest of the island is composed of alluvial plains below 100 m in elevation. Ninety percent of the population lives on these alluvial plains. The distance from the mountaintops to the sea is very short, less than 70 km on average. Most of the riverbed slopes exceed 1/100 in the upstream reaches and are between 1/200 and 1/500 in the downstream reaches, which results in average rainfall-runoff concentration times of between 6 and 72 h in the townships (Jang, 2015) and a lag time between observed peak precipitation and flooding of between 2 and 10 h (Jang et al., 2012) time and high density of the population in the plains areas further increase the damage caused by floods. Taiwan is one of the most disaster-prone countries in the world; thus, it has been selected as the study area here for the development of an urban inundation warning system.

3.2 Observed inundation alerts

Records such as the time of occurrence, depth, and extent of inundation are used to calibrate and validate early warning systems. Collecting accurate information is thus incredibly important. However, data collection during major floods is challenging. For example, identifying the occurrence time of an inundation is always an issue because of the lack of in situ monitoring devices. This study used urban inundation alerts issued by the WRA as a reference to evaluate the system's performance. The WRA issues alerts following the Common Alerting Protocol (CAP), which was first published by the OASIS Emergency Management Technical Committee in 2005 (OASIS Emergency Management Technical Committee, 2005). The WRA updates its alerts every 10 min and uploads the information to an open-source platform operated by the National Science and Technology Center for Disaster Reduction (Lee et al., 2014). The CAP data include observed flood warning information, such as the flood warning's location and duration. Information on seven typhoons, including Soulik (2013), Trami (2013), Matmo (2014), Fung-wong (2014), Linfa (2015), Soudelor (2015), and Dujuan (2015), was collected to evaluate the system's performance. Five of these typhoons made landfall and resulted in heavy rainfall and floods. For example, Soudelor dropped more than 1100 mm of precipitation within 24 h and had wind gusts of up to 66.1 m s^{-1} in northern Taiwan (i.e., Suao Township, Yilan County). Detailed information on these seven typhoons is listed in Table 1. The landfall time was identified when the eye of the typhoon made landfall. Of these typhoons, the eyes of Trami and Linfa did not

Table 1. Information of seven typhoons during 2013–2015 used to evaluate the system performance.

Year	Name	Warning period (LST)	Landfall time (LST)
2013	Soulik	2013/07/11 08:30–2013/07/13 23:30	2013/07/13 03:00
2013	Trami	2013/08/20 11:30–2013/08/22 08:30	2013/08/21/ 18:00*
2014	Matmo	2014/07/21 17:30–2014/07/23 23:30	2014/07/23 00:10
2014	Fung-wong	2014/09/19 08:30–2014/09/22 08:30	2014/09/21 10:00
2015	Linfa	2015/07/06 08:30–2015/07/09 05:30	2015/07/08 15:00*
2015	Soudelor	2015/08/06 11:30–2015/08/09 08:30	2015/08/08 04:40
2015	Dujuan	2015/09/27 08:30–2015/09/29 17:30	2015/09/28 17:40

* For typhoons that did not make landfall, this study defined the landfall time while the minimum observational station pressure was observed when typhoon was closest to Taiwan.

Table 2. Contingency table used for the system performance evaluation.

		CAP records from WRA	
		Issued	Not issued
Forecasted by the proposed system	Issued	Hit	False alarm
	Not issued	Miss	No event

make landfall. For reference, this study selected the minimum observed atmospheric pressure at a weather station to define the time when these two typhoons were closest to Taiwan. The selected weather stations were the Taipei station for Trami and the Kaohsiung station for Linfa.

4 Results and discussion

This study relied on the contingency information shown in Table 2 to evaluate the performance of the proposed system. Hits and misses were associated with the observed records and determined based on whether the system's warning forecasts were consistent with the observations. A false alarm was associated with forecasts that did not correlate with observed data. "No event" was assigned to a township when neither the CAP records nor the model indicated flooding. Because floods are not frequent events, the no event (no flooding) scenario typically had a higher frequency than the other three fields. Different measures that have been broadly adopted by previous studies (e.g., Nguyen et al., 2015; Yang et al., 2015; Zhang et al., 2015) were used to evaluate the system's performance:

$$\text{probability of detection (POD)} = \frac{\text{Hit}}{\text{Hit} + \text{Miss}}, \tag{3}$$

$$\text{false alarm ratio (FAR)} = \frac{\text{False alarm}}{\text{Hit} + \text{false alarm}}, \tag{4}$$

$$\text{success ratio (SR)} = \frac{\text{Hit}}{\text{Hit} + \text{false alarm}}, \tag{5}$$

$$\text{threat score (TS)} = \frac{\text{Hit}}{\text{Hit} + \text{Miss} + \text{false alarm}}. \tag{6}$$

Both POD and TS are sensitive to hits and range from 0 to 1. The only difference between these two values is that POD ignores false alarms and TS does not. POD has the ability to be artificially improved by the issuance of additional alarms, which would increase the number of hits. TS is also known as the critical success index (CSI) and usually results in poorer scores for rare events. SR and FAR are the success ratio and false alarm ratio, respectively. FAR is used in conjunction with POD. If FAR equals 0.5 or less, the performance is considered tolerable (Coughlan de Perez et al., 2016). The sum of SR and FAR equals 1, and both indices ignore misses. This study combined SR and FAR into one index (SR–FAR) that had a range from -1 to 1. A positive value (> 0) for SR–FAR was expected given that the likelihood of correct warnings is acceptable. Rare events such as floods result in extremely large numbers of no events, which could greatly affect the forecast results. In this study, a no event forecast can provide information to decision makers that allows them to allocate resources to those townships with a higher inundation risk. Equations (3)–(6) do not consider the "no event" scenario in their formulas. The accuracy (ACC) of the model, which is shown in Eq. (7) and is also called the proportion of correct forecasts (Wilks, 2005), is simple and intuitive, and it served as a valuable reference in this study.

$$\text{Accuracy (ACC)} = \frac{\text{Hit} + \text{No event}}{\text{Hit} + \text{False alarm} + \text{Miss} + \text{No event}} \tag{7}$$

The next section presents the performance evaluation of the proposed system and then modifies the forecasting results using a hybrid of real-time observation and rainfall forecasts to improve the first 24 h inundation forecasts. This study used the time the typhoon made landfall as a reference point to define the evaluation period. The time needed to generate a rainfall forecast is 6 h, noted as one date–time group (dtg). The evaluation period was plus–minus three dtg's (18 h) relative to the time at which a typhoon made landfall. For example, Table 1 shows that Trami made landfall at 18:00, 21 August 2013. The landfall dtg is at 14:00, 21 August for 1–24 h; 14:00, 20 August for 25–48 h; and 14:00, 19 August for 49–72 h. The -1 dtg is 08:00, 21 August for 1–24 h; 08:00, 20

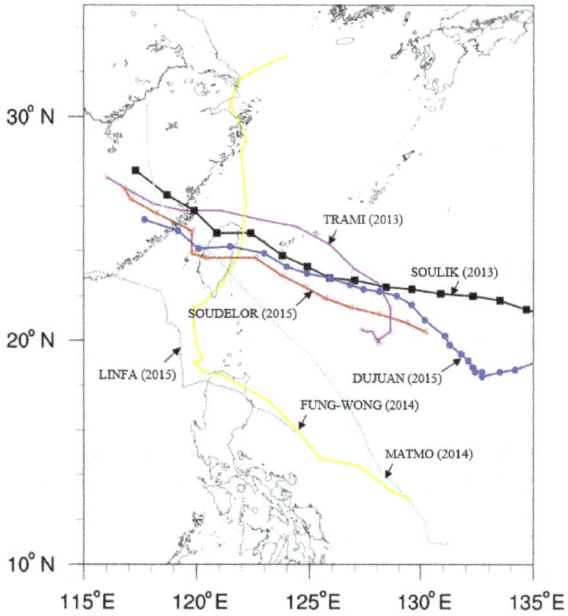

Figure 5. Location of the island of Taiwan, seven typhoons during 2013–2015, and their observed tracks.

August for 25–48 h; and 08:00, 19 August for 49–72 h. The average impact duration of a typhoon in Taiwan is 73.68 h (Huang et al., 2012). A typhoon has the most impact during the evaluation period (a total of 36 h).

4.1 Original forecast results without a data assimilation technique

Both the typhoon tracks and geography affected the performance of the rainfall forecasts. Figure 5 shows the observed typhoon tracks, and Fig. 6 compares the forecasted and observed tracks for Soulik, Soudelor, and Matmo. The models of the first two typhoons were consistent with the observed tracks, while the third was not; as a result, the performance of rainfall forecasts during the first two typhoons exceeded that of the third typhoon. The causes of the track forecast errors are beyond the discussion of this study. Use of ensemble rainfall forecasts as inputs to produce flood warning forecasts should take into account uncertainties such as track and rainfall forecast errors in numerical weather predictions. Figures 7–9 show the differences between the observed and forecasted flood warnings without a data assimilation technique over three lead-time periods (1–24, 25–48, and 49–72 h). Tables 3–5 summarize the average ACC, POD and SR–FAR results for different lead-time lengths during the evaluation period. The proposed system provides probabilistic forecasts. For example, 50 % flood probability means that at least 10 out of 20 TAPEX members produced rainfall forecasts that met or exceeded the rainfall thresholds. The appropriate probability threshold that initiated response actions was discussed. Six probability thresholds (10, 30, 50, 70, 80,

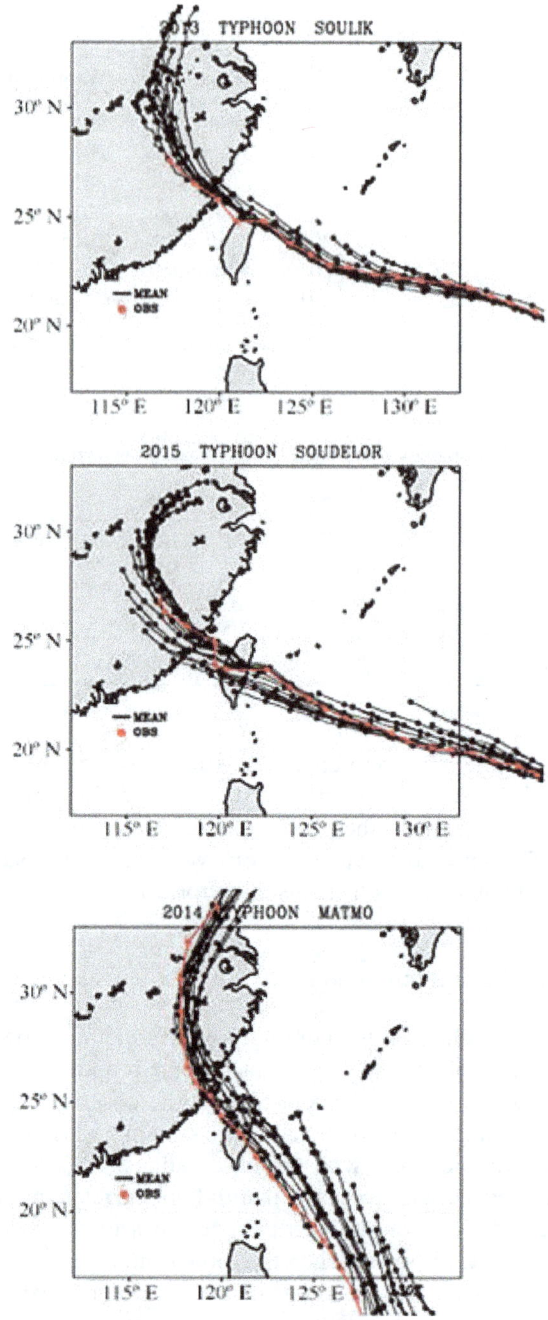

Figure 6. Comparisons of forecasted and observed typhoon tracks for Soulik (top), Soudelor (middle), and Matmo (bottom): black lines are TAPEX's ensemble mean forecasted tracks and each black line's forecasting length is 72 h.

and 100 %) were selected. The results showed that forecasts with lower possibility thresholds had higher TSs (Figs. 7–9). For example, Fig. 7 shows that the TSs of Soudelor are 0.1–0.4 for the 10 % probability threshold, which are higher than those for the 70 % probability threshold. All tables showed that the average performance of low-possibility thresholds over the evaluation period resulted in better TS and POD

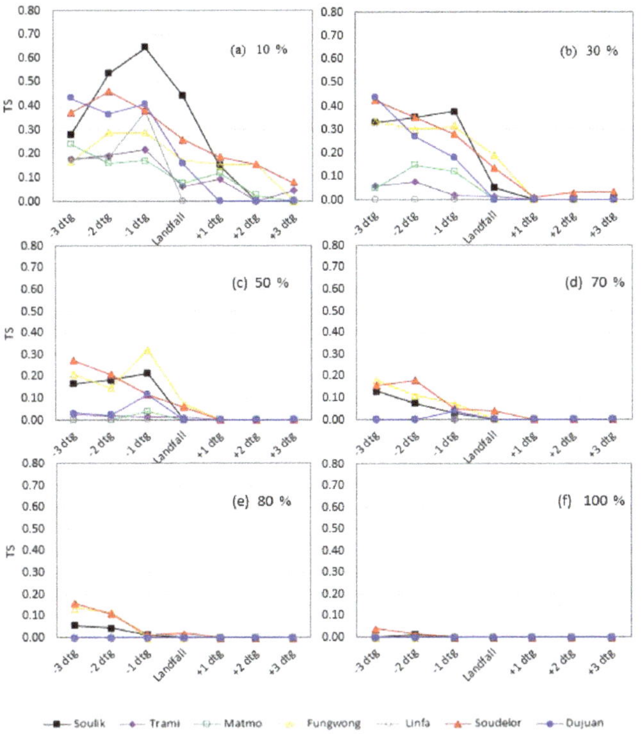

Figure 7. Comparisons of TS performance with a 1 to 24 h lead time considering various probability thresholds without a data assimilation technique.

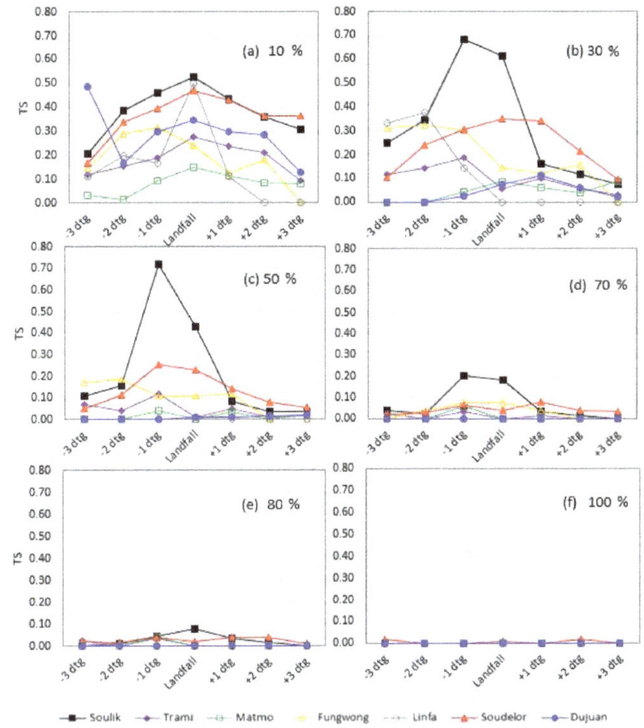

Figure 8. Comparisons of TS performance with a 25 to 48 h lead time considering various probability thresholds without a data assimilation technique.

scores. A lower probability threshold means a lower inundation threshold. Thus, the number of hits increased, but the number of false alarms increased as well. Decision makers generally consider an increased number of actions "in vain" when taking emergency measures based on a low probability threshold. The higher probability thresholds (e.g., a probability threshold > 50 %) had lower TSs and indicated that TAPEX ensemble rainfall forecasts were usually underestimated in this study. TAPEX's forecasted tracks had an impact on the rainfall forecasts, which affected the accuracy of the inundation forecasting. Soudelor and Soulik had the best performance in terms of TSs. The results for these typhoons were consistent with the track forecasts' performance (Fig. 5). The results also showed that the TS performance decreased after the typhoons made landfall. The period from −3 dtg to landfall is shown in Figs. 7–9. The steep terrain of Taiwan poses a challenge to the vortex initialization in numerical weather prediction models. Most current techniques are unable to properly initiate a typhoon vortex near complex terrain, when in reality the typhoons are already well developed at the time of landfall. The typhoons, due to their proximity to Taiwan by the time of model initiation, are not well developed in the models because of the terrain. The vortex is initialized near the complex terrain, and the current technique in TAPEX may not perform as well as it does when the vortex is in the open ocean. This introduces errors into the consequent precipitation forecast. This observation explains the decreased system performance when the TAPEX model initialization involves a typhoon close to or making landfall on Taiwan, even if the forecast time is as small as 1 to 24 h. The same issue does not create problems when the lead time is greater (or the typhoon is farther away). However, due to the complexity of the atmosphere, other issues, such as lack of observations, can cause the initial field degradation. Consequently, the typhoon tracks, rainfall, and related inundation forecasts were inevitably influenced. In the tables, the majority of ACC values exceeded 0.7. The less likely the inundation, the higher the ACC value. For example, only a few inundation alerts were issued during Linfa; the system's corresponding ACC scores were above 0.9. However, the POD and SR–FAR values were not as good as the ACC values in this case. The POD scores were zero. The SR–FAR values could not be calculated because there were zero hits and false alarms. When the system produced less accurate forecasts, the performance of the POD and SR–FAR functions decreased, resulting in a lower number of observed inundation alerts. A large number of inundation alerts were issued by the WRA during Soudelor and Soulik. The ACC numbers were below 0.8. The POD and SR–FAR numbers were relatively better than those in Linfa. A lower possibility threshold indicated that more hits and false alarms occurred; this resulted in negative SR–FAR scores.

Table 3. Average performance with 1 to 24 h lead time during the evaluation period for all possibility thresholds.

%	Soulik			Trami			Matmo			Fung-wong			Linfa			Soudelor			Dujuan		
	ACC	POD	SR–FAR	ACC	POD	SR–FAR	ACC	POD	SR–FAR	ACC	POD	SR–FAR	ACC	POD	SR–FAR	ACC	POD	SR–FAR	ACC	POD	SR–FAR
10	0.787	0.517	0.033	0.776	0.186	−0.433	0.778	0.241	−0.626	0.914	0.348	−0.341	0.976	0.196	−0.419	0.758	0.396	−0.050	0.839	0.252	0.083
30	0.795	0.201	0.204	0.823	0.021	−0.250	0.867	0.046	−0.472	0.944	0.219	0.360	0.981	0.000	n/a*	0.792	0.166	0.462	0.844	0.100	0.519
50	0.787	0.092	0.220	0.827	0.009	0.333	0.874	0.007	−0.667	0.944	0.129	0.818	0.981	0.000	n/a*	0.780	0.077	0.517	0.834	0.027	0.222
70	0.781	0.031	0.133	0.826	0.000	n/a*	0.876	0.000	−1.000	0.941	0.058	1.000	0.981	0.000	n/a*	0.775	0.049	0.514	0.832	0.007	−0.143
80	0.779	0.015	−0.059	0.826	0.000	n/a*	0.877	0.000	−1.000	0.940	0.039	1.000	0.981	0.000	n/a*	0.772	0.031	0.500	0.832	0.000	−1.000
100	0.780	0.002	1.000	0.826	0.000	n/a*	0.877	0.000	n/a*	0.937	0.000	n/a*	0.981	0.000	n/a*	0.767	0.005	0.000	0.833	0.000	n/a*

* n/a means that either FAR or POD had zero values in the denominator and cannot be calculated.

Table 4. Average performance with 25 to 48 h lead time during the evaluation period for all probability thresholds.

%	Soulik			Trami			Matmo			Fung-wong			Linfa			Soudelor			Dujuan		
	ACC	POD	SR–FAR	ACC	POD	SR–FAR	ACC	POD	SR–FAR	ACC	POD	SR–FAR	ACC	POD	SR–FAR	ACC	POD	SR–FAR	ACC	POD	SR–FAR
10	0.732	0.746	−0.127	0.621	0.494	−0.544	0.620	0.297	−0.779	0.913	0.374	−0.337	0.923	0.271	−0.856	0.756	0.616	−0.038	0.842	0.383	0.075
30	0.821	0.453	0.258	0.817	0.117	−0.187	0.811	0.086	−0.757	0.947	0.245	0.490	0.978	0.063	−0.625	0.796	0.291	0.265	0.833	0.053	0.023
50	0.825	0.274	0.594	0.830	0.044	0.357	0.857	0.023	−0.781	0.944	0.116	0.895	0.981	0.000	−1.000	0.787	0.147	0.400	0.832	0.010	−0.111
70	0.795	0.081	0.725	0.827	0.009	0.333	0.877	0.010	−0.143	0.940	0.039	1.000	0.982	0.000	n/a*	0.772	0.047	0.256	0.831	0.000	−1.000
80	0.784	0.029	0.600	0.826	0.007	0.200	0.877	0.007	0.333	0.937	0.000	n/a*	0.982	0.000	n/a*	0.769	0.028	0.143	0.831	0.000	−1.000
100	0.779	0.000	−1.000	0.826	0.000	n/a*	0.877	0.000	n/a*	0.937	0.000	n/a*	0.982	0.000	n/a*	0.769	0.007	1.000	0.833	0.000	n/a*

* n/a means that either FAR or POD had zero values in the denominator and cannot be calculated.

Table 5. Average performance with 49 to 72 h lead time during the evaluation period for all probability thresholds.

%	Soulik			Trami			Matmo			Fung-wong			Linfa			Soudelor			Dujuan		
	ACC	POD	SR–FAR	ACC	POD	SR–FAR	ACC	POD	SR–FAR	ACC	POD	SR–FAR	ACC	POD	SR–FAR	ACC	POD	SR–FAR	ACC	POD	SR–FAR
10	0.671	0.576	−0.299	0.514	0.573	−0.610	0.483	0.317	−0.835	0.860	0.303	−0.668	0.910	0.188	−0.913	0.761	0.531	−0.024	0.826	0.129	−0.138
30	0.762	0.092	−0.301	0.715	0.219	−0.593	0.771	0.099	−0.814	0.935	0.071	−0.1544	0.978	0.000	−1.000	0.782	0.154	0.248	0.829	0.002	−0.818
50	0.782	0.022	0.333	0.784	0.105	−0.531	0.866	0.017	−0.737	0.937	0.000	−1.000	0.982	0.000	n/a*	0.768	0.040	0.022	0.831	0.000	−1.000
70	0.780	0.000	n/a*	0.821	0.047	−0.245	0.876	0.003	−0.500	0.937	0.000	n/a*	0.982	0.000	n/a*	0.768	0.016	0.059	0.833	0.000	n/a*
80	0.780	0.000	n/a*	0.826	0.021	0.000	0.877	0.003	1.000	0.937	0.000	n/a*	0.982	0.000	n/a*	0.769	0.014	0.333	0.833	0.000	n/a*
100	0.780	0.000	n/a*	0.826	0.000	n/a*	0.877	0.000	n/a*	0.937	0.000	n/a*	0.982	0.000	n/a*	0.767	0.000	n/a*	0.833	0.000	n/a*

* n/a means that either FAR or POD had zero values in the denominator and cannot be calculated.

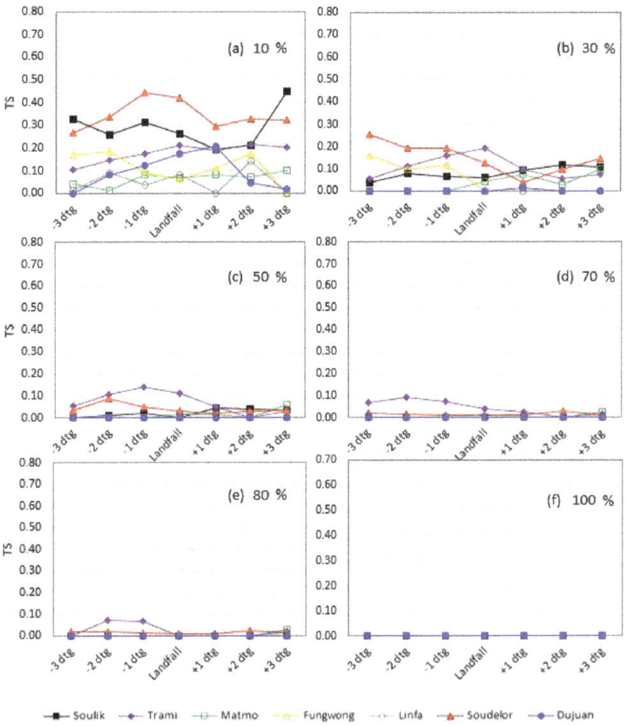

Figure 9. Comparisons of TS performance with a 49 to 72 h lead time considering various probability thresholds without a data assimilation technique.

In general, the SR–FAR scores decreased when the forecast lead time increased. However, the results for Soulik were opposite for the 50 % probability threshold and below. The TS was higher when the probability increased by up to 50 % prior to the typhoon making landfall (i.e., −1 dtg). The number of false alarms decreased when the probability threshold increased. This helped improve the TS at −1 dtg. However, this finding did not hold true when the probability threshold was above 70 %. Typhoon Matmo performed worst in terms of SR–FAR scores for the three different lead-time lengths. Figure 5 shows that the forecasted tracks did not coincide with the observed track. When a typhoon made landfall, the topography affected the performance of the numerical weather models, worsening the performance of the inundation warning forecasts. All of the results above indicate that the greatest uncertainty in the forecasts appears in the numerical weather predictions, which also has an important impact on other related disaster forecasts.

4.2 Modified forecasts using the data assimilation technique

To decrease the uncertainty of numerical weather predictions and improve the performance of inundation alert forecasting, this study applied a data assimilation technique that combined real-time observed and forecasted rainfall amounts to modify the forecasts. The data assimilation technique de-

Table 6. Average performance with 1 to 24 h lead time during the evaluation period for all probability thresholds using the data assimilation technique.

%	Soulik			Trami			Matmo			Fung-wong			Linfa			Soudelor			Dujuan		
	POD	FAR	TS	POD	FAR	TS	POD	FAR	TS	POD	FAR	TS	POD	FAR	TS	POD	FAR	TS	POD	FAR	TS
10	0.783	0.401	0.513	0.247	0.610	0.178	0.399	0.727	0.194	0.406	0.640	0.236	0.239	0.667	0.162	0.604	0.437	0.411	0.381	0.375	0.310
30	0.508	0.260	0.431	0.079	0.404	0.075	0.198	0.429	0.172	0.290	0.274	0.262	0.022	0.000	0.022	0.316	0.181	0.295	0.211	0.155	0.203
50	0.359	0.223	0.326	0.063	0.250	0.062	0.129	0.250	0.123	0.187	0.065	0.185	0.022	0.000	0.022	0.220	0.106	0.214	0.131	0.156	0.128
70	0.285	0.193	0.267	0.047	0.200	0.046	0.096	0.147	0.094	0.129	0.000	0.129	0.022	0.000	0.022	0.194	0.090	0.190	0.095	0.152	0.093
80	0.265	0.186	0.250	0.042	0.182	0.042	0.086	0.103	0.085	0.110	0.000	0.110	0.022	0.000	0.022	0.164	0.069	0.162	0.075	0.184	0.074
100	0.245	0.153	0.235	0.040	0.150	0.039	0.069	0.087	0.069	0.058	0.000	0.058	0.022	0.000	0.022	0.115	0.057	0.114	0.068	0.097	0.067

Table 7. Performance of all typhoons with the data assimilation technique for 1 to 24, 25 to 48, and 49 to 72 h lead time

%	1–24 h		25–48 h		49–72 h	
	FAR	TS	FAR	TS	FAR	TS
10	0.51	0.33	0.67	0.25	0.76	0.18
30	0.26	0.25	0.49	0.19	0.75	0.08
50	0.18	0.18	0.34	0.11	0.72	0.03
70	0.15	0.15	0.29	0.03	0.59	0.01
80	0.14	0.13	0.38	0.01	0.42	0.01
100	0.12	0.11	0.20	0.00	–	0.00

creased the temporal uncertainty of numerical rainfall forecasts and improved the accuracy of early warning notifications. The longest rainfall threshold duration to trigger an inundation alerts is 24 h in this study. The technique was used to address the gap in forecasted rainfall data with observed rainfall information. The absence of forecasted rainfall values occurred in the first warning period (i.e., 1–24 h). Therefore, this study used the data assimilation technique to improve the 1 to 24 h forecasts. Table 6 shows the modified forecast results compared to the original forecasts. Compared to the results without the hybrid technique, all performance measures' scores improved significantly. For example, when all typhoons were tested using the original forecasts, the system performed best during Soulik. Using the hybrid technique, the POD scores improved from 0.517 to 0.783 and from 0.002 to 0.245 for the 10 and 100 % probability thresholds, respectively. The TSs improved from 0.293 to 0.513 and from 0.002 to 0.235 for the 10 and 100 % probability thresholds, respectively. The probability threshold represents the number of ensemble members' forecasted rainfall events that met or exceeded the rainfall thresholds. The hybrid technique forecasts thus support the idea that a higher probability threshold indicates lower uncertainty in terms of forecasting. The FAR and POD scores decreased when the probability threshold increased. Decision-making confidence increases when the probability threshold increases and the FAR decreases. Coughlan de Perez et al. (2016) concluded that the likelihood of taking a necessary action when the FAR is lower than 0.5 would satisfy the decision maker's requirements for not taking action potentially in vain. Table 6 shows that most of the FAR scores improved to below 0.5 using the hybrid technique. Though these values improved compared to previous results, all of the POD scores were still low and continued to decrease when the probability threshold increased. The low POD score implies a lower hit rate. To improve these values, identifying the accuracy and uncertainty of rainfall forecasts is necessary.

Table 7 shows the overall performance of the system for seven typhoons in terms of FAR and TS scores. The overall results indicate that the FAR score decreases when the possibility threshold increases. The FAR score is smaller than

0.5 when the possibility is greater than 30 % with a lead time of 48 h. Therefore, the system performance meets the requirements of decision makers to take action during typhoon events (Coughlan de Perez et al., 2016). However, the system cannot provide acceptable forecasts with a lead time greater than 48 h, regardless of which possibility threshold is selected. This finding limits the use of the system when the lead time is greater than 48 h. The system integrates TAPEX data to obtain forecasted typhoon tracks and rainfall amounts. However, for some local convections, such as afternoon thunderstorms, the current 5 km spatial resolution of TAPEX might not be sufficient to resolve these weather phenomena as well as it does for much larger-scale weather systems, such as typhoons. These small-scale weather systems pose another limitation to the use of this system.

5 Conclusions

This study proposed an early inundation warning system that integrates ensemble rainfall forecasts and rainfall thresholds. Five rainfall thresholds with different durations were applied. Seven typhoon events during the period 2013–2015 and real inundation alert records from the WRA were used to evaluate the performance of the system. Five performance measures and a period of 18 h (3 dtg's) before and after a typhoon made landfall were considered. The system applied ensemble rainfall forecasts and provided probabilistic forecasts. Therefore, six different probability thresholds were considered to trigger the issuance of inundation alerts and calculate various performance scores. An appropriate probability threshold helps decision makers take fewer actions in vain. The results showed that a lower probability threshold had a higher POD score, which is associated with a higher inundation alert detection rate. The downside of a lower probability threshold is a higher FAR score. If the FAR is above 0.5, the system is considered impractical (Coughlan de Perez et al., 2016). Although the system performed better before a typhoon made landfall, particularly in terms of TSs, it was still unable to identify the most useful probability threshold for identifying when emergency responders should take various actions. Numerical weather predictions were the dominant input influencing the forecast results. The system's performance varied according to the different typhoons tested. In other words, the system cannot maintain a constant level of performance due the temporal and spatial uncertainties in the numerical rainfall forecasts. Taiwan's steep terrain also poses a challenge to the vortex initialization in numerical weather prediction models and contributes to the uncertainty inherent in the rainfall forecasts. In conclusion, the findings of this study suggest that a better forecast is usually produced (1) when the forecasted typhoon tracks are consistent with the observed tracks and (2) before a typhoon makes landfall.

Finally, the authors developed a data assimilation technique that combined real-time observed and forecasted rainfall to decrease the uncertainty of numerical weather predictions and to improve 24 h inundation forecasts. The results showed that the FAR scores decreased when the probability threshold increased. All FAR scores were below 0.5 or less when the probability threshold was 30 % or above. This technique improved the appeal of the early warning system and generated more valuable forecasts that allowed decision makers to take fewer actions in vain. To further decrease the uncertainty of numerical weather predictions and improve the performance of inundation forecasts, advanced techniques, such as radar observations and associated data assimilation systems, could be considered in the future. A greater number of extreme weather events are likely in the future due to global climate change. These extreme events will bring high-intensity rainfalls over very short time spans. Radar observations efficiently improve very short-range rainfall forecasts, which are essential for accurate inundation forecasts. Rainfall thresholds need to be updated to meet the present flood capacity, such as when a new storm sewage system is put in place. After all, decision makers use forecasted rainfall and threshold-based early warning systems for a high-level overview of flood risk only. Given its advantage of an extended lead time and rapid estimation process, the model presented here is beneficial for emergency deployment to prepare large areas in advance of flooding. For small-area forecasts during a disaster, a complex physics-based model is recommended to replace the threshold-based model and provide detailed information.

6 Data availability

The TAPEX's rainfall forecast data were provided by the Taiwan Typhoon and Flood Research Institute (TTFRI, 2016). The government's law prohibits the data to be freely distributed. An application is needed before receiving the raw data. The rainfall thresholds for all townships were identified by the Water Resources Agency in Taiwan (WRA, 2016). National Science and Technology Center for Disaster Reduction in Taiwan provided the historical flood warnings of various typhoons (NCDR, 2016).

Appendix A: List of abbreviations

ACC	accuracy
bias	bias score
CAP	Common Alerting Protocol
CCU	Chinese Culture University
CWB	Central Weather Bureau
DEMs	digital elevation models
dtg	date–time group
dur	durations
GFS	Global Forecast System
EFAS	European Flood Alert System
FAR	false alarm ratio

FFG	flash-flood guidance
h	hour
km^2	square kilometer
LST	local standard time
m	meter
mm	millimeter
NCDR	National Science and Technology Center for Disaster Reduction
NCEP	National Centers for Environment Prediction
NCHC	National Center for High-Performance Computing
NCU	National Central University
NTNU	National Taiwan Normal University
NTU	National Taiwan University
NWP	numerical weather prediction
NWS	US National Weather Service
POD	probability of detection
QPFs	quantitative precipitation forecasts
SR	success ratio
TAPEX	TAiwan cooperative Precipitation Ensemble forecast eXperiment
TS	threat score
TTFRI	Taiwan Typhoon and Flood Research Institute
WRA	Water Resources Agency, Ministry of Economic Affairs

Acknowledgements. The authors would like to thank the Water Resources Agency, Taiwan, for providing the rainfall threshold values and the Central Weather Bureau, Taiwan, for providing rainfall observation data.

Edited by: H. Cloke

References

Alfieri, L., Salamon, P., Pappenberger, F., Wetterhall, F., and Thielen, J.: Operational early warning systems for water-related hazards in Europe, Environ. Sci. Policy, 21, 35–49, 2012.

Alfieri, L., Pappenberger, F., and Wetterhall, F.: The extreme runoff index for flood early warning in Europe, Nat. Hazards Earth Syst. Sci., 14, 1505–1515, doi:10.5194/nhess-14-1505-2014, 2014.

Badrzadeh, H., Sarukkalige, R., and Jayawardena, A. W.: Hourly runoff forecasting for flood risk management: application of various computational intelligence models, J. Hydrol., 529, 1633–1643, 2015.

Barredo, J. I.: Normalised flood losses in Europe: 1970–2006, Nat. Hazards Earth Syst. Sci., 9, 97–104, doi:10.5194/nhess-9-97-2009, 2009.

Carpenter, T. M., Sperfslage, J. A., Georgakakos, K. P., Sweeney, T., and Fread, D. L.: National threshold runoff estimation utilizing GIS in support of operational flash flood warning systems, J. Hydrol., 224, 21–44, 1999.

Chang, L., Shen, H., Wang, Y., Huang, J., and Lin, Y.: Clustering-based hybrid inundation model for forecasting flood inundation depths, J. Hydrol., 385, 257–268, 2010.

Cheng, C. and Liao, W.: Current situation and sustainability of water resource in Taiwan, in: Proceedings of 1st Asian Water Saving Council Conference, 141–148, 2011.

Cirbus, J. and Podhoranyi, M.: Cellular automata for the flow simulations on the earth surface, optimization computation process, Appl. Math. Inf. Sci., 7, 2149–2158, 2013.

Cloke, H. L. and Pappenberger, F.: Ensemble flood forecasting: a review, J. Hydrol., 375, 613–626, 2009.

Coughlan de Perez, E., van den Hurk, B., van Aalst, M. K., Jongman, B., Klose, T., and Suarez, P.: Forecast-based financing: an approach for catalyzing humanitarian action based on extreme weather and climate forecasts, Nat. Hazards Earth Syst. Sci., 15, 895–904, doi:10.5194/nhess-15-895-2015, 2015.

Coughlan de Perez, E., van den Hurk, B., van Aalst, M. K., Amuron, I., Bamanya, D., Hauser, T., Jongma, B., Lopez, A., Mason, S., Mendler de Suarez, J., Pappenberger, F., Rueth, A., Stephens, E., Suarez, P., Wagemaker, J., and Zsoter, E.: Action-based flood forecasting for triggering humanitarian action, Hydrol. Earth Syst. Sci., 20, 3549–3560, doi:10.5194/hess-20-3549-2016, 2016.

Crosta, G. B. and Frattini, P.: Distributed modelling of shallow landslides triggered by intense rainfall, Nat. Hazards Earth Syst. Sci., 3, 81–93, doi:10.5194/nhess-3-81-2003, 2003.

Dale, M., Wicks, J., Mylne, K., Pappenberger, F., Laeger, S., and Taylor, S.: Probabilistic flood forecasting and decision-making: an innovative risk-based approach, Nat. Hazards, 70, 159–172, 2014.

De Kleermaeker, S., Verlaan, M., Kroos, J., and Zijl, F.: A new coastal flood forecasting system for the Netherlands, Proceedings of Hydro 2012 "Taking Care of the Sea", Rotterdam, the Netherlands, 13–15 November 2012, 135–140, 2012.

Doong, D.-J., Chuang, L. Z.-H., Wu, L.-C., Fan, Y.-M., Kao, C. C., and Wang, J.-H.: Development of an operational coastal flooding early warning system, Nat. Hazards Earth Syst. Sci., 12, 379–390, doi:10.5194/nhess-12-379-2012, 2012.

Georgakakos, K.: Modern operational flash flood warning systems based on flash flood guidance theory: performance evaluation, International Conference on Innovation Advances and Implementation of Flood Forecasting Technology, Bergen-Tromsø, Norway, 9–13 October 2005, 1–10, 2005.

Georgakakos, K. P.: Analytical results for operational flash flood guidance, J. Hydrol., 317, 81–103, 2006.

Georgakakos, K., Graham, R., Jubach, R., Modrick, T., Shamir, E., Spencer, C., and Sperfslage, J.: Global flash flood guidance system, phase I, Hydrological Research Center, San Diego, CA, HRC Technical Report No. 9, 2013.

Gourley, J. J., Flamig, Z. L., Hong, Y., and Howard, K. W.: Evaluation of past, present and future tools for radar-based flash-flood prediction in the USA, Hydrol. Sci. J., 59, 1377–1389, 2014.

Gruntfest, E. and Handmer, J. (Eds.): Coping with Flash Floods, NATO Science Series, Vol. 77, Springer Verlag Science, Dordrecht, the Netherlands, 243 pp., 2001.

Guzzetti, F., Peruccacci, S., Rossi, M., and Stark, C. P.: Rainfall thresholds for the initiation of landslides in central and southern Europe, Meteorol. Atmos. Phys., 98, 239–267, 2007.

Hallegatte, S.: A cost effective solution to reduce disaster losses in developing countries: hydro-meteorological services, early warning, and evacuation, World Bank Policy Research Working Paper 6058, 2012.

Hallegatte, S., Green, C., Nicholls, R. J., and Corfee-Morlot, J.: Future flood losses in major coastal cities, Nat. Clim. Change, 3, 802–806, 2013.

Hirabayashi, Y., Mahendran, R., Koirala, S., Konoshima, L., Yamazaki, D., Watanabe, S., Kim, H., and Kanae, S.: Global flood risk under climate change, Nat. Clim. Change, 3, 816–821, 2013.

Hoedjes, J., Kooiman, A., Maathuis, B., Said, M., Becht, R., Limo, A., and Su, B.: A conceptual flash flood early warning system for Africa based on terrestrial microwave links and flash flood guidance, ISPRS Int. J. Geoinf., 3, 584–598, 2014.

Hsiao, L., Chen, D., Kuo, Y., Guo, Y., Yeh, T., Hong, J., Fong, C., and Lee, C.: Application of WRF 3DVAR to operational typhoon prediction in Taiwan: impact of outer loop and partial cycling approaches, Weather Forecast., 27, 1249–1263, 2012.

Hsiao, L., Yang, M., Lee, C., Kuo, H., Shih, D., Tsai, C., Wang, C., Chang, L., Chen, D. Y., Feng, L., Hong, J., Fong, C., Chen, D., Yeh, T., Huang, C., Guo, W., and Lin, G.: Ensemble forecasting of typhoon rainfall and floods over a mountainous watershed in Taiwan, J. Hydrol., 506, 55–68, 2013.

Huang, J., Yu, C., Lee, J., Cheng, L., Lee, T., and Kao, S.: Linking typhoon tracks and spatial rainfall patterns for improving flood lead time predictions over a mesoscale mountainous watershed, Water Resour. Res., 48, W09540, doi:10.1029/2011WR011508, 2012.

Huthoff, F., Remo, J. W. F., and Pinter, N.: Improving flood preparedness using hydrodynamic levee-breach and inundation modelling: middle Mississippi River, USA, J. Flood Risk Manage., 8, 2–18, 2015.

Jang, J.: An advanced method to apply multiple rainfall thresholds for urban flood warnings, Water, 7, 6056–6078, 2015.

Jang, J., Yu, P., Yeh, S., Fu, J., and Huang, C.: A probabilistic model for realtime flood warning based on deterministic flood inundation mapping, Hydrol. Process., 26, 1079–1089, 2012.

Lee, C., Wu, S., and Chang, T.: Implementing common alerting protocol for disaster reduction in Taiwan, 2014 TGIS GeoInformatics – UGIS Joint International Conference, Kaohsiung, Taiwan, 25–27 June, 2014.

Lin, G. and Jhong, B.: A real-time forecasting model for the spatial distribution of typhoon rainfall, J. Hydrol., 521, 302–313, 2015.

Lin, G., Lin, H., and Chou, Y.: Development of a real-time regional-inundation forecasting model for the inundation warning system, J. Hydroinform., 15, 1391–1407, 2013.

Liu, L., Liu, Y., Wang, X., Yu, D., Liu, K., Huang, H., and Hu, G.: Developing an effective 2-D urban flood inundation model for city emergency management based on cellular automata, Nat. Hazards Earth Syst. Sci., 15, 381–391, doi:10.5194/nhess-15-381-2015, 2015.

López-Trujillo, D.: Real Time Flood Alert System (RTFAS) for Puerto Rico, US Department of Interior, US Geological Survey, 2010.

Martina, M. L. V., Todini, E., and Libralon, A.: A Bayesian decision approach to rainfall thresholds based flood warning, Hydrol. Earth Syst. Sci., 10, 413–426, doi:10.5194/hess-10-413-2006, 2006.

McBride, J. L. and Ebert, E. E.: Verification of quantitative precipitation forecasts from operational numerical weather prediction models over Australia, Weather Forecast., 15, 103–121, 2000.

National Science and Technology Center for Disaster Reduction, Taiwan (NCDR): Observed flood warnings for typhoons from 2013 to 2015, retrieved from: https://alerts.ncdr.nat.gov.tw/AlertdataDownload.aspx, 2016.

Nguyen, P., Thorstensen, A., Sorooshian, S., Hsu, K., and AghaKouchak, A.: Flood forecasting and inundation mapping using HiResFlood-UCI and near-real-time satellite precipitation data: the 2008 Iowa flood, J. Hydrometeorol., 16, 1171–1183, 2015.

OASIS Emergency Management Technical Committee: Common alerting protocol, v.1.1., Organization for the Advancement of Structured Information Standards (OASIS), Boston, MA, 2005.

Pappenberger, F., Beven, K. J., Hunter, N. M., Bates, P. D., Gouweleeuw, B. T., Thielen, J., and de Roo, A. P. J.: Cascading model uncertainty from medium range weather forecasts (10 days) through a rainfall-runoff model to flood inundation predictions within the European Flood Forecasting System (EFFS), Hydrol. Earth Syst. Sci., 9, 381–393, doi:10.5194/hess-9-381-2005, 2005.

Posner, A. J. and Georgakakos, K. P.: Soil moisture and precipitation thresholds for real-time landslide prediction in El Salvador, Landslides, 12, 1179–1196, 2015.

Raynaud, D., Thielen, J., Salamon, P., Burek, P., Anquetin, S., and Alfieri, L.: A dynamic runoff co-efficient to improve flash flood early warning in Europe: evaluation on the 2013 central European floods in Germany, Meteorol. Appl., 22, 410–418, 2015.

Sampson, C. C., Smith, A. M., Bates, P. D., Neal, J. C., Alfieri, L., and Freer, J. E.: A high-resolution global flood hazard model,

Water Resour. Res., 51, 7358–7381, 2015.

Shamir, E., Georgakakos, K. P., Spencer, C., Modrick, T. M., Murphy Jr., M. J., and Jubach, R.: Evaluation of real-time flash flood forecasts for Haiti during the passage of hurricane Tomas, 4–6 November, 2010, Nat. Hazards, 67, 459–482, 2013.

Shao, Q., Weatherley, D., Huang, L., and Baumgartl, T.: RunCA: a cellular automata model for simulating surface runoff at different scales, J. Hydrol., 529, 816–829, 2015.

Shi, H., Li, T., Liu, R., Chen, J., Li, J., Zhang, A., and Wang, G.: A service-oriented architecture for ensemble flood forecast from numerical weather prediction, J. Hydrol., 527, 933–942, 2015.

Taiwan Typhoon and Flood Research Institute (TTFRI): TAPEX's rainfall forecasts for typhoons during 2013 to 2015, retrieved from: https://tf-rat.narlabs.org.tw/member/login/?next=/, 2016.

Thielen, J., Bartholmes, J., Ramos, M.-H., and de Roo, A.: The European Flood Alert System – Part 1: Concept and development, Hydrol. Earth Syst. Sci., 13, 125–140, doi:10.5194/hess-13-125-2009, 2009.

Water Resources Agency, Ministry of Economic Affairs, Taiwan (WRA): Rainfall thresholds for townships in Taiwan, retrieved from: http://fhy.wra.gov.tw/Pub_Web_2011/Page/SurveyStationInfos.aspx, 2016.

Wilks, D.: Statistical Methods in the Atmospheric Sciences, 2nd Edn. Elsevier, Burlington, 2005.

Wu, D.: Rainfall threshold values for flood inundation, Workshop on Flood and Drought Prevention and Mitigation, Taipei, Taiwan, 12 April 2012, 2013.

Wu, D. and Wang, Y.: A research on using critical precipitation value for taiwan inundation warning system, in: Proceedings of the 2009 Conference for Disaster Management in Taiwan, Taipei, Taiwan, 19 December, 2009.

Wu, S., Hsu, C., Lien, H., and Chang, C.: Modeling the effect of uncertainties in rainfall characteristics on flash flood warning based on rainfall thresholds, Nat. Hazards, 75, 1677–1711, 2015.

Yang, T., Yang, S., Ho, J., Lin, G., Hwang, G., and Lee, C.: Flash flood warnings using the ensemble precipitation forecasting technique: a case study on forecasting floods in Taiwan caused by typhoons, J. Hydrol., 520, 367–378, 2015.

Zhang, Y., Reed, S., Gourley, J. J., Cosgrove, B., Kitzmiller, D., Seo, D., and Cifelli, R.: The impacts of climatological adjustment of quantitative precipitation estimates on the accuracy of flash flood detection, J. Hydrol., 541, 387–400, 2015.

7

Framework for developing hybrid process-driven, artificial neural network and regression models for salinity prediction in river systems

Jason M. Hunter[1], **Holger R. Maier**[1], **Matthew S. Gibbs**[1,2], **Eloise R. Foale**[1], **Naomi A. Grosvenor**[1], **Nathan P. Harders**[1], **and Tahali C. Kikuchi-Miller**[1]

[1]School of Civil, Environmental and Mining Engineering, The University of Adelaide, Adelaide, 5005 SA, Australia
[2]Department of Environment and Water, GPO Box 2384, Adelaide, 5001 SA, Australia

Correspondence: Jason M. Hunter (jason.hunter@adelaide.edu.au)

Abstract. Salinity modelling in river systems is complicated by a number of processes, including in-stream salt transport and various mechanisms of saline accession that vary dynamically as a function of water level and flow, often at different temporal scales. Traditionally, salinity models in rivers have either been process- or data-driven. The primary problem with process-based models is that in many instances, not all of the underlying processes are fully understood or able to be represented mathematically. There are also often insufficient historical data to support model development. The major limitation of data-driven models, such as artificial neural networks (ANNs) in comparison, is that they provide limited system understanding and are generally not able to be used to inform management decisions targeting specific processes, as different processes are generally modelled implicitly. In order to overcome these limitations, a generic framework for developing hybrid process and data-driven models of salinity in river systems is introduced and applied in this paper. As part of the approach, the most suitable sub-models are developed for each sub-process affecting salinity at the location of interest based on consideration of model purpose, the degree of process understanding and data availability, which are then combined to form the hybrid model. The approach is applied to a 46 km reach of the Murray River in South Australia, which is affected by high levels of salinity. In this reach, the major processes affecting salinity include in-stream salt transport, accession of saline groundwater along the length of the reach and the flushing of three waterbodies in the floodplain during overbank flows of various magnitudes. Based on trade-offs between the degree of process understanding and data availability, a process-driven model is developed for in-stream salt transport, an ANN model is used to model saline groundwater accession and three linear regression models are used to account for the flushing of the different floodplain storages. The resulting hybrid model performs very well on approximately 3 years of daily validation data, with a Nash–Sutcliffe efficiency (NSE) of 0.89 and a root mean squared error (RMSE) of 12.62 mg L^{-1} (over a range from approximately 50 to 250 mg L^{-1}). Each component of the hybrid model results in noticeable improvements in model performance corresponding to the range of flows for which they are developed. The predictive performance of the hybrid model is significantly better than that of a benchmark process-driven model (NSE = −0.14, RMSE = 41.10 mg L^{-1}, G_{bench} index = 0.90) and slightly better than that of a benchmark data-driven (ANN) model (NSE = 0.83, RMSE = 15.93 mg L^{-1}, G_{bench} index = 0.36). Apart from improved predictive performance, the hybrid model also has advantages over the ANN benchmark model in terms of increased capacity for improving system understanding and greater ability to support management decisions.

1 Introduction

Models are being used increasingly for the management of hydrological systems such as streamflow (e.g. Hsu et al., 2002; Shamseldin et al., 2002; Dessie et al., 2014; Yaseen et al., 2016; Gibbs et al., 2018), reservoir inflow (e.g. Tsai et al., 2014; Gragne et al., 2015; Chang and Tsai, 2016), floods (e.g. Quiroga et al., 2013; Alvarez-Garreton et al., 2015; Kasiviswanathan et al., 2016), baseflow (e.g. Corzo and Solomatine, 2007; Li et al., 2014), water level (e.g. Chang and Chang, 2006; Shiri et al., 2016), groundwater level (Adamowski and Chan, 2011; Mohanty et al., 2015; Chang et al., 2016), evaporation (e.g. Parasuraman et al., 2007; Guo et al., 2016; Kisi and Demir, 2016), stream temperature (e.g. Gallice et al., 2015), ecosystem services and response (e.g. Chang et al., 2013; Duku et al., 2015), raw water quality (Zhang and Stanley, 1997) and a range of other water quality parameters (e.g. Pulido-Velazquez et al., 2015; Kisi and Parmar, 2016), such as suspended sediment (e.g. Mount et al., 2012; Duan et al., 2015), phosphate (e.g. Chang et al., 2016) and salinity (e.g. Maier and Dandy, 1966; Bowden et al., 2005b). Such models can take different forms, ranging from fully process-driven (Habib et al., 2007; Liu et al., 2007), to conceptual (Clark et al., 2011; Fenicia et al., 2011; Kavetski and Fenicia, 2011), to data-driven (Maier and Dandy, 1996; Bowden et al., 2005a). Process-driven models are developed from the known physical process(es) in a system, which are represented mathematically. Conceptual models represent the key elements of a system and the hypothesised relationships between them and data-driven models are developed purely on available data with limited or no knowledge of the physical process represented in the model structure (Maier et al., 2010). As pointed out by Mount et al. (2016), all of these model types are part of a spectrum based on the degree of influence of derived hypothetic knowledge or empirical data in their development. Hypothetic influence can be high if the underlying processes are well understood, as processes can be both described mathematically and incorporated into system models. In contrast, poorly-understood processes may not be able to be described explicitly in mathematical form, so the degree of hypothetic influence is necessarily lower. This is due to the fact that information contained within measured data of the underlying system behaviour has to be relied upon to a greater degree for model development (e.g. for determination of the functional form of the model, as well as model calibration).

Apart from the degree to which underlying process are understood and can be represented mathematically, the most appropriate modelling approach is also a function of the extent to which available data can adequately support model development (Grayson and Blöschl, 2000). For example, the structure of models that are based on well-known underlying processes can be highly complex with a large number of parameters to accommodate existing understanding; therefore these models require significant volumes of data, which may not be available, for calibration. However, it should be noted that system understanding and/or analytical techniques designed to identify dominant processes can be useful in reducing model complexity and the amount of data needed for calibration (e.g. Gibbs et al., 2012; Markstrom et al., 2016). Conversely, and somewhat counter-intuitively, data-driven models might be more suitable in such situations, as they are generally designed to make the best use of existing data. In other words, in more hypothetically influenced models, data requirements are dictated by model structure, which is generally derived based on system understanding. In contrast, in data-driven models, model structure is a function of available data. This means that best use can be made of existing data without the need to collect additional data to meet the requirements of a predetermined model structure, as is generally the case with process-driven models (Mount et al., 2016). However, this has the disadvantage that not all dominant processes might be represented in the resulting model.

The final factor that can affect the suitability of different model types for a particular application is the intended purpose of the model. For example, if the primary purpose of the model is to improve system understanding, a model with a higher degree of hypothetic influence is likely to be of more value, although data-driven models have also been shown to be able to provide some insight into underlying processes (e.g. Jain et al., 2004; Kingtson et al., 2005; Jain and Kumar, 2009; Mount et al., 2013). If the primary purpose of the model is the evaluation of various management options, care needs to be taken that these options are represented as inputs to the model and that the impacts of the management options are represented as model outputs, which can be achieved using a variety of model types. In other words, input–output relationships of the relevant processes need to be included in the model, but this can be achieved with or without explicit modelling of the underlying physical processes. Finally, if the primary purpose of the model is forecasting, explicit representation of the underlying processes is generally less important when compared with model accuracy. If a purely data-driven model is used, however, care needs to be taken to ensure that the model is updated when faced with input patterns that lie outside those used for initial model calibration (see Bowden et al., 2012; Zheng et al., 2018).

When dealing with complex integrated hydrological systems, overall system behaviour is likely to be affected by a number of different sub-processes. While some of these might be well understood and supported by sufficient data to enable them to be modelled explicitly, others might not. Consequently, a model that falls on a single point of the hypothetically-driven – data-driven hydrological modelling spectrum might not be best suited to addressing the problem under consideration. In order to allow the relative strengths of different types of models on the hydrological modelling spectrum to be utilised fully, the use of hybrid models has been suggested (Corzo and Solomatine, 2007; Corzo et al., 2009; Robertson and Sharp, 2013; Mount et al., 2016; Humphrey et al., 2016). Such models combine modelling approaches that fall on different points of the modelling spectrum to enable the most appropriate degree of hypothetic and data influence to be utilised in the modelling of each sub-process.

While a number of studies have illustrated the potential benefit of hybrid models, they have generally been confined to rainfall–runoff/streamflow modelling (e.g. Hsu et al., 2002; Wang et al., 2006; Noori and Kalin 2016; Zhang et al., 2016). In addition, these studies have focused on a particular hybrid model structure, rather than a generic framework that can be used to develop the most suitable hybrid model in different settings. In order to overcome the above shortcomings, the objectives of this paper are as follows:

- To introduce a generic, high level conceptual framework for the development of hybrid models for modelling salinity in river systems that uses a combination of model purpose, knowledge of underlying system processes and the type and amount of available data to provide guidance for the selection of the suitable sub-models. This should thereby enable the most appropriate balance between hypothetic and data influence to be struck in model development. While the proposed approach is specific to salinity modelling, the underlying principles presented are likely to be more widely applicable. The modelling of salinity in river systems is selected as the focus of the approach for the following reasons:

  a. High levels of salinity are a major concern in many river systems around the world (Rengasamy, 2006), due to their potentially detrimental impacts on the growth of agricultural crops, vegetation, bacteria and algae (Hart et al., 1991; Maier and Dandy, 1996) and adverse effects on water quality, as well as the stability of freshwater and neighbouring ecosystems. In addition, high salinity levels can have significant negative financial consequences stemming from the ongoing expense of treating drinking water, pumping at groundwater interception wells and from diminishing agricultural returns (Moxey, 2012).

  b. Salinity in river systems is generally affected by a number of complex processes (Williams, 2001; Goss, 2003), and the degree to which these processes are understood varies significantly (Maier and Dandy 1996; Woods, 2015). For example, there is generally a good understanding of the processes involved in the transport of salt with discharge, as salt is a conservative constituent. However, understanding of the complex processes associated with the accession of additional salt loads into the main river channel is often limited, as they are generally influenced by multiple interacting factors (e.g. land use, historical inundation regime, surface water-groundwater interactions, abstraction/recharge processes). In addition, the data needed to support the development of different types of models is highly variable.

  c. Current efforts directed towards the modelling of salinity in river systems has generally relied on either process-driven (Banerjee et al., 2011; Habib et al., 2007; Liu et al., 2007; Woods, 2015) or data-driven (Maier and Dandy, 1996; Huang and Foo, 2002; Suen and Lai, 2013; Bowden et al., 2005a; Bowden et al., 2002; Kingston et al., 2005; Rath et al., 2017) approaches. This has resulted in a number of limitations, such as difficulties in modelling the accession of salt via groundwater, wetlands and floodplains (e.g. groundwater regime shifts and flushing) explicitly (Harrington et al., 2006), which in turn makes it difficult to understand the relative importance of the different sources of saline accessions and to assess the potential utility of some of the management options mentioned earlier.

  d. To illustrate the application and test the utility of the framework by applying it to a reach of the Murray River in South Australia, as this is an area where improved salinity modelling for management purposes would be of significant benefit (Beecham et al., 2003).

The remainder of this paper is organised as follows: the proposed framework providing guidance for developing hybrid salinity models for river systems is introduced in Sect. 2, followed by the application of the approach to a case study in a South Australian reach of the Murray River in Sect. 3. Details of the development of the relevant models are given in Sect. 4, while the corresponding results are presented and discussed in Sect. 5. A summary and conclusions are provided in Sect. 6.

2 Proposed framework

2.1 Overview

An outline of the proposed high level conceptual framework for supporting the development of hybrid models of salinity in complex river systems is given in Fig. 1. As can be seen, a hybrid model that consists of a number of sub-models representing the different processes affecting river salinity (e.g. in-channel salt transport, overland flow, flushing of anabranches and saline groundwater accession) is suggested. The key feature of the proposed framework is that it uses a combination of model purpose, process understanding, and data availability and suitability to identify the most appropriate sub-models for the various processes under consideration. This includes the degree to which a particular process can be described mathematically and the suitability of the available data for the development of different types of models (e.g. the degree to which the available data are able to support the development of a process-driven model, even when the underlying processes are well understood and can be described

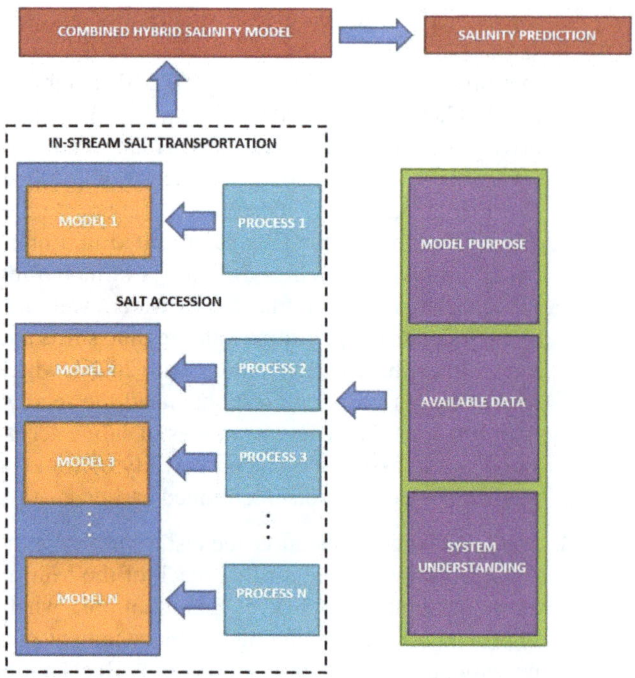

Figure 1. Conceptualisation of proposed hybrid modelling approach.

mathematically). This enables the models representing the various sub-processes to be developed by considering the degree of hypothetic and data influence that is most appropriate, thereby tailoring the model development process to the system under consideration. Given the conceptual nature of the framework, it provides high-level guidance and there is some subjectivity in its application to a particular case study. Details of the various components of the proposed framework are given in the following sub-sections.

2.2 Identification of relevant sub-processes

In order to apply the proposed approach, a suitable conceptual understanding of the processes that affect salinity in the system under consideration is required. In general, a distinction can be made between in-stream salt transport and the accession/addition of salt into the main channel via a variety of mechanisms that are affected by whether the river is operating under normal or flood conditions (Tefler et al., 2012), as illustrated in Fig. 2. The transport of in-stream salt occurs along the main channel, from the top of the figure flowing through the river toward the bottom. Under normal conditions, the only potential source of salt accession is generally via the inflow of saline groundwater, provided the stream is gaining (i.e. if the height of the water table is above the river level) (Fig. 3). However, during these conditions, salt from the groundwater store can also mobilise into various floodplain elements, such as wetlands and anabranches. In addition, the salt load in these floodplain elements increases as a result of evapo-concentration. Under flood conditions (Fig. 2), saline inflow from groundwater is likely to cease, as the river level is likely to be higher than the water table, resulting in a reversal of the flow direction (Fig. 3). Consequently, instead of the river gaining water (and salt) from groundwater, there will be a loss of fresh water from the river to recharge the groundwater system.

However, wetlands and anabranches that are disconnected under normal flow conditions may connect to the main river channel under flood conditions, adding salt that has been building up in these systems since the last flood when water levels recede. The amount of salt added is a function of the magnitude of the flood and the time and conditions (e.g. degree of evapo-concentration) since the occurrence of the last flood. Broader inundation of the floodplain results in recharge to the groundwater system, and leads to increased flux, and hence salt load, from the groundwater system to the river once the river level returns to normal conditions. As part of the proposed approach, all of the specific sub-processes that contribute to salinity for the case study under consideration need to be identified.

2.3 Identification of the most suitable model types

As part of this step, the model types that are most suitable for modelling the relevant sub-processes identified in the previous step are determined based on joint consideration of model purpose, system understanding and data availability and suitability (Fig. 1). Model purpose has an influence on which of the relevant processes have to be modelled explicitly. For example, if the overall purpose of the modelling exercise is to obtain salinity forecasts at downstream locations, there might not be a need to model all contributing processes explicitly (Maier and Dandy, 1996). In contrast, if the purpose of the model is to gain increased system understanding or to enable the impacts of various salt management options to be considered, all sub-processes will most likely have to be modelled explicitly. This also has an impact on which potential model inputs are considered. For example, if forecasting is the primary model purpose, auto-regressive values of the model output should be considered as potential inputs (e.g. Bowden et al., 2005b), as this is likely to improve the quality of the forecasts. In contrast, if the purpose of the model is to assess the impact of different management options on salinity, auto-regressive values of the model output cannot be considered as potential model inputs, as the model output has to be independent of the model input(s) in such cases.

Once the processes for which sub-models are required have been identified, the most appropriate model type has to be selected for each of these. As previously mentioned, this requires an appropriate balance between hypothetic and data influence. The degree to which the selected processes are understood and can be described mathematically can be highly variable, as can the state of the available data to support different modelling approaches. For example, the trans-

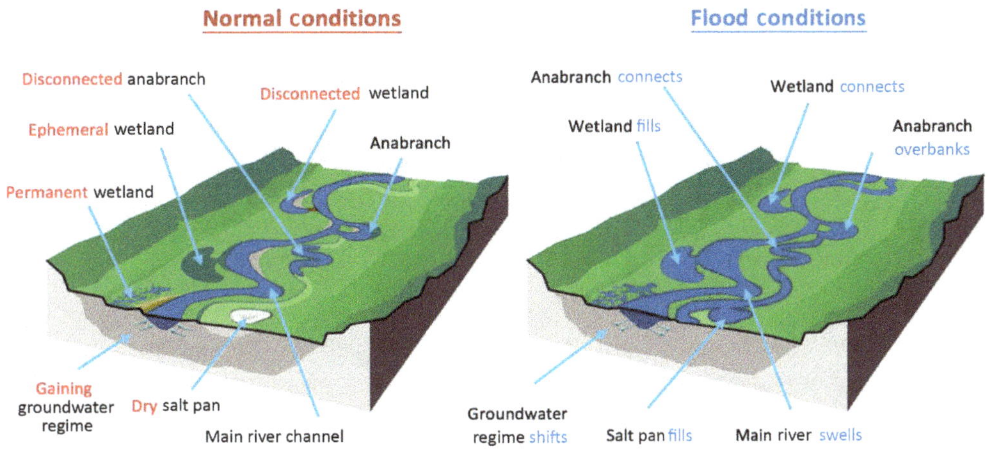

Figure 2. Processes affecting saline accessions during normal and flood conditions.

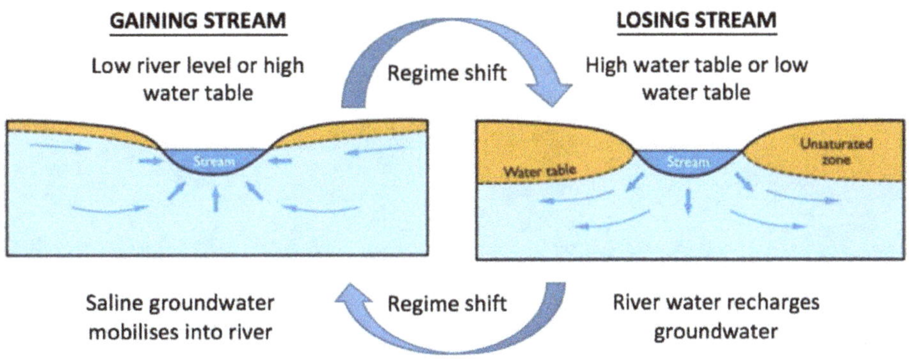

Figure 3. Processes affecting saline groundwater accessions for gaining and losing streams.

port of a conservative constituent with discharge, i.e. the process of in-stream salt transport, is generally well understood and requires relatively little data to be modelled explicitly, as the main processes consist of flow routing and storage. Consequently, the use of process models might be most appropriate. However, the same is unlikely to be true when modelling different processes of salt accession, as these are generally more complex, site specific (depending on soil types and groundwater conditions) and less well understood, making more hypothetically-influenced models an attractive alternative. If sufficient data are available, the use of universal function approximators, such as artificial neural networks (ANNs), might be best (Mount et al., 2016). However, a scarcity of data representing rare events, such as the large flood events that might be required to flush the salt stores in the wetlands and anabranches adjacent to the main river channel, might make models with a smaller number of parameters, such as regression, a better option.

It is important to note that the proposed framework is conceptual in nature and designed to provide high-level guidance. Consequently, its implementation for particular case studies is subjective. For example, how much data is required to support a particular modelling approach is case study dependent and relies on the judgement of the model developer. Consequently, this stage of the process may be iterative. Following the application of the developed sub-models, results may assist in identifying limitations in understanding or missing information.

Hence, understanding gained from the application of the sub-model combinations developed may be applied to optimise the number and type(s) of sub-models included in the final hybrid model.

2.4 Development of required sub-models and hybrid model

Once the most appropriate model types have been determined for each of the sub-processes to be modelled explicitly, the corresponding models have to be developed. This process should follow state-of-the-art approaches for the development of the different types of models (see Jakeman et al., 2006; Maier et al., 2010; Wu et al., 2014; Hamilton et al., 2015; Galelli et al., 2014; Humphrey et al., 2017). Finally, the various sub-models have to be combined to develop the desired hybrid model (Kelly et al., 2013).

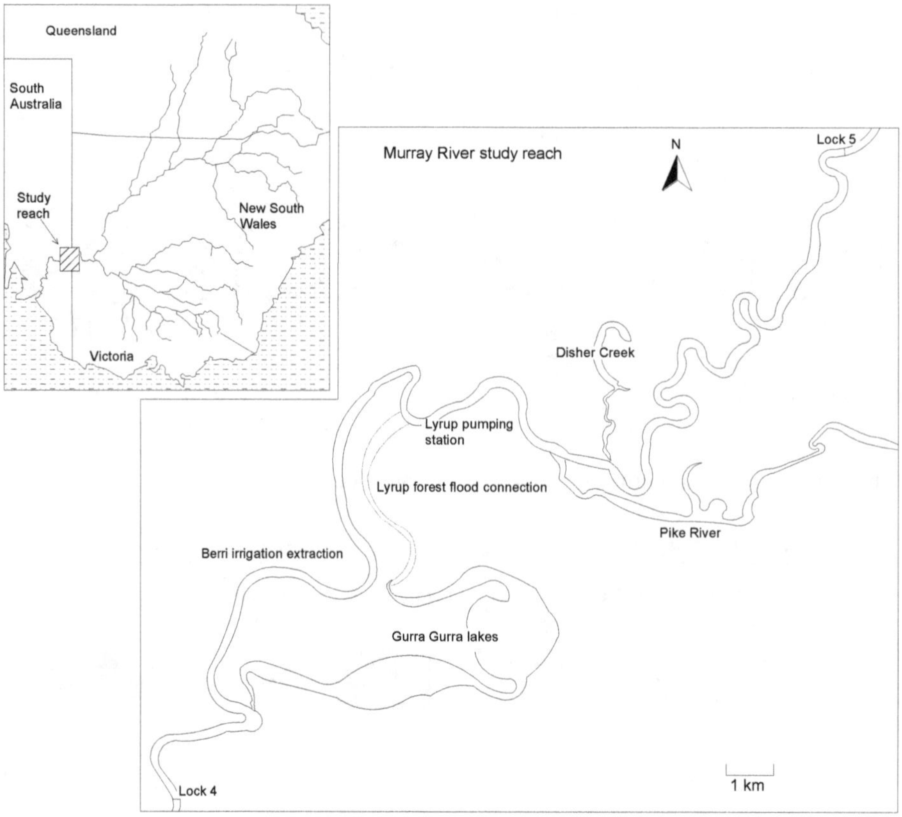

Figure 4. Map of case study reach.

3 Case study

3.1 Background

In order to illustrate and test the utility of the conceptual high-level framework introduced in Sect. 2, it is applied to a case study of the 46 km reach between Lock 5 and Lock 4 on the Murray River in South Australia (Fig. 4). This particular location is chosen due to the following:

- The reach is underlain by highly saline regional groundwater systems that provide significant salt accession into the river along this reach. As such, there are substantial saline accessions in this reach that are not well understood.
- This reach exemplifies a range of processes that are known to facilitate salt accession into a reach of a river, such as groundwater gain, inflow from streams, creeks and wetlands, and in-channel transport.
- Conditions along this reach have been monitored for many years, so there are suitable datasets for model development.
- The reach has had minimal changes in salinity management over time, so external influences on the underlying processes represented by the historical data are minimal.
- A number of changes are occurring in the Murray River system that will result in changes in the flow and inundation regime:

  a. The Murray–Darling Basin Plan, as per parts 1A and 2 of the Commonwealth Water Act (2007), will return some water previously allocated to consumptive use to the environment, with the aim of increasing ecosystem health through processes such as increased frequency of inundation. The ability to predict the effects on salinity resulting from these changes is currently very limited.

  b. Large-scale floodplain regulators are proposed in this reach to further increase inundation frequency to improve ecosystem health. A greater understanding of, and the ability to, predict salt accession under different conditions along this reach will improve the ability to manage and operate this infrastructure once constructed.

3.2 Identification of relevant sub-processes

The main processes affecting salinity in the reach of interest include in-channel salt transport, as well as accession of salt via groundwater and flushing of several large floodplains and wetlands. There are also several anabranches and backwaters, which include Pike River, the Gurra Gurra lakes and Disher Creek (Fig. 4). In-channel salt transport is primarily driven by advection with flow in the main river channel, as well as the storage mixing volume behind the lock, which may change as water levels are manipulated.

Saline accession via groundwater is a function of the relative levels of the groundwater adjacent to the river channel and the water levels in the main river channel. As discussed in Sect. 2.2, if the water levels in the main river channel are below the adjacent groundwater levels, groundwater flows into the main river. As the salinity of the groundwater that flows into the main river channel in the reach of interest is very high (up to 70 000 $\text{mg}\,\text{L}^{-1}$) (Barnett, 2007), this can be a significant source of saline accession. Analysis of the available data from groundwater wells (WaterConnect, 2016) showed that groundwater levels are relatively constant and that saline groundwater accession occurs at flows below approximately 40 000 $\text{ML}\,\text{day}^{-1}$, after which overbank flow commences. This is useful in that it demonstrates a link between groundwater accession and river flow rate, and implies that models of such processes are unhelpful at higher flows.

There are three large salt sources that connect to the river at different flow rates: Disher Creek, Gurra Gurra lakes and the Pike River. Disher Creek is an evaporation basin for irrigation runoff and salt interception scheme flows. Water is held in the basin to be disposed of through evaporation or, if this is not sufficient, water can be pumped to the Noora drainage disposal basin 20 km away from the river. The management rules for Disher Creek allow for the release of water from the evaporation basin to the river at flows greater than 15 000 $\text{ML}\,\text{day}^{-1}$, which is assumed to be sufficient to dilute the highly saline releases from the basin. It should be noted that releases are not always triggered at this flow, and the release rate can be modified depending on the measured salinity in the river at the time.

At lower river flows, the Gurra Gurra lakes are a terminal wetland system, with one connection to the river. Under these conditions, evaporation removes water from the wetland, which is then replaced from the river, and this process results in naturally higher salinities than in the main river channel. At flows of approximately 30 000 $\text{ML}\,\text{day}^{-1}$ the water level in the river rises high enough to connect the flow path to the north of the lakes (through Lyrup Forest), resulting in a through-flowing system. When this is the case, the lakes are flushed, and the saline water from the wetlands contributes to the main river channel, thus increasing its salt content.

The Pike River is an anabranch that loops around Lock 5 and the upper area of the reach. The lock provides a 3 m head difference, and due to this, the Pike River can flow around the lock and back into the river below it. There are two inlets to the Pike River above Lock 5, which are both regulated; this was historically to supply flows for irrigation purposes. At flows approaching 50 000 $\text{ML}\,\text{day}^{-1}$, the main river channel begins to overflow, and connects a number of temporary flow paths to the Pike River, resulting in wash-off and transport of salt that may have been deposited on the floodplain in that locality. This flow of 50 000 $\text{ML}\,\text{day}^{-1}$ is representative of when overbank flows start to occur along the reach between locks 4 and 5 (and the lower Murray River more broadly), where this process occurs along the river, as well as the longer-term process of recharge to groundwater and increased flux once river levels recede.

It should be noted that the amount of salt flushed into the main river channel from these systems is difficult to predict, as it is not only a function of flow, but also of the time between flushing events, the duration of an event and the nature of the flow regime at the time. Moreover, the accession processes are very different at different flow regimes. For example, at 15 000 $\text{ML}\,\text{day}^{-1}$, the salt contribution from Disher Creek may be represented by a point load into the main channel. However, at flows greater than 40 000 $\text{ML}\,\text{day}^{-1}$, Disher Creek disappears entirely from the map, as it is swallowed by an extensive floodplain. When this occurs, the creek does not contribute any additional salt to the system: it has already been entirely flushed and the only water in the area is flowing downstream as part of the flooded main river channel.

3.3 Identification of relevant sub-models

The purpose of the hybrid model is to quantify salinity responses to proposed managerial changes to the flow and inundation regime in the reach of the river under consideration under the Murray–Darling Basin Plan. These changes will be enacted by the construction of additional control structures, and by selective releases or routing of volumetric flow. The salinity response to such changes is of interest in that the consequences of poor water quality can be high, and the modelling of different processes of accession is poorly understood. There is consequently value in identifying and modelling the main processes of accession separately, so that future management may determine the best locations of control options in addition to assessing the magnitude of their effects on salinity.

The data available for model development are shown in Table 1, including flow, temperature, stage height and salinity, which are measured at various points along the reach of interest. All historical data are available as daily readings. The recording and management of these data are undertaken by the South Australian Department of Water, Environmental and Natural Resources (DEW), largely funded through the Murray–Darling Basin Authority (MDBA). The time periods for which data are available vary, depending on when a measuring station was constructed, and when the instru-

Table 1. Details of available model data. All data are recorded at a daily resolution and have been sourced from DEW (Department of Environment, Water and Natural Resources).

Parameter	Location (Station Number)	Time Period
Flow rate (ML day^{-1})	Lock 5 (downstream, A4260513)	23 Jan 1981–25 May 2016
Flow rate (ML day^{-1})	Lock 4 (downstream, A260515)	1 Jul 1983–30 Jun 2012
Flow rate (ML day^{-1})	Lyrup pumping station (A4260663)	12 Nov 1993–18 Jun 2017
Temperature (°C)	Berri irrigation extraction (A4260537)	29 Mar 2001–2 Aug 2016
Temperature (°C)	Pike River outlet (downstream, A260645)	26 Sep 1991–18 Jun 2017
Salinity (mg^{-1})	Lock 5 (upstream, A4260512)	4 Jul 1972–1 May 2013
Salinity (mg L^{-1})	Lock 4 (upstream, A1260514)	18 Jan 1994–20 Mar 2017
Salinity (mg L^{-1})	Pike River outlet (downstream, A260645)	26 Sep 1991–18 Jun 2017
Salinity (mg L^{-1})	Berri irrigation extraction (A4260537)	17 Oct 1942–20 Mar 2017
Water level (m)	Lock 5 (downstream, A4260513)	1 Apr 1924–1 May 2013
Water level (m)	Lyrup pumping station (A4260663)	11 Nov 1993–18 Jun 2017
Water level (m)	Lock 4 (upstream, A1260514)	1 Apr 1927–1 May 2013
Water level (m)	Berri irrigation extraction (A4260537)	1 Jan 1974–20 Mar 2017
Salt load (kg day^{-1})	Lock 4 (downstream, A260515)	1 Jul 1983–30 Jun 2012

ments to measure certain parameters were commissioned. Some datasets are deemed not suitable for model development and are therefore excluded from this study, often due to short data records of only a few years in length (e.g. measurements taken at the mouth of Gurra Gurra lakes). Key locations include the two locks that define the extent of the reach considered, as well as the Pike River anabranch. Berri river extraction and Lyrup pumping station, which are downstream of the Pike River, also provide useful information for salt transport and accession along the reach.

As discussed in Sect. 2.3, which modelling approach is most suitable for a particular process is a combination of the degree of process understanding and data availability. The relative degree with which these two factors are satisfied for the processes to be modelled in this case study (see Sect. 3.2), based on a subjective assessment of available information, is summarised in Table 2. As can be seen, the processes affecting in-stream salt transport are well understood, able to be represented mathematically and supported by sufficient data to enable a process-driven model to be developed. However, the processes associated with the various modes of saline accession are not considered to be well understood, making data-driven models the best option. In relation to groundwater accession, the degree of available data is high, as this occurs during non-flood events and relevant data are measured daily. Consequently, an artificial neural network is considered the most appropriate modelling approach due to its universal function approximation ability and its successful application to the prediction of salinity in the Murray River in previous studies (e.g. Maier and Dandy, 1996). However, as the saline accessions corresponding to overbank flow and flushing only occur during flood events, which occur infrequently, the data available on these processes is considered insufficient to support the development of a model with a potentially large number of parameters, such as an artificial neural network. Instead, a linear regression model is considered most appropriate to represent these processes due to the combination of a low degree of process understanding and a low degree of data availability.

A conceptual representation of the resulting hybrid model is given in Fig. 5. As can be seen, the models corresponding to the four main sub-processes associated with saline accession identified in Sect. 3.2 are conceptualised as being applicable at different flow rates. This was determined based on a preliminary analysis of the available data, with one model being applied to only one discrete bracket of flow. As a result, while each model primarily represents the process with which it is associated, it might also represent other processes that occur during the range of flows for which each model is developed. As shown in Fig. 5, the four models of saline accession (models 2 to 5) feed into the in-stream salt transport model (model 1).

3.4 Development of required sub-models and hybrid model

In this section, details of the development of the five submodels (Fig. 5) are given. The modelling data and performance metrics are described first, as they are common to all models, followed by details on the development of the three different modelling types (i.e. process-driven (model 1), artificial neural network (model 2) and linear regression (models 3 to 5)). This is followed by details of the model development process for each of these model types, as well as how they are combined to form the hybrid model.

3.4.1 Model development data

The quality of the available data (see Sect. 3.1) is checked by visual inspection and a number of small periods of missing

Table 2. Identified processes and the degree of data and understanding that are available for each.

Process	Degree of data availability	Degree of process understanding	Selected model type
In-stream salt transport	Medium	High	Process-driven
Overbank flow & flushing of Pike River	Low	Low	Linear regression
Overbank flow & flushing of Gurra Gurra lakes	Low	Low	Linear regression
Overbank flow & flushing of Disher Creek	Low	Low	Linear regression
Groundwater accession	High	Low	Artificial neural network

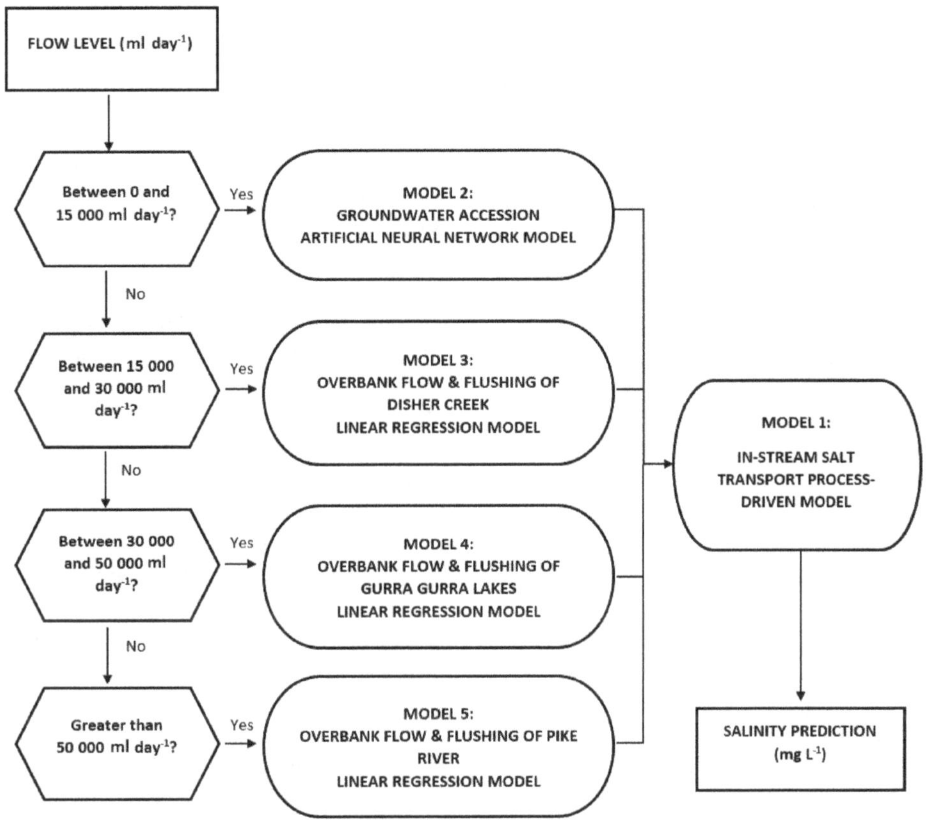

Figure 5. Conceptual representation of components of hybrid model and how they are connected.

data (1–3 days) are filled in using linear interpolation, which is common practice for gaps in data of this duration and is unlikely to result in any significant loss of information (Kornelson and Coulibaly, 2014). Two longer periods of missing flow data at Lock 5 in the period 2011–2012 (132 and 72 days, respectively) are filled in by correlation with corresponding water level data. To ensure consistency between sub-models, the longest common period of available data is used for the development of all models, which is from 18 January 1994 to 30 June 2012. The available data are split so that the first 80 % (i.e. from 18 January 1994 to 21 October 2008) are used for model calibration and the subsequent 20 % (i.e. from 22 October 2008 to 30 June 2012) are used for validation for all models. It should be noted that a regional drought event from 2001 to 2010, which is reflected in almost a decade of low flows (< 15 000 ML day^{-1}), is the most significant unusual feature in the data and is purposely split between both calibration and validation datasets. To ensure all inputs into the ANN and regression models span the same ranges and can thus be combined during the modelling process, all data are standardised to have a mean of zero and a standard deviation of one, as per Eq. (1).

$$y = \frac{(x - \bar{x})}{\sigma}, \tag{1}$$

where y is the standardised input (dimensionless), x is the original input (with units), $\bar{x}$ is the mean of original inputs and σ is the standard deviation of original inputs.

3.4.2 Model performance assessment

All data are calibrated and validated against the salinity at Lock 4. The root mean squared error (RMSE, Eq. 2) and the Nash–Sutcliffe Efficiency (NSE, Eq. 3) are used as metrics to judge the fit of the predicted variables to the observed data when calibrating the parameters of the models:

$$\mathrm{RMSE} = \sqrt{\frac{1}{n}\sum_{i=1}^{n}\left(x_{\mathrm{m}}^{i} - x_{\mathrm{o}}^{i}\right)^2} \tag{2}$$

$$\mathrm{NSE} = 1 - \frac{\sum_{i=1}^{n}\left(x_{\mathrm{m}}^{i} - x_{\mathrm{o}}^{i}\right)^2}{\sum_{i=1}^{n}\left(x_{\mathrm{o}}^{i} - \bar{x}_{\mathrm{o}}\right)^2}, \tag{3}$$

where n is the number of points in the series, x_{m} is modelled points, x_{o} is observed points and $\bar{x}_{\mathrm{o}}$ is the mean of observed points.

The goodness of fit of the hybrid model is evaluated against two benchmark models, one process-driven and one data-driven, with the metric G_{bench} introduced in Seibert (2001) (Eq. 4):

$$G_{\mathrm{bench}} = 1 - \frac{\sum_{i=1}^{n}\left(x_{\mathrm{o}}^{i} - x_{\mathrm{m}}^{i}\right)^2}{\sum_{i=1}^{n}\left(x_{\mathrm{o}}^{i} - x_{\mathrm{b}}^{i}\right)^2}, \tag{4}$$

where x_{b} are the benchmark model data points.

The G_{bench} index is structured similar to the NSE, but replaces the mean of the observed time series with the time series of a benchmark model: therefore an index of zero indicates that model performance is equal to that of the benchmark model, negative values indicate that the performance of the model under consideration is inferior to that of the benchmark model and positive values indicate the opposite.

3.4.3 Process-driven salt transport sub-model (model 1)

The purpose of this model is to simulate the in-stream transport of salt from Lock 5 to Lock 4, thereby predicting salinity at Lock 4 as a function of the upstream salt load, without considering saline accessions due to groundwater inflow or the flushing of anabranches and backwaters along the reach. The model is developed using eWater Source (Welsh et al., 2013). Routing is represented using a piecewise linear lookup table, where the travel times for key flow rates are calculated based on travel times of flow peaks in the historical record. A dead storage volume is used to represent the mixing time for salinity, as the travel time for solutes is much slower than the wave celerity travel time resulting from the analysis of flow peaks. The routing or transportation method is a fully mixed water quality constituent. This technique has been used since the 1970s in various water quality models and hence is well developed. The key principles are that the mass balance of the modelled constituents (e.g. salt) is maintained in all divisions of all links (which represent a reach). Calculations take place for every time step, which is daily.

Development of the travel times and dead storage volumes for all areas of the river were calibrated in a previous study, as outlined in MDBC (2002). The salt constituent volumes in the upstream reaches were also determined as part of this previous work, however, the salt accession within the study reach itself is considered as part of the calibration process. As this is a process-driven model, the required inputs are predetermined due to the mathematical specification of the model. These include upstream flow and salinity at Lock 5, and knowledge of the physical characteristics of the reach, such as its total length and the location of various extraction points (e.g. Lyrup pumping station and Berri irrigation extraction). The transport model routes the upstream salt through the reach down to Lock 4, and converts it into salinity by multiplying the salt load with the rate of flow. The difference between this transported salinity and the measured salinity at Lock 4 (i.e. the residual salinity of model 1) represents the salt that is gained by the river due to the accession processes that occur within the reach itself (Sect. 3.2), and is the salinity that is predicted by the remaining hybrid component models (i.e. models 2, 3, 4 & 5).

3.4.4 Artificial neural network (ANN) groundwater accession model (model 2)

The purpose of this model is to predict accession of salt for flows that are less than 15 000 ML day^{-1}, which is sourced primarily from groundwater inflows and small increases in water level at the upstream end of the reach between locks 5 and 4. These tend to occur due to the same mechanisms along large stretches of the river during a single dry or wet event. However, their response times can be vastly different, ranging from days for responses to changing water levels, to months for the slower responses to saline groundwater accessions.

The model is implemented using the "Validann" R package (Humphrey et al., 2017) using the steps in the ANN model development process outlined in Maier et al. (2010) and Wu et al. (2014). Potential input variables considered include salinity, temperature, water level and flow rates, measured at various time lags and locations along the length of the reach (Table 1). It should be noted that past values of the model output are not considered as potential model inputs, as the purpose of the model is to assess the impact of different management options on river salinity rather than forecasting future salinity values (see Sect. 2.3). The model output is the residual salinity between the process-driven in-channel salt transport model (model 1) and the measured salinity at Lock 4, for the days for which the measured flow at Lock 5 is less than 15 000 ML day^{-1}. Appropriate model inputs are determined with the aid of correlation analyses between potential model inputs and the model output, resulting in the selection of two inputs: (1) salinity at Lock 5, and (2) the water level measured at Lyrup pumping station (Fig. 6), which are lagged by 5 and 3 days, respectively. The selected inputs reflect that saline accessions at low flows are primarily driven by upstream salinities (primarily affecting water level) and water levels (primarily affecting groundwater inflow).

Multilayer perceptrons (MLPs) are used as the model architecture, as this is by far the most commonly used architecture in ANN applications in hydrology and water resources (Maier et al., 2010; Wu et al., 2014). Different combinations of standard activation functions (linear, sigmoidal and hyperbolic tangent) are trialled on the calibration data, resulting in the selection of the hyperbolic tangent function applied to the single hidden layer and the linear function being applied to the output layer. An ANN with three hidden nodes performs best on the calibration data, based on trials with one to four hidden nodes. The ANN is fully connected, so there are nine weighted connections (i.e. nine parameters to calibrate). The development data contain 5391 points (i.e. 5391 days over the calibration period), so there are almost 600 data points available to calibrate each parameter, making overfitting unlikely. These parameters are calibrated using the Broyden–Fletcher–Goldfarb–Shanno (BFGS) algorithm, as this method generally performs well when simulating hydrological phenomena (Zounemat-Kermani et al., 2016). Optimal values of the learning rate and momentum are obtained using trial and error on the calibration dataset, and the initial weights are selected randomly on the range [−0.5,0.5]. To test against overfitting, the model is validated against the residual salinity between the transport model (model 1) and the measured salinity at Lock 4, over the period 22 October 2008 to 30 June 2012 for all data points corresponding to flow rates at Lock 5 of less than 15 000 ML day^{-1}.

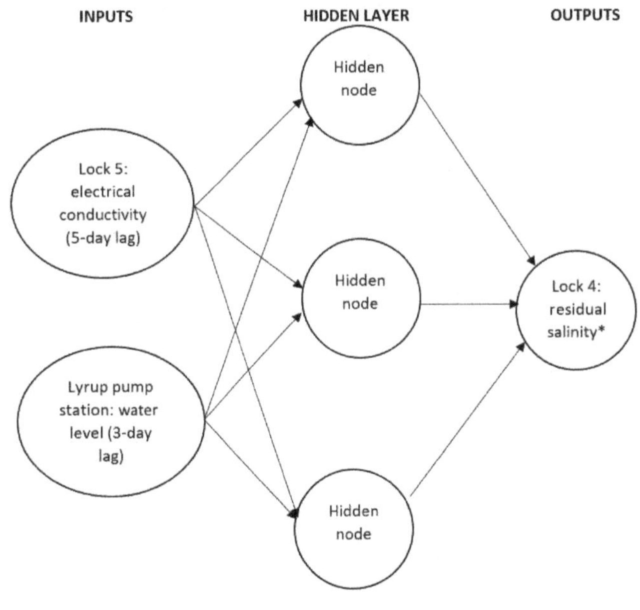

Figure 6. Structure and inputs of ANN model for predicting saline groundwater accession as part of the hybrid model (model 2). The residual salinity output is the difference between the measured salinity and the output from the process-driven salt transport model for all flow rates that are less than 15 000 ML day^{-1}.

3.4.5 Linear regression models for Disher Creek, Pike River and Gurra Gurra lakes accession (models 3 to 5)

The purpose of the linear regression models (models 3 to 5, Fig. 5) is to predict accessions of salt to the river at flows in excess of 15 000 ML day^{-1}. These flows are dominated by Disher Creek and Pike River at a point downstream from Lock 5 (see Fig. 4), and the inflow from the Gurra Gurra lakes at a point upstream from Lock 4 (Fig. 4); they are also likely to include saline accessions from groundwater and increases in water level at flows closer to 15 000 ML day^{-1}. The data in Table 1 are available as potential inputs for model development, with the most relevant inputs determined by trial and error during the calibration process. Because of the scarcity of peak flow data and the complexity of the physical processes being modelled, it is valuable to incorporate as much process understanding into these models as possible. This is achieved by ensuring that some of the more important known process drivers, such as peak duration, time since last peak flow and historical flow volumes, are represented as potential model inputs, as follows:

- The 5-year historical volume of water (ML) is extracted from the daily flow rate measurements at Lock 5 (Q_2) because residual waters from receding historical peaks or overbank flows can create concentrated pockets of salt once the water has evaporated, which is then available to be accessed by the next overbank flow.

- The duration of the peak flow event (D) is extracted as a count of days, which begins incrementing at the commencement of a peak flow event. This is due to the fact that longer floods allow the extended floodplain more time to connect with groundwater aquifers at a distance from the river, and to react with the salt content of the soil.

- The time since last peak flow (T) is extracted as a count of days, which begins incrementing when the daily flow rate falls below the defined peak flow. This is because a longer time since the last peak allows for a greater amount of saline groundwater to seep into depressions and shallow reservoirs, which may be some distance from the main channel, thereby increasing the amount of salt available for overbank flow.

For the purpose of this work, a peak flow event begins when the flow rate at Lock 5 is larger than the lower flow bound of an individual model. For example, a peak flow event for model 3 is any daily flow that is 15 000 ML day^{-1} or higher, while a peak flow event for model 5 is any daily flow that is 50 000 ML day^{-1} or higher.

The models are developed in Microsoft Excel, using the Solver function (i.e. a gradient method) to optimise the coefficients from a range of starting positions and to minimise the chance of identifying locally optimal parameter values. The models are calibrated against the residual salinity at Lock 4 obtained from the process-driven in-stream salt transport model (model 1), over the time period from 18 January 1994 to 21 October 2008, using the NSE as the objective function. Data from 22 October 2008 to 30 June 2012 are used for validation, however, flow data from 18 January 1989 are also used in order to calculate the summed, volumetric flow from 5 years previous (Q_2), which is considered as a potential input for these models, as mentioned above.

The resulting equations for models 3, 4, and 5 are given by Eqs. (5), (6) and (7), respectively:

model 3

$$\mathrm{RS_p} = 0.4\mathrm{EC} + 0.35Q_2 + 0.1D \quad \text{if} \quad 15000 \le Q_1 < 30000; \tag{5}$$

model 4

$$\mathrm{RS_p} = 0.4\mathrm{WL} + 0.3T + 0.35D \quad \text{if} \quad 30000 \le Q_1 < 50000; \tag{6}$$

model 5

$$\mathrm{RS_p} = 0.3\mathrm{WL} - 0.45Q_1 \quad \text{if} \quad Q_1 > 50000, \tag{7}$$

where $\mathrm{RS_p}$ are the residuals between the measured salinity and the output salinity from the process-driven salt transport model for all peak flow rates greater than or equal to 15 000 ML day^{-1} (mg L^{-1}); Q_1 is the flow rate downstream of Lock 5 (ML day^{-1}); Q_2 is the 5-year historical volume of water at Lock 5 (ML); WL is the water level at Lyrup pump station (m); T is the time since last peak flow (days); D is the duration of peak flow event (days); and EC is the salinity downstream from Lock 5 (mg L^{-1}).

The selected inputs for model 3 indicate a positive correlation with salinity at Lock 5 (EC), the 5-year historical flow volume at Lock 5 (Q_2) and peak flow event duration (D). The positive correlation with EC is most likely related to shallow overbank flow, as discussed in Sect. 3.5.4. Q_2 is likely to increase salt load, as greater volumes of historical flow indicate a greater likelihood that events that generate saline accessions have occurred in the previous 5 years. Furthermore, D is likely to be positively correlated with saline accessions, as longer flood durations provide more time for groundwater recharge during an event and therefore later discharge on the recession of a flood, which in turn results in slow response saline accessions.

The selected inputs for model 4 indicate a positive correlation with higher water levels at Lyrup pump station (WL), time since last peak flow (T) and peak flow event duration (D). The positive correlation with WL is most likely due to the fact Lyrup pump station is just upstream of the connection that flushes Gurra Gurra lakes at high flows, with higher water levels at Lyrup pump station providing an indication of an increased ability to flush the lakes. Higher values of time since last peak flow are likely to increase saline accessions, as longer periods of time between floods provide more time for evapo-concentration to occur, as well as saline groundwater to flow into the Gurra Gurra lakes system, which is very shallow. Finally, as is the case for model 3, longer flood durations provide more time for stored saline water to be flushed into the main river channel.

The selected inputs for model 5 indicate a positive correlation with higher water levels at Lyrup pump station (WL) and a negative correlation with flow downstream of Lock 5 (Q_1). The positive correlation with WL is most likely related to the fact that larger areas of the floodplain are inundated at higher water levels, increasing overall salt load. The negative correlation with flow is most likely due to the increased dilution of the salt load at the high flows to which this model caters.

3.4.6 Hybrid model

A schematic of the resulting hybrid model is shown in Fig. 7. As can be seen, the process-driven in-stream salt transport model (model 1) forms the basis of the hybrid model, with the different types of accessions, modelled using the ANN and regression models, added at appropriate locations. Specifically, the first and second regression models (models 3 and 4, respectively) are added downstream of Lock 5, at the approximate location of the Pike River and Disher Creek outlets. The third regression model (model 5) is added upstream of Lock 4, near the entrance of Gurra Gurra lakes. The output from the groundwater accession ANN (model 2) is added to model 1 at Lock 4. Although groundwater accession occurs along the length of the reach under consideration, these inflows are impractical to segregate. The outputs from models 2 to 5 are only added to model 1 when triggered by the corresponding flow rate in a given time step: there is only one model besides model 1 that describes the salinity levels on any given day.

3.5 Development of benchmark models

To enable the predictive ability of the hybrid model to be assessed in an objective manner it is compared with that of two benchmark models, which represent commonly used paradigms for modelling salinity in rivers in previous studies, including data-driven ANN and process-driven models. As summarised in Table 3, although the process-driven benchmark model does account for saline accession along the reach

Table 3. Method by which different processes are represented by the hybrid and benchmark data- and process-driven models. The inputs are represented either explicitly (by separate processes within the model) or implicitly. The outputs are either dynamic (the salt load varies in response to some time-dependent environmental changes) or static.

Process I Model	Data-driven benchmark model		Process-driven benchmark model		Hybrid model	
In-stream salt transport	Implicit	Dynamic	Explicit	Dynamic	Explicit	Dynamic
Groundwater accession	Implicit	Dynamic	Explicit	Static	Explicit	Dynamic
Pike River inflow	Implicit	Dynamic	Explicit	Static	Explicit	Dynamic
Gurra Gurra lakes inflow	Implicit	Dynamic	Explicit	Static	Explicit	Dynamic
Disher Creek inflow	Implicit	Dynamic	Explicit	Static	Explicit	Dynamic

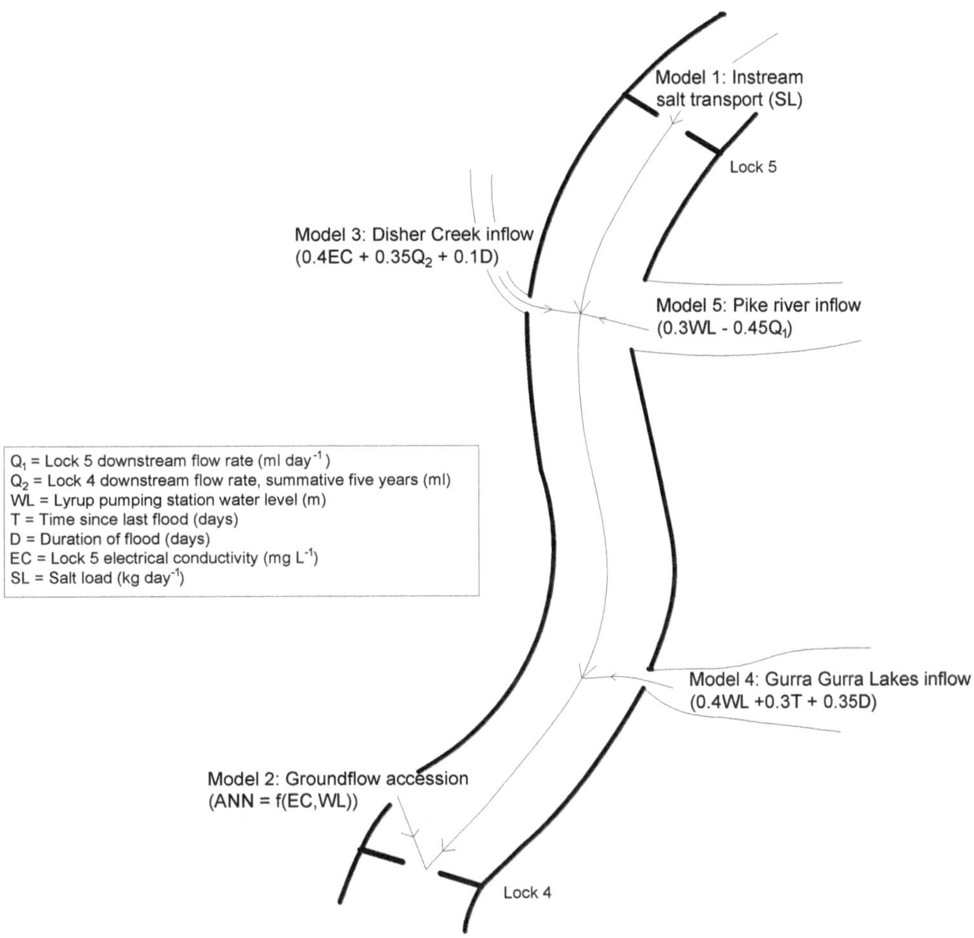

Figure 7. Schematic of the hybrid model on a stylised view of the river (not to scale). The regression models are input as point loads, while the groundwater accession is added to the transport model at Lock 4.

of interest explicitly, this is done via the addition of average historical accessions, as this represents the best available information. Consequently, unlike the hybrid model, which accounts for saline accession in a dynamic fashion, the benchmark process-driven model does so in static fashion. However, in-stream salt transport processes are represented in an explicit and dynamic manner. In contrast, in the benchmark ANN model, all processes are represented implicitly, as it predicts salinity at Lock 4 as a function of available data along the length of the reach of interest without explicit consideration of any of the underlying processes; however, this is done in a dynamic fashion. For the sake of consistency with the development of the hybrid model, the calibration and validation data, the model development processes and the way model performance is assessed are identical to those used for the development of model 1 (process-driven benchmark model) and model 2 (ANN benchmark model). Further de-

Table 4. Performance statistics for the hybrid model and its component models for the validation data.

Model	NSE	RMSE (mg L^{-1})
In-stream salt transport (model 1)	−1.56	61.70
In-stream salt transport + groundwater accession (models 1 + 2)	0.38	30.46
In-stream salt transport + groundwater accession + flushing of Disher Creek, Gurra Gurra lakes & Pile River (models 1–5 – complete hybrid model)	0.89	12.62

tails of the development of the two benchmark models are given in the subsequent sections.

3.5.1 ANN model (benchmark)

The purpose of this ANN model is to predict the total salinity in the river at Lock 4 directly. This is in contrast to the ANN model that forms part of the hybrid model (model 2), which predicts the residual salinity between the process-driven in-stream salt transport model predictions and the measured salinity at Lock 4 for flows up to 15 000 ML day^{-1}. The latter is aimed at representing groundwater accession and low flow processes only. As mentioned above, the benchmark ANN model is developed using the same methodology as that used for model 2 (see Sect. 3.5.4). A summary of the resulting ANN model is given in Fig. 8.

3.5.2 Process-driven model (benchmark)

The benchmark process-driven model is identical to the one used in the hybrid model. However, as shown in Table 3 and described above, while salt accession within the study reach is modelled dynamically using ANN and regression models in the hybrid model, in the benchmark model salt accession is represented as an average of historical salt loads. This is done by calculating the average daily salt load at Lock 4 for the calibration period of 18 January 1994 to 21 October 2008, which is then applied as two constant point loads. Most (approximately 82 %) of this load is applied upstream of the Berri irrigation extraction, as this area forms a longer part of the study reach than the area downstream of this location. The remainder of the constant salt load is applied downstream from Berri.

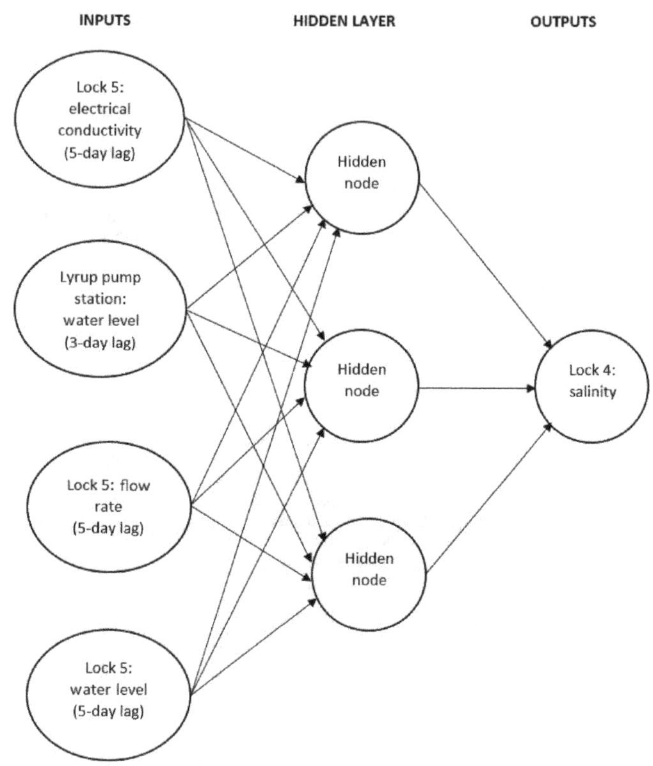

Figure 8. Structure and inputs of the benchmark ANN.

4 Results and discussion

The time series plots of actual versus predicted salinities for the hybrid model and its component models for the validation data are given in Fig. 9, with the corresponding performance statistics given in Table 4. As can be seen, the hybrid model performs very well, with a NSE of 0.89 and an RMSE of 12.62 mg L^{-1} (for data ranging from approximately 50 to 250 mg L^{-1}). The time series plot shows that the model has captured all variations in salinity very well, with only small over- or under-predictions (Fig. 9c).

The performance of the process-based in-stream salt transport model (model 1) on its own is not very good, with a NSE of −1.56 and an RMSE of 61.70 mg L^{-1}. This is because this model does not include any of the saline accessions within the reach, and therefore under-predicts salinity values significantly. However, the model is able to capture most major variations in salinity (Fig. 9a). When the ANN model designed to primarily capture saline groundwater accessions (model 2) is added, the NSE increases to 0.38 and the RMSE reduces to 30. 46 mg L^{-1}, which is due to improved performance during periods of low flow, as expected (Fig. 9b, d). When the linear regression models designed to capture saline accessions from the flushing of Disher Creek, Gurra Gurra

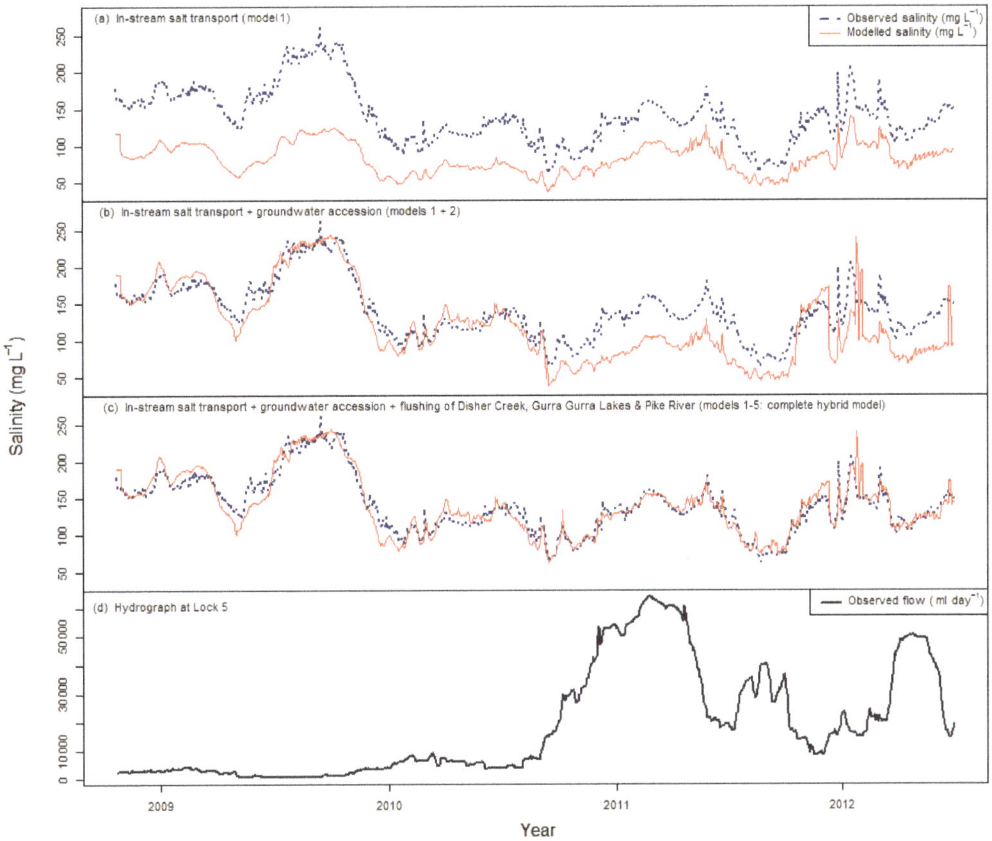

Figure 9. Measured versus modelled salinity at Lock 4 for the hybrid model and its components for the validation data, as well as corresponding hydrograph for Lock 5.

lakes and Pike River (models 3 to 5) are added, an NSE value of 0.89 and an RMSE value of 12.62 mg L^{-1} are achieved as a result of increased performance during high flow periods. This indicates the value of the proposed hybrid approach, as each of the models of the different sub-processes improve model performance significantly.

The hybrid model also performs favourably compared with the two benchmark models, as shown in Table 5 and Fig. 10. The process-driven benchmark model performs significantly worse than the hybrid model, with NSE and RMSE values of −0.14 and 41.10 mg L^{-1}, respectively, compared with corresponding values of 0.89 and 12.62 mg L^{-1} for the hybrid model. As can be seen in Fig. 10, this is due to an over-prediction of saline accessions by the benchmark process-driven model, as these are based on static historical values rather than being modelled dynamically as a function of changes in flow, water levels and upstream salinity, as is the case for the hybrid model. However, the addition of the average values of the historical saline accessions results in an improvement in model performance compared with that of the process-driven in-stream salt transport model used in the hybrid model (model 1) – an increase in NSE from −1.56 to −0.14 and a reduction in RMSE from 61.70 to 41.10 mg L^{-1}.

In contrast to the process-driven benchmark model, the data-driven benchmark model performs only slightly worse than the hybrid model, with NSE and RMSE values of 0.83 and 15.93 mg L^{-1}, respectively, compared with corresponding values of 0.89 and 13.58 mg L^{-1} for the hybrid model. As can be seen in Fig. 10, the primary differences between the hybrid and data-driven benchmark models are that the benchmark model under-predicts saline accessions during low-flow periods (e.g. towards the end of 2009) and over-predicts saline accessions during high-flow periods (e.g. in the first half of 2011 and 2012). This highlights the benefits of the hybrid model in being able to tailor models to accessions during low- and high-flow periods. The superior performance of the hybrid model is reinforced by positive values of the G_{bench} index of 0.36 and 0.90 for the data-driven and process-driven benchmark models, respectively.

In addition to resulting in improved predictive performance, a major benefit of the hybrid model is that it, unlike both benchmark models, can be used to assist in dealing with some of the proposed changes in river management, which are in part driven by the Murray–Darling Basin Plan. Historically, river management has focused on ensuring a supply of water for consumptive demands. However, more recently,

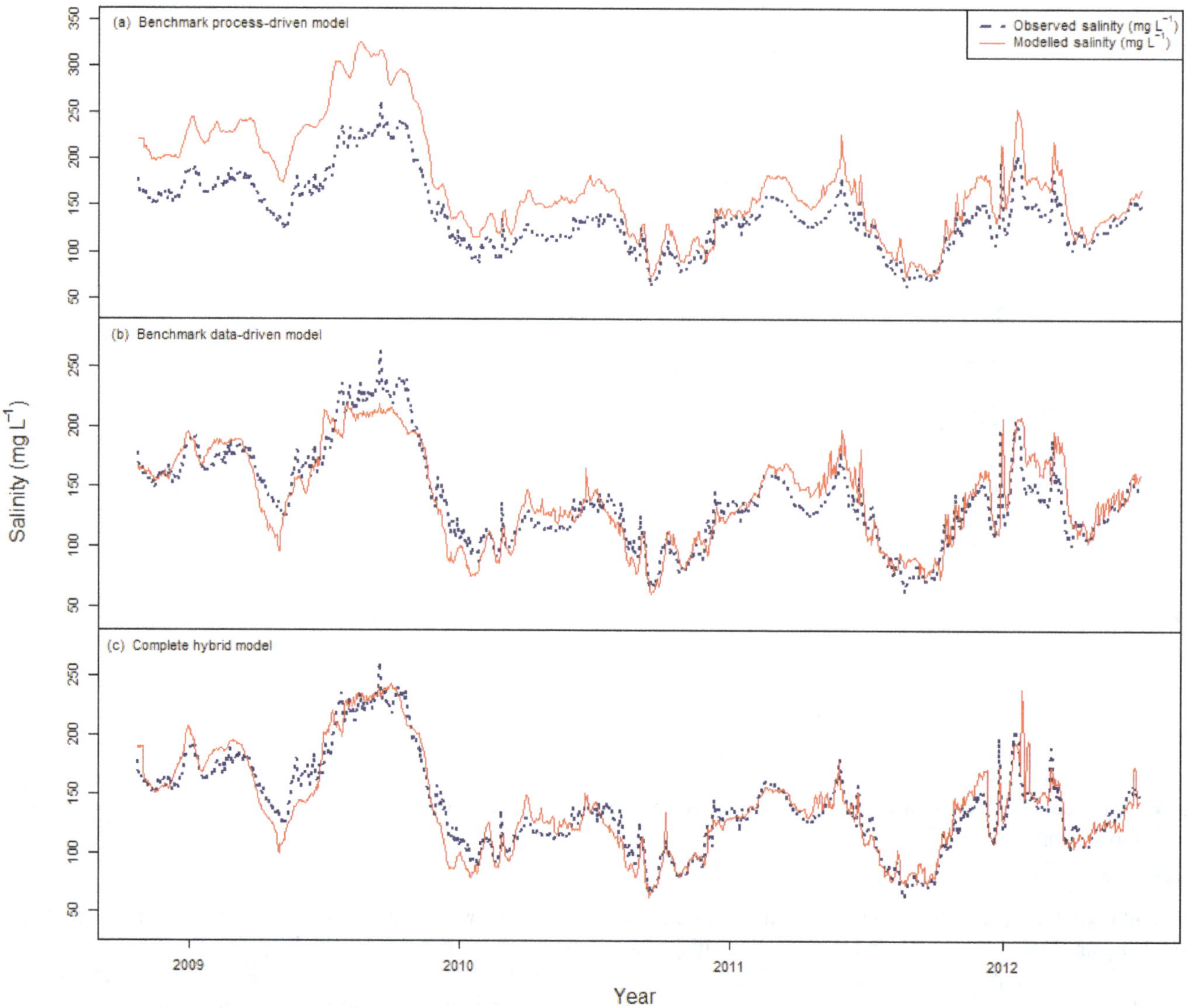

Figure 10. Measured versus modelled salinity at Lock 4 for the hybrid model and the two benchmark models for the validation data.

Table 5. Performance statistics for the two benchmark models and the hybrid model for the validation data.

Model	NSE	RMSE (mg L^{-1})
Process-driven benchmark model	−0.14	41.10
Data-driven benchmark model	0.83	15.93
Hybrid model	0.89	13.58

river management is being expanded to include the improvement of environmental outcomes. This includes changing the flow regime via the delivery of environmental water, and by constructing control infrastructure on the floodplain to increase inundation frequency and duration. As the variables that affect the flow regime are included as inputs in the hybrid model, this model can be used to assess the impact of some of the proposed management options on river salinity. In addition, the hybrid model contributes to an increased understanding of the underlying processes.

Overall, the results of the illustrative case study highlight the potential benefits of the proposed framework. By considering the relevant processes affecting river salinity at the site of interest, how well they are understood and can be represented mathematically, how much data there is to support model development and what the primary purpose of the model is, a hybrid model was able to be developed that not only results in better predictive performance than the corresponding benchmark process- and data-driven models, but is also more useful from a management perspective. However, given the conceptual nature of the proposed framework and the level of subjectivity required to implement it, it is not possible to tell if an even better model could have been developed had different decisions been made with regard to model types.

5 Summary and conclusions

This paper introduces a framework for the development of hybrid models for the prediction of salinity in rivers. As part of the framework, relevant sub-processes contributing to river salinity are identified, followed by the selection and development of the most appropriate sub-models for each of these based on model purpose, degree of process understanding and data availability, which are then combined to form the hybrid model.

The approach is illustrated for a reach of the Murray River in South Australia. The resulting model consists of five sub-models, including a process-driven in-stream salt transport model, an ANN model to primarily cater to saline groundwater accessions and three linear regression models to account for the flushing of three different waterbodies in the floodplain. Results show that the hybrid model performs very well and is able to capture all variations in salinity with high levels of accuracy. The value of using a hybrid approach is demonstrated by the incremental increase in model performance when different sub-models are added. The superior performance of the hybrid model compared with that of two benchmark models, based on commonly used methods for modelling salinity in rivers including a process-driven and a data-driven ANN model, further highlights the model's value. In addition to superior predictive performance, the hybrid model results in the development of increased process understanding and is able to be used to assist with the evaluation of various river management options.

Overall, the proposed hybrid approach shows significant promise, although there would be value in applying it to different river systems where different processes dominate and different types of data are available. While the approach has been developed specifically for the modelling of salinity in rivers, there is no reason why some of the underlying principles cannot be applied successfully for other types of hydrological models.

Competing interests. The authors declare that they have no conflict of interest.

Acknowledgements. The authors would like to thank the two anonymous reviewers for their comments, which improved and clarified the manuscript.

Edited by: Dimitri Solomatine

References

Adamowski, J. and Chan, H. F.: A wavelet neural network conjunction model for groundwater level forecasting, J. Hydrol, 407, 28–40, https://doi.org/10.1016/j.jhydrol.2011.06.013, 2011.

Alvarez-Garreton, C., Ryu, D., Western, A. W., Su, C.-H., Crow, W. T., Robertson, D. E., and Leahy, C.: Improving operational flood ensemble prediction by the assimilation of satellite soil moisture: comparison between lumped and semi-distributed schemes, Hydrol. Earth Syst. Sci., 19, 1659–1676, https://doi.org/10.5194/hess-19-1659-2015, 2015.

Banerjee, P., Singh, V. S., Chatttopadhyay, K., Chandra, P. C., and Singh, B.: Artificial neural network model as a potential alternative for groundwater salinity forecasting, J. Hydrol., 398, 212–220, https://doi.org/10.1016/j.jhydrol.2010.12.016, 2011.

Barnett, S.: Gurra Gurra Wetland Complex – Groundwater Data Review, Dept. of Water, Land and Biodiversity Conservation, 4, 2007.

Beecham, R., Arranz, P., Boddy, J., Burrell, M., Gilmore, R., Javam, A., Martin, J., O'Neill, R., and Salbe, I.: Implementing daily salinity models in the NSW Murray Darling Basin tributaries, in: Modsim 2003, International Congress on Modelling and Simulation, Vol 1–4: Vol 1: Natural Systems, Pt 1; Vol 2: Natural Systems, Pt 2; Vol 3: Socio-Economic Systems; Vol 4: General Systems, 362–367, 2003.

Bowden, G. J., Maier, H. R., and Dandy, G. C.: Optimal division of data for neural network models in water resources applications, Water Resour. Res., 38, 1010, https://doi.org/10.1029/2001wr000266, 2002.

Bowden, G. J., Maier, H. R., and Dandy, G. C.: Input determination for neural network models in water resources applications. Part 1. Background and methodology, J. Hydrol., 301, 75–92, https://doi.org/10.1016/j.jhydrol.2004.06.021, 2005a.

Bowden, G. J., Maier, H. R., and Dandy, G. C.: Input determination for neural network models in water resources applications. Part 2. Case study: forecasting salinity in a river, J. Hydrol., 301, 93–107, https://doi.org/10.1016/j.jhydrol.2004.06.020, 2005b.

Bowden, G. J., Maier, H. R., and Dandy, G. C.: Real-time deployment of artificial neural network forecasting models: Understanding the range of applicability, Water Resour. Res., 48, W10549, https://doi.org/10.1029/2012WR011984, 2012.

Chang, F.-J. and Chang, Y.-T.: Adaptive neuro-fuzzy inference system for prediction of water level in reservoir, Adv. Water Resour., 29, 1–10, https://doi.org/10.1016/j.advwatres.2005.04.015, 2006.

Chang, F.-J. and Tsai, M.-J.: A nonlinear spatio-temporal lumping of radar rainfall for modeling multi-step-ahead inflow forecasts by data-driven techniques, J. Hydrol., 535, 228–236, https://doi.org/10.1016/j.jhydrol.2016.01.056, 2016.

Chang, F.-J., Tsai, W.-P., Chen, H.-K., Yam, R. S.-W., and Herricks, E. E.: A self-organizing radial basis network for estimating riverine fish diversity, J. Hydrol., 476, 280–289, https://doi.org/10.1016/j.jhydrol.2012.10.038, 2013.

Chang, F.-J., Chen, P.-A., Chang, L.-C., and Tsai, Y.-H.: Estimating spatio-temporal dynamics of stream total phosphate concentration by soft computing techniques, Sci. Total Environ., 562, 256–269, https://doi.org/10.1016/j.scitotenv.2016.03.219, 2016.

Clark, M. P., Kavetski, D., and Fenicia, F.: Pursuing the method of multiple working hypotheses for hydrological modeling, Water Resour. Res., 47, W09301, https://doi.org/10.1029/2010WR009827, 2011.

Commonwealth of Australia: Water Act (An act to make provision for the management of the water resources of the Murray-Darling Basin, and to make provision for other matters of national interest in relation to water and water information, and for related purposes), Commonwealth Consolidated Acts, Sects. 28–32, 2007.

Corzo, G. and Solomatine, D.: Baseflow separation techniques for modular artificial neural network modelling in flow forecasting, Hydrol. Sci. J., 52, 491–507, https://doi.org/10.1623/hysj.52.3.491, 2007.

Corzo, G. A., Solomatine, D. P., Hidayat, de Wit, M., Werner, M., Uhlenbrook, S., and Price, R. K.: Combining semi-distributed process-based and data-driven models in flow simulation: a case study of the Meuse river basin, Hydrol. Earth Syst. Sci., 13, 1619–1634, https://doi.org/10.5194/hess-13-1619-2009, 2009.

Dessie, M., Verhoest, N. E. C., Pauwels, V. R. N., Admasu, T., Poesen, J., Adgo, E., Deckers, J., and Nyssen, J.: Analyzing runoff processes through conceptual hydrological modeling in the Upper Blue Nile Basin, Ethiopia, Hydrol. Earth Syst. Sci., 18, 5149–5167, https://doi.org/10.5194/hess-18-5149-2014, 2014.

Duan, W. L., He, B., Takara, K., Luo, P. P., Nover, D., and Hu, M. C.: Modeling suspended sediment sources and transport in the Ishikari River basin, Japan, using SPARROW, Hydrol. Earth Syst. Sci., 19, 1293–1306, https://doi.org/10.5194/hess-19-1293-2015, 2015.

Duku, C., Rathjens, H., Zwart, S. J., and Hein, L.: Towards ecosystem accounting: a comprehensive approach to modelling multiple hydrological ecosystem services, Hydrol. Earth Syst. Sci., 19, 4377–4396, https://doi.org/10.5194/hess-19-4377-2015, 2015.

Fenicia, F., Kavetski, D., and Savenije, H. H. G.: Elements of a flexible approach for conceptual hydrological modeling: 1. Motivation and theoretical development, Water Resour. Res., 47, W11510, https://doi.org/10.1029/2010wr010174, 2011.

Galelli, S., Humphrey, G. B., Maier, H. R., Castelletti, A., Dandy, G. C., and Gibbs, M. S.: An evaluation framework for input variable selection algorithms for environmental data-driven models, Environ. Modell. Softw., 62, 33–51, https://doi.org/10.1016/j.envsoft.2014.08.015, 2014.

Gallice, A., Schaefli, B., Lehning, M., Parlange, M. B., and Huwald, H.: Stream temperature prediction in ungauged basins: review of recent approaches and description of a new physics-derived statistical model, Hydrol. Earth Syst. Sci., 19, 3727–3753, https://doi.org/10.5194/hess-19-3727-2015, 2015.

Gibbs, M. S., Maier, H. R., and Dandy, G. C.: A generic framework for regression regionalization in ungauged catchments, Environ. Modell. Softw., 27–28, 1–14, https://doi.org/10.1016/j.envsoft.2011.10.006, 2012.

Gibbs, M. S., McInerney, D., Humphrey, G., Thyer, M. A., Maier, H. R., Dandy, G. C., and Kavetski, D.: State updating and calibration period selection to improve dynamic monthly streamflow forecasts for an environmental flow management application, Hydrol. Earth Syst. Sci., 22, 871–887, https://doi.org/10.5194/hess-22-871-2018, 2018.

Goss, K. F.: Environmental flows, river salinity and biodiversity conservation: managing trade-offs in the Murray & Darling basin, Aust. J. Bot., 51, 619–625, https://doi.org/10.1071/BT03003, 2003.

Government of South Australia, Department of Environment, Water and Natural Resources (DEWNR): WaterConnect groundwater data, available at: www.waterconnect.sa.gov.au (last access: 9 November 2017), 2015.

Gragne, A. S., Sharma, A., Mehrotra, R., and Alfredsen, K.: Improving real-time inflow forecasting into hydropower reservoirs through a complementary modelling framework, Hydrol. Earth Syst. Sci., 19, 3695–3714, https://doi.org/10.5194/hess-19-3695-2015, 2015.

Grayson, R. B. and Blöschl, G.: Spatial patterns in catchment hydrology: Observations and modelling, Cambridge University Press, UK, https://doi.org/10.1002/esp.378, 2000.

Guo, D., Westra, S., and Maier, H. R.: An R package for modelling actual, potential and reference evapotranspiration, Environ. Modell. Softw., 78, 216–224, https://doi.org/10.1016/j.envsoft.2015.12.019, 2016.

Habib, E., Nuttle, W. K., Rivera-Monroy, V. H., Gautam, S., Wang, J., Meselhe, E., and Twilley, R. R.: Assessing effects of data limitations on salinity forecasting in Barataria basin, Louisiana, with a Bayesian analysis, J. Coastal Res., 23, 749–763, https://doi.org/10.2112/06-0723.1, 2007.

Hamilton, S. H., ElSawah, S., Guillaume, J. H. A., Jakeman, A. J., and Pierce, S. A.: Integrated assessment and modelling: Overview of salient dimensions, Environ. Modell. Softw., 64, 215–229, https://doi.org/10.1016/j.envsoft.2014.12.005, 2015.

Harrington, N., Van den Akker, J., and Brown, K.: Padthaway Salt Accession Study Volume Three: Conceptual Models, Government of South Australia, Dept. of Water, Land and Biodiversity Conservation, Adelaide, 11, 15–16, 38, 112, 2006.

Hart, B. T., Bailey, P., Edwards, R., Hortle, K., James, K., McMahon, A., Meredith, C., and Swadling, K.: A Review of the Salt Sensitivity of the Australian Fresh-Water Biota, Hydrobiologia, 210, 105–144, https://doi.org/10.1007/bf00014327, 1991.

Hsu, K. L., Gupta, H. V., Gao, X. G., Sorooshian, S., and Imam, B.: Self-organizing linear output map (SOLO): An artificial neural network suitable for hydrologic modeling and analysis, Water Resour. Res., 38, 1302, https://doi.org/10.1029/2001wr000795, 2002.

Huang, W. R. and Foo, S.: Neural network modeling of salinity variation in Apalachicola River, Water Res., 36, 356–362, https://doi.org/10.1016/s0043-1354(01)00195-6, 2002.

Humphrey, G. B., Gibbs, M. S., Dandy, G. C., and Maier, H. R.: A hybrid approach to monthly streamflow forecasting: Integrating hydrological model outputs into a Bayesian artificial neural network, J. Hydrol., 540, 623–640, https://doi.org/10.1016/j.jhydrol.2016.06.026, 2016.

Humphrey, G. B., Maier, H. R., Wu, W., Mount, N. J., Dandy, G. C., Abrahart, R. J., and Dawson, C. W.: Improved validation framework and R-package for artificial neural network models, Environ. Modell. Softw., 92, 82–106, https://doi.org/10.1016/j.envsoft.2017.01.023, 2017.

Jain, A. and Kumar, S.: Dissection of trained neural network hydrological models for knowledge extraction, Water Resour. Res., 45, W07420, https://doi.org/10.1029/2008WR007194, 2009.

Jain, A., Sudheer, K. P., and Srinivasulu, S.: Identification of physical processes inherent in artificial neural network rainfall runoff models, Hydrol. Process., 18, 571–581, https://doi.org/10.1002/hyp.5502, 2004.

Jakeman, A. J., Letcher, R. A., and Norton, J. P.: Ten iterative steps in development and evaluation of environmental models, Environ. Modell. Softw., 21, 602–614, https://doi.org/10.1016/j.envsoft.2006.01.004, 2006.

Kasiviswanathan, K. S., He, J., Sudheer, K. P., and Tay, J.-H.: Potential application of wavelet neural network ensemble to forecast streamflow for flood management, J. Hydrol., 536, 161–173, https://doi.org/10.1016/j.jhydrol.2016.02.044, 2016.

Kavetski, D. and Fenicia, F.: Elements of a flexible approach for conceptual hydrological modeling: 2. Application and experimental insights, Water Resour. Res., 47, W11511, https://doi.org/10.1029/2011wr010748, 2011.

Kelly, R. A., Jakeman, A. J., Barreteau, O., Borsuk, M. E., ElSawah, S., Hamilton, S. H., Henriksen, H. J., Kuikka, S., Maier, H. R., Rizzoli, A. E., van Delden, H., and Voinov, A. A.: Selecting among five common modelling approaches for integrated environmental assessment and management, Environ. Modell. Softw., 47, 159–181, https://doi.org/10.1016/j.envsoft.2013.05.005, 2013.

Kingston, G. B., Lambert, M. F., and Maier, H. R.: Bayesian training of artificial neural networks used for water resources modeling, Water Resour. Res., 41, W12409, https://doi.org/10.1029/2005wr004152, 2005.

Kisi, O. and Demir, V.: Evapotranspiration estimation using six different multi-layer perceptron algorithms, Irrigation & Drainage Systems Engineering, 5, 164, https://doi.org/10.4172/2168-9768.1000164, 2016.

Kisi, O. and Parmar, K. S.: Application of least square support vector machine and multivariate adaptive regression spline models in long term prediction of river water pollution, J. Hydrol., 534, 104–112, https://doi.org/10.1016/j.jhydrol.2015.12.014, 2016.

Kornelsen, K. and Coulibaly, P.: Comparison of Interpolation, Statistical, and Data-Driven Methods for Imputation of Missing Values in a Distributed Soil Moisture Dataset, J. Hydrol. Eng., 19, 26–43, https://doi.org/10.1061/(asce)he.1943-5584.0000767, 2014.

Li, L., Maier, H. R., Partington, D., Lambert, M. F., and Simmons, C. T.: Performance assessment and improvement of recursive digital baseflow filters for catchments with different physical characteristics and hydrological inputs, Environ. Modell. Softw., 54, 39–52, https://doi.org/10.1016/j.envsoft.2013.12.011, 2014.

Liu, W. C., Chen, W. B., Cheng, R. T., Hsu, M. H., and Kuo, A. Y.: Modeling the influence of river discharge on salt intrusion and residual circulation in Danshuei River estuary, Taiwan, Cont. Shelf Res., 27, 900–921, https://doi.org/10.1016/j.csr.2006.12.005, 2007.

Maier, H. R. and Dandy, G. C.: The use of artificial neural networks for the prediction of water quality parameters, Water Resour. Res., 32, 1013–1022, 1996.

Maier, H. R., Jain, A., Dandy, G. C., and Sudheer, K. P.: Methods used for the development of neural networks for the prediction of water resource variables in river systems: Current status and future directions, Environ. Modell. Softw., 25, 891–909, https://doi.org/10.1016/j.envsoft.2010.02.003, 2010.

Markstrom, S. L., Hay, L. E., and Clark, M. P.: Towards simplification of hydrologic modeling: identification of dominant processes, Hydrol. Earth Syst. Sci., 20, 4655–4671, https://doi.org/10.5194/hess-20-4655-2016, 2016.

MDBC: Setting up of MSM-BIGMOD modelling Suite for the River Murray System, Technical, Murray-Darling Basin Commission, 2002.

Mohanty, S., Jha, M. K., Raul, S. K., Panda, R. K., and Sudheer, K. P.: Using artificial neural network approach for simultaneous forecasting of weekly groundwater levels at multiple sites, Water Resour. Manag., 29, 5521–5532, https://doi.org/10.1007/s11269-015-1132-6, 2015.

Mount, N. J., Abrahart, R. J., Dawson, C. W., and Ab Ghani, N.: The need for operational reasoning in data-driven rating curve prediction of suspended sediment, Hydrol. Process., 26, 3982–4000, https://doi.org/10.1002/hyp.8439, 2012.

Mount, N. J., Maier, H. R., Toth, E., Elshorbagy, A., Solomatine, D., Chang, F. J., and Abrahart, R. J.: Data-driven modelling approaches for socio-hydrology: opportunities and challenges within the Panta Rhei Science Plan, Hydrolog. Sci. J., 61, 1182–1208, https://doi.org/10.1080/02626667.2016.1159683, 2016.

Moxey, A.: Agriculture and Water Quality: Monetary Costs and Benefits across OECD Countries, 11–12, 2012.

Noori, N. and Kalin, L.: Coupling SWAT and ANN models for enhanced daily streamflow prediction, J. Hydrol., 533, 141–151, https://doi.org/10.1016/j.jhydrol.2015.11.050, 2016.

Parasuraman, K., Elshorbagy, A., and Carey, S. K.: Modelling the dynamics of the evapotranspiration process using genetic programming, Hydrol. Sci. J., 52, 563–578, https://doi.org/10.1623/hysj.52.3.563, 2007.

Pulido-Velazquez, M., Peña-Haro, S., García-Prats, A., Mocholi-Almudever, A. F., Henriquez-Dole, L., Macian-Sorribes, H., and Lopez-Nicolas, A.: Integrated assessment of the impact of climate and land use changes on groundwater quantity and quality in the Mancha Oriental system (Spain), Hydrol. Earth Syst. Sci., 19, 1677–1693, https://doi.org/10.5194/hess-19-1677-2015, 2015.

Quiroga, V. M., Popescu, I., Solomatine, D. P., and Bociort, L.: Cloud and cluster computing in uncertainty analysis of integrated flood models, J. Hydroinform., 15, 55–70, https://doi.org/10.2166/hydro.2012.017, 2013.

Rath, J. S., Hutton, P. H., Chen, L., and Roy, S. B.: A hybrid empirical-Bayesian artificial neural network model of salinity in the San Francisco Bay-Delta estuary, Environ. Modell. Softw., 93, 193–208, https://doi.org/10.1016/j.envsoft.2017.03.022, 2017.

Rengasamy, P.: World salinization with emphasis on Australia, J. Exp. Bot., 57, 1017–1023, https://doi.org/10.1093/jxb/erj108, 2006.

Robertson, W. M. and Sharp, J. M.: Estimates of recharge in two arid basin aquifers: a model of spatially variable net infiltration and its implications (Red Light Draw and Eagle Flats, Texas, USA), Hydrogeol. J., 21, 1853–1864, https://doi.org/10.1007/s10040-013-1018-8, 2013.

Seibert, J.: On the need for benchmarks in hydrological modelling, Hydrol. Process., 15, 1063–1064, https://doi.org/10.1002/hyp.446, 2001.

Shamseldin, A. Y., Nasr, A. E., and O'Connor, K. M.: Comparison of different forms of the Multi-layer Feed-Forward Neural Network method used for river flow forecasting, Hydrol. Earth Syst. Sci., 6, 671–684, https://doi.org/10.5194/hess-6-671-2002, 2002.

Shiri, J., Shamshirband, S., Kisi, O., Karimi, S., Bateni, S. M., Nezhad, S. H. H., and Hashemi, A.: Prediction of water-level in the Urima Lake using the extreme learning machine approach, Water Resour. Res., 30, 5217–5229, https://doi.org/10.1007/s11269-016-1480-x, 2016.

Suen, J. P. and Lai, H. N.: A salinity projection model for determining impacts of climate change on river ecosystems in Taiwan, J. Hydrol., 493, 124–131, https://doi.org/10.1016/j.jhydrol.2013.04.020, 2013.

Tefler, A., Burnell, R., and Charles, A.: Salt interception schemes and instream processes, in Mallee Salinity Workshop, Mallee Catchment Management Authority, 30 May 2012.

Tsai, M.-J., Abrahart, R. J., Mount, N. J., and Chang, F.-J.: Including spatial distribution in a data-driven rainfall-runoff model to improve reservoir inflow forecasting in Taiwan, Hydrol. Process., 28, 1055–1070, https://doi.org/10.1002/hyp.9559, 2014.

Wang, W., Van Gelder, P., Vrijling, J. K., and Ma, J.: Forecasting daily streamflow using hybrid ANN models, J. Hydrol., 324, 383–399, https://doi.org/10.1016/j.jhydrol.2005.09.032, 2006.

Welsh, D. W., Vaze, J., Dutta, D., Rassam, D., Rahman, J. M., Jolly, I. D., Wallbrink, P., Podger, G. M., Bethune, M., Hardy, M. J., Teng, J., and Lerat, J.: An integrated modelling framework for regulated river systems, Env. Modell. Softw., 39, 81–102, https://doi.org/10.1016/j.envsoft.2012.02.022, 2013.

Williams, W. D.: Anthropogenic salinisation of inland waters, Hydrobiologia, 466, 329–337, https://doi.org/10.1023/a:1014598509028, 2001.

Woods, J.: Modelling salt dynamics on the River Murray floodplain in South Australia: Conceptual model, data review and salinity risk approaches, Goyder Institute for Water Research Technical Report Series No. 15/9, 2015.

Wu, W., Dandy, G. C., and Maier, H. R.: Protocol for developing ANN models and its application to the assessment of the quality of the ANN model development process in drinking water quality modelling, Environ. Modell. Softw., 54, 108–127, https://doi.org/10.1016/j.envsoft.2013.12.016, 2014.

Yaseen, Z. M., Jaafar, O., Deo, R. C., Kisi, O., Adamowski, J., Quilty, J., and El-Shafie, A.: Stream-flow forecasting using extreme learning machines: A case study in a semi-arid region in Iraq, J. Hydrol., 542, 603–614, https://doi.org/10.1016/j.jhydrol.2016.09.035, 2016.

Zhang, H. B., Singh, V. P., Bin Wang, B., and Yu, Y. H.: CEREF: A hybrid data-driven model for forecasting annual streamflow from a socio-hydrological system, J. Hydrol., 540, 246–256, https://doi.org/10.1016/j.jhydrol.2016.06.029, 2016.

Zhang, Q. and Stanley, S. J.: Forecasting raw-water quality parameters for the North Saskatchewan River by neural network modeling, Water Res., 31, 2340–2350, https://doi.org/10.1016/s0043-1354(97)00072-9, 1997.

Zheng, F., Maier, H. R., Wu, W., Dandy, G. C., Gupta, H. V., and Zhang, T.: On lack of robustness in hydrological model development due to absence of guidelines for selecting calibration and evaluation data: Demonstration for data driven models, Water Resour. Res., 54, 1013–1030, https://doi.org/10.1002/2017WR021470, 2018.

Zounemat-Kermani, M., Kişi, Ö., Adamowski, J., and Ramezani-Charmahineh, A.: Evaluation of data driven models for river suspended sediment concentration modeling, J. Hydrol., 535, 457–472, https://doi.org/10.1016/j.jhydrol.2016.02.012, 2016.

8

Skilful seasonal forecasts of streamflow over Europe?

Louise Arnal[1,2], Hannah L. Cloke[1,3,4,5], Elisabeth Stephens[1], Fredrik Wetterhall[2], Christel Prudhomme[2,6,7], Jessica Neumann[1], Blazej Krzeminski[2], and Florian Pappenberger[2]

[1]Department of Geography and Environmental Science, University of Reading, Reading, RG6 6AB, UK
[2]European Centre for Medium-Range Weather Forecasts, Shinfield Park, Reading, RG6 9AX, UK
[3]Department of Meteorology, University of Reading, Reading, RG6 6BB, UK
[4]Department of Earth Sciences, Uppsala University, Uppsala, 752 36, Sweden
[5]Centre of Natural Hazards and Disaster Science, CNDS, Uppsala, 752 36, Sweden
[6]Department of Geography, Loughborough University, Loughborough, LE11 3TU, UK
[7]NERC Centre for Ecology & Hydrology, Wallingford, OX10 8BB, UK

Correspondence: Louise Arnal (l.l.s.arnal@pgr.reading.ac.uk, louise.arnal@ecmwf.int)

Abstract. This paper considers whether there is any added value in using seasonal climate forecasts instead of historical meteorological observations for forecasting streamflow on seasonal timescales over Europe. A Europe-wide analysis of the skill of the newly operational EFAS (European Flood Awareness System) seasonal streamflow forecasts (produced by forcing the Lisflood model with the ECMWF System 4 seasonal climate forecasts), benchmarked against the ensemble streamflow prediction (ESP) forecasting approach (produced by forcing the Lisflood model with historical meteorological observations), is undertaken. The results suggest that, on average, the System 4 seasonal climate forecasts improve the streamflow predictability over historical meteorological observations for the first month of lead time only (in terms of hindcast accuracy, sharpness and overall performance). However, the predictability varies in space and time and is greater in winter and autumn. Parts of Europe additionally exhibit a longer predictability, up to 7 months of lead time, for certain months within a season. In terms of hindcast reliability, the EFAS seasonal streamflow hindcasts are on average less skilful than the ESP for all lead times. The results also highlight the potential usefulness of the EFAS seasonal streamflow forecasts for decision-making (measured in terms of the hindcast discrimination for the lower and upper terciles of the simulated streamflow). Although the ESP is the most potentially useful forecasting approach in Europe, the EFAS seasonal streamflow forecasts appear more potentially useful than the ESP in some regions and for certain seasons, especially in winter for almost 40 % of Europe. Patterns in the EFAS seasonal streamflow hindcast skill are however not mirrored in the System 4 seasonal climate hindcasts, hinting at the need for a better understanding of the link between hydrological and meteorological variables on seasonal timescales, with the aim of improving climate-model-based seasonal streamflow forecasting.

1 Introduction

Seasonal streamflow forecasts predict the likelihood of a difference from normal conditions in the following months. Unlike forecasts at shorter timescales, which aim to predict individual events, seasonal streamflow forecasts aim at predicting long-term (i.e. weekly to seasonal) averages. The predictability in seasonal streamflow forecasts is driven by two components of the Earth system, the initial hydrological conditions (IHC; i.e. of snowpack, soil moisture, streamflow and reservoir levels, etc.) and large-scale climate patterns, such as the El Niño–Southern Oscillation (ENSO), the North Atlantic Oscillation (NAO), the Pacific-North American (PNA) pattern and the Indian Ocean Dipole (IOD) (Yuan et al., 2015b).

The first seasonal streamflow forecasting method, based on a regression technique developed around 1910–1911 in the United States, harnessed the predictability from accurate IHC of snowpacks to derive streamflow for the following summer (Church, 1935). This statistical method recognized antecedent hydrological conditions and land surface memory as key drivers of streamflow generation for the following months.

Alongside the physical understanding of streamflow generation processes came technical developments, such as the creation of the first hydrological models and the acquisition of longer observed meteorological time series, which led to the creation of the first operational model-based seasonal streamflow forecasting system. This system, called extended streamflow prediction (ESP; i.e. note that ESP nowadays stands for ensemble streamflow prediction, although it refers to the same forecasting method), was developed by the United States National Weather Service (NWS) in the 1970s (Twedt et al., 1977; Day, 1985). The ESP forecasts are produced by forcing a hydrological model, initialized with the current IHC, with the observed historical meteorological time series available. The output is an ensemble streamflow forecast (where each year of historical data is a streamflow trace) for the following season(s) (Twedt et al., 1977; Day, 1985). The quality of the ESP forecasts can be high in basins where the IHC dominate the surface hydrological cycle for several months (the exact forecast quality depending on the time of year and the basin's physiographic characteristics; Wood and Lettenmaier, 2008).

In basins where the meteorological forcings drive the predictability, however, the lack of information on the future climate is a limitation of the ESP forecasting method and might result in unskilful ESP forecasts. This drawback led to the investigation of the use of seasonal climate forecasts, in place of the historical meteorological inputs, to feed hydrological models and extend the predictability of hydrological variables on seasonal timescales (Pagano and Garen, 2006). This investigation was made possible by technical and scientific advances. Scientifically, seasonal climate forecasts were improved greatly by the understanding of ocean–atmosphere–land interactions and the identification of large-scale climate patterns as drivers of the hydro-meteorological predictability (Goddard et al., 2001; Troccoli, 2010). This was technically implementable with the increase in computing resources, making it possible to run dynamical coupled ocean–atmosphere–land general circulation models on the global scale at high spatial and temporal resolutions (Doblas-Reyes et al., 2013). An additional technical challenge, the coarse spatial resolution of seasonal climate forecasts compared to the finer resolution of hydrological models, had to be addressed. To tackle this issue, many authors have explored different ways of downscaling climate variables for hydrological applications (Maraun et al., 2010, and references therein).

While climate-model-based seasonal streamflow forecasting experiments are more common outside of Europe, for example for the United States (Wood et al., 2002, 2005; Mo and Lettenmaier, 2014), Australia (Bennett et al., 2016), or Africa (Yuan et al., 2013), they remain limited in Europe, with a few examples in France (Céron et al., 2010; Singla et al., 2012; Crochemore et al., 2016), in central Europe (Demirel et al., 2015; Meißner et al., 2017), in the United Kingdom (Bell et al., 2017; Prudhomme et al., 2017) and at the global scale (Yuan et al., 2015a; Candogan Yossef et al., 2017). This is because, although the quality of seasonal climate forecasts has increased over the past decades, there remains limited skill in seasonal climate forecasts for the extra-tropics, particularly for the variables of interest for hydrology, notably precipitation and temperature (Arribas et al., 2010; Doblas-Reyes et al., 2013).

In Europe, the NAO is one of the strongest predictability sources of seasonal climate forecasts; it is associated with changes in the surface westerlies over the North Atlantic and Europe, and hence with changes in temperature and precipitation patterns over Europe (Hurrell, 1995; Hurrell and Van Loon, 1997). It was shown to affect streamflow predictability, especially during winter (Dettinger and Diaz, 2000; Bierkens and van Beek, 2009; Steirou et al., 2017), in addition to the IHC and the land surface memory. It was furthermore shown to be an indicator of flood damage and occurrence in parts of Europe (Guimarães Nobre et al., 2017).

As the quality and usefulness of seasonal streamflow forecasts increase, their usability for decision-making has lagged behind. Translating the quality of a forecast into an added value for decision-making and incorporating new forecasting products into established decision-making chains are not easy tasks. This has been explored for many water-related applications, such as navigation (Meißner et al., 2017), reservoir management (Viel et al., 2016; Turner et al., 2017), drought-risk management (Sheffield et al., 2013; Yuan et al., 2013; Crochemore et al., 2017), irrigation (Chiew et al., 2003; Li et al., 2017), water resource management (Schepen et al., 2016) and hydropower (Hamlet et al., 2002), but seasonal streamflow forecasts have yet to be adopted by the flood preparedness community.

The European Flood Awareness System (EFAS) is at the forefront of seasonal streamflow forecasting, with one of the first operational pan-European seasonal hydrological forecasting systems. The aim of this paper is to bridge the current gap in pan-European climate-model-based seasonal streamflow forecasting studies. Firstly, the setup of the newly operational EFAS climate-based seasonal streamflow forecasting system is presented. A Europe-wide analysis of the skill of this forecasting system compared to the ESP forecasting approach is then presented, in order to identify whether there is any added value in using seasonal climate forecasts instead of historical meteorological observations for forecasting streamflow on seasonal timescales over Europe. Subsequently, the potential usefulness of the EFAS seasonal streamflow forecasts for decision-making is assessed.

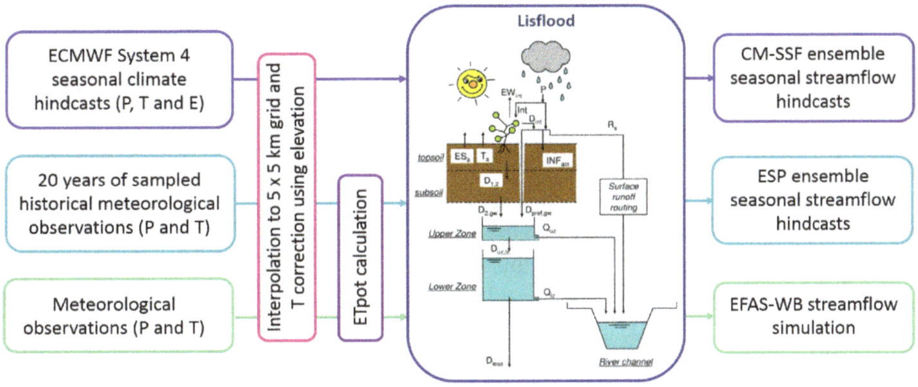

Figure 1. Schematic of the EFAS-WB streamflow simulation and of the CM-SSF and ESP seasonal streamflow hindcast generation, where P is precipitation, T is temperature, E is evaporation and ETpot is potential evapotranspiration. The Lisflood model diagram was taken from Burek et al. (2013).

2 Data and methods

2.1 EFAS hydrological simulation and seasonal hindcasts

The data used in this paper include a streamflow simulation and two seasonal streamflow hindcasts (Fig. 1). Further information on these datasets is given below.

2.1.1 Hydrological modelling and streamflow simulation

The Lisflood model was used to produce all the simulations and hindcasts used in this paper. Lisflood is a GIS-based hydrological rainfall–runoff–routing distributed model written in the PCRaster Dynamic Modelling Language, which enables it to use spatially distributed maps (i.e. both static and dynamic) as input (De Roo et al., 2000; Van Der Knijff et al., 2010). The Lisflood model was calibrated to produce pan-European parameter maps. The calibration was performed for 693 basins from 1994 to 2002 using the Standard Particle Swarm Optimisation 2011 (SPSO-2011) algorithm. The calibration was carried out for parameters controlling snowmelt, infiltration, preferential bypass flow through the soil matrix, percolation to the lower groundwater zone, percolation to deeper groundwater zones, residence times in the soil and subsurface reservoirs, river routing and reservoir operations for a few basins. The results were validated with the Nash–Sutcliffe efficiency (NSE) for the validation period 2003–2012. In validation (calibration), Lisflood obtained a median NSE of 0.57 (0.62). Basins with large discrepancies between the observed and simulated flow statistics were situated mainly on the Iberian Peninsula and on the Baltic coasts (see Zajac et al., 2013, and Smith et al., 2016, for further details).

The Lisflood model is run operationally in EFAS, with the simulation domain covering Europe at a 5×5 km resolution. A reference simulation, called the EFAS water balance (EFAS-WB), is available on a daily time step starting from February 1990. Lisflood simulates the hydrological processes within a basin (most of which are mentioned above), starting from the previous day's IHC (e.g. snow cover, storage in the upper and lower zones, soil moisture, initial streamflow, reservoir filling) and forced with the most recent observed meteorological fields (i.e. of precipitation, potential evapotranspiration and temperature; provided by the EFAS meteorological data collection centres). The observed meteorological fields are daily maps of spatially interpolated point measurements of precipitation (from more than 6000 stations) and temperature (from more than 4000 stations) at the surface level. These same data are used to produce interpolated potential evapotranspiration maps from the Penman–Monteith method (Alfieri et al., 2014). All meteorological variables are interpolated on a 5×5 km grid using an inverse distance weighting scheme and the temperature is first corrected using the elevation (Smith et al., 2016).

The EFAS-WB is the best estimate of the hydrological state at a given time and for a given grid point in EFAS and is thus used as initial conditions from which the seasonal hydrological forecasts are started.

2.1.2 Ensemble seasonal streamflow hindcasts

In this paper, two types of ensemble seasonal streamflow hindcasts are used: the ensemble streamflow prediction (ESP) hindcast (hereafter referred to as ESP) and the System 4-driven seasonal streamflow hindcast (hereafter referred to as CM-SSF (climate-model-based seasonal streamflow forecast), following the notation from Yuan et al. (2015b)).

They are both initialized from the EFAS-WB, on the first day of each month, to produce a new ensemble streamflow forecast up to a lead time of 7 months (215 days), with a daily time step. Both hindcasts are generated from February 1990 for the same European domain as the EFAS-WB, at the same 5×5 km resolution. The unique difference between the

ESP and the CM-SSF is the meteorological forcing used to drive the hydrological model, described below.

The ESP is produced by driving the Lisflood model with 20 (the number of years of data available at the time the hindcast was produced) randomly sampled years of historical meteorological observations (i.e. the same as the meteorological observations used to produce the EFAS-WB, excluding the year of meteorological observations corresponding to the year that is being forecasted). A new 20-member ESP is thus generated at the beginning of each month and for the next 7 months.

The CM-SSF is produced by driving the Lisflood model with the ECMWF System 4 seasonal climate hindcast (Sys4, i.e. of precipitation, evaporation and temperature). Sys4 has a spatial horizontal resolution of about 0.7° (approximately 70 km). It is re-gridded to the Lisflood spatial resolution using an inverse distance weighting scheme and the temperature is first corrected using the elevation. Sys4 is made up of 15 ensemble members, extended to 51 every 3 months (Molteni et al., 2011). From 2011 onwards the Sys4 forecasts were run in real time and all contained 51 ensemble members. A new 15- to 51-member CM-SSF is hence produced at the beginning of each month and for the next 7 months. Operationally, the CM-SSF forecasts are currently used in EFAS to generate a seasonal streamflow outlook for Europe at the beginning of every month.

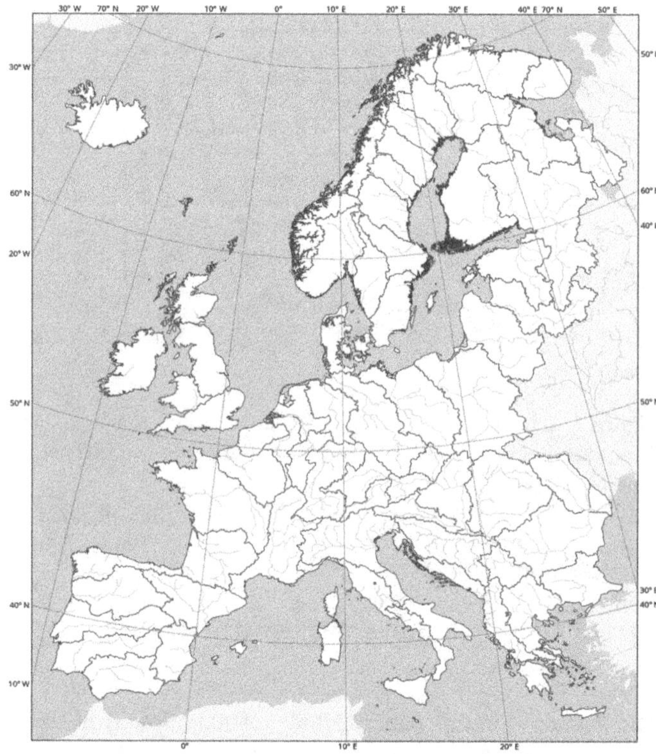

Figure 2. Map of the 74 European regions (dark blue outlines) selected for the analysis of the CM-SSF and the ESP.

2.2 Hindcast evaluation strategy

For this study, monthly region specific discharge averages of the hindcasts (CM-SSF and ESP) and EFAS-WB were used. The specific discharge is the discharge per unit area of an upstream basin. For this paper, the gridded daily specific discharge was calculated by dividing the gridded daily discharge output maps (of the hindcasts and the EFAS-WB) by the Lisflood gridded upstream area static map. Subsequently, the gridded daily specific discharge maps were used to calculate daily region averaged specific discharges (for each region in Fig. 2) by summing up the daily specific discharge values of each grid cell within a region, divided by the number of grid cells in that region. Finally, monthly specific discharge region averages were calculated for each calendar month.

The regions displayed in Fig. 2 were created by merging several basins together (basins used operationally in EFAS for the shorter timescale forecasts), while respecting hydro-climatic boundaries. They were chosen for the analysis presented in this paper for two main reasons. Firstly, they are the regions used operationally to display the EFAS seasonal streamflow outlook. Secondly, they were created in order to capture large-scale variability in the weather.

The analysis of the hindcasts was performed on monthly specific discharge (hereafter referred to as streamflow) region averages for hindcast starting dates spanning February 1990 to November 2016 (included; approximately 27 years of data), with 1 to 7 months of lead time. In this paper, 1 month of lead time refers to the first month of the forecast (e.g. the January 2017 streamflow for a forecast made on 1 January 2017). Two months of lead time is the second month of the forecast (e.g. the February 2017 streamflow for a forecast made on 1 January 2017), etc. Monthly averages were selected for the analysis presented in this paper as it is a valuable aggregation time step for decision-makers for many water-related applications (as shown in the literature for applications such as, for example, navigation (Meißner et al., 2017), reservoir management (Viel et al., 2016; Turner et al., 2017), drought-risk management (Yuan et al., 2013), irrigation (Chiew et al., 2003; Li et al., 2017) and hydropower (Hamlet et al., 2002)).

Several verification scores were selected in order to assess the hindcasts' quality. These verification scores were chosen to cover a wide range of hindcast attributes (i.e. accuracy, sharpness, reliability, overall performance and discrimination). All of these verification scores, except for the verification score selected to look at hindcast discrimination, are the same as chosen in Crochemore et al. (2016), and are described below. The EFAS-WB streamflow simulations were used as a proxy for observation against which the seasonal streamflow hindcasts were evaluated, hence minimizing the impact of model errors on the hindcasts' quality.

2.2.1 Hindcast accuracy

Both hindcasts (CM-SSF and ESP) were assessed in terms of their accuracy, the magnitude of the errors between the hindcast ensemble mean and the "truth" (i.e. the EFAS-WB). For this purpose, the mean absolute error (MAE) was calculated for each region, target month (i.e. the month that is being forecast) and lead time (i.e. 1 to 7 months). The lower the MAE, the more accurate the hindcast.

2.2.2 Hindcast sharpness

Both hindcasts were also assessed in terms of their sharpness, an attribute of the hindcast only, which is a measure of the spread of the ensemble members of a hindcast. In this paper, the 90 % interquantile range (IQR; i.e. the difference between the 95th and 5th percentiles of the hindcast distribution) was calculated for each region, target month and lead time. The lower the IQR, the sharper the hindcast.

2.2.3 Hindcast reliability

Both hindcasts were additionally assessed in terms of their reliability, the statistical consistency between the hindcast probabilities and the observed frequencies. For this purpose, the probability integral transform (PIT) diagram was calculated for each region, target month and lead time (Gneiting et al., 2007). The PIT diagram is the cumulative distribution of the PIT values as a function of the PIT values. The PIT values measure where the "truth" (i.e. EFAS-WB) falls relative to the percentiles of the hindcast distribution. For a perfectly reliable hindcast, the "truth" should fall uniformly in each percentile of the hindcast distribution, giving a PIT diagram that falls exactly on the 1-to-1 diagonal. A hindcast that systematically under- (over-) predicts the "truth" will have a PIT diagram below (above) the diagonal. A hindcast that is too narrow (i.e. underdispersive; hindcast distribution smaller than the distribution of the observations) (large (i.e. overdispersive; hindcast distribution greater than the distribution of the observations)) will have a transposed S-shaped (S-shaped) PIT diagram (Laio and Tamea, 2007).

In order to compare the reliability across all regions, target months and lead times, the area between the PIT diagram and the 1-to-1 diagonal was computed for all PIT diagrams (Renard et al., 2010). The smaller this area, the more reliable the hindcast.

Furthermore, to disentangle the causes of poor reliability, the spread and bias of the hindcasts were calculated for all PIT diagrams, using two measures first introduced by Keller and Hense (2011): ß-score and ß-bias, respectively. By definition, a perfectly reliable hindcast (with regards to its spread) will have a β-score of zero (to which a tolerance interval of ± 0.09 was added), whereas a hindcast that is too narrow (large) will have a negative (positive) β-score (outside of the tolerance interval). A perfectly reliable hindcast (with regards to its bias) will have a β-bias of zero (to which a tolerance interval of ± 0.09 was added), whereas a hindcast that systematically under- (over-) predicts the "truth" will have a negative (positive) β-bias (outside of the tolerance interval).

2.2.4 Hindcast overall performance

The hindcasts were furthermore assessed in terms of their overall performance from the continuous rank probability score (CRPS), calculated for each region, target month and lead time (Hersbach, 2000). The CRPS is a measure of the difference between the hindcast and the observed (i.e. EFAS-WB) cumulative distribution functions. The lower the CRPS, the better the overall performance of the hindcast.

In this paper, the skill of the CM-SSF is benchmarked with respect to the ESP in order to identify whether there is any added value in using Sys4 instead of historical meteorological observations for forecasting the streamflow on seasonal timescales over Europe. To this end, skill scores were calculated for the MAE, IQR, PIT diagram area and CRPS, using the following equation:

$$\text{Skill score} = 1 - \frac{\text{score}_{\text{CM-SSF}}}{\text{score}_{\text{ESP}}}. \tag{1}$$

Skill scores were calculated for each region, target month and lead time and will be referred to as MAESS, IQRSS, PITSS and CRPSS, respectively. Skill scores larger (smaller) than zero indicate more (less) skill in the CM-SSF compared to the ESP. A skill score of zero means that the CM-SSF is as skilful as the ESP. Note that as the ESP is not a "naive" forecast, using it as a benchmark might lead to lower skill than benchmarking the CM-SSF against, for example, climatology.

2.2.5 Hindcast potential usefulness

For decision-making, the ability of a seasonal forecasting system to predict the right category of an event (e.g. above or below normal conditions) months ahead is of great importance (Gobena and Gan, 2010). In this paper, the potential usefulness of the CM-SSF and the ESP to forecast lower and higher than normal streamflow conditions within their hindcasts is assessed.

To do so, the relative operating characteristic (ROC) score, a measure of hindcast discrimination (Mason and Graham, 1999), was calculated. The thresholds selected to calculate the ROC are the lower and upper terciles of the EFAS-WB climatology for each season. They were calculated for the simulation period (February 1990 to May 2017), by grouping together EFAS-WB monthly streamflows for each month falling in a season (SON: September–October–November, DJF: December–January–February, MAM: March–April–May and JJA: June–July–August). For each season and each region a lower and upper tercile streamflow value was obtained, subsequently used as thresholds against which to calculate the probability of detection (POD) and the false alarm rate (FAR; with 0.1 probability bins) for both hindcasts, and for each region, season and lead time. Finally, the area under the ROC curve, i.e. the ROC score, was calculated for both hindcasts, for each region, season and lead time. The ROC score ranges from 0 to 1, with a perfect score of 1. A hindcast with a ROC score ≤ 0.5 is unskilful, i.e. less good than the long-term average climatology which has a ROC of 0.5, and is therefore not useful.

Because the ROC score was calculated from a low number of events (i.e. approximately 27 years $\times$ 3 months in each season $\times 1/3$ (lower or upper tercile) $= 27$ simulated events), the hindcasts were judged skilful and useful when their ROC score ≥ 0.6 instead of 0.5. Moreover, the CM-SSF was categorized as more useful than the ESP when the CM-SSF's ROC score was at least 10 % larger than the ESP's ROC score.

3 Results

3.1 Overall skill of the CM-SSF

In the first part of the results, the skill of the CM-SSF (benchmarked with respect to the ESP) is presented, in terms of the accuracy (MAESS), sharpness (IQRSS), reliability (PITSS) and overall performance (CRPSS) in the hindcast datasets. This will benchmark the added value of using Sys4 against the use of historical meteorological observations for forecasting the streamflow on seasonal timescales over Europe.

As shown by the MAESS boxplots (Fig. 3), the CM-SSF appears on average more accurate than the ESP for the first month of lead time only, for all seasons excluding spring (MAM). Beyond 1 month of lead time, the CM-SSF becomes on average as or less accurate than the ESP. There are however noticeable differences between the different seasons. The CM-SSF shows the largest improvements in the average accuracy compared to the ESP in winter (DJF) and for the first month of lead time. For longer lead times (i.e. 2 to 7 months), the accuracy of the CM-SSF is on average quite similar to that of the ESP in autumn (SON) and winter, and on average lower in spring and summer (JJA). The boxplots for the CRPSS look very similar to the MAESS boxplots, the main difference being the lower average scores for 2 to 7 months of lead time in autumn and winter (Fig. 3).

The boxplots of the IQRSS show that the CM-SSF predictions are on average as sharp as those of the ESP for the first month of lead time (slightly sharper in autumn; Fig. 3). For 2 to 7 months of lead time, in autumn and winter, the CM-SSF predictions are on average sharper than those of the ESP, whereas in spring and summer, the CM-SSF predictions are on average slightly less sharp than the ESP predictions.

As shown by the boxplots of the PITSS (Fig. 3), the CM-SSF predictions are less reliable than the ESP predictions for all seasons and months of lead time. For the first month of lead time and all seasons, 10–20 % of the ESP hindcasts and less than 5 % of the CM-SSF hindcasts are reliable (Fig. 4). About 40–60 % of the ESP hindcasts are not reliable for the first month of lead time and all seasons due to the ensemble spread. Approximately half of these hindcasts are too large, while the other half (slightly more in autumn and winter) are too narrow. Furthermore, 50–80 % of the ESP hindcasts under-predict the simulated streamflow for the first month of lead time and all seasons. The percentage of reliable (unreliable) ESP hindcasts increases (decreases) with lead time, as the effect of the IHC fades away. About 70–90 % of the CM-SSF hindcasts are too narrow for the first month of lead time and all seasons. With increasing lead time, the percentage of CM-SSF hindcasts that are too narrow (large) decreases (increases), especially in spring. Approximately 40–50 % of the CM-SSF hindcasts over-predict the simulated streamflow in spring and summer for the first month of lead time (and increasingly over-predict with longer lead times). In autumn and winter, about 70 % of the CM-SSF hindcasts under-predict the simulated streamflow for the first month of lead time (and increasingly under-predict with longer lead times).

For all verification scores, the boxplots for autumn and winter are slightly smaller than for spring and summer, hinting at a smaller variability in the verification scores amongst regions and target months in autumn and winter than in spring and summer. Furthermore, the presence of the boxplots above the zero line (i.e. no skill line) for all lead times suggests that the CM-SSF is more skilful than the ESP for some regions and target months, beyond the first month of lead time.

3.2 Potential usefulness of the CM-SSF

In the second part of the results, the potential usefulness of the CM-SSF compared to the ESP is described for decision-making. Here, potential usefulness is defined as the ability of the forecasting systems to predict lower or higher streamflows than normal, as measured with the ROC score.

Generally, either of the two forecasting systems (CM-SSF or ESP) is capable of predicting skilfully whether the streamflow will be anomalously low or high in the coming months (Fig. 5). However, for a few seasons and regions, none of the two forecasting systems is skilful at predicting lower and/or higher streamflows than normal. This is especially noticeable in winter.

For most seasons and regions, the ESP is more skilful than the CM-SSF at predicting lower and higher streamflows than normal. However, in winter for most regions and during other seasons for several regions, the CM-SSF appears more skilful than the ESP. Regions where the CM-SSF best predicts lower and higher streamflows than normal at most lead times are summarized in Table 1 for all four seasons and the lower and upper terciles of the simulated streamflow.

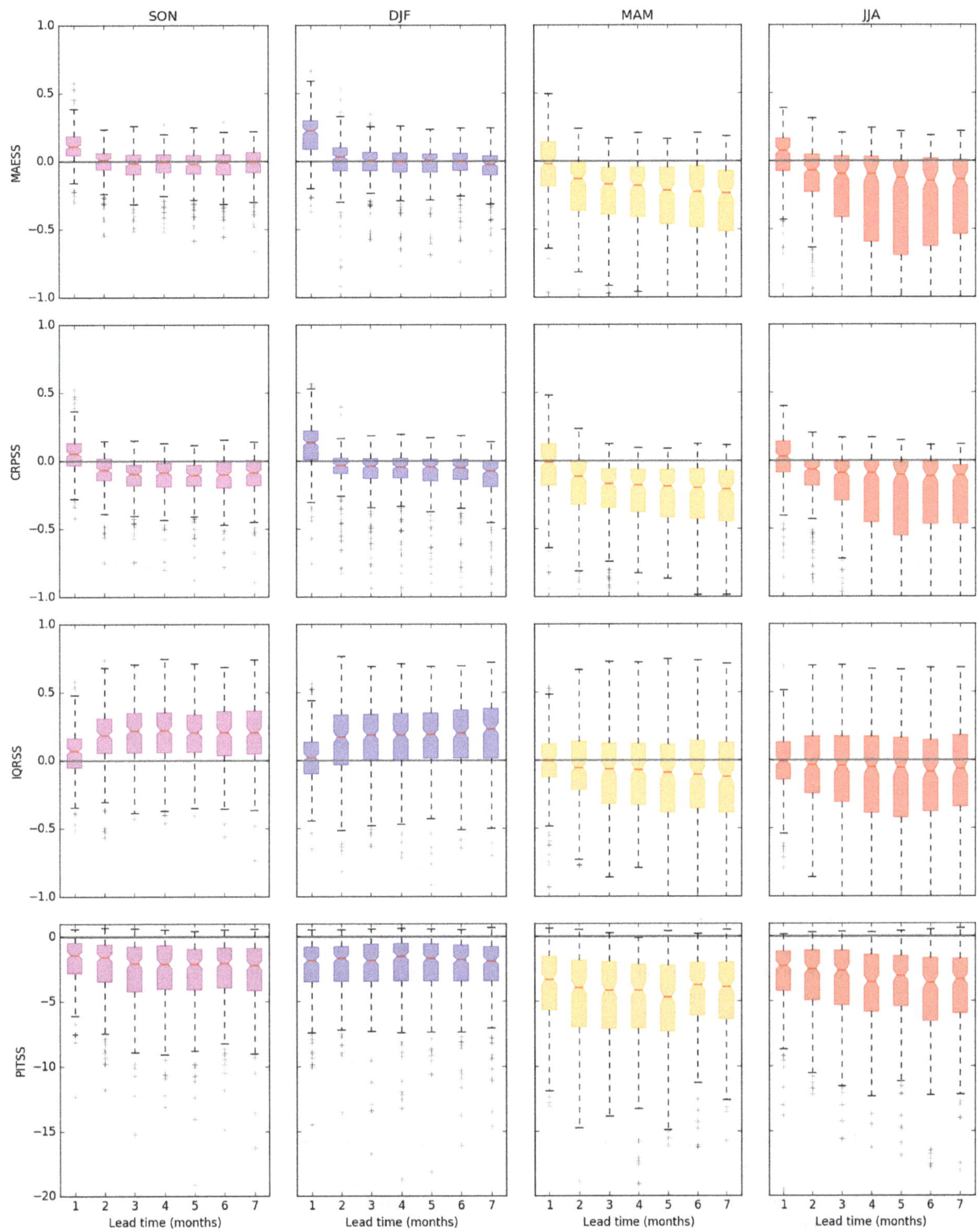

Figure 3. Boxplots of the MAESS, CRPSS, IQRSS and PITSS (from the top to bottom rows) for all four seasons (SON, DJF, MAM and JJA from the left-most to right-most columns) as a function of lead time (i.e. 1 to 7 months). The boxplots contain the scores for all target months falling in a given season and all 74 European regions. For all scores, values larger (smaller) than zero indicate that the CM-SSF is more (less) skilful than the ESP (benchmark). Where the skill is zero, the CM-SSF is as skilful as the ESP for the hindcast period. Note that the PITSS plots have a different y-axis scale.

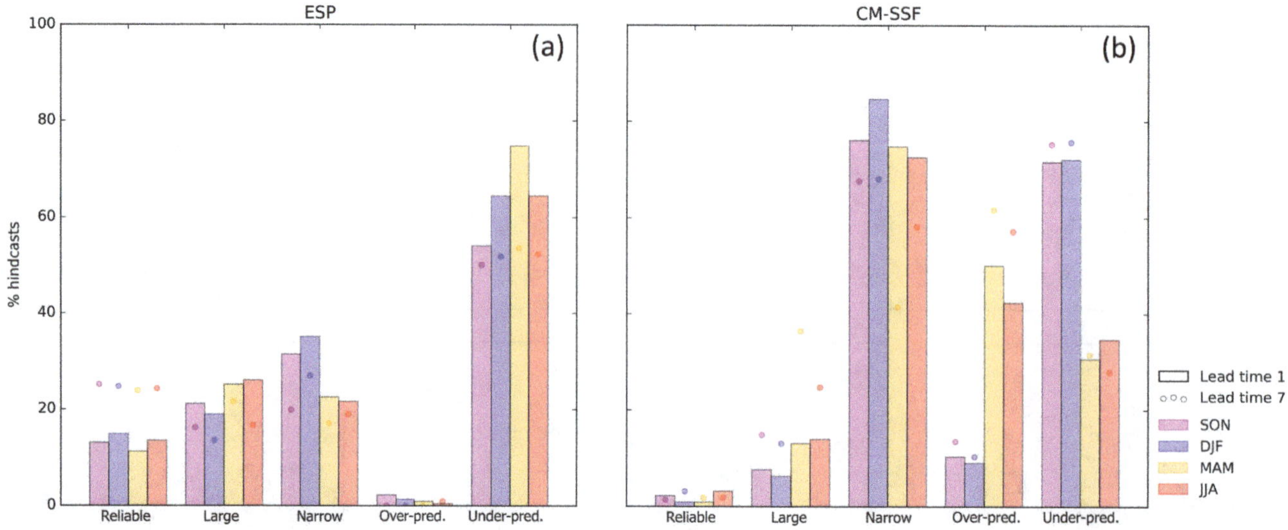

Figure 4. Plots of the percentage of the ESP **(a)** and the CM-SSF **(b)** hindcasts falling into each reliability category (reliable – in terms of both spread and bias, too large, too narrow, over-predicting and under-predicting) for all four seasons (SON, DJF, MAM and JJA from the left-most to right-most bars in each reliability category). The results are shown as bar charts for the first month of lead time and as circles for the seventh month of lead time. These lead times were selected for display to highlight the evolution of reliability between the first and last months of the hindcast. The percentages were calculated from hindcasts for all target months falling in a given season and all 74 European regions.

Table 1. Regions where the CM-SSF is more skilful than the ESP at predicting anomalously low (lower tercile; first column) or high (upper tercile; second column) streamflows for all four seasons (SON, DJF, MAM and JJA from the top to bottom rows). This is a summary of the information displayed in Fig. 5.

	Lower tercile	Upper tercile
SON	– Few regions in Fennoscandia – Po River basin (northern Italy) – Elbe River basin (south of Denmark) – Upstream of the Rhine River basin – Upstream of the Danube River basin – Duero River basin (Iberian Peninsula)	– Few regions in Fennoscandia – Iceland – Parts of the Danube River basin – Segura River basin (Iberian Peninsula)
DJF	Many regions except – in most of Fennoscandia north of the Baltic Sea, – parts of central Europe.	Same as lower tercile.
MAM	– Few regions on the Iberian Peninsula – Few regions in the western part of central Europe	Same as lower tercile.
JJA	– Few regions in the United Kingdom (UK) – Ireland – North-western edge of the Iberian Peninsula – Regions in Fennoscandia around the Baltic Sea – Regions south of the North Sea	– Northern part of the UK – Ireland – North-western edge of the Iberian Peninsula – Regions in Fennoscandia around the Baltic Sea – Around the Elbe River basin – Upstream of the Danube River basin – Along the Adriatic Sea in Italy

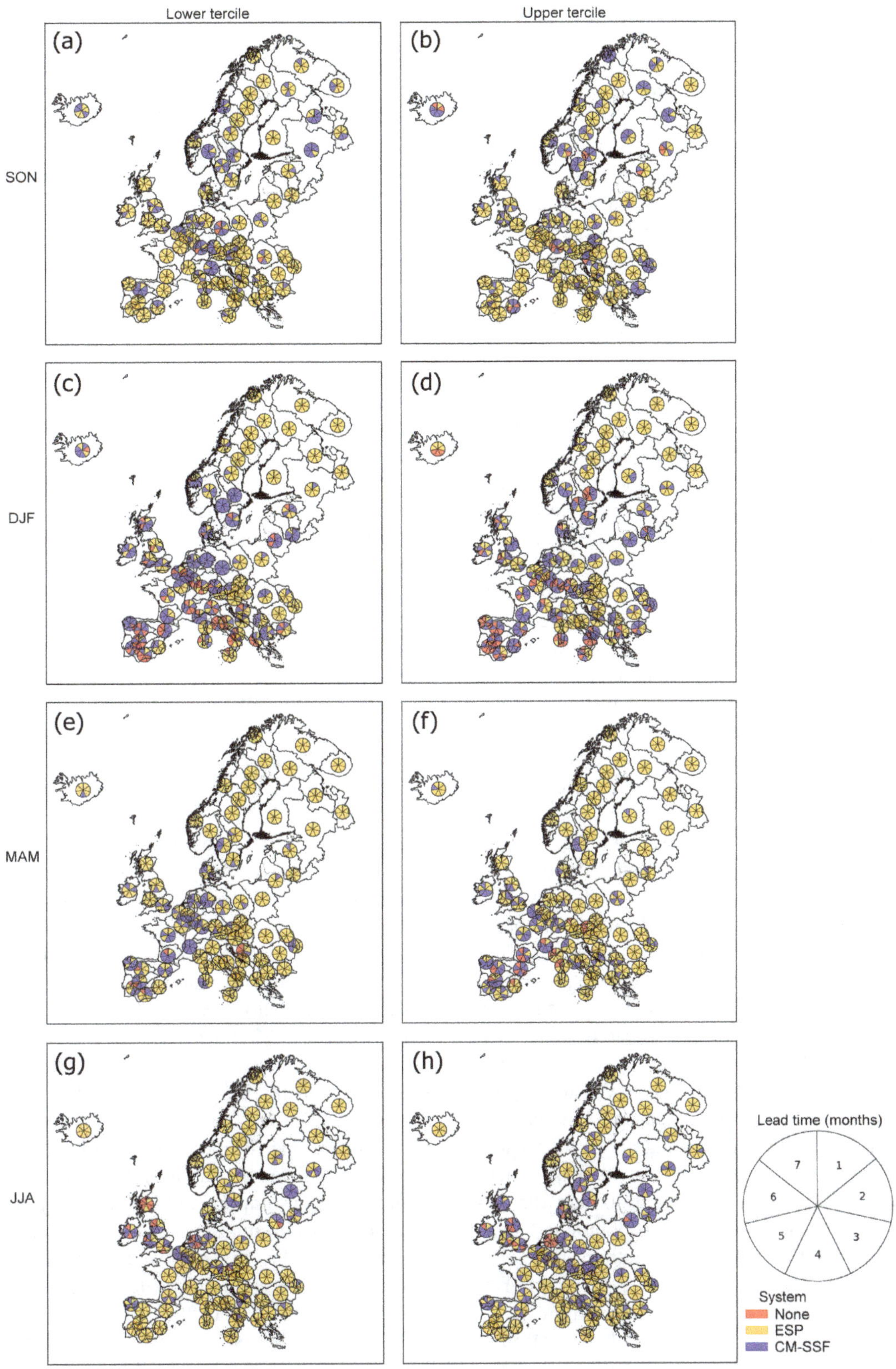

Figure 5. Maps of the best system (as measured with the ROC score) for all four seasons (SON, DJF, MAM and JJA) and the lower and upper simulated streamflow seasonal terciles (left-most and right-most columns, respectively) in each region from **(a)** to **(h)**. The pie charts display the best system for each lead time (i.e. 1 to 7 months), as shown in the example pie chart on the bottom right of this figure. There are three possible cases: (1) neither the ESP nor the CM-SSF is skilful (red colours), (2) the ESP is skilful and better than the CM-SSF (yellow colours), and (3) the CM-SSF is skilful and better than the ESP (blue colours).

4 Discussion

4.1 Does seasonal climate information improve the predictability of seasonal streamflow forecasts over Europe?

On average over Europe and across all seasons, the CM-SSF is skilful (in terms of hindcast accuracy, sharpness and overall performance, using the ESP as a benchmark) for the first month of lead time only. This means that, on average, Sys4 improves the predictability over historical meteorological information for pan-European seasonal streamflow forecasting for the first month of lead time only. At longer lead times, historical meteorological information becomes as good as or better than Sys4 for seasonal streamflow forecasting over Europe. Crochemore et al. (2016) and Meißner et al. (2017) similarly found positive skill in the seasonal streamflow forecast (Sys4 forced hydrological model compared to an ESP) for the first month of lead time, after which the skill faded away for basins in France and central Europe, respectively. Additionally, on average over Europe and across all seasons, the CM-SSF is less reliable than the ESP for all lead times. This is due to a combination of too narrow and biased CM-SSF hindcasts, where the bias depends on the season that is being forecasted. As mentioned in the methods section of this paper, the ESP is not a "naive" benchmark, which might partially explain the limited predictability gained from Sys4.

The predictability varies per season and the CM-SSF predictions are on average sharper than and as accurate as the ESP predictions in autumn and winter beyond the first month of lead time (and increasingly sharper with longer lead times). The CM-SSF however tends to systematically under-predict the autumn and winter simulated streamflow (and increasingly under-predicts with longer lead times). In spring and summer, the CM-SSF predictions are on average less sharp and less accurate than the ESP predictions, and they tend to systematically over-predict the simulated streamflow (and increasingly over-predict with longer lead times).

The added predictability gained from Sys4 was shown to lead to skilful CM-SSF predictions of lower and higher streamflows than normal for specific seasons and regions. The CM-SSF is more skilful at predicting anomalously low and high streamflows than the ESP in certain seasons and regions, and noticeably in winter in almost 40 % of the European regions, mostly clustered in rainfall-dominated areas of western and central Europe. Several authors have discussed the higher winter predictability over (parts of) Europe, with examples in basins in France (Crochemore et al., 2016), central Europe (Steirou et al., 2017), the UK (Bell et al., 2017) and the Iberian Peninsula (Lorenzo-Lacruz et al., 2011). Bierkens and van Beek (2009) additionally showed that there was a higher winter predictability in Scandinavia, the Iberian Peninsula and around the Black Sea. Our results are mostly consistent with these findings, except for Scandinavia, where the ESP is more skilful than the CM-SSF in winter. Bierkens and van Beek (2009) produced the seasonal streamflow forecast analysed in their paper by forcing a hydrological model with resampled years of historical meteorological information based on their winter NAO index. However, Sys4 has difficulties in forecasting the NAO over Europe (Kim et al., 2012), which could have led to these inconsistent results with the ones presented by Bierkens and van Beek (2009).

In spring, the CM-SSF is more skilful than the ESP at predicting lower and higher streamflows than normal beyond 1 month of lead time in approximately 15 % of the European regions, and mostly in regions of western Europe. This could be due to a persistence of the skill from the previous winter through the land surface memory (i.e. groundwater-driven streamflow or snowmelt-driven streamflow), as highlighted by Bierkens and van Beek (2009) for Europe, Singla et al. (2012) for parts of France, Lorenzo-Lacruz et al. (2011) for the Iberian Peninsula and Meißner et al. (2017) for the Rhine. Moreover, it could be that most of the gained predictability occurs in March, a transition month between the more predictable winter (as mentioned above) and spring, as discussed by Steirou et al. (2017). The ESP is overall more skilful than the CM-SSF at predicting the spring streamflow in snow-dominated regions (e.g. most of Fennoscandia and parts of central and eastern Europe). This hints at the importance of the IHC (i.e. of snowpack) and the land surface memory for forecasting the spring streamflow in snow-dominated regions in Europe.

The added predictability from Sys4 for forecasting lower and higher streamflows than normal is limited in summer and autumn for most regions. The CM-SSF is more skilful at predicting anomalously low and high streamflows than the ESP in about 10–20 % of the European regions during those seasons. Other studies have found similar patterns for (parts of) Europe; these include less skill in summer than in winter overall for basins in France (Crochemore et al., 2016), less skill for the low flow season (July to October) for basins in central Europe (Meißner et al., 2017), negative correlations in summer and autumn seasonal streamflow forecasts in central Europe as the influence of the winter NAO fades away (Steirou et al., 2017), and less skill overall in summer than in winter in Europe (Bierkens and van Beek, 2009). The lower CM-SSF skill for predicting lower and higher streamflows than normal in summer could additionally be due to the convective storms in summer over Europe, which are hard to predict, and to the fact that it is the dry season in most of Europe, where rivers are groundwater fed. Therefore, in this season, the quality of the IHC controls the streamflow predictability.

While the CM-SSF is most skilful (in terms of hindcast accuracy, sharpness and overall performance, using the ESP as a benchmark) in autumn and winter and most potentially useful in winter, this does not appear to correlate with high performance in the Sys4 precipitation and temperature hindcasts (as seen on the maps of correlation for Sys4 precipitation and temperature for all four seasons (SON, DJF,

MAM and JJA) and with 2 months of lead time (as identified in this paper); available at https://meteoswiss.shinyapps.io/skill_metrics/, Forecast skill metrics, 2017). Over Europe, the Sys4 precipitation and temperature hindcasts are the most skilful in summer and the least skilful in autumn and winter. Moreover, the regions of high CM-SSF skill for predicting lower and upper streamflows than normal do not clearly correspond to regions of high performance in the Sys4 precipitation and temperature hindcasts. These differences could be partially induced by the different benchmark used to evaluate the skill of the CM-SSF (i.e. the ESP) compared to the one used to look at the performance of the Sys4 precipitation and temperature hindcasts (i.e. ERA-Interim). However, these results clearly indicate that looking at the performance of the Sys4 precipitation and temperature hindcasts only does not give a good indication of the skill and potential usefulness of the seasonal streamflow hindcasts over Europe, and that marginal performance in seasonal climate forecasts can translate through to more predictable seasonal streamflow forecasts, and vice versa. The added predictability in the CM-SSF could be due to the combined predictability in the precipitation and temperature hindcasts, as well as a lag in the predictability from the land surface memory.

In most regions and for most seasons, at least one of the two forecasting systems (CM-SSF or ESP) is able to predict lower or higher streamflows than normal. However, in winter, the number of regions and lead times for which none of the forecasting systems are skilful increases. This could be because in winter, many regions experience weather-driven high streamflows and the performance of Sys4 is limited at this time of year (as mentioned above). In those regions, the seasonal streamflow forecasts could be improved either by improving the IHC, through for example data assimilation, or by improving the seasonal climate forecasts.

Overall, the ESP appears very skilful at forecasting lower or higher streamflows than normal, showing the importance of IHC and the land surface memory for seasonal streamflow forecasting (Wood and Lettenmaier, 2008; Bierkens and van Beek, 2009; Yuan et al., 2015b).

4.2 What is the potential usefulness and usability of the EFAS seasonal streamflow forecasts for flood preparedness?

What appears like little added skill does not necessarily mean no skill for the forecast users and can in fact be a large added value for decision-making (Viel et al., 2016). The ability of a seasonal streamflow forecasting system to predict the right category of an event months ahead is valuable for many water-related applications (e.g. navigation, reservoir management, drought-risk management, irrigation, water resource management, hydropower and flood preparedness). From the results presented in this paper, it appears that either of the two forecasting systems (CM-SSF or ESP) is capable of predicting lower or higher streamflows than normal months in advance, thanks to the predictability gained from the IHC, the land surface memory and the seasonal climate hindcast in some regions and for certain seasons.

However, as highlighted by White et al. (2017), there is currently a gap between usefulness and usability of seasonal information. What is a useful scientific finding does not automatically translate into usable information which will fit into any user's decision-making chain (Soares and Dessai, 2016). While several authors have already investigated the usability of seasonal streamflow forecasts for applications such as navigation (Meißner et al., 2017), reservoir management (Viel et al., 2016; Turner et al., 2017), drought-risk management (Sheffield et al., 2013; Yuan et al., 2013; Crochemore et al., 2017), irrigation (Chiew et al., 2003; Li et al., 2017), water resource management (Schepen et al., 2016) and hydropower (Hamlet et al., 2002), its application to flood preparedness is still left mostly unexplored. One exception being Neumann et al. (in review), who look at the use of the CM-SSF to predict the 2013/14 Thames basin floods. This is partially due to the complex nature of flood generating mechanisms, still poorly studied on seasonal timescales beyond snowmelt-driven spring floods, as well as the fact that seasonal forecasts reflect the likelihood of abnormal seasonal streamflow totals, but without much skilful information on the exact timing, location and severity of the impact of individual flood events within that season. Coughlan de Perez et al. (2017) looked at the usefulness of seasonal rainfall forecasts for flood preparedness in Africa and highlighted the complexities behind using these forecasts as a proxy for floodiness (for a discussion on floodiness, see Stephens et al., 2015). Furthermore, decision-makers in the navigation, reservoir management, drought-risk management, irrigation, water resource management and hydropower sectors are familiar with working on long timescales (i.e. several weeks to months ahead). In contrast, the flood preparedness community is currently mostly used to working on timescales of hours to a couple of days.

The Red Cross Red Crescent Climate Centre has recently designed a new approach that harnesses the usefulness of seasonal climate information for decision-making for disaster management. This approach, called "Ready-Set-Go!", is made up of three stages. The "Ready" stage is based on seasonal forecasts, where they are used as monitoring information to drive contingency planning (e.g. volunteer training). The "Set" stage is triggered by sub-seasonal forecasts, used as early-warning information to alert volunteers. Finally, the "Go!" stage is based on short-range forecasts and consists in the evacuation of people and the distribution of aid (White et al., 2017). Using a similar approach, seasonal streamflow forecasts could complement existing forecasts at shorter timescales and provide monitoring and early-warning information for flood preparedness. Such an approach however requires the use of consistent forecasts from short to seasonal timescales. In this context, moving to seamless forecasting is becoming vital (Wetterhall and Di Giuseppe, in review).

Soares and Dessai (2016) also identified the accessibility to the information, enhanced by collaborations and ongoing relationships between users and producers, as a key enabler of the usability of seasonal information. International projects, such as the Horizon 2020 IMPREX (IMproving PRedictions and management of hydrological EXtremes) project (van den Hurk et al., 2016), alongside promoting scientific progress on hydrological extremes forecasting from short to seasonal timescales over Europe, gather together forecasters and decision-makers and can effectively demonstrate the added value of the integration of seasonal information in decision-making chains. The Hydrologic Ensemble Prediction EXperiment (HEPEX) is another international initiative that brings together researchers and practitioners in the field of ensemble prediction for water-related applications. It is an ideal environment for collaboration and fosters communication and outreach on topics such as the usefulness and usability of seasonal information for decision-making.

4.3 Aspects for future work

In this paper, terciles of the simulated streamflow are used. However, and because the application of the EFAS seasonal streamflow forecasts is of particular relevance for flood preparedness, the evaluation of the hindcasts for lower and higher streamflow extremes (for example the 5th and 95th percentiles, respectively) would be more relevant and might give very different results. This was not done in this paper as the time period covered by the seasonal streamflow hindcasts (i.e. approximately 27 years) was not long enough for statistically reliable results for lower and higher streamflow extremes. The limited hindcast length is a common problem in seasonal predictability studies. Increasing the hindcast length back in time could lead to more stable Sys4 hindcasts and hence to more stable and potentially skilful seasonal streamflow hindcasts (Shi et al., 2015).

Furthermore, in this paper, the hindcasts were analysed against simulated streamflow, used as a proxy for observed streamflow. This is necessary because it enables an analysis of the quality of the hindcasts over the entire computation domain, rather than at non-evenly spaced stations over the same domain (Alfieri et al., 2014). Further work could however include carrying out a similar analysis for selected river stations in Europe, in order to account for model errors in the hindcast evaluation.

The calculation of the verification scores (excluding the ROC) was made by randomly selecting 15 ensemble members from the 51 ensemble members of the CM-SSF hindcasts, for starting dates for which the ensemble varies between 15 and 51 members (i.e. hindcasts made on 1 January, March, April, June, July, September, October and December; this is due to the split between 15 and 51 ensemble members in the Sys4 hindcasts, as described in Sect. 2.1.2 of this paper). In order to investigate the potential impact of this evaluation strategy on the results presented in this paper, the

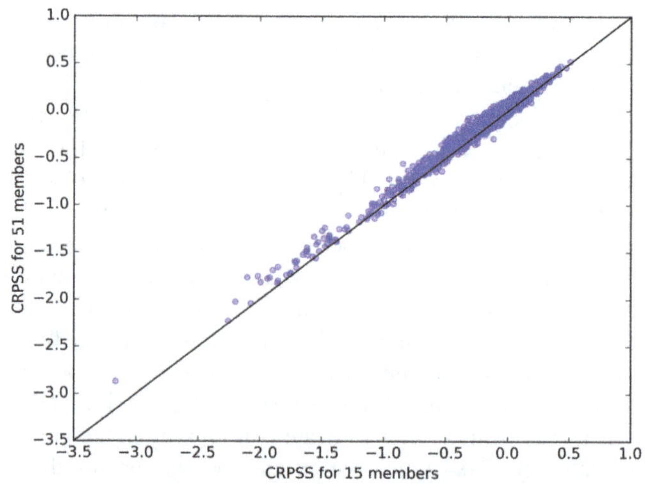

Figure 6. CRPSS calculated for the CM-SSF against the ESP (benchmark) for hindcasts made on 1 February, May, August and November, all lead times (i.e. 1 to 7 months) and all 74 European regions. The x-axis (y-axis) contains the CRPSS calculated from 15 (all 51) ensemble members of the CM-SSF.

CRPSS was calculated for 15 and 51 ensemble members of the CM-SSF hindcasts for starting dates for which 51 ensemble members are available for the full hindcast period (i.e. hindcasts made on 1 February, May, August and November). This is displayed in Fig. 6 for all hindcast starting dates, lead times (i.e. 1 to 7 months) and regions combined. Overall, it is apparent that the impact of this evaluation strategy on the results presented in this paper should be minimal, as all points align themselves approximately with the 1-to-1 diagonal.

The next version of the ECMWF seasonal climate forecast, SEAS5, was released in November 2017. Future work could include forcing the Lisflood model with SEAS5 and comparing the obtained seasonal streamflow hindcasts to the CM-SSF presented in this paper. This should indicate whether developments to the seasonal climate forecast translate through to better pan-European seasonal streamflow forecasts, which is of particular interest for regions and seasons when neither the ESP nor the CM-SSF is currently skilful.

The operational EFAS medium-range streamflow forecasts are currently post-processed as a means to improve their reliability (Smith et al., 2016, and references therein). Results from this paper have shown that the CM-SSF is mostly unreliable (with regards to the EFAS-WB) and could hence benefit from post-processing of the seasonal climate forecast. However, post-processing techniques used for the EFAS medium-range streamflow forecasts might not be suitable for the CM-SSF, as the seasonal climate forecast used for the latter should be post-processed in terms of its seasonal anomalies rather than for errors in the timing, volume and magnitude of specific events. This is currently being considered for operational implementation within EFAS and is an active area of discussion within the EFAS user community.

For the analysis presented in this paper, the CM-SSF was benchmarked against the ESP. Several other techniques exist for seasonal streamflow forecasting, such as statistical methods using predictors ranging from climate indices to antecedent observed precipitation and crop production metrics, to mention a few (e.g. Mendoza et al., 2017; Slater et al., 2017). Further analysis could include benchmarking the CM-SSF against one or multiple statistical methods, to assess the relative benefits of various seasonal streamflow forecasting techniques.

In this paper, the ability of both systems (CM-SSF and ESP) to forecast lower and higher streamflows than normal was explored, with several hypotheses made to link the streamflow predictability to regions' hydro-climatic processes. This includes the higher potential usefulness of the ESP in forecasting the spring streamflow in snow-dominated regions and the summer streamflow in regions where rivers are groundwater fed. In these regions and for these seasons, the IHC and the land surface memory drive the predictability. The CM-SSF provides an added potential usefulness in winter in the rainfall-dominated regions of central and western Europe, where the skill appears to persist through to spring due to the land surface memory (i.e. groundwater-driven streamflow and snowmelt-drive streamflow). While further exploration of these hypotheses is outside of the scope of this paper, future work is required to disentangle the links between the added predictability from Sys4 and the basins' hydro-climatic characteristics, for example, understanding the predictability in snow-dominated basins, arid regions and temperate groundwater-fed basins.

In this context, additional work to further disentangle and quantify the contribution of both predictability sources (seasonal climate forecasts versus IHC) to seasonal streamflow forecasting quality over Europe could be carried out by using the EPB (end point blending) method (Arnal et al., 2017).

5 Conclusions

In this paper, the newly operational EFAS seasonal streamflow forecasting system (producing the CM-SSF forecasts by forcing the Lisflood model with the ECMWF System 4 seasonal climate forecasts (Sys4)) was presented and benchmarked against the ESP forecasting approach (ESP forecasts produced by forcing the Lisflood model with historical meteorological observations) for the hindcast period 1990 to 2017. On average, Sys4 improves the predictability over historical meteorological information for pan-European seasonal streamflow forecasting for the first month of lead time only (in terms of hindcast accuracy, sharpness and overall performance). However, the predictability varies per season and the CM-SSF is more skilful on average at predicting autumn and winter streamflows than spring and summer streamflows. Additionally, parts of Europe exhibit a longer predictability, up to 7 months of lead time, for certain months within a season. In terms of hindcast reliability, the CM-SSF is on average less skilful than the ESP for all lead times, due to a combination of too narrow and biased CM-SSF hindcasts, where the bias depends on the season that is being forecasted.

Subsequently, the potential usefulness of the two forecasting systems (CM-SSF and ESP) was assessed by analysing their skill in predicting lower and higher streamflows than normal. Overall, at least one of the two forecasting systems is capable of predicting those events months in advance. The ESP appears the most skilful on average, showing the importance of IHC and the land surface memory for seasonal streamflow forecasting. Nevertheless, for certain regions and seasons the CM-SSF is the most skilful at predicting anomalously low or high streamflows beyond 1 month of lead time, noticeably in winter for almost 40 % of the European regions. This potential usefulness could be harnessed by using seasonal streamflow forecasts as complementary information to existing forecasts at shorter timescales, to provide monitoring and early-warning information for flood preparedness.

Overall, patterns in skill in the CM-SSF are however not mirrored in the Sys4 precipitation and temperature hindcasts. This suggests that using seasonal climate forecast performance as a proxy for seasonal streamflow forecasting skill is not adequate and that more work is needed to understand the link between meteorological and hydrological variables on seasonal timescales over Europe.

Competing interests. The authors declare that they have no conflict of interest.

Special issue statement. This article is part of the special issue "Sub-seasonal to seasonal hydrological forecasting". It is not associated with a conference.

Acknowledgements. Louise Arnal, Hannah L. Cloke and Jessica Neumann gratefully acknowledge financial support from the Horizon 2020 IMPREX project (grant agreement 641811) (project IMPREX: www.imprex.eu). Louise Arnal's time was additionally partly funded by a University of Reading PhD scholarship. Fredrik Wetterhall, Christel Prudhomme and Blazej Krzeminski's work was supported by the EFAS computational centre in support to the Copernicus Emergency Management Service/Early Warning Systems (Flood) (contract no. 198702 from JRC-IES). Elisabeth Stephens is thankful for support from the Natural Environment Research Council and Department for International Development (grant number NE/P000525/1) under the Science for Humanitarian Emergencies and Resilience (SHEAR) research programme.

Edited by: Ilias Pechlivanidis

References

Alfieri, L., Pappenberger, F., Wetterhall, F., Haiden, T., Richardson, D., and Salamon, P.: Evaluation of ensemble streamflow predictions in Europe, J. Hydrol., 517, 913–922, https://doi.org/10.1016/j.jhydrol.2014.06.035, 2014.

Arnal, L., Wood, A. W., Stephens, E., Cloke, H. L., and Pappenberger, F.: An Efficient Approach for Estimating Streamflow Forecast Skill Elasticity, J. Hydrometeorol., 18, 1715–1729, https://doi.org/10.1175/JHM-D-16-0259.1, 2017.

Arribas, A., Glover, M., Maidens, A., Peterson, K., Gordon, M., MacLachlan, C., Graham, R., Fereday, D., Camp, J., Scaife, A. A., Xavier, P., McLean, P., and Colman, A.: The GloSea4 Ensemble Prediction System for Seasonal Forecasting, Mon. Weather. Rev., 139, 1891–1910, https://doi.org/10.1175/2010MWR3615.1, 2010.

Bell, V. A., Davies, H. N., Kay, A. L., Brookshaw, A., and Scaife, A. A.: A national-scale seasonal hydrological forecast system: development and evaluation over Britain, Hydrol. Earth Syst. Sci., 21, 4681–4691, https://doi.org/10.5194/hess-21-4681-2017, 2017.

Bennett, J. C., Wang, J. Q., Li, M., Robertson, D. E., and Schepen, A.: Reliable long-range ensemble streamflow forecasts: Combining calibrated climate forecasts with a conceptual runoff model and a staged error model, Water Resour. Res., 52, 8238–8259, https://doi.org/10.1002/2016WR019193, 2016.

Bierkens, M. F. and van Beek, L. P.: Seasonal Predictability of European Discharge: NAO and Hydrological Response Time, J. Hydrometeorol., 10, 953–968, https://doi.org/10.1175/2009JHM1034.1, 2009.

Burek, P., Van Der Knijff, J. M., and De Roo, A.: LISFLOOD – Distributed Water Balance and Flood Simulation Model – Revised User Manual 2013, EUR – Scientific and Technical Research Reports, Publications Office of the European Union, Luxembourg, 150 pp., https://doi.org/10.2788/24719, 2013.

Candogan Yossef, N., van Beek, R., Weerts, A., Winsemius, H., and Bierkens, M. F. P.: Skill of a global forecasting system in seasonal ensemble streamflow prediction, Hydrol. Earth Syst. Sci., 21, 4103–4114, https://doi.org/10.5194/hess-21-4103-2017, 2017.

Céron, J.-P., Tanguy, G., Franchistéguy, L., Martin, E., Regimbeau, F., and Vidal, J.-P.: Hydrological seasonal forecast over France: feasibility and prospects, Atmos. Sci. Lett., 11, 78–82, https://doi.org/10.1002/asl.256, 2010.

Chiew, F. H., Zhou, S. L., and McMahon, T. A.: Use of Seasonal Streamflow Forecasts in Water Resources Management, J. Hydrol., 270, 135–144, https://doi.org/10.1016/S0022-1694(02)00292-5, 2003.

Church, J. E.: Principles of snow surveying as applied to forecasting stream flow, edited by: Merrill, M. C., J. Agric. Res., Washington, D. C., Vol. 51, no. 2, 97–130, 1935.

Coughlan de Perez, E., Stephens, E., Bischiniotis, K., van Aalst, M., van den Hurk, B., Mason, S., Nissan, H., and Pappenberger, F.: Should seasonal rainfall forecasts be used for flood preparedness?, Hydrol. Earth Syst. Sci., 21, 4517–4524, https://doi.org/10.5194/hess-21-4517-2017, 2017.

Crochemore, L., Ramos, M.-H., and Pappenberger, F.: Bias correcting precipitation forecasts to improve the skill of seasonal streamflow forecasts, Hydrol. Earth Syst. Sc., 20, 3601–3618, https://doi.org/10.5194/hess-2016-78, 2016.

Crochemore, L., Ramos, M.-H., Pappenberger, F., and Perrin, C.: Seasonal streamflow forecasting by conditioning climatology with precipitation indices, Hydrol. Earth Syst. Sci., 21, 1573–1591, https://doi.org/10.5194/hess-21-1573-2017, 2017.

Day, G. N.: Extended streamflow forecasting using NWSRFS, J. Water Res. Plan. Man., 111, 157–170, https://doi.org/10.1061/(ASCE)0733-9496(1985)111:2(157), 1985.

De Roo, A. P., Wesseling, C. G., and Van Deursen, W. P.: Physically based river basin modelling within a GIS: the LISFLOOD model, Hydrol. Process., 14, 1981–1992, https://doi.org/10.1002/1099-1085(20000815/30)14:11/12<1981::AID-HYP49>3.0.CO;2-F, 2000.

Demirel, M. C., Booij, M. J., and Hoekstra, A. Y.: The skill of seasonal ensemble low-flow forecasts in the Moselle River for three different hydrological models, Hydrol. Earth Syst. Sci., 19, 275–291, https://doi.org/10.5194/hess-19-275-2015, 2015.

Dettinger, M. D. and Diaz, H. F.: Global characteristics of stream flow seasonality and variability, J. Hydrometeorol., 1, 289–310, https://doi.org/10.1175/1525-7541(2000)001<0289:GCOSFS>2.0.CO;2, 2000.

Doblas-Reyes, F. J., García-Serrano, J., Lienert, F., Biescas, A. P., and Rodrigues, L. R.: Seasonal climate predictability and forecasting: status and prospects, WIRES Clim. Change., 4, 245–268, https://doi.org/10.1002/wcc.217, 2013.

Forecast skill metrics: https://meteoswiss.shinyapps.io/skill_metrics/, last access: 3 October 2017.

Gneiting, T., Balabdaoui, F., and Raftery, A. E.: Probabilistic forecasts, calibration and sharpness, J. Roy. Stat. Soc. B, 69, 243–268, https://doi.org/10.1111/j.1467-9868.2007.00587.x, 2007.

Gobena, A. K. and Gan, T. Y.: Incorporation of seasonal climate forecasts in the ensemble streamflow prediction system, J. Hydrol., 385, 336–352, https://doi.org/10.1016/j.jhydrol.2010.03.002, 2010.

Goddard, L., Mason, S. J., Zebiak, S. E., Ropelewski, C. F., Basher, R., and Cane, M. A.: Current approaches to seasonal to interannual climate predictions, Int. J. Climatol., 21, 1111–1152, https://doi.org/10.1002/joc.636, 2001.

Guimarães Nobre, G., Jongman, B., Aerts, J., and Ward, P. J.: The role of climate variability in extreme floods in Europe, Environ. Res. Lett., 12, 084012, https://doi.org/10.1088/1748-9326/aa7c22, 2017.

Hamlet, A. F., Huppert, D., and Lettenmaier, D. P.: Economic Value of Long-Lead Streamflow Forecasts for Columbia River Hydropower, J. Water Res. Plan. Man., 128, 91–101, https://doi.org/10.1061/(ASCE)0733-9496(2002)128:2(91), 2002.

Hersbach, H.: Decomposition of the Continuous Ranked Probability Score for Ensemble Prediction Systems, Weather Forecast., 15, 559–570, https://doi.org/10.1175/1520-0434(2000)015<0559:DOTCRP>2.0.CO;2, 2000.

Hurrell, J. W.: Decadal trends in the North Atlantic oscillation: Regional temperatures and precipitation, Science, 269, 676–679, https://doi.org/10.1126/science.269.5224.676, 1995.

Hurrell, J. W. and Van Loon, H.: Decadal Variations in Climate Associated with the North Atlantic Oscillation, in: Climatic Change at High Elevation Sites, edited by: Diaz, H. F., Beniston, M., and Bradley, R. S., Springer, Dordrecht, 69–94, https://doi.org/10.1007/978-94-015-8905-5_4, 1997.

Keller, J. D. and Hense, A.: A new non-Gaussian evaluation method for ensemble forecasts based on analysis rank histograms, Meteorol. Z., 20, 107–117, https://doi.org/10.1127/0941-2948/2011/0217, 2011.

Kim, H.-M., Webster, P. J., and Curry, J. A.: Seasonal prediction skill of ECMWF System 4 and NCEP CFSv2 retrospective forecast for the Northern Hemisphere Winter, Clim. Dynam., 39, 2957–2973, https://doi.org/10.1007/s00382-012-1364-6, 2012.

Laio, F. and Tamea, S.: Verification tools for probabilistic forecasts of continuous hydrological variables, Hydrol. Earth Syst. Sci., 11, 1267–1277, https://doi.org/10.5194/hess-11-1267-2007, 2007.

Li, Y., Giuliani, M., and Castelletti, A.: A coupled human-natural system to assess the operational value of weather and climate services for agriculture, Hydrol. Earth Syst. Sci., 21, 4693–4709, https://doi.org/10.5194/hess-21-4693-2017, 2017.

Lorenzo-Lacruz, J., Vicente-Serrano, S. M., López-Moreno, J. I., González-Hidalgo, J. C., and Morán-Tejeda, E.: The response of Iberian rivers to the North Atlantic Oscillation, Hydrol. Earth Syst. Sci., 15, 2581–2597, https://doi.org/10.5194/hess-15-2581-2011, 2011.

Maraun, D., Wetterhall, F., Ireson, A. M., Chandler, R. E., Kendon, E. J., Widmann, M., Brienen, S., Rust, H. W., Sauter, T., Themessl, M., Venema, V. K. C., Chun, K. P., Goodess, C. M., Jones, R. G., Onof, C., Vrac, M., and Thiele-Eich, I.: Precipitation downscaling under climate change: Recent developments to bridge the gap between dynamical models and the end user, Rev. Geophys., 48, Rg3003, https://doi.org/10.1029/2009rg000314, 2010.

Mason, S. J. and Graham, N. E.: Conditional Probabilities, Relative Operating Characteristics, and Relative Operating Levels, Weather Forecast., 14, 713–725, https://doi.org/10.1175/1520-0434(1999)014<0713:CPROCA>2.0.CO;2, 1999.

Meißner, D., Klein, B., and Ionita, M.: Development of a monthly to seasonal forecast framework tailored to inland waterway transport in central Europe, Hydrol. Earth Syst. Sci., 21, 6401–6423, https://doi.org/10.5194/hess-21-6401-2017, 2017.

Mendoza, P. A., Wood, A. W., Clark, E., Rothwell, E., Clark, M. P., Nijssen, B., Brekke, L. D., and Arnold, J. R.: An intercomparison of approaches for improving operational seasonal streamflow forecasts, Hydrol. Earth Syst. Sci., 21, 3915–3935, https://doi.org/10.5194/hess-21-3915-2017, 2017.

Mo, K. C. and Lettenmaier, D. P.: Hydrologic Prediction over the Conterminous United States Using the National Multi-Model Ensemble, J. Hydrometeorol., 15, 1457–1472, https://doi.org/10.1175/JHM-D-13-0197.1, 2014.

Molteni, F., Stockdale, T., Balmaseda, M., Balsamo, G., Buizza, R., Ferranti, L., Magnusson, L., Mogensen, K., Palmer, T., and Vitart, F.: The new ECMWF seasonal forecast system (System 4), ECMWF Tech. Memorandum, 656, 1–49, 2011.

Neumann, J. L., Arnal, L., Magnusson, L., and Cloke, H.: The 2013/14 Thames basin floods: Do improved meteorological forecasts lead to more skilful hydrological forecasts at seasonal timescales?, J. Hydrometeorol., in review, 2018.

Pagano, T. C. and Garen, D. C.: Integration of climate information and forecasts into western US water supply forecasts, Climate variations, climate change, and water resources engineering, edited by: Garbrecht, J. D. and Piechota, T. C., American Society of Civil Engineers location, Reston, Virginia, US, 86–103, 2006.

Prudhomme, C., Hannaford, J., Harrigan, S., Boorman, D., Knight, J., Bell, V., Jackson, C., Svensson, C., Parry, S., Bachiller-Jareno, N., Davies, H. N., Davis, R., Mackay, J., Mackenzie, A., Rudd, A. C., Smith, K., Bloomfield, J., Ward, R., and Jenkins, A.: Hydrological Outlook UK: an operational streamflow and groundwater level forecasting system at monthly to seasonal time scales, Hydrolog. Sci. J., 62, 2753–2768, https://doi.org/10.1080/02626667.2017.1395032, 2017.

Renard, B., Kavetski, D., Kuczera, G., Thyer, M., and Franks, S. W.: Understanding predictive uncertainty in hydrologic modeling: The challenge of identifying input and structural errors, Water Resour. Res., 46, W05521, https://doi.org/10.1029/2009WR008328, 2010.

Schepen, A., Zhao, T., Wang, Q. J., Zhou, S., and Feikema, P.: Optimising seasonal streamflow forecast lead time for operational decision making in Australia, Hydrol. Earth Syst. Sci., 20, 4117–4128, https://doi.org/10.5194/hess-20-4117-2016, 2016.

Sheffield, J., Wood, E. F., Chaney, N., Guan, K., Sadri, S., Yuan, X., Olang, L., Amani, A., Ali, A., Demuth, S., and Ogallo, L.: A Drought Monitoring and Forecasting System for Sub-Sahara African Water Resources and Food Security, B. Am. Meteorol. Soc., 95, 861–882, https://doi.org/10.1175/BAMS-D-12-00124.1, 2013.

Shi, W., Schaller, N., MacLeod, D., Palmer, T. N., and Weisheimer, A.: Impact of hindcast length on estimates of seasonal climate predictability, Geophys. Res. Lett., 42, 1554–1559, https://doi.org/10.1002/2014GL062829, 2015.

Singla, S., Céron, J.-P., Martin, E., Regimbeau, F., Déqué, M., Habets, F., and Vidal, J.-P.: Predictability of soil moisture and river flows over France for the spring season, Hydrol. Earth Syst. Sci., 16, 201–216, https://doi.org/10.5194/hess-16-201-2012, 2012.

Slater, L. J., Villarini, G., Bradley, A. A., and Vecchi, G. A.: A dynamical statistical framework for seasonal streamflow forecasting in an agricultural watershed, Clim. Dynam., 1–17, https://doi.org/10.1007/s00382-017-3794-7, 2017.

Smith, P., Pappenberger, F., Wetterhall, F., Thielen, J., Krzeminski, B., Salamon, P., Muraro, D., Kalas, M., and Baugh, C.: On the operational implementation of the European Flood Awareness System (EFAS), ECMWF Tech. Memorandum, 778, 1–34, 2016.

Soares, M. B. and Dessai, S.: Barriers and enablers to the use of seasonal climate forecasts amongst organisations in Europe, Climatic Change, 137, 89–103, https://doi.org/10.1007/s10584-016-1671-8, 2016.

Steirou, E., Gerlitz, L., Apel, H., and Merz, B.: Links between large-scale circulation patterns and streamflow in Central Europe: A review, J. Hydrol., 549, 484–500, https://doi.org/10.1016/j.jhydrol.2017.04.003, 2017.

Stephens, E., Day, J. J., Pappenberger, F., and Cloke, H.: Precipitation and floodiness, Geophys. Res. Lett., 42, 10316–10323, https://doi.org/10.1002/2015GL066779, 2015.

Troccoli, A.: Seasonal climate forecasting, Meteorol. Appl., 17, 251–268, https://doi.org/10.1002/met.184, 2010.

Turner, S. W. D., Bennett, J. C., Robertson, D. E., and Galelli, S.: Complex relationship between seasonal streamflow forecast skill and value in reservoir operations, Hydrol. Earth Syst. Sci., 21, 4841–4859, https://doi.org/10.5194/hess-21-4841-2017, 2017.

Twedt, T. M., Schaake, J. C., and Peck, E. L.: National Weather Service extended streamflow prediction, Proceedings Western Snow

Conference, Albuquerque, New Mexico, 52–57, April 1977.
van den Hurk, B. J. J. M., Bouwer, L. M., Buontempo, C., Döscher, R., Ercin, E., Hananel, C., Hunink, J., Kjellström, E., Klein, B., Manez, M., Pappenberger, F., Pouget, L., Ramos, M.-H., Ward, P. J., Weerts, A., and Wijngaard, J.: Improving predictions and management of hydrological extremes through climate services: www.imprex.eu, Climate Services, 1, 6–11, https://doi.org/10.1016/j.cliser.2016.01.001, 2016.
Van Der Knijff, J. M., Younis, J., and De Roo, A. P.: LISFLOOD: a GIS-based distributed model for river basin scale water balance and flood simulation, Int. J. Geogr. Inf. Sci., 24, 189–212, https://doi.org/10.1080/13658810802549154, 2010.
Viel, C., Beaulant, A.-L., Soubeyroux, J.-M., and Céron, J.-P.: How seasonal forecast could help a decision maker: an example of climate service for water resource management, Adv. Sci. Res., 13, 51–55, https://doi.org/10.5194/asr-13-51-2016, 2016.
Wetterhall, F. and Di Giuseppe, F.: The benefit of seamless forecasts for hydrological predictions over Europe, Hydrol. Earth Syst. Sci. Discuss., https://doi.org/10.5194/hess-2017-527, in review, 2017.
White, C. J., Carlsen, H., Robertson, A. W., Klein, R. J., Lazo, J. K., Kumar, A., Vitart, F., Coughlan de Perez, E., Ray, A. J., Murray, V., Bharwani, S., MacLeod, D., James, R., Fleming, L., Morse, A. P., Eggen, B., Graham, R., Kjellström, E., Becker, E., Pegion, K. V., Holbrook, N. J., McEvoy, D., Depledge, M., Perkins-Kirkpatrick, S., Brown, T. J., Street, R., Jones, L., Remenyi, T., Hodgson-Johnston, I., Buontempo, C., Lamb, R., Meinke, H., Arheimer, B., and Zebiak, S. E.: Potential applications of subseasonal-to-seasonal (S2S) predictions, Meteorol. Appl., 24, 315–325, https://doi.org/10.1002/met.1654, 2017.
Wood, A. W., Kumar, A., and Lettenmaier, D. P.: A retrospective assessment of National Centers for Environmental Prediction climate model-based ensemble hydrologic forecasting in the western United States, J. Geophys. Res.-Atmos., 110, D04105, https://doi.org/10.1029/2004JD004508, 2005.
Wood, A. W. and Lettenmaier, D. P.: An ensemble approach for attribution of hydrologic prediction uncertainty, Geophys. Res. Lett., 35, L14401, https://doi.org/10.1029/2008GL034648, 2008.
Wood, A. W., Maurer, E. P., Kumar, A., and Lettenmaier, D. P.: Long-range experimental hydrologic forecasting for the eastern United States, J. Geophys. Res.-Atmos., 107, 4429, https://doi.org/10.1029/2001JD000659, 2002.
Yuan, X., Roundy, J. K., Wood, E. F., and Sheffield, J.: Seasonal forecasting of global hydrologic extremes: system development and evaluation over GEWEX basins, B. Am. Meteorol. Soc., 96, 1895–1912, https://doi.org/10.1175/BAMS-D-14-00003.1, 2015a.
Yuan, X., Wood, E. F., and Ma, Z.: A review on climate-model-based seasonal hydrologic forecasting: physical understanding and system development, Wiley Interdisciplinary Reviews: Water, 2, 523–536, https://doi.org/10.1002/wat2.1088, 2015b.
Yuan, X., Wood, E. F., Chaney, N. W., Sheffield, J., Kam, J., Liang, M., and Guan, K.: Probabilistic Seasonal Forecasting of African Drought by Dynamical Models, J. Hydrometeorol., 14, 1706–1720, https://doi.org/10.1175/JHM-D-13-054.1, 2013.
Zajac, Z., Zambrano-Bigiarini, M., Salamon, P., Burek, P., Gentile, A., and Bianchi, A.: Calibration of the lisflood hydrological model for europe – calibration round 2013, Joint Research Centre, European Commission, 2013.

9

Assimilation of passive microwave AMSR-2 satellite observations in a snowpack evolution model over northeastern Canada

Fanny Larue[1,2,3], **Alain Royer**[1,2], **Danielle De Sève**[3], **Alexandre Roy**[1,2,4,a], **and Emmanuel Cosme**[5]

[1]CARTEL, Université de Sherbrooke, Québec, Canada
[2]Centre d'Études Nordiques, Québec, Canada
[3]IREQ, Hydro-Québec, Québec, Canada
[4]Département de Géographie, Université de Montréal, Québec, Canada
[5]Institut des Géosciences de l'Environnement, IGE, UGA-CNRS, Grenoble, France
[a]now at: Université du Québec à Trois-Rivière, Québec, Canada

Correspondence: Fanny Larue (fanny.larue@usherbrooke.ca)

Abstract. Over northeastern Canada, the amount of water stored in a snowpack, estimated by its snow water equivalent (SWE) amount, is a key variable for hydrological applications. The limited number of weather stations driving snowpack models over large and remote northern areas generates great uncertainty in SWE evolution. A data assimilation (DA) scheme was developed to improve SWE estimates by updating meteorological forcing data and snowpack states with passive microwave (PMW) satellite observations and without using any surface-based data. In this DA experiment, a particle filter with a Sequential Importance Resampling algorithm (SIR) was applied and an inflation technique of the observation error matrix was developed to avoid ensemble degeneracy. Advanced Microwave Scanning Radiometer 2 (AMSR-2) brightness temperature (T_B) observations were assimilated into a chain of models composed of the Crocus multilayer snowpack model and radiative transfer models. The microwave snow emission model (Dense Media Radiative Transfer – Multi-Layer model, DMRT-ML), the vegetation transmissivity model (ω-τ_{opt}), and atmospheric and soil radiative transfer models were calibrated to simulate the contributions from the snowpack, the vegetation, and the soil, respectively, at the top of the atmosphere. DA experiments were performed for 12 stations where daily continuous SWE measurements were acquired over 4 winters (2012–2016). Best SWE estimates are obtained with the assimilation of the T_{Bs} at 11, 19, and 37 GHz in vertical polarizations. The overall SWE bias is reduced by 68 % compared to the original SWE simulations, from 23.7 kg m^{-2} without assimilation to 7.5 kg m^{-2} with the assimilation of the three frequencies. The overall SWE relative percentage of error (RPE) is 14.1 % (19 % without assimilation) for sites with a fraction of forest cover below 75 %, which is in the range of accuracy needed for hydrological applications. This research opens the way for global applications to improve SWE estimates over large and remote areas, even when vegetation contributions are up to 50 % of the PMW signal.

1 Introduction

In Quebec, eastern Canada, snowmelt runoff has become a major economic issue and plays a considerable role in flood events (Perry, 2000). Good forecasting of this water supply is essential in optimizing hydroelectric dam management. The amount of water stored in a snowpack is estimated by the snow water equivalent (SWE). Accurately predicting the evolution of the SWE is challenging over large and remote areas due to the high spatial and temporal variability of the snowpack and to the lack of in situ data, which are time-consuming and expensive to measure. Current operational hydrological forecasting models used by Hydro-Québec, one of the larger energy producers in North America, rely on the interpolation of surface snow survey measurements (Tapsoba et al., 2005; Brown et al., 2018). It has been shown that the highest uncertainties in hydrological forecasting related to snow result from a lack of accurate estimates of the amount of snow accumulated over a large area during the winter season (Turcotte et al., 2010). To better determine the spatial distribution of the SWE, many approaches use snowpack models to simulate the evolution of the snow cover in response to meteorological conditions (Brun et al., 1989; Jordan, 1991; Lehning et al., 2002). However, using these models is challenging due to the incomplete meteorological forcing data for remote areas where weather stations are scarce (Raleigh et al., 2015) and the snow physics simplifications used in the models (Foster et al., 2005).

The assimilation of satellite observations is a promising approach for reducing uncertainties related to the lack of in situ data (Pietroniro and Leconte, 2005; Durand et al., 2009; Touré et al., 2011; De Lannoy et al., 2012; DeChant and Moradkhani, 2011; Andreadis and Lettenmaier, 2012; Kwon et al., 2017). In particular, passive microwave (PMW) satellite observations, which measure brightness temperatures ("T_B"), are sensitive to the volume of snow and provide information at a good temporal and spatial coverage (Hallikainen, 1984; Chang et al., 1996; Tedesco et al., 2004). It has been shown that the assimilation of PMW satellite data into snow models adds valuable information to compensate for initialization errors and improve SWE simulated by snow models (Sun et al., 2004). These approaches appear to be very promising to evaluate and predict water resources but are still under development for further use in operational hydrological applications (Xu et al., 2014). Larue et al. (2017) showed that the GlobSnow-2 SWE product (Takala et al., 2011), which assimilates both T_B satellite data and local snow depth observations, was not accurate enough for hydrological modeling, mainly because of its dependence on in situ data in remote areas.

The main difficulty in the assimilation of PMW satellite observations in boreal forest areas is the quantification of all the contributions that affect the measured signal. PMW satellite observations have a low spatial resolution ($\sim 10 \times 10\,\text{km}^2$) and satellite sensors measure many contributions in addition to the PMW emission from the volume of the snowpack (vegetation canopy, ice crust, frozen/unfrozen soil, lakes, moisture in the snow, topography, etc.) (Kelly et al., 2003; Koenig and Forster, 2004). In boreal areas, the PMW emission from the forest canopy within a pixel can contribute up to half of the PMW signal measured by satellite sensors (Roy et al., 2012, 2016). This contribution does not only depend on the fraction of forest cover, but also on the biomass (liquid water content, LWC), the vegetation volume, and the canopy structure (stem, leaf, trunk) (Franklin, 1987). To adjust snowpack model simulations, several studies suggest using radiative transfer models, coupled to a snowpack model, to take into account the different contributions to the PMW signal at the top of the atmosphere and to directly assimilate PMW satellite observations (Brucker et al., 2011; Durand et al., 2011; Langlois et al., 2012; Roy et al., 2016). However, the assimilation of PMW must be used with care, and a good understanding of the interactions between the properties and microwave emission of the snowpack is crucial to avoid degradation of the SWE estimates. For instance, the assimilation of passive microwave in wet snow conditions can introduce large uncertainties since the presence of liquid water in the snowpack increases T_{Bs}, whereas increases in snow grain size decrease the brightness temperature independent of any change in SWE (Klehmet et al., 2013). The assimilation of PMW thus can help to adjust the modeled snowpack states during the winter, but it cannot be used at the beginning and at the end of the season (snowmelt periods).

This paper aims at developing and validating the assimilation of PMW satellite observations for SWE improvements over Quebec by adjusting meteorological forcing data and simulated snowpack states without using any surface-based observations. Advanced Microwave Scanning Radiometer 2 (AMSR-2) satellite sensors provide the T_B observations at 11, 19, and 37 GHz. The data assimilation scheme (DA) is a Sequential Importance Resampling particle filter (referred to as PF-SIR) (Van Leeuwen, 2009, 2014). The PMW emission from the snowpack is computed by using the Crocus snowpack model (Brun et al., 1989) coupled to a microwave snow emission model, the Dense Media Radiative Transfer – Multi-Layer model (DMRT-ML) (Picard et al., 2013). This scheme is further referred as the Crocus/DMRT-ML chain and was previously calibrated over Quebec (Larue et al., 2018). As a first step, the previous study of Larue et al. (2018) tested the feasibility of the DA scheme in a controlled environment by using synthetic T_{Bsnow} observations, obtained by running the Crocus/DMRT-ML chain with perturbed meteorological forcings. The results showed SWE root mean square error (RMSE) reduced by 82 % with the multivariate assimilation of T_{Bs} at 37, 19, and 11 GHz in vertical polarizations, compared to SWE RMSE without assimilation. In the present study, the same DA setup as described in Larue et al. (2018) was implemented, except that real satellite observations were used. For the assimilation of satellite data, the challenge is to accurately simulate the T_B measured at the top of the atmosphere ($T_{B\,TOA}$) by including contributions other than snow (i.e., soil, vegetation, and atmosphere). The vegetation transmissivity model (ω-τ_{opt}), the Wegmüller and Mätzler (1999) soil emission model, and the Liebe (1989) atmospheric emission model were added and calibrated to simulate the PMW emission of satellite observations (Roy et al., 2016).

The specific objectives of this paper were thus to (1) calibrate the soil and the vegetation radiative transfer models coupled with the Crocus/DMRT-ML chain to simulate $T_{B\,TOA}$ over several years (2012 to 2016); and (2) evaluate the performance of the assimilation of PMW data in Crocus using SWE measurements obtained over 12 reference nivometric stations from 2012 to 2016 (43 winters). This paper opens the way to a functional spatialized method for improving SWE estimates over large and remote areas without using surface-based data.

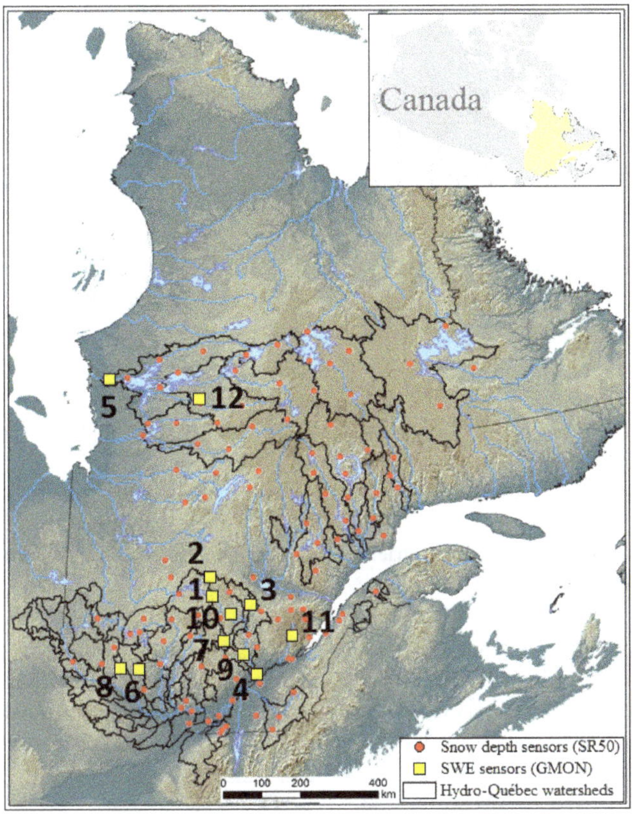

Figure 1. SWE measurement stations with the "GMON" SWE sensors (yellow squares; see Table 1 for details) in the province of Quebec. The red circles are the snow depth sensors ("SR50") used by Hydro-Québec for hydrological purposes, overlaid on a relief map (from blue – low – to brown – higher altitudes) and watershed contours (black lines).

2 General framework

2.1 Study area and evaluation database

Figure 1 shows the region of interest located in the province of Quebec, eastern Canada (46–56° N). This area includes the La Grande (LG) watershed, in north-central Quebec (below 56° N), and the Outaouais and Saint-Maurice watersheds, in southwestern and south-central Quebec, respectively (46–48° N), which are equipped with SWE and snow depth sensors for hydrological purposes. Quebec is characterized by different ecoclimatic conditions, a high percentage of forested area (dense boreal forests and mixed coniferous and deciduous), and a flat topography.

To evaluate SWE simulations, SWE measurements were acquired from 2012 to 2016 by 12 nivometric stations (see numbered stations on Fig. 1), located through a north–south gradient in Quebec. This SWE database (coordinates, sensors, operating period, etc.) was fully described in Larue et al. (2018). Table 1 describes the main station characteristics, including the mean maximum SWE values over operating periods. Daily SWE measurements were derived from gamma ray SWE sensors (Campbell Scientific CS725, "GMON") with an average error of +5 % (Choquette et al., 2008). Two stations (nos. 5 and 12) were located in the subarctic ecoclimatic zone (53–54° N, James Bay area), eight in the coniferous boreal zone (46–48° N), and two (Nos. 4 and 11) in a mixed forest area in southern Quebec (45.3° N). Sensors were calibrated by Hydro-Québec from numerous field measurement campaigns during the first year following their installation.

A total of 43 winters were studied (Table 1). These winters were all very different. Winter 2012–2013 had the lowest snow accumulation in 10 years (165 cm), whereas winter 2013–2014 was very snowy (379 cm) compared to the average snow accumulation (217 cm). Winter 2014–2015 was unusually cold (3° below average temperatures), and winter 2015–2016 was the warmest in 60 years (statistics can be found at http://www.mddep.gouv.qc.ca, last access: 18 October 2018).

2.2 General setup

Figure 2 shows the general methodology developed to simulate and to assimilate AMSR-2 satellite observations into the snowpack model.

To simulate the signal measured by satellite sensors at the top of the atmosphere ($T_{\mathrm{B\,TOA}}$), a chain of models was implemented and calibrated over eastern Canada. The 3-hourly continuous atmospheric forcing database provided by the Global Environmental Multiscale weather prediction model (referred to as "GEM"; Coté et al., 1998) was used to drive the multilayer Crocus snowpack model (described in Sect. 3.2.1). Each GEM grid cell has a spatial resolution of $10 \times 10\,\mathrm{km}^2$, which is on the same order as the observation scale. The Crocus model updates the snowpack every 15 min by interpolating meteorological inputs, but in this study we used daily Crocus outputs (SWE, snow depth, density, etc.) computed at 14:00 local time (19:00 UTC), in agreement with the AMSR-2 pass (Sect. 3.1.1). The DMRT-ML radiative transfer model (Sect. 3.2.1), driven with Crocus outputs, was used to simulate the PMW emission from the modeled snowpack (referred to as "T_{Bsnow}") at 11, 19, and 37 GHz, at vertical and horizontal polarizations ("V-pol" and "H-pol", respectively). The contribution of the atmosphere was estimated by using an atmospheric model (Liebe, 1989) driven with the total precipitable water integrated over 28 atmospheric layers and provided by GEM (Sect. 3.3). The surface emissivity for a rough soil was deduced by calibrating the Wegmüller and Mätzler (1999) soil model, and vegetation contributions were quantified with the (ω-τ_{opt}) radiative transfer model (Sect. 3.3). To take canopy emissivity variability into account, the inversions of the (ω, τ_{opt}) parameters were linked to the 4-day leaf area index (LAI) product from MODIS data ($1 \times 1\,\mathrm{km}^2$), averaged for each AMSR-2 grid cell ($10 \times 10\,\mathrm{km}^2$) (Sect. 3.3). These inversions of soil

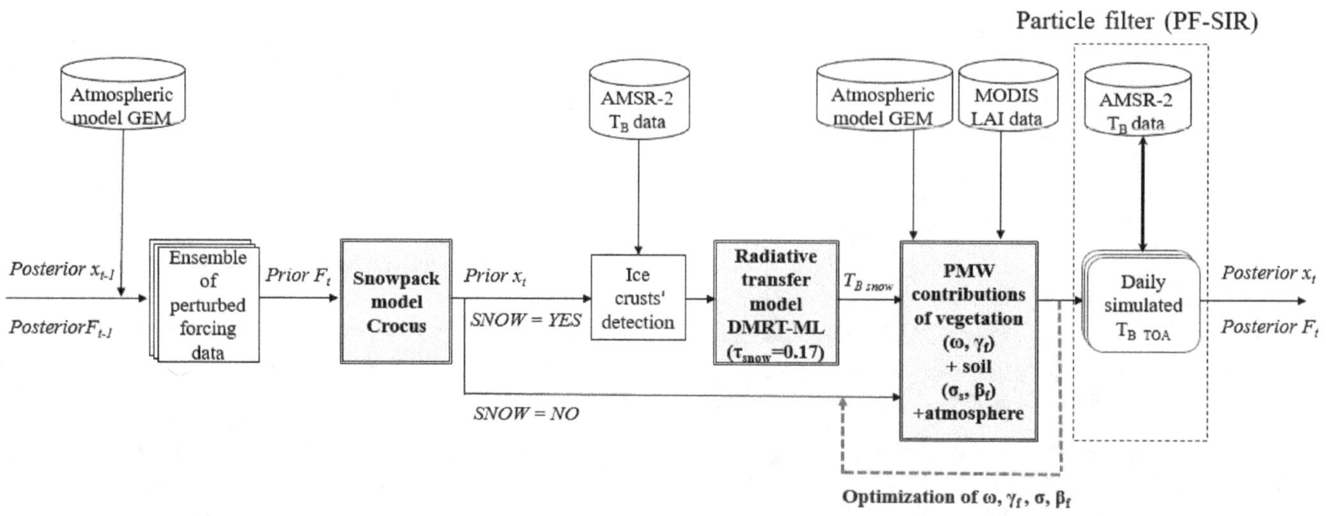

Figure 2. Methodological scheme describing the DA scheme in the chain of models for SWE retrievals by updating perturbed atmospheric forcing data and snowpack states ("F_t" and "x_t", respectively; see Sect. 3.4).

Table 1. Characteristics of the nivometric SWE stations: site number, latitude (Lat.), longitude (Long.) and elevation (El., a.s.l. in meters) of stations. Dist. GEM–station is the distance between the station and the center of the associated GEM grid cell (with GEM, Global Environmental Multiscale weather prediction model; Sect. 2.2). The time period of observations, average of the maximum observed data over the studied period, and data provider are given (HQ: Hydro-Québec, U. Sherb: University of Sherbrooke, U. Laval: University of Laval).

Sites no.	Lat.	Long.	El.	Dist. GEM–station (km)	Time period	Mean maximum SWE value (kg m^{-2})	Data provider
1	48.3	−74.1	100	3.4	2012–2016	272	HQ
2	48.9	−74.2	100	4.9	2012–2016	277	HQ
3	47.9	−72.9	100	4.7	2012–2016	252	HQ
4	46.6	−72.8	136	4.2	2012–2016	253	HQ
5	53.7	−78.2	103	4.2	2012–2016	213	HQ
6	46.7	−76.0	229	2.3	2012–2016	161	HQ
7	47.0	−74.3	469	3.3	2012–2016	235	HQ
8	46.9	−76.4	330	1.8	2012–2016	212	HQ
9	46.9	−73.7	372	1.9	2012–2016	180	HQ
10	47.7	−73.6	398	3.5	2012–2016	202	HQ
11	47.3	−71.2	669	2.6	2015–2016	396	U. Laval
12	53.4	−75.0	389	4.0	2014–2016	211	U. Sherb
Mean				3.4	2012–2016	237	

and vegetation parameters were performed over the summer period to avoid bias due to the presence of the snowpack.

The brightness temperatures (T_{Bs}) measured by AMSR-2 satellite sensors were assimilated in a DA scheme (see Sect. 3.4). Raleigh et al. (2015) have shown that meteorological forcing data are the major sources of errors in snow model simulations. Hence, we assume here that the uncertainties of GEM meteorological forcing data are the only sources of errors in the T_B modeling. It is very difficult to quantify modeling errors due to physical simplifications inside the model due to the spatial scale of the observations. Further studies are needed to estimate these errors over the study area and to take them into account in the DA experiment. The observation error was assumed to be known and the modeling errors were estimated by perturbing selected meteorological forcing variables. An ensemble of 150 T_B simulations was obtained and the distribution of these *prior* estimates represents the modeling error in response to GEM uncertainties. A particle filter with an SIR algorithm was used to update the simulated $T_{B\,TOA}$ over the winter by adjusting meteorological forcing data and snowpack states (*posterior* estimates) when an observation was available (Fig. 2).

Several configurations of the DA scheme were tested over three evaluation sites representing different environmental conditions. The best configuration was evaluated over the validation reference sites from 2012 to 2016 (for 43 winters; Sect. 3.4).

Comparing data simulated at the station against model cells involves uncertainty due to spatial variations of the snowpack and land cover. This is a well-known problem for model validation studies and we assume here that the high number of sites (12 SWE stations or 43 snowpack simulations) provides a useful assessment of simulations. It is also known that the spatial localization of measurements can lead to some biases (Molotch and Bales, 2005). To diversify its measurements, Hydro-Québec has installed two SWE sensors in the forest, and not in a clearing as is the usual practice for ease of maintenance.

3 Materials and methods

3.1 Data

3.1.1 AMSR-2 observations

AMSR-2 satellite sensors (Imaoka et al., 2010) provide PMW satellite observations on the 11 (10.7), 19, and 37 GHz channels at V-pol and H-pol. Images produced by AMSR-2 are freely available on the Japan Aerospace Exploration Agency (JAXA) website. This study used the Level 3 Version 2 product, which provides daily T_{Bs} normalized on a North Hemisphere polar stereographic projection with a spatial resolution of $10 \times 10\,\mathrm{km}^2$ (see https://gportal.jaxa.jp/, last access: 18 October 2018, for the specifications of the projection), from 1 August 2012 to 1 July 2016. T_{Bs} from AMSR-2 are computed twice a day: around 13:30 local time, or 17:30 UTC (ascending pass), and around 01:30 local time, or 05:30 UTC (descending pass). Only the ascending pass was used in this study since the snowpack was computed once a day at 14:00 (local time). The use of the ascending pass allowed the nighttime refreezing process to be avoided. To reduce observation errors due to the daytime melting process, the approach was evaluated during the dry snow period, from December to mid-March. This aspect is further discussed in Sect. 5.1.

3.1.2 LAI MODIS data

The 4-day LAI product provided by MODIS TERRA data (MOD15A3; Myneni et al., 2002) was used to characterize the vegetation contributions to the total emissivity (Fig. 2). The product has a spatial resolution of $1 \times 1\,\mathrm{km}^2$ and was resampled on the AMSR-2 grid of $10 \times 10\,\mathrm{km}^2$ by averaging all LAI data within each AMSR-2 grid cell (referred to as "$\mathrm{LAI}_{\mathrm{AMSR-2}}$"). For each site, Table 2 describes the summer and winter average values ("$\mathrm{LAI}_{\mathrm{summer}}$" and "$\mathrm{LAI}_{\mathrm{winter}}$") calculated using $\mathrm{LAI}_{\mathrm{AMSR\text{-}2}}$ from 1 July to 31 August and from 1 January to 1 March over the 2012 to 2016 time period, respectively (Roy et al., 2014).

3.1.3 Land cover map of Canada

The land cover map of Canada Circa 2000 (available at http://www.geobase.ca/geobase/en/data/landcover/index.html, last access: 18 October 2018) (referred to as "LCC") was used to extract the fraction of forest cover ("f_{cover}") within each AMSR-2 grid cell. This product provides the percentage of coniferous, herbaceous, deciduous, and water areas with a spatial resolution of $1 \times 1\,\mathrm{km}^2$ and was resampled to generate average values within each $10 \times 10\,\mathrm{km}^2$ AMSR-2 grid cell. Table 2 shows the fractions of forest cover provided by the LCC and resampled over AMSR-2 grid cells for each site. As expected, Sites 5 and 12, which are located in the subarctic area (Fig. 1), have a low f_{cover} (below 32 %). The other sites in boreal areas have an f_{cover} of up to 60 %. Sites 6 and 9 are in particularly densely forested areas, with a high f_{cover} (up to 80 %). The measured T_{B} signal can be significantly affected by the forest and the signature of the underlying snow is attenuated during the winter period in such densely forested areas. The sensitivity of the DA scheme to the f_{cover} was analyzed for sites with an f_{cover} above and below 75 % (Sect. 4.2.1).

Moreover, the presence of lakes can affect the PMW signal. Lake ice (when snow cover is absent) increases the PMW signal at high frequencies, and at low frequencies, the contribution of water bodies acts as a reflector and the emissivity remains low (De Sève et al., 1999). With snow cover on lakes, the different snow states on the lakes compared to snow cover under forest also modified the emitted signal (see Derksen et al., 2012, 2014). Nevertheless, we made the hypothesis that these impacts were negligible over our studied sites, which have lake water fractions under 7 % within their AMSR-2 grid cells (Table 2) (masks are generally applied for water fractions of up to 20 %; Takala et al., 2011).

3.2 Simulation of the PMW emission from the snowpack

3.2.1 Coupling of Crocus and DMRT-ML

The chain of models developed to simulate T_{Bsnow} is identical to that of Larue et al. (2018), so only a brief description of the approach is detailed here (see Fig. 2).

The Crocus snowpack evolution model (Brun et al., 1989, 1992; Vionnet et al., 2012) is coupled with the ISBA land surface model within the SURFEX interface (Surface Externalisée, in French) (Decharme et al., 2011; Masson, 2013). SURFEX/ISBA/Crocus (hereafter referred to as "Crocus") computes the evolution of the physical properties of the snowpack and the underlying ground (soil). In particular, the snow layers are modeled with a set of variables representing the morphological properties of snow grains (shape and size), including the specific surface area (SSA), which is one of the most sensitive variables for snowpack emission simulations. The snow microstructure evolves in time according

Table 2. LAI_{summer} is the mean of the LAI provided by MODIS for the summer period (1 July to 31 August) and averaged over the AMSR-2 grid cell ($10 \times 10\,km^2$), LAI_{winter} is the mean LAI for the winter period (1 January to 1 March). f_{cover} is the fraction of forest cover within the AMSR-2 grid cell extracted from the land cover map Circa 2000 (see Sect. 3.1.3). The percentages of coniferous, deciduous, and water areas are the percentages distributed within the f_{cover}. Sites are ranked in the increasing order of f_{cover}. The three highlighted sites (gray cells) are the sites selected to test the configuration of the DA scheme in Sect. 3.4.3.

Site no.	LAI_{summer}	LAI_{winter}	f_{cover} (%)	Coniferous (%)	Deciduous (%)	Water (%)
No. 12	1.07	0.04	24.2	77.6	14.4	4.9
No. 5	1.07	0.08	31.5	66.5	25.9	7.0
No. 4	2.63	0.06	47.6	8.5	70.3	1.4
No. 7	3.13	0.28	59.3	49.9	45.8	4.0
No. 10	2.47	0.17	61.8	67.3	30.1	2.4
No. 1	2.96	0.28	63.7	41.6	55.8	2.2
No. 3	3.69	0.25	65.5	44.6	52.1	3.3
No. 2	1.99	0.12	66.6	79.4	16.6	3.5
No. 8	4.11	0.22	72.1	15.5	80.2	4.3
No. 11	2.43	0.19	74.5	52.5	46.6	0.5
No. 6	2.82	0.11	81.5	18.1	75.3	6.5
No. 9	3.65	0.43	84.0	60.9	36.1	2.9

to semi-empirical laws (Vionnet et al., 2012). Crocus is the only model able to simulate the SSA as a prognostic variable (rather than as a diagnostic variable) by using the formulations of Carmagnola et al. (2014). The number of snow layers is dynamic and evolves with physical properties updated at each time step. The maximum number of simulated snow layers was fixed at 15 in this study as a compromise between accuracy and computing time (not shown). Configuration and initialization of the Crocus snowpack model are the same as described in Larue et al. (2018).

$T_{B\,snow}$ was computed by driving the radiative transfer model DMRT-ML with Crocus outputs. The DMRT-ML model is well detailed in the literature (Tsang et al., 1992; Tsang and Kong, 2001; Picard et al., 2013, Royer et al., 2017), so only the calibration is described here. Snow grain size, and more generally snow microstructure, are factors that most affect the accuracy of simulated PMW emission from a snowpack as they determine the strength of scattering mechanisms in the snowpack at the high frequencies used (Roy et al., 2013; Leppänen et al., 2015; Sandells et al., 2017, Larue et al., 2018). In DMRT-ML, snow grains are represented as spheres of ice with variable interactions between them. The potential formation of clusters of grains, which increases the effective snow grain size, is not taken into account, generating uncertainties (Picard et al., 2013). Several studies have shown that DMRT-ML needed an effective scaling factor to represent the stickiness between snow grains and to correct the snow microstructure representation (Brucker et al., 2011; Roy et al., 2013; Royer et al., 2017). Larue et al. (2018) have shown that a mean snow stickiness parameter (τ_{snow}) of 0.17 was optimal to simulate T_{Bsnow} over boreal snow in Quebec (RMSE of 27 K) when DMRT-ML is driven by Crocus snow profiles. This constant τ_{snow} value was thus used in the implemented chain of models (Sect. 3.4.3; Experiment A and B). Nevertheless, this effective parameter could change with snow type (Royer et al., 2017; Larue et al., 2018). Hence, the quality of the DA scheme with the use of the τ_{snow} parameter as a free variable was studied (Sect. 3.4.3, Experiment C).

3.2.2 Ice lens detection algorithm

Since ice lenses ("ILs") within a snowpack significantly reduce T_B mainly at H-pol (Montpetit et al., 2013; Roy et al., 2016), ice layers must be detected and added in the simulated Crocus snow profiles to improve T_{Bsnow} simulations. T_B in H-pol are much more attenuated by the presence of an IL than T_B in V-pol, since the coefficient of reflectivity is stronger in H-pol (Montpetit et al., 2013). Therefore, by following the daily evolution of the PMW emission from the snowpack with AMSR-2 observations, the formation of an IL can be detected by using a threshold on the polarization ratio (PR) defined by Cavalieri et al. (1984) for a given frequency (ν):

$$\mathrm{PR}(\nu) = \frac{T_B(\nu, \text{V-pol}) - T_B(\nu, \text{H-pol})}{T_B(\nu, \text{V-pol}) + T_B(\nu, \text{H-pol})}. \quad (1)$$

In this study, an IL was added on the top of the simulated snowpack if the AMSR-2 PR(11) was above 0.06 (Roy, 2014). This IL was represented as a 1 cm layer with a density of $900\,kg\,m^{-3}$ and with the snow grain radius set to zero (Roy et al., 2016). The difficulty is to know how to evolve this IL in the snowpack. The Crocus snowpack model has not yet been adapted to integrate the formation of ILs and evolve them in a coherent way (Quéno et al., 2016). Nevertheless, it was shown in Larue et al. (2018) (from field measurements) that an IL of 1 cm located at 4 cm from the surface of the simulated snowpack minimized the bias of DMRT-ML simulations due to the presence of an IL (regardless of its real location in the snow profile). Hence, the IL first added at the surface of the snowpack was moved to 4 cm from the surface as soon as snowfall was detected with GEM precipitation data or, if not, after 5 days to take into account the snowpack transformations (percolations, sublimations, etc.). The maximum number of detected IL was fixed at two. When a second IL was detected (IL2), IL2 was added at the surface while the first detected IL (IL1) was left at 4 cm. After the next snowfall (or after 5 days otherwise), IL1 was moved to 8 cm from the surface and IL2 to 4 cm. For instance, during winter 2014–2015, one IL was detected at Sites 1 and 12 (22 and 15 December 2014). At Site 9, two ILs were detected: one on 10 December 2014 and another on 1 January 2015.

This is a simplified way to take into account the presence of ILs, and further studies are needed to dynamically evolve these ILs in the snowpack and to model the impact on the neighboring layers. This work is particularly complex, and no solution has yet been found (D'Ambroise et al., 2017), in particular because measurements are difficult to take.

3.3 Simulation of the PMW emission at the top of the atmosphere

The PMW brightness temperature ($T_{\mathrm{B,\,TOA}}$) emitted at the scale of the AMSR-2 product can be written for each grid cell as

$$T_{\mathrm{B,\,TOA}} = f_{\mathrm{season}} \cdot T_{\mathrm{B\,forest}} + (1 - f_{\mathrm{season}})\, T_{\mathrm{B\,open}} + T_{\mathrm{B\,atm}\uparrow}, \quad (2)$$

where $T_{\mathrm{B\,atm}\uparrow}$ is the ascending atmospheric contribution, estimated using the Liebe (1989) model implemented in the Helsinki University of Technology (HUT) snow emission model (Pulliainen et al., 1999). The model considers radiative transfer through the atmospheric layers and provides $T_{\mathrm{B\,atm}\uparrow}$ values at the satellite sensor level (Liebe, 1989) according to the precipitable water integrated for all atmospheric layers provided by GEM. f_{season} is the seasonal (winter or summer) fraction of forest cover in the AMSR-2 grid cell, $T_{\mathrm{B\,forest}}$ is the PMW emission with vegetation contributions, and $T_{\mathrm{B\,open}}$ is the PMW emission without vegetation contributions.

The f_{cover} values provided by the LCC map were constants, whereas these fractions of forest evolve with the season. To take into account the temporal evolution of the forest cover for the winter and summer periods (defined as the time period with and without snow, respectively) and to estimate the f_{season} used in Eq. (2), f_{cover} was linked respectively to $\mathrm{LAI_{winter}}$ and to $\mathrm{LAI_{summer}}$ by comparing the f_{cover} map to the two resampled maps (both resampled on the AMSR-2 projection) throughout Quebec (not shown). The seasonal fractions of f_{cover} were related to seasonal LAIs with Eqs. (3) and (4) for summer and winter, respectively:

$$f_{\mathrm{summer}} = 0.9 \cdot (1 - \exp(-2.7 \cdot \mathrm{LAI_{summer}}))^{3.2} \quad (3)$$

$$f_{\mathrm{winter}} = 0.9 \cdot (1 - \exp(-16.0 \cdot \mathrm{LAI_{winter}}))^{0.3}. \quad (4)$$

The linear correlation between the f_{summer} values estimated from the LCC and the f_{summer} values fitted to LAI data with the Eq. (3) had a coefficient correlation R equal to 0.94 and a p value below 0.01. For the LCC f_{winter} values and the f_{winter} values fitted to the LAI data (see Eq. 4), the coefficient correlation R was equal to 0.95 and the p value was below 0.01.

3.3.1 Vegetation contributions

The PMW emission from the vegetation varies with the forest characteristics, such as the biomass, the structure of the vegetation, or the liquid water content of the canopy. In this study, the vegetation contribution was modeled with the simplified radiative transfer model (ω-τ_{opt}) (Mo et al., 1982), in which the parameters should be estimated by fitting the simulated T_{Bs} with observations (Grant et al., 2008; Roy et al., 2012). The ω is the single scattering factor of the albedo. Given the incidence angle $\theta = 55°$ of AMSR-2 satellite sensors, the optical thickness of the vegetation τ_{opt} was a function of the forest transmissivity (γ) such that $\gamma = \exp(-\tau_{\mathrm{opt}}/\cos\theta)$. The forest transmissivity varies with the frequency (ν) used and is further called γ_ν. At the satellite sensor, the expression of $T_{\mathrm{B\,TOA}}$ in boreal areas was described by Eq. (2), which can be detailed with Eqs. (5) and (6) (see Roy et al., 2012):

$$\begin{aligned} T_{\mathrm{B\,forest}} = \big[&\gamma_\nu \cdot e_{\mathrm{surf}} \cdot T_{\mathrm{surf}} + (1-\omega)\cdot(1-\gamma_\nu)\cdot T_{\mathrm{veg}} \\ &+ \gamma_\nu \cdot (1-e_{\mathrm{surf}})\cdot(1-\omega)\cdot(1-\gamma_\nu)\cdot T_{\mathrm{veg}} \\ &+ (1-e_{\mathrm{surf}})\cdot {\gamma_\nu}^2 \cdot T_{\mathrm{B\,atm}\downarrow} + (1-\gamma_\nu)\cdot\omega \\ &\cdot T_{\mathrm{B\,atm}\downarrow}\big]\,\gamma_{\mathrm{atm}} \end{aligned} \quad (5)$$

$$T_{\mathrm{B\,open}} = \big[e_{\mathrm{surf}} \cdot T_{\mathrm{surf}} + (1-e_{\mathrm{surf}})\cdot T_{\mathrm{B\,atm}\downarrow}\big]\,\gamma_{\mathrm{atm}}, \quad (6)$$

where T_{surf} is the surface temperature, e_{surf} is the surface emissivity under the canopy (with or without snow) for a given frequency, and T_{veg} is the temperature of the vegetation (taken as equal to the air temperature at 2 m, provided by GEM). $T_{\mathrm{B\,atm}\downarrow}$ is the descending atmospheric contributions and γ_{atm} is the transmittance of the atmosphere. These atmospheric contributions were modeled using the Liebe (1989) model, as were the $T_{\mathrm{B\,atm}\uparrow}$ values. Thus, for snow-free conditions, only forest (ω, γ_ν) and soil (e_{surf}) parameters were unknown and needed to be adjusted for each site by fitting the model output to the observations.

3.3.2 Soil contributions

To deduce the surface emissivity for rough soil ($e_{\mathrm{surf,\,p}}$ for a given polarization p), the Wegmüller and Mätzler (1999) soil model was used to calculate the surface reflectivity of rough soil under the canopy ($r_{\mathrm{surf,\,p}}$ for a given polarization p), with or without snow by using Eqs. (7) and (8):

$$r_{\mathrm{surf,\,H}} = 1 - e_{\mathrm{surf,\,H}} = \Gamma_{\mathrm{Fresnel,\,H}} \cdot \exp(-\sigma_{\mathrm{s}}^{\sqrt{\alpha \cdot \cos\theta}}) \quad (7)$$

$$r_{\mathrm{surf,\,V}} = 1 - e_{\mathrm{surf,\,V}} = r_{\mathrm{surf,\,H}} \cos\theta^{\beta_\nu}, \quad (8)$$

where $r_{\mathrm{surf,\,p}}$ mainly depends on the surface roughness and Fresnel coefficients ($\Gamma_{\mathrm{Fresnel,\,H}}$). In Eq. (7), the simplified parameter $\sigma_{\mathrm{s}} = k\sigma$ was used, where k is the wave number and σ the standard deviation of the surface height (in meters); α is a constant parameter fixed to -0.1 (Wegmüller and Mätzler (1999). For frozen soil, parameters derived from Montpetit et al. (2018) were used (see Sect. 4.1). For thawed soil, $\Gamma_{\mathrm{Fresnel,\,H}}$ was estimated from the dielectric constant calculated with the Dobson (1985) equations according to the soil moisture and soil temperature. These variables were computed with the Crocus model, coupled to the ISBA land surface model, and extracted daily (at 14:00, as the other variables). The soil reflectivity in vertical polarization also depends on a parameter β_ν (Montpetit et al., 2018), which describes the polarization of the signal and is frequency-dependent. Note that we will often use the "ν" subscript hereafter to denote quantities that are dependent on frequency.

Hence, the soil parameter e_{surf} was linked to the set of values (σ_{s}, β_ν) and mainly evolved with soil moisture and soil temperature.

3.3.3 Inversions of vegetation and soil parameters

The inversion of forest (ω, γ_ν) and soil (σ_{s}, β_ν) parameters was carried out in summer to avoid the bias due to the presence of a snowpack. Forest parameters (ω, γ_ν) depend on the forest characteristics, such as the biomass and the structure of the canopy at each site. They also depend on LAI, which allows the season forest emission cycle to be accounted for. Using the vegetation water content equation defined by Pampaloni and Paloscia (1986), the parameter γ_ν was related to the 4-day LAI for a given frequency ν with the Eq. (9):

$$\gamma_\nu = e^{-b \cdot k^a \cdot \left(\exp\left(-\frac{\mathrm{LAI}}{3}\right) - 1\right)/\cos\theta}, \quad (9)$$

where a and b are two constants to calibrate. To reduce the number of unknown variables, Eq. (9) was simplified to use only one constant η_ν such as $\eta_\nu = e^{-b \cdot k^a}$.

The vegetation and soil parameters were inverted by minimizing the difference between $T_{\mathrm{B\,TOA}}$ simulations and $T_{\mathrm{B\,TOA}}$ measured with AMSR-2 sensors at 11, 19, and 37 GHz in vertical polarizations. We used the same approach developed by Roy et al. (2014) since it was well adapted for PMW emission in boreal areas: the two frequency-dependent parameters (η_ν, β_ν) and two frequency-invariant parameters (ω, σ_{s}) were inverted with a two-stage calibration by permuting all possible combinations of the two frequency invariant parameters. Specifically, ω values varied from 0.02 to 0.16 in steps of 0.01, and σ_{s} varied from 0.01 to 1.1 in steps of 0.05. This yields a total of 300 possible combinations of the frequency invariant parameters. Then, for each possible combination of the frequency-invariant parameters, a calibration of the frequency-dependent parameters, η_ν and β_ν, was performed for each frequency. A total of 900 frequency-dependent calibrations were thus computed. Finally, for each possible combination of the frequency-invariant parameters, the total post-calibration T_{B} RMSE across all three frequencies was computed. The combination of frequency-invariant parameters resulting in the lowest T_{B} RMSE was chosen.

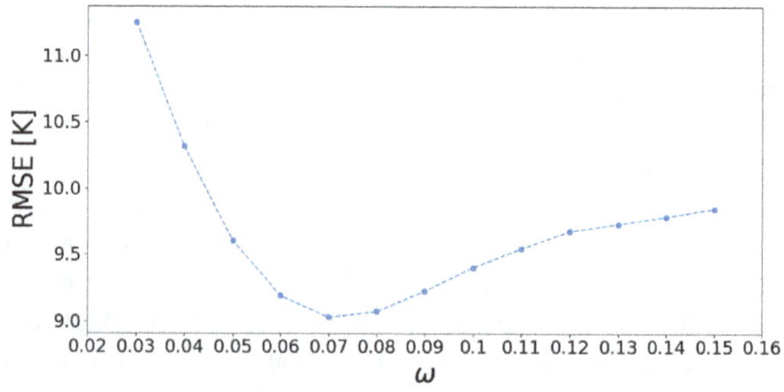

Figure 3. Overall T_{B} RMSE (at 11, 19, and 37 GHz, for the 12 sites and for the summer period) between the simulated and measured $T_{\mathrm{B\,TOA}}$ as a function of the values of ω. A σ_{s} value at 0.2 cm gives the best results but T_{B} RMSE is not very sensitive to this variable. The parameters β_ν and η_ν were optimized for each (ω, σ_{s}) couple according to the frequency used.

$T_{\mathrm{B\,TOA}}$ were simulated from 2012 to 2016. The inversion was not very sensitive to σ_{s} (not shown) and Fig. 3 shows the optimal overall $T_{\mathrm{B\,TOA}}$ RMSE between simulated and measured $T_{\mathrm{B\,TOA}}$ for the 12 sites and for the summer period according to ω values. Over the summer period, a ω value at 0.07 and a σ_{s} value at 0.2 cm gave best results for $T_{\mathrm{B\,TOA}}$ simulations, with a minimum overall RMSE equal to 9.0 K. These parameters were previously optimized over the same study area by Montpetit et al. (2018). A value of $\omega = 0.07$ was coherent with the literature for dense boreal forest areas (Pellarin et al., 2006; Meissner and Wentz, 2010; Roy et al., 2012). For this optimal (ω, σ_{s}) set of values, the mean optimal values of the η_ν and β_ν factors were estimated, by considering the soil contribution constant if the soil was frozen.

3.4 Data assimilation setup

The DA setup is the same as the one developed in Larue et al. (2018) except that we added an inflation technique of the covariance matrix of observation errors (**R** matrix) to avoid ensemble degeneracy, i.e., when an ensemble collapses to a unique particle (Arulampalam et al., 2002).

3.4.1 DA framework

The DA scheme is a particle filter with a Sequential Importance Resampling algorithm (PF-SIR) that is well documented in Van Leeuwen (2009, 2014) and Gordon et al. (1993) and relatively easy to implement with a snowpack model (Dechant and Moradkhani, 2011; De Lannoy et al., 2012; Charrois et al., 2016; Larue et al., 2018). The PF-SIR represents the probability density function (pdf) of the model state with an ensemble of states (called particles), which is updated when an observation is available. An ensemble approach was preferred because of the nonlinearity of the system. Moreover, the particle filter approach can cope with the variable number of state variables resulting from the changing number of snow layers in Crocus. The created ensemble represents uncertainty in SWE and in T_B simulations due to the uncertainties of meteorological inputs (Fig. 2).

Assuming that the meteorological forcing data were the only source of uncertainties, the ensemble of T_{Bs} was created by running the chain of models described in Fig. 2 with an ensemble of perturbed inputs. The assimilation was performed daily (at 13:00) and the ensemble was composed of 150 members, which was found to be an adequate size (Larue et al., 2018). The daily ensemble of meteorological forcing data was created by perturbing selected GEM data (air temperature, wind speed, precipitation, and short- and long-wave radiation) with Gaussian noise according to their respective uncertainties (estimated in Larue et al., 2018). Meteorological forcing perturbations were propagated in time following a first-order autoregressive process to simulate their realistic temporal variations (Charrois et al., 2016). Precipitation, wind speed, and short-wave radiation ("SW_{down}") were perturbed by a multiplicative factor centered at 1. Perturbation boundaries were fixed at -0.9 and 0.9. The air temperature was perturbed by an additive factor, with boundaries fixed at -3 and $+3$ K. Perturbed long-wave radiation ("LW_{down}") was estimated with perturbed T_{air} from a linear regression estimated in Larue et al. (2018). In order to maintain physical consistency in the simulations, SW_{down} was limited to 200 W m^{-2} when there was precipitation (presence of clouds) (Charrois et al., 2016).

The observation error standard deviation associated with AMSR-2 observations was assumed to be 2 K (Durand and Margulis, 2006, 2007). Note that in reality it was probably larger since it represents all mismatches between observations and simulations obtained if the model was run with "correct" inputs. This observation error cannot be easily estimated (low spatial resolution, representativeness, etc.), but it is only a sort of initial value here, since we used a covariance inflation to adjust it.

DA experiments were applied between 1 November and 1 May. To avoid wet snow conditions, which increase the emissivity of the snowpack, whereas the SWE does not change, the DA was not performed when liquid water content was observed in the modeled snowpack. This variable was computed by Crocus, driven with original meteorological forcing data. SWE values were evaluated over both the dry snow period (from 1 December to 15 March) and the whole winter (when a snowpack was detected).

3.4.2 DA and inflation technique

The snowpack prior state x_t at time t is computed with the updated past state of snowpack simulations at time $t-1$ (posterior state x_{t-1}) and the prior perturbed meteorological forcing data F_t from time $t-1$ to t (see Fig. 2). The predicted observation is computed with

$$y_t^i = h\left(x_t^i\right), \tag{10}$$

where y_t^i is $T_{B\,TOA}$ predicted from particle i ($i = 0 \ldots N$, with N the ensemble size). The observation operator h is the τ_{snow}-calibrated DMRT-ML model and the calibrated radiative transfer models estimating soil, atmosphere, and vegetation contributions. In the analysis step, the new posterior distribution is updated by weighting each particle x_t^i according to the distance between y_t^i and the AMSR-2 T_B observation (with the weight we_t^i). With the SIR algorithm, the pdf is resampled by duplicating particles with high weights (i.e., close to observations) and dropping with negligible weights (far from observations). With Arakawa's procedure used here for ensemble resampling (Arakawa, 1996; same as Charrois et al., 2016), a particle is definitely selected if its weight is higher than or equal to the inverse of the ensemble size ($we_t^i > 1/N$, with $N = 150$).

Ensemble resampling considerably reduces the risk of degeneracy but does not eliminate it. Degeneracy starts when only a few particles have significant weights. These particles are selected many times, leading to a loss of diversity of the posterior ensemble. After several assimilation steps, the ensemble quickly reduces to a single particle. Ensemble degeneracy can be detected when the number of selected particles (those with high weights) is below an effective limit number N_{keep}, here fixed at 25 as a compromise between the quality of the DA scheme and the size of the ensemble (not shown). In this study, we developed a new technique to avoid a degeneracy problem, which consists in the online adjustment of the **R** matrix (i.e., observation standard deviation squared times the identity matrix) such that the weight of the 25th selected particle (we_{keep}) is at least equal to $1/N$. The rationale here is that, because the weights are nonlinear functions of the observation error covariance matrix, a larger matrix tends to flatten the distribution of weights and favors the selection of more particles. This adjustment is performed with an inflation of the initial matrix, and the detailed algorithm is provided in Appendix A.

Ensemble degeneracy is often caused by extreme precipitation events resulting in very high T_B values difficult to represent with the model. The online adjustment technique mitigates the consequences of this model deficiency on the snow

Table 3. Experiment setup information. Exp. is the experiment identifier (see text). T_{Bs} are in V-pol.

Exp.	State variables	Sites	Time period	Assimilated frequencies	Acronyms
A	T_{air}, wind, precipitation SW_{down}, LW_{down}	1, 9, 12, one winter	1 November 2014 to 1 May 2015	$\Delta T_{B\,19\text{-}37}$ $\Delta T_{B\,19\text{-}37}$ and $\Delta T_{B\,11\text{-}19}$ $T_{B\,11}$, $T_{B\,19}$, $T_{B\,37}$	DA1_DTB19-37 DA2_DTB19-37, DTB11-19 DA3_TB_11,19,37
B	T_{air}, wind, precipitation, SW_{down}, LW_{down}	12 sites, 43 winter simulations	From 1 November to 1 May, winters 2012 to 2016	$T_{B\,11}$, $T_{B\,19}$, $T_{B\,37}$	DA_b_TB_11,19,37
C	T_{air}, wind, precipitation, SW_{down}, LW_{down}, ω, τ_{snow}	12 sites, 43 winter simulations	From 1 November to 1 May, winters 2012 to 2016	$T_{B\,11}$, $T_{B\,19}$, $T_{B\,37}$	DA_c_TB_11,19,37

simulations over the rest of the season. The other side of the coin is that a "good" observation can be ruled out if the model is not able to reproduce it, thereby reducing the accuracy of the snowpack estimation.

3.4.3 Experimental setup

To study the sensitivity and the quality of T_B assimilation for SWE improvements, three experiments were performed.

(a) Experiment A: to test the feasibility of the DA scheme for several environmental conditions, and to find the best DA configuration to apply, T_B assimilation for three representative sites was performed in a preliminary experiment for winter 2014–2015. Following a north–south gradient, we selected Site 12 ($f_{cover} = 24.2\,\%$, northern coniferous area), Site 1 ($f_{cover} = 63.7\,\%$, coniferous area), and Site 9 ($f_{cover} = 84.0\,\%$, mixed forest area), each representing a different environmental condition. Over these three sites, we estimated the quality of the DA scheme according to the assimilated frequencies: (a) assimilation of the T_B difference between 19 and 37 GHz (referred to as $\Delta T_{B\,19\text{-}37}$); (b) assimilation of the $\Delta T_{B\,19\text{-}37}$ and the T_B difference between 11 and 19 GHz, in V-pol (referred to as "$\Delta T_{B\,11\text{-}19}$"); and (c) assimilation of the three T_{Bs} at 11, 19, and 37 GHz in V-pol ($T_{B\,11}$, $T_{B\,19}$, $T_{B\,37}$). Table 3 summarizes the experiment setup information (acronyms of the experiments, sites, time period). We used V-pol T_B because H-pol T_B is more sensitive to the stratigraphy of the snowpack and to the presence of ILs (Mätzler, 1987). While the DA of T_{Bs} at 11, 19, and 37 GHz in V-pol should give best results since this combination of frequencies imposes more constraints, the risk of encountering a degeneracy problem is higher. The combination of both $\Delta T_{B\,19\text{-}37}$ and $\Delta T_{B\,11\text{-}19}$ is commonly used in the literature for SWE retrievals (Chang et al., 1987; Tedesco et al., 2004; Tedesco and Nervekar, 2010). The assimilation of the $\Delta T_{B\,19\text{-}37}$ only was also studied to analyze the sensitivity of T_B assimilation for deep snowpack when $T_{B\,37}$ saturates for a SWE up to about 150 mm (Mätzler et al., 1994) and to evaluate the supply of information from 11 GHz in the assimilation of both $\Delta T_{B\,19\text{-}37}$ and $\Delta T_{B\,11\text{-}19}$ for SWE improvements.

To quantify the performance of the DA scheme, the daily RMSEs of ensembles of simulated SWE obtained with and without the DA scheme were compared by using Eq. (11):

$$\mathrm{RMSE}_t = \sqrt{\left(\frac{1}{N}\sum_{i=1}^{N}\left(X_{\mathrm{sim}\,i,t} - X_{\mathrm{Obs}\,t}\right)^2\right)}, \tag{11}$$

where N is the ensemble size, $X_{\mathrm{sim}\,i,t}$ is the simulated variable from the member i at time t, and $X_{\mathrm{Obs}\,t}$ is the diagnostic variable at time t obtained from AMSR-2 observations.

(b) Experiment B: the best configuration of the DA scheme (DA of the three T_{Bs} at 11, 19, and 37 GHz in V-pol, called the "DA_b_TB_11, 19, 37" experiment in Table 3) was applied over the 43 winters. To estimate the accuracy for hydrological applications, the median of the resampled SWE ensemble obtained with the DA_b_TB_11, 19, 37 experiment (called "SWE_{DA}"further) was compared to SWE measurements. The median was used instead of the mean to reduce the potential impact of extreme perturbations. The evaluation of this experiment was performed by comparing SWE_{DA} RMSE and the relative percentage of error ("RPE") values to the original SWE simulations (SWE_{Crocus}), obtained by driving Crocus with original meteorological forcing data. The RPE is defined as

$$\mathrm{RPE} = 100 \cdot \frac{|\mathrm{Bias}|}{\overline{\mathrm{MEAN}_{\mathrm{obs}}}}. \tag{12}$$

The mean biases of SWE estimates obtained without and with assimilation were also compared. Performance was estimated for SWE higher than 48 kg m^{-2}(about 20 cm of snow

Table 4. Effective parameters calibrated for the 12 studied sites to quantify soil contributions e_{surf} (calibrated surface roughness "cal. σ_s" and calibrated polarization ratio "cal. β_v") and vegetation contributions (controlled by the calibrated (ω, η_v) parameters "cal. ω" and "cal. η_v" according to the daily LAI) measured at the top of the atmosphere. The parameterization of frozen ground was estimated by Montpetit et al. (2018). ε_{eff} is the effective dielectric constant estimated with the permittivity of frozen and unfrozen soils derived from Dobson's equations (1985). Annual and seasonal $T_{B\,TOA}$ RMSEs estimated for the summer and the winter period ($RMSE_{summer}$ and $RMSE_{winter}$) are calculated from 2012 to 2016 with the calibrated parameters.

Frequency (GHz)	Frozen soil			Unfrozen soil		Cal. ω	Cal. η_v	Mean $RMSE_{summer}$ (K)	Mean annual RMSE (K)
	ε_{eff}	σ_s (cm)	β_v	Cal. σ_s (cm)	Cal. β_v				
11	3.18-0.006134i		1.08		0.69		0.01	8.6	8.4
19	3.42-0.00508i	0.19	0.72	0.2	0.60	0.07	0.05	8.7	9.1
37	4.47-0.32643i		0.42		0.67		0.23	10.1	26.0

depth), derived from measurements, to attenuate problems of shallow snow cover variability or heterogeneity in the AMSR-2 grid cells. To analyze the impact of the vegetation, results were separated according to the fraction of f_{cover}: moderate f_{cover} ($f_{cover} < 75\,\%$, 10 sites) and high f_{cover} ($f_{cover} > 75\,\%$, 2 sites) (see Table 2 for f_{cover} site information).

The accuracy needed for hydrological applications is a SWE RPE lower than 15 % (Vachon, 2009; Larue et al., 2017), which is the same performance objective as the CoreH2O project (10 % for SWE >30 cm and 3 cm for SWE <30 cm, Rott et al., 2010) and the GlobSnow2 product (Luojus et al., 2014). This error threshold corresponds to an RMSE of about 45 kg m^{-2} for a measured average Quebec snowpack about 300 kg m^{-2} of SWE. The ability to accurately estimate the annual SWE maximum (SWE_{max}) was also studied since it is one of the most important variables for hydrological applications. It allows the amount of water stored in the snowpack before the spring snowmelt to be described. To avoid extreme values, the SWE_{max} is estimated as the average of the SWE for a time period of ± 2 days around the detected SWE_{max}.

(c) Experiment C: the quality of the DA scheme could strongly depend upon the choice of state variables. In the A and B experiments, we chose to pre-calibrate forest and soil parameters and to use a constant snow stickiness parameter (τ_{snow}) fixed at 0.17 (Larue et al., 2018). Nevertheless, these calibrations are empirical and should be adjusted for each site. It depends on several parameters that are difficult to measure at a 10×10 km^2 spatial scale (snow grains, canopy, biomass, etc.). The forest parameter ω strongly affects the PMW emission from the vegetation, which can represent more than 60 % of the signal measured by satellite sensors (see discussion in Sect. 5.2). Kwon et al. (2017) have shown that the contribution of $T_{B\,Veg}$ to $T_{B\,TOA}$ was better represented by considering ω free in the DA scheme, and improvements in the resulting SD were evident for the forest land cover type (about 5 % with DMRT-ML). In Experiment C, the DA scheme was thus tested using ω and τ_{snow} as free variables in the assimilation process (called the "DA_c_TB_11,19,37" experiment). The DA_c_TB_11,19,37 experiment is identical to the DA_b_TB_11,19,37 experiment (over the 43 winters); only the state variables were changed. The ω parameter was perturbed with Gaussian noise, centered on 0.07 (as calibrated), with a standard deviation of 0.02 and bounded by 0.05 and 0.12 (reasonable range of $T_{B\,TOA}$ RMSE values; Fig. 3). The snow stickiness parameter was perturbed by Gaussian noise, centered on 0.17, with a standard deviation of 0.15 and bounded by 0.1 and 0.46. These limits correspond to the range of τ_{snow} values extracted from Larue et al. (2018) over the same study area. The ensemble size was kept to 150 members.

4 Results

4.1 Results of model inversions

The mean optimal values of the η_v and β_v factors were estimated for the optimal (ω, σ_s) set of values (0.07 and 0.2, respectively; see Table 4). The (σ_s, β_v) soil parameters are given in Table 4 and are used to estimate the $T_{B\,TOA}$ RMSE obtained with the calibrated chain of models. Without parameter inversions, the annual mean RMSE of the original T_B simulations varies from 12.9 to 47.1 K for the three frequencies (not shown). With parameter inversions over the summer period, the three frequencies have a similar T_B $RMSE_{summer}$ (8.6–10.1 K, Table 4), while over the year (using parameters inverted over the summer period) the annual $T_{B\,TOA}$ RMSE significantly increases at 37 GHz due to the presence of the snowpack (26.0 K). The inversions make it possible to reduce the annual $T_{B,37}$ RMSE by 21.1 K. Figure 4a, b, and c show the multiyear $T_{B\,TOA}$ variations for Sites 12, 1, and 9, respectively, from 2012 to 2016 and at 37 GHz. At this

Table 5. Averaged SWE ensemble RMSE (see Eq. 11) obtained with and without DA, according to the experiment (see Sect. 3.4.3, Table 3 for acronyms) for each tested site. $RMSE_{dry\text{-}snow}$ is the SWE ensemble RMSE obtained from 1 December to 15 March. $RMSE_{annual}$ is estimated over the whole winter (when snowpack is detected). No. corresponds to the site (see Table 1).

	Experiments	SWE ensemble RMSE ($kg\,m^{-2}$)			
		No. 1	No. 12	No. 9	Overall
$RMSE_{dry\ snow}$ ($kg\,m^{-2}$)	Without assimilation (prior estimates)	50.7	28.6	47.8	42.4
	DA1_DTB19-37	21.1	18.1	32.4	23.8
	DA2_DTB19-37, DTB11-19	14.0	26.9	28.2	23.0
	DA3_TB_11,19,37	10.5	19.7	21.0	17.1
$RMSE_{annual}$ ($kg\,m^{-2}$)	Without assimilation (prior estimates)	47.2	28.9	48.9	41.7
	DA1_DTB19-37	21.6	22.1	40.1	27.9
	DA2_DTB19-37,DTB11-19	16.7	28.1	38.5	27.8
	DA3_TB_11,19,37	15.1	21.9	29.1	22.1

frequency, the simulated $T_{B\,TOA}$ is strongly underestimated when a snowpack is observed. This is likely due to an overestimation of the SWE or snow grain sizes since $T_{B,37}$ are attenuated in the snowpack as snow grains act as diffusers while the $T_{B,19}$ and $T_{B,11}$ are relatively unaffected by snow grains ($RMSE_{summer}$ similar to $RMSE_{winter}$ at 11 and 19 GHz; Table 2). Simulated SWE values were overestimated by 16 % and 20.2 % compared to SWE measurements for Sites 1 and 9, respectively, for winter 2014–2015. The objective of T_B assimilation is to reduce these overestimations. Note that the SWE simulated at Site 12 is underestimated by 19 %. The underestimation of $T_{B,37}$ can also be caused by an underestimation of the vegetation contributions. This aspect is further discussed in Sect. 5.2.

By integrating ILs within the snowpack when the PR11 is above 0.06, the annual $T_{B\,TOA}$ RMSE at 37 GHz is reduced and goes from 28.9 to 26.0 K.

In winter, the overall $T_{B\,TOA}$ RMSE (all frequencies) is equal to 17.4 K from 2012 to 2016 (not shown), similar to the overall RMSE estimated for the τ_{snow}-calibrated DMRT-ML driven by in situ measurements in an open area and equal to 19.9 K compared to surface-based radiometric measurements in Quebec (Larue et al., 2018).

4.2 Results of AMSR-2 data assimilation (DA)

4.2.1 Experiment A

Figure 5 shows variations of the daily RMSE of the SWE ensemble (see Eq. 11) obtained without and with DA (prior and posterior estimates) according to the combination of frequencies used as observations (DA1_DTB19-37, DA2_DTB19-37, DTB11-19 and DA3_TB_11,19,37 experiments; see Table 3). Table 5 summarizes these averaged RMSEs for the studied period (dry snow period and whole winter) for tested sites.

Over the three sites and for the dry snow period, the DA reduced the overall SWE ensemble RMSE by 43.9 %, 45.8 %, and 59.7 % with the DA1_DTB19-37, DA2_DTB19-37,DTB11-19, and DA3_TB_11,19,37 experiments, respectively, compared to the SWE ensemble RMSE obtained with prior estimates (Table 5). The assimilation of the three frequencies (DA3_TB_11,19,37) helps to improve SWE simulations, giving the lowest RMSE compared to other scenarios. The same trend is observed over the whole winter and the assimilation of the three frequencies reduces the overall SWE ensemble RMSE by 47.0 % (SWE ensemble RMSE of 22.1 $kg\,m^{-2}$) compared to the SWE ensemble RMSE of prior estimates (SWE ensemble RMSE of 41.7 $kg\,m^{-2}$).

In our previous work (Larue et al., 2018), we have shown a reduction of 82 % of the SWE RMSE by assimilating both the $\Delta T_{B\,19\text{-}37}$ and $\Delta T_{B\,11\text{-}19}$ and using synthetic observation data over a dry snow period. The differences between results using synthetic and real data in DA experiments are likely due to two aspects. Firstly, the snow model does not resolve the intra-pixel surface variability. We assumed homogeneous snow cover within the pixel in open areas, thus with no interactions between snow and vegetation. Even if we compare simulations with surface-based measurements in open areas, this could introduce large uncertainties (Roy et al., 2016). Secondly, the land cover variability and heterogeneity within each pixel also induce uncertainties in the mean T_B simulation over a pixel (T_B weighted by the fraction of forest cover; see Eq. 2).

Figure 6 illustrates the comparison between SWE measurements, the original SWE Crocus simulations (SWE_{Crocus}), and the median of the SWE ensemble obtained with the DA3_TB_11,19,37 experiment. The yellow envelope corresponds to the SWE ensemble obtained without DA (prior estimates) and shows a large ensemble spread in response to meteorological forcing uncertainties. The gray envelope is the resampled SWE ensemble (posterior estimates). SWE simulations are very sensitive to the uncertainties of

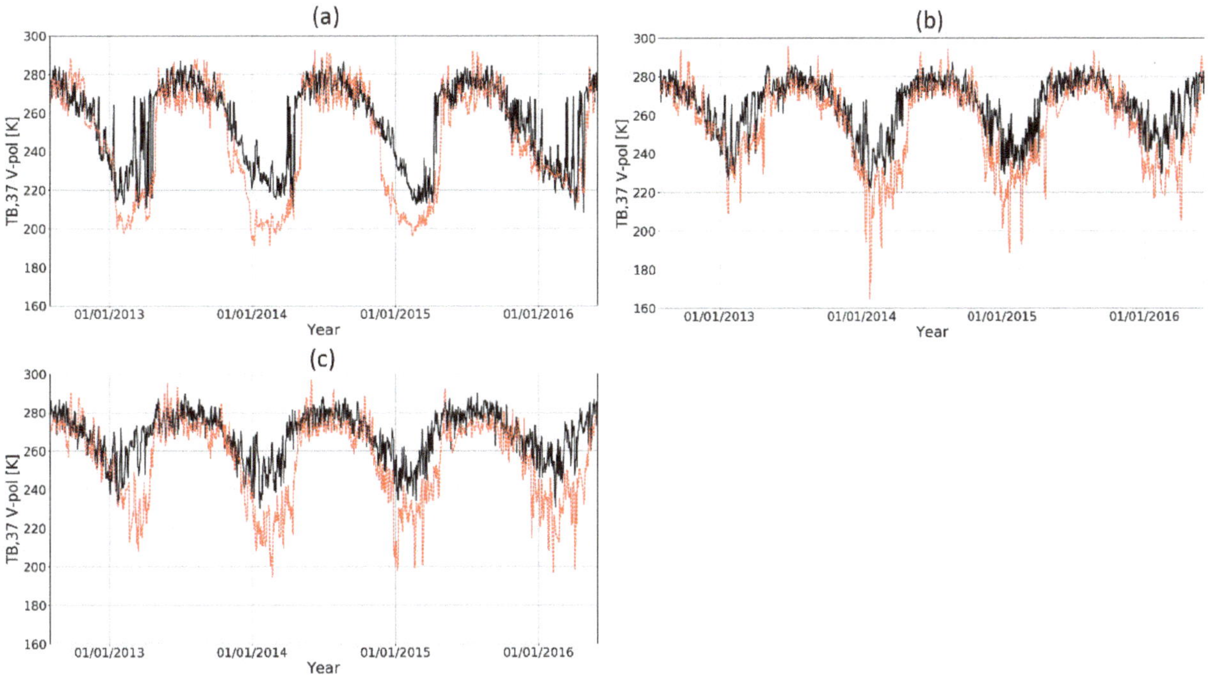

Figure 4. Multi-year variations of simulated $T_{B\,TOA}$ (red dotted lines) and measured $T_{B\,TOA}$ (black full lines) from 2012 to 2016 at 37 GHz in vertical polarization: **(a)** Site 12 (f_{cover} of 24 %), **(b)** Site 1 (f_{cover} of 64 %), and **(c)** Site 9 (f_{cover} of 84 %) (see Table 2).

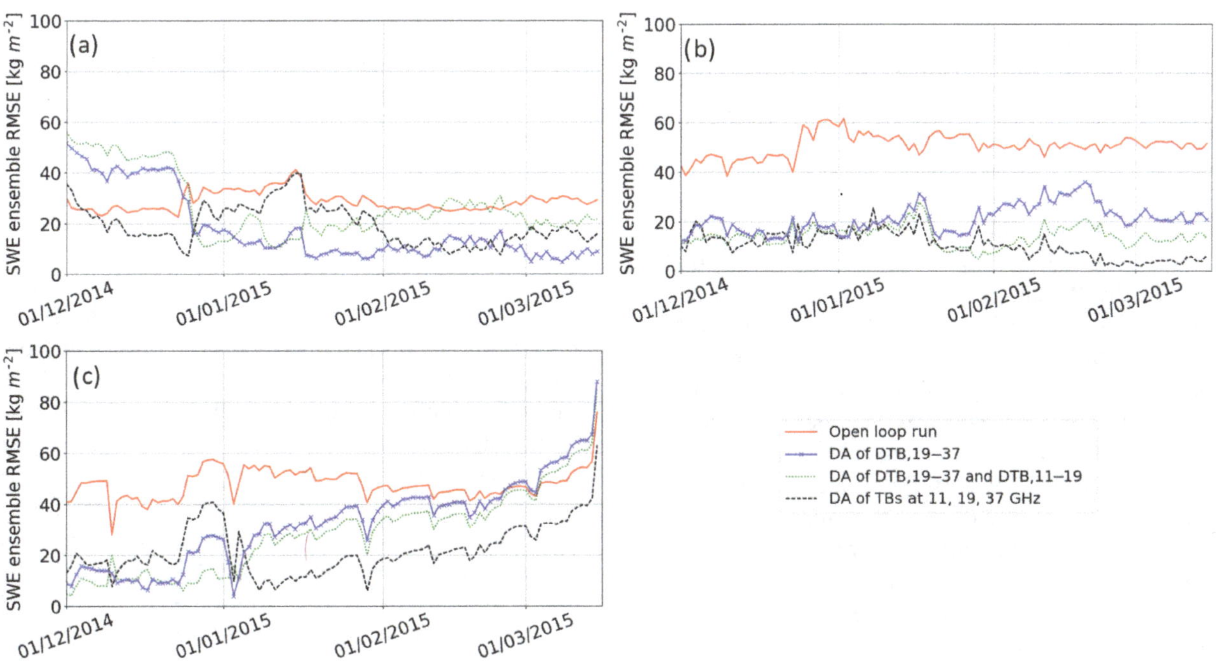

Figure 5. Variations of the SWE ensemble RMSE (Eq. 11) obtained with and without DA for the dry snow period (from 1 December to 15 March). Experiments are performed for **(a)** Site 12, **(b)** Site 1, and **(c)** Site 9, over the winter 2014–2015. The red line is the SWE ensemble RMSE obtained without DA (open loop runs), the blue line is the RMSE obtained with the DA1_TB19-37 experiment, the green dashed line the RMSE with the DA2_TB19-37,TB11-19 experiment, and the black dotted line the RMSE with the DA3_TB_11,19,37 experiment.

Table 6. Averaged SWE RMSE, bias, and RPE (Eq. 12) over the 43 winters for original SWE simulation (SWE_{Crocus}) and assimilated SWE_{DA}(Experiment B). Statistical performances were estimated for SWE_{obs}>48 kg m^{-2} (snow depth higher than ~ 20 cm). $\overline{SWE_{obs}}$ and $\overline{SWE_{sim}}$ are the averaged observed and simulated SWE, respectively.

		SWE_{Crocus}				SWE_{DA} with the DA_b_TB_11,19,37			
	$\overline{SWE_{obs}}$	RMSE (kg m^{-2})	Bias (kg m^{-2})	RPE %	$\overline{SWE_{sim}}$	RMSE (kg m^{-2})	Bias (kg m^{-2})	RPE %	$\overline{SWE_{sim}}$
f_{cover}<75 %	162.2	42.5	17.3	19.0	179.4	37.1	−1.2	14.1	161.0
f_{cover}>75 %	139.0	62.0	47.8	33.9	186.8	68.0	40.2	38.0	179.2
Mean	157.3	45.0	23.7	22.1	41.2	7.5	18.4	164.8	

meteorological forcing data at the beginning of the winter season. If an event (melting or precipitation) is missed, a constant bias on SWE estimates is kept throughout the winter. For Sites 1 and 9, the DA scheme allows the correction of these uncertainties at the beginning of the season: the SWE ensemble RMSEs of posterior estimates are reduced by about 30 kg m^{-2} at the beginning of the season, compared to the RMSE of prior estimates (Fig. 5). For these two sites, the SWE ensemble RMSE obtained with the DA1_DTB19-37 experiment increases as the snowpack becomes deeper, especially from mid-January when the snowpack becomes deeper than 100 kg m^{-2} (Fig. 6). The PMW signal from the snowpack at 37 GHz saturates for such deep snowpack (Mätzler et al., 1982; Mätzler, 1994; De Sève et al., 1997, 2007) and the assimilation of $\Delta T_{Bv,19-37}$ only does not give enough information to significantly improve SWE retrievals. For Site 9, posterior estimates are deteriorated at the end of the season compared to prior estimates with the DA1_DTB19-37 experiment. By adding $\Delta T_{Bv,11-19}$ (DA2_DTB19-37,DTB11-19), this effect is reduced but stays sensitive to the depth of the snowpack (Fig. 5).

Note that the gray envelope does not always include the observations (Fig. 6a, c). This could be due to an underestimation of the **R** matrix. In the developed approach, the inflation technique of the **R** matrix is limited by a threshold on the α factor fixed at 5 since the simulations are limited by the simplifications of physical parameters in the models and we may introduce a bias if we force them to follow the observation by perturbing meteorological forcings only. Further work is needed to quantify the model errors in order to consider it in the DA scheme and to improve the representativeness of the simulations. To represent the uncertainties about the physical processes simulated with the Crocus snow model, a new system based on snow model ensembles could be an alternative. Such an approach was recently developed by implementing different configurations estimating the physical parameters of the Crocus snow model (ESCROC, Lafaysse et al., 2017).

4.2.2 Experiment B

The median of the resampled ensemble of SWE obtained with the DA of the three frequencies (SWE_{DA}) is used to estimate the global performance of the DA scheme for SWE improvements. Table 6 details the statistical performance of simulated SWE_{DA} compared to measurements and to the original SWE Crocus simulations (SWE_{Crocus}) over the 43 winters. Figure 7 compares the SWE_{DA}, SWE_{Crocus}, and SWE measurements (SWE_{obs}) from 2012 to 2016 for four sites with different f_{cover} taken as an example: Site 5 ($f_{cover} = 31.5$ %), 10 ($f_{cover} = 61.8$ %), 1 ($f_{cover} = 63.7$ %), and 9 ($f_{cover} = 84.0$ %). In this section, we first analyze the overall SWE improvements obtained with T_B assimilation compared to original SWE simulations, and the impact of the vegetation on the quality of the DA scheme is discussed.

Overall SWE improvements compared to original Crocus simulations

The overall SWE_{Crocus} RMSE, bias, and RPE are of 45.0 kg m^{-2}, 23.7 kg m^{-2}, and 22.1 %, respectively (Table 6). In comparison, the overall SWE_{DA} RMSE, bias, and RPE are improved and equal to 41.2 kg m^{-2}, 7.5 kg m^{-2}, and 18.4 %, respectively. The overall bias is reduced by 16.2 kg m^{-2} (68 % of SWE_{Crocus} bias) with the DA scheme. The DA of the three frequencies thus helps to improve SWE estimates over Quebec. Correlation between SWE_{DA} simulations and SWE measurements gives a similar R coefficient to the one obtained with SWE_{Crocus} simulations ($R = 0.79$ and $R = 0.78$, respectively), but the offset is significantly reduced with SWE_{DA} compared to SWE_{Crocus} (offset = 10 and 29 kg m^{-2}, respectively). We analyzed the number of cases with significant improvements for the total of 43 simulations studied by considering a 5 % threshold on the bias and RMSE differences before and after assimilation. The SWE_{DA} bias is significantly reduced for 26 winters (60 % of cases) compared to original SWE simulations. However, the RMSE is significantly improved for only 35 % of simulations, and in 35 % of cases, RMSEs are similar.

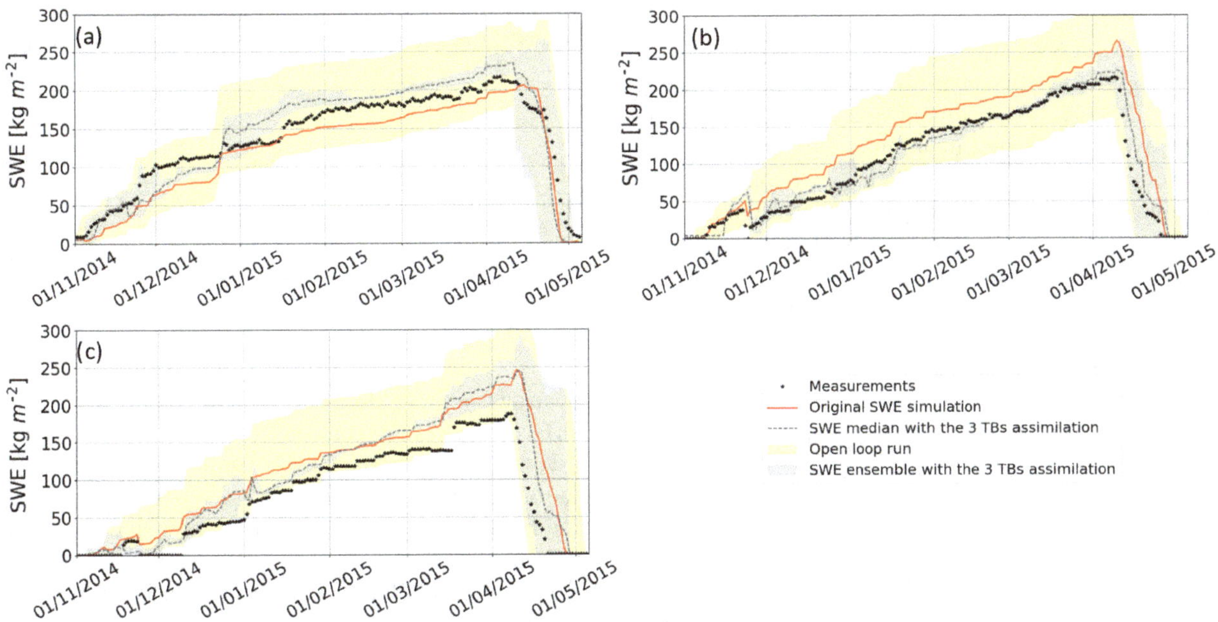

Figure 6. Evolution of SWE measurements (black points) and SWE simulations. The SWE_{Crocus} is the red line, and the SWE obtained with the DA3_TB_11,19,37 experiment is the gray dotted line. The yellow envelope is the spread of the SWE ensemble obtained with open loop runs (prior estimates). The gray envelope is the spread of the SWE ensemble obtained with the assimilation of the three frequencies (posterior estimates). Both spreads are delimited by the 5th and the 95th percentiles. Experiments are computed for **(a)** Site 12, **(b)** Site 1, and **(c)** Site 9, over the winter 2014–2015.

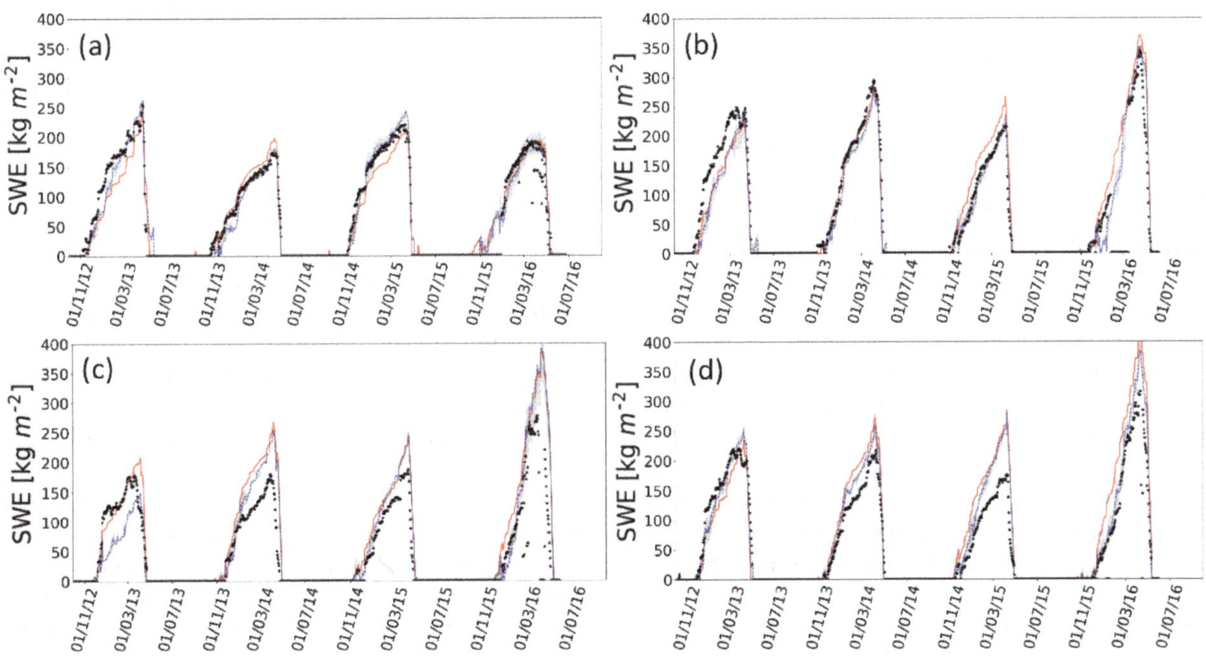

Figure 7. Evolution of SWE measurements (black points), original SWE simulations (red full line), and the median of the SWE ensemble obtained with the DA_b_TB_11,19,37 experiment (SWE_{DA}) (blue dotted line). The gray envelope is the spread of the SWE_{DA} ensemble (posterior estimates). Experiments are computed for **(a)** Site 5 ($f_{cover} = 31.5\,\%$), **(b)** Site 1 ($f_{cover} = 63.7\,\%$), **(c)** Site 9 ($f_{cover} = 84\,\%$), and **(d)** Site 10 ($f_{cover} = 61.8\,\%$), from 2012 to 2016.

Evaluation of SWE_{max}

The mean observed SWE_{max} is equal to 235.6 kg m^{-2} from 2012 to 2016, and the mean simulated SWE_{max} is equal to 278.3 and 266.8 kg m^{-2} without and with the assimilation, respectively. Compared to original SWE simulations, the DA scheme improves 63 % of SWE_{max} simulations with an overall improvement of 12.2 kg m^{-2}, corresponding to 8 % of SWE measurements (Table 6). Such an uncertainty extended over the whole territory could have a strong impact, considering that 1 mm of SWE in the LG watershed could represent USD 1 million in hydroelectric power production (Brown and Tapsoba, 2007).

SWE accuracy for sites according to the f_{cover}

The overall RPE obtained with the DA scheme is below 15 % (RPE = 14.1 %) for sites with an f_{cover} below 75 % (Table 6), which is the accuracy required for hydrological applications (Larue et al., 2017). Hence, the accuracy of SWE_{DA} estimates, obtained without the use of any surface-based data, is very encouraging for hydrological needs in remote areas. In comparison, the GlobSnow-2 SWE product (Takala et al., 2011), which assimilates both T_{Bs} and in situ snow depth measurements, has a SWE RPE equal to 35.9 % over the same area in Quebec (Larue et al., 2017), twice the uncertainty of SWE_{DA}. Figure 7a and b (Sites 5 and 1) show that for a single site, the original SWE_{Crocus} simulation works well for some years but can be underestimated or overestimated over other years. The DA scheme allows a more stable solution when the overall f_{cover} is under 75 % (not the case for Site 9, for example).

Nevertheless, even if the overall RMSE is improved, the DA scheme does not help to improve SWE estimates for sites with an f_{cover} above 75 % (RMSE of 66 kg m^{-2}) compared to original SWE simulations (RMSE of 62.0 kg m^{-2}). The presence of vegetation is a major source of uncertainty in $T_{B\,TOA}$ simulations. The emission of the trees is superimposed on the signal emitted by the underlying snowpack and increases the T_B measured at the satellite level (Chang et al., 1996; Brown et al., 2003). At same time, the canopy also attenuates the surface emission toward the satellite. These contributions are complex to quantify since it depends not only on the tree fraction within the pixel but also on the tree species and states which emit/attenuate a different PMW signal depending on their biomass (liquid water content), volume, and structure (stem, leaf, trunk) (Franklin, 1987). Also, the presence of trees modifies snow accumulation on the ground, depending on interception, shade, and sublimation effects (Dutra et al., 2011; Wang et al., 2009), which increases the spatial variability of the snowpack within the same pixel. These interactions between the vegetation and the snowpack are not taken into account in Crocus, and this might induce uncertainties due to model errors. Note that SWE sensors are mostly installed in clearings, which reduces this impact in comparisons against surface-based measurements.

Table 7. Same as Table 6 but using the forest parameter ω and the snow stickiness parameter (τ_{snow}) as free variables in the DA scheme (Experiment C) to improve SWE retrievals ($SWE_{DA,\,\omega,\,\tau s}$).

		$SWE_{DA,\,\omega,\,\tau s}$ with the DA of the three frequencies			
	$\overline{SWE_{obs}}$	RMSE (kg m^{-2})	Bias (kg m^{-2})	RPE %	$\overline{SWE_{sim}}$
f_{cover}<75 %	162.2	45.6	−14.8	21.6	147.4
f_{cover}>75 %	139.0	45.1	−7.1	17.5	131.9
Mean	157.3	45.5	−13.2	20.7	144.1

Kwon et al. (2016) used a similar snow radiance assimilation system to correct SD by updating the Community Land Model, version 4 (CLM4), snow and soil states, and radiative transfer model with the assimilation of the 19 and 37 GHz of AMSR-E. Over North America, it produced significant improvements of SD for the tundra type, but also produced degradations for taiga snow class and forest land cover (7.1 % and 7.3 % degradations, respectively). In the present study, the use of a multilayer snowpack model makes it possible to represent PMW emission from the snowpack with DMRT-ML well, and to improve overall snowpack simulations with T_B assimilation in boreal areas when the f_{cover} is below 75 %. Kwon et al. (2017) obtained better results for areas with a high f_{cover} in comparison to their previous study (Kwon et al., 2016) over North America by using the vegetation parameter ω as a free variable in the DA scheme, instead of pre-calibrating it as we chose to do. This aspect is further studied with the experiment C.

4.2.3 Experiment C

Table 7 shows the statistical SWE performances obtained with the DA_c_TB_11,19,37 experiment (see Table 3 for definitions), where ω and τ_{snow} are taken as free variables in the DA scheme ("$SWE_{DA,\,\omega,\,\tau s}$") over the 43 winters.

The overall $SWE_{DA,\,\omega,\,\tau s}$ RMSE, bias, and RPE are equal to 45.5 kg m^{-2}, −13.2 kg m^{-2}, and 20.7 %, respectively, very close to the statistical performances of the original SWE_{Crocus} simulations. The use of ω and τ_{snow} as free variables does not help to improve SWE_{Crocus} simulations for sites with an f_{cover} below 75 %, but the bias is significantly reduced for sites with an f_{cover} above 75 % (−7.1 kg m^{-2} and a RPE of 17.5 %). In addition, the simulated SWE_{max} is improved for 86 % of the 43 simulations (37 cases), with a reduction of the SWE_{max} bias of 26.6 kg m^{-2} (17 % of SWE measurements) compared to the original SWE_{Crocus} simulation.

The use of pre-calibrated parameters is justified because the parameters ω and τ_{snow} were not measurable and could not be directly validated. Furthermore, if parameters are added as state variables in the DA scheme, a larger ensemble size in the DA scheme would be needed to improve the representativeness of T_B uncertainties and to ensure the solution's stability (or at least to prevent a degeneracy problem). The ensemble size was kept to 150 here but this DA_c_TB_11,19,37 experiment should produce improved results with a larger ensemble size. However, this would require a significant computational effort. This study is a preliminary step of a PMW DA implementation for operational hydrological applications, so there was a need to limit computing time. These results suggest that the developed approach using pre-calibrated ω and τ_{snow} parameters helps to improve the retrievals for sites with an f_{cover} below 75 %, and the use of ω and τ_{snow} parameters as free variables in the DA scheme should be investigated in further work for sites with more than 75 % forest cover.

5 Discussion

In this section, we discuss (a) the sensitivity of wet snow conditions for T_B assimilation, and (b) the percentage of surface, vegetation, and atmospheric contributions in the PMW signal measured by satellite sensors.

5.1 Wet snow conditions

In wet snow conditions, water droplets act as emission sources (especially at frequencies above 30 GHz), and the snowpack becomes close to a blackbody (Brucker et al., 2011; Picard et al., 2013; Klehmet et al., 2013). The PMW observations are thus complex to use for SWE retrievals, especially at the end of the season before the spring snowmelt when the SWE is maximal. Figure 8 illustrates the SWE_{DA} obtained with the DA of the three frequencies applied over the whole winter and when the snow is dry only (with an LWC equal to 0 kg m^{-2}), for Site 3 (winter 2013/2014). SWE estimates are strongly deteriorated when T_B assimilation is performed in wet snow conditions. For this example, the SWE_{DA} RMSE is equal to 31.1 kg m^{-2} with a DA performed over the dry snow period only and significantly increases to 70.2 kg m^{-2} by assimilating T_{Bs} over the whole winter (dry and wet snow conditions).

Here we used the LWC simulated by Crocus to detect wet and dry snow. This variable is subject to model errors and is linked to the original atmospheric forcing data and to their uncertainties. Further studies are needed to automatically detect wet snow events by using direct satellite observations. Previous studies have shown the potential to use the gradient ratio (GR $= T_{B\,37} - T_{B\,19}/T_{B\,37} + T_{B\,19}$) to detect rain-on-snow events in arctic areas (Langlois et al., 2017; Dolant et al., 2016), and this approach should be investigated for boreal forest areas in further work. The use of active microwave observations is also a promising approach with a good spatial resolution (Roy et al., 2010).

5.2 Land cover contributions within the simulated $T_{B\,TOA}$

To properly assimilate PMW satellite observations, all contributions that affect the observed signal need to be well identified and quantified. The estimation of $T_{B\,TOA}$ (see Eqs. 5 and 6) can be written as the sum of the PMW contributions of the open surface ($T_{B\,surf}$), vegetation ($T_{B\,veg}$), and atmosphere ($T_{B\,atm}$) according to the fraction of forest (f_{cover}, estimated with the LAI as in Eqs. 2 and 3) and open area ($1 - f_{cover}$) with Eqs. (13), (14), and (15) as

$$T_{B\,veg} = f_{cover} \cdot \big[(1-\omega)\cdot(1-\gamma_v)\cdot T_{veg} + \gamma_v \cdot (1-e_{surf})\cdot(1-\omega)\cdot(1-\gamma_v)\cdot T_{veg}\big]\cdot \gamma_{atm} \quad (13)$$

$$T_{B\,surf} = f_{cover}\cdot\big[\gamma_v \cdot e_{surf}\cdot T_{surf}\big]\cdot\gamma_{atm} + (1-f_{cover})\cdot[e_{surf}\cdot T_{surf}]\cdot\gamma_{atm} \quad (14)$$

$$T_{B\,atm} = f_{cover}\cdot\Big(\big[(1-e_{surf})\cdot{\gamma_v}^2\cdot T_{B\,atm\downarrow} + (1-\gamma_v)\cdot\omega\cdot T_{B\,atm\downarrow}\big]\cdot\gamma_{atm} + T_{B\,atm\uparrow}\Big) + (1-f_{cover})\cdot\big((1-e_{surf})\cdot T_{B\,atm\downarrow}\gamma_{atm} + T_{B\,atm\uparrow}\big). \quad (15)$$

Figure 9 illustrates the percentage of each contribution before DA at 11, 19, and 37 GHz in V-pol from 2012 to 2016, for the summer and for the winter periods (defined when snowpack is detected) for Site 12 (f_{cover} of 24.2 %), Site 1 (f_{cover} of 63.7 %), and Site 9 (f_{cover} of 84.0 %). The percentages of each contribution are similar at 11 and 19 GHz. The contributions from the atmosphere are weak. As expected for all frequencies, the surface contributions increase for the winter period with the presence of the snowpack, while the vegetation contributions decrease as the LAI decreases, especially at 37 GHz. For Site 12, the surface contributions represent more than 80 % of the PMW signal in winter when the vegetation contributions represent less than 10 % of the PMW signal (same magnitude as atmosphere contributions). For Site 1, the surface and the vegetation contributions are similar in winter (40 %–55 %), whereas the vegetation contributions were more than 60 % of the PMW signal in summer. For Site 9, the vegetation contributions remain the main contribution to the PMW signal in comparison to the surface contributions, even in winter (50 %–70 % of the PMW signal for 37–10 GHz). In this dense boreal forest area, the measured snowpack emission represents less than 30 % of the measured signal, and SWE improvements using only T_B observations are challenging. This high vegetation contribution (emission and attenuation) explains why the developed DA scheme does not succeed in significantly improving SWE estimates for these sites with an f_{cover} exceeding 75 %.

6 Summary and conclusion

An ensemble data assimilation (DA) scheme was implemented in a calibrated chain of models (Crocus/DMRT-ML,

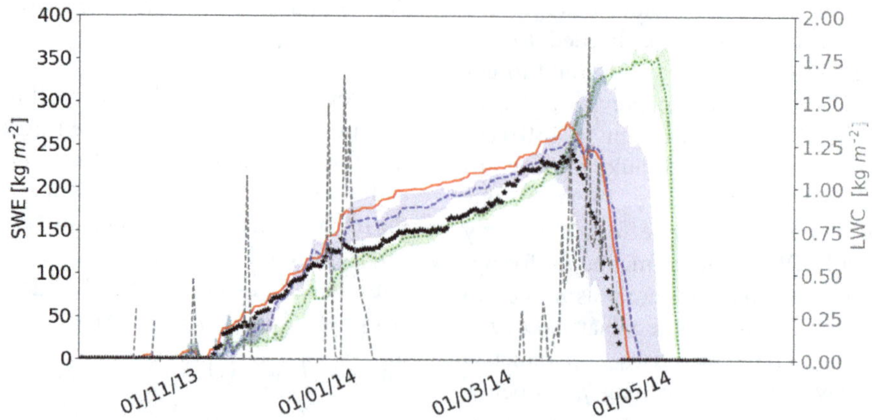

Figure 8. Evolutions of measured SWE (black points) for Site 3 from 2013 to 2014, original SWE Crocus simulation (red full line), and SWE_{DA} obtained with a DA of the three frequencies applied for the entire winter (green dotted line) and when LWC = 0 only (blue full line). The simulated total liquid water content (LWC) in the snowpack (dotted gray lines) is also shown.

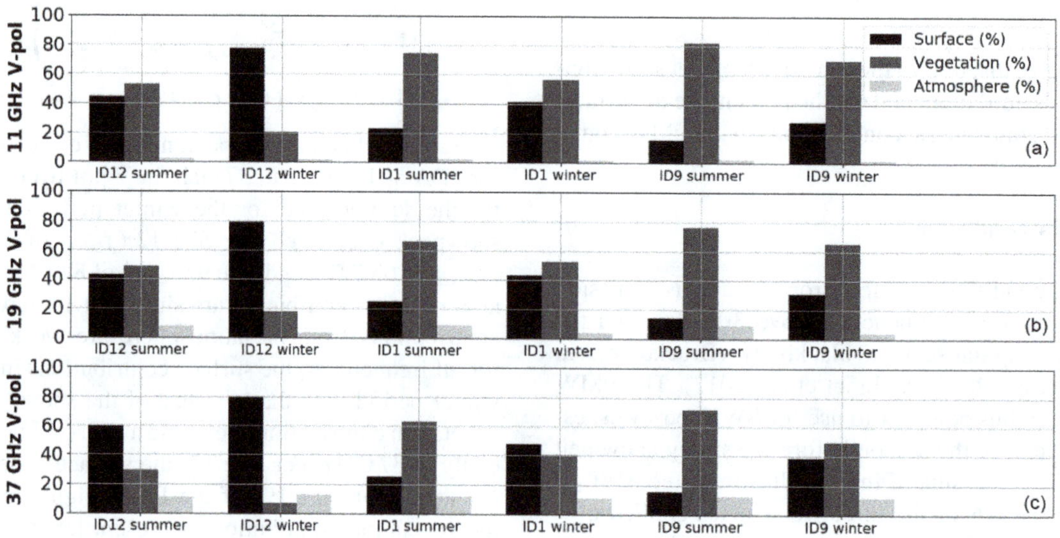

Figure 9. Percentage of surface (black), vegetation (dark gray), and atmosphere (light gray) contributions to the simulated PMW signal at the top of the atmosphere (before DA) at the three frequencies 11 **(a)**, 19 **(b)**, and 37 GHz **(c)**. ID12, ID1, and ID9 are Site 12 (f_{cover} of 24.2 %), 1 (f_{cover} of 63.7 %), and 9 (f_{cover} of 84.0 %), respectively (see Table 2). Summer and winter periods are defined as periods when snowpack is observed or not.

soil, vegetation, and atmosphere radiative transfer models) to improve SWE estimates by updating forcing data and snowpack states with the AMSR-2 satellite observations. The developed approach does not use any surface-based data and was tested over a boreal forest area in Quebec (eastern Canada). The proposed DA scheme is a particle filter with a resampled SIR algorithm, using an inflation technique of the **R** matrix to avoid degeneracy problems. The multilayer snowpack model, Crocus, coupled to the surface land model ISBA, was used to simulate the evolution of the snowpack. The DMRT-ML, the (ω-τ_{opt}) model, an atmospheric model, and the Wegmüller and Mätzler (1999) radiative transfer model were pre-calibrated to simulate the PMW contributions from the snowpack, the vegetation, and the soil, respectively, at the top of the atmosphere. The DA scheme was performed over 43 winters (12 sites from 2012 to 2016; Table 1), only in the presence of dry snow. Ice lenses were detected and integrated in the snowpack by using a thresholding approach on the polarization ratio at 11 GHz. The study shows the following.

1. $T_{B\,TOA}$can be well simulated with the developed chain of models. By calibrating soil and forest parameters ($\omega = 0.07$ and $\sigma_s = 0.2$ cm), the annual $T_{B\,TOA}$ RMSE (all frequencies) is equal to 14.5 K from 2012 to 2016. This RMSE is similar to the overall RMSE estimated with the τ_{snow}-calibrated DMRT-ML model driven by

in situ measurements in an open area (19.9 K compared to surface-based radiometric measurements in Quebec; Larue et al., 2018).

2. The assimilation of T_{Bs} at 11, 19, and 37 GHz (V-pol) improves the SWE estimates compared to the assimilation of $\Delta T_{\mathrm{B}\,19-37}$ only (sensitive to snowpack depth) or to the assimilation of both $\Delta T_{\mathrm{B}\,19-37}$ and $\Delta T_{\mathrm{B}\,11-19}$. For three sites (with different f_{cover}), the SWE ensemble RMSE of posterior estimates is reduced by 47 % over the whole winter compared to the SWE ensemble RMSE of prior estimates (open loop runs).

3. By using pre-calibrated ω and τ_{snow} parameters in the DA scheme, the overall bias (for 43 winters) of the original $\mathrm{SWE}_{\mathrm{Crocus}}$ simulations is significantly reduced by assimilating T_{Bs} at 11, 19, and 37 GHz (from 23.7 to 7.5 kg m^{-2}). The bias on $\mathrm{SWE}_{\mathrm{max}}$ is reduced by 12.2 kg m^{-2} (8 % of SWE measurements). The overall RPE goes from 22.1 % to 18.4 %, which is close to the range of accuracy needed for hydrological applications (SWE RPE < 15 %). This accuracy is achieved with the T_{B} assimilation for sites with an f_{cover} below 75 %, but the DA deteriorates SWE simulations for sites with an f_{cover} above 75 %. However, by using ω and τ_{snow} as free variables, the DA significantly improves SWE simulations for sites with an f_{cover} above 75 % (RPE = 17.5 %).

Even with the difficulties associated with quantifying all the different factors that contribute to the PMW signal measured by satellite sensors in remote boreal areas (canopy, ice crust, frozen/unfrozen ground, presence of lakes, moisture in the snow, topography, etc.) (Kelly et al., 2003; Koenig and Forster, 2004), and even when vegetation contributions are 50 % of the PMW signal, the implementation of a DA scheme in a well-calibrated chain of models reduces SWE uncertainties without using any surface-based data. This assimilation scheme can be easily implemented in an operational system using real satellite-borne observations, despite the relatively significant computing time required. This research opens the way for global applications to obtain more accurate SWE estimates over large and remote areas where few meteorological weather stations are present.

Data availability. The daily SWE data provided by Hydro-Québec are used for hydrological purposes and are not available to the public due to legal constraints on the data's availability. For the SWE and SD data, and field campaign measurements provided by the University of Sherbrooke, please contact the coauthor Alain Royer (Alain.Royer@USherbrooke.ca). Meteorological GEM data are freely available on the government of ECCC's website (Environment and Climate Change Canada) at https://weather.gc.ca/grib/grib2_reg_10km_e.html, last access: 10 October 2018). Other data used are listed in the references.

Appendix A: Online adjustment of the observation error covariance matrix R

Online adjustment of covariance matrices in data assimilation is quite a common approach with the ensemble Kalman filter (Dee, 1995; Miyoshi, 2011; Brankart et al., 2010) but not with the particle filter. However, in many implementations of the particle filter, the measurement pdf is considered Gaussian, so the particle weights are computed using the observation error covariance matrix **R**. This matrix can therefore also be subject to adjustment in the context of the particle filter. Online adjustment can be and is often performed by tuning a simple inflation of the initial covariance matrix. This is the approach chosen here.

Noting $\delta_i = y - h(x_i)$, the innovation for particle i, the weight of this particle, is

$$\widetilde{\mathrm{we}}_i = \frac{\mathrm{we}_i}{\sum\limits_j \mathrm{we}_j}, \tag{A1}$$

where

$$\mathrm{we}_i = \exp\left(-\frac{1}{2}\delta_i^T \mathbf{R}^{-1}\delta_i\right). \tag{A2}$$

An inflation of matrix **R** by a factor $1/\alpha$ (larger than 1) can be interpreted as an exponent α (smaller than 1) to we_i. Because the weights we_i are nonlinear functions of **R**, inflating **R** tends to flatten their distribution. Online adjustment consists in finding a value for α that flattens the distribution of weights to the point at which N_{keep} particles are selected with certainty, N_{keep} being a number to be prescribed. If the number N_{keep} is fixed, when the resampling step is performed using Arakawa's procedure (Arakawa, 1997), the weight of the $N_{\mathrm{keep-th}}$ particle to be selected, $\widetilde{\mathrm{we}}_{\mathrm{keep}}$, must become equal to $\widetilde{\mathrm{we}}_{\mathrm{ref}} = 1/N_{\mathrm{keep}}$. Consequently,

$$\widetilde{\mathrm{we}}_{\mathrm{keep}} = \frac{\left(\mathrm{we}_{\mathrm{keep}}\right)^{\alpha}}{\sum\limits_j \left(\mathrm{we}_j\right)^{\alpha}} = \widetilde{\mathrm{we}}_{\mathrm{ref}}, \tag{A3}$$

or, written differently after taking the logarithm:

$$\alpha = \left(\log(\widetilde{\mathrm{we}}_{\mathrm{ref}}) + \log\left(\sum_i \left(\mathrm{we}_j\right)^{\alpha}\right)\right) / \log(\mathrm{we}_{\mathrm{keep}}). \tag{A4}$$

This equation for α is not solvable analytically. Instead, we find α after the convergence of the series:

$$\alpha_n = \left(\log(\widetilde{\mathrm{we}}_{\mathrm{ref}}) + \log\left(\sum_i \mathrm{we}_j^{\alpha_{n-1}}\right)\right) / \log(\mathrm{we}_{\mathrm{keep}}). \tag{A5}$$

The result of this adjustment is illustrated in Fig. A1. The blue dots show the first 20 weights of a sorted distribution for an ensemble of 50 particles strongly prone to degeneracy: only 4 particles have a weight larger than $1/50 = 0.02$.

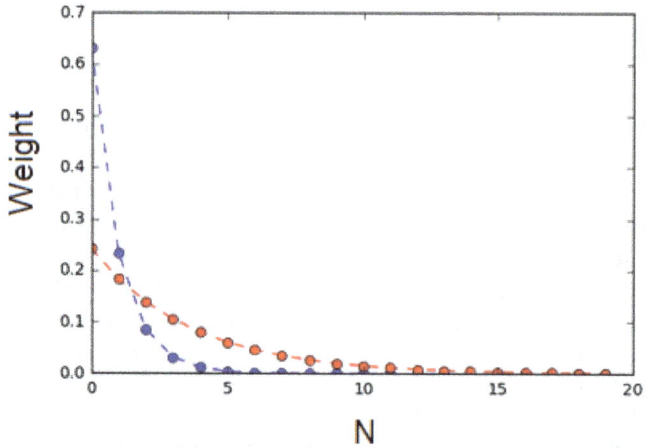

Figure A1. Weight distribution of the first 20 weights of a sorted distribution for an ensemble of 50 particles: distribution before the adjustment (blue dotted points), showing a degeneracy problem, and distribution after the adjustment procedure (red dotted points), in which weight distribution is flattened and significant weights are distributed around N_{keep} particles (10 particles for this example).

The minimum number of particles to be selected is fixed to $N_{\text{keep}} = 10$. After the adjustment procedure, the identified inflation factor for matrix **R** is 3.6 ($\alpha = 0.277$), and the weight we_{keep} of the 10th particle is exactly equal to 0.02.

Obviously, this procedure is only used if the number of selected particles is below the N_{keep} threshold with the initial weights.

Author contributions. FL, AR, DS, and EC contributed to the conception and design of the work. All authors contributed to the acquisition, analysis, and interpretation of data. FL created new software used in the work. FL and AR wrote the manuscript, and all authors contributed to revisions of the manuscript.

Competing interests. The authors declare that they have no conflict of interest.

Special issue statement. This article is part of the special issue "Integration of Earth observations and models for global water resource assessment". It is not associated with a conference.

Acknowledgements. The authors would like to thank the data providers: Hydro-Québec, Environment Canada (CMC-ECCC), and the University of Laval. This project was supported by financial contributions from NSERC, Canada, FRQ-NT and MITACS Quebec, and the CFQCU France–Quebec collaboration program.

Edited by: Jaap Schellekens

References

Andreadis, K. M. and Lettenmaier, D. P.: Implications of representing snowpack stratigraphy for the assimilation of passive microwave satellite observations, J. Hydrometeorology, 13, 1493–1506, https://doi.org/10.1175/JHM-D-11-056.1, 2012.

Arakawa, A.: Adjustment mechanisms in atmospheric motions, J. Meteor. Soc. Japan, Special issue of collected papers, 75, 155–179, 1997.

Arulampalam, M. S., Maskell, S., Gordon, N., and Clapp, T.: A tutorial on particle filters for online nonlinear/non-Gaussian Bayesian tracking, IEEE T. Signal Proces., 50, 174–188, 2002.

Brankart, J.-M., E. Cosme, C.-E. Testut, P. Brasseur, and J. Verron: Efficient Adaptive Error Parameterizations for Square Root or Ensemble Kalman Filters: Application to the Control of Ocean Mesoscale Signals, Mon. Weather Rev., 138, 932–950, 2010.

Brown, R. and Tapsoba, D.: Improved mapping of snow water equivalent over Quebec, 64th Eastern Snow Conference, St. John's, Newfoundland, Canada, 2007.

Brown, R. D., Brasnett, B., and Robinson, D.: Gridded North American monthly snow depth and snow water equivalent for GCM evaluation, Atmos. Ocean, 41, 1–14, https://doi.org/10.3137/ao.410101, 2003.

Brown, R., Tapsoba, D., and Derksen, C.: Evaluation of snow water equivalent datasets over the Saint-Maurice river basin region of southern Québec, Hydrol. Process., 32, 2748–2764, https://doi.org/10.1002/hyp.13221, 2018.

Brucker, L., Royer, A., Picard, G., Langlois, A., and Fily, M.: Hourly simulations of the microwave brightness temperature of seasonal snow in Quebec, Canada, using a coupled snow evolution-emission model, Remote Sens. Environ., 115, 1966–1977, 2011.

Brun, E., Martin, E., Simon, V., Gendre, C., and Coleou, C.: An energy and mass model of snow cover suitable for operational avalanche forecasting, J. Glaciol., 35, 333–342, 1989.

Brun, E., David, P., Sudul, M., and Brunot, G.: A numerical model to simulate snow cover stratigraphy for operational avalanche forecasting, J. Glaciol., 38, 13–22, 1992.

Carmagnola, C. M., Morin, S., Lafaysse, M., Domine, F., Lesaffre, B., Lejeune, Y., Picard, G., and Arnaud, L.: Implementation and evaluation of prognostic representations of the optical diameter of snow in the SURFEX/ISBA-Crocus detailed snowpack model, The Cryosphere, 8, 417–437, https://doi.org/10.5194/tc-8-417-2014, 2014.

Cavalieri, D. J., Gloersen, P., and Campbell, W. J.: Determination of sea ice parameters with the Nimbus 7 SMMR, J. Geophys. Res., 89, 5355–5369, 1984.

Chang, A. T. C., Foster, J. L., and Hall, D. K.: Nimbus-7 derived global snow cover parameters, Ann. Glaciol., 9, 39–44, 1987.

Chang, A. T. C., Foster, J. L., and Hall, D. K.: Effects of forest on the snow parameters derived from microwave measurements during the BOREAS winter field campaign, Hydrol. Process., 10, 1565–1574, 1996.

Charrois, L., Cosme, E., Dumont, M., Lafaysse, M., Morin, S., Libois, Q., and Picard, G.: On the assimilation of optical reflectances and snow depth observations into a detailed snowpack model, The Cryosphere, 10, 1021–1038, https://doi.org/10.5194/tc-10-1021-2016, 2016.

Choquette, Y., Lavigne, P., Nadeau, M., Ducharme, P., Martin, J. P., Houdayer, A., and Rogoza, J.: GMON, a new sensor for snow water equivalent via gamma monitoring, Proceedings Whistler 2008 International Snow Science Workshop, 21–27 September 2008, Whistler, B. C., 2008.

Côté, J., Gravel, S., Méthot, A., Patoine, A., Roch, M., and Staniforth, A.: The operational CMC-MRB Global Environmental Multiscale model: Part I – Design considerations and formulation, Mon. Weather Rev., 126, 1373–1395, 1998.

DeChant, C. and Moradkhani, H.: Radiance data assimilation for operational snow and streamflow forecasting, Adv. Water Res., 34, 351–364, 2011.

Decharme, B., Boone, A., Delire, C., and Noilhan, J.: Local evaluation of the Interaction between Soil Bio- sphere Atmosphere soil multilayer diffusion scheme using four pedotransfer functions, J. Geophys. Res., 116, D20, https://doi.org/10.1029/2011JD016002, 2011.

Dee, D. P.: On-line Estimation of Error Covariance Parameters for Atmospheric Data Assimilation, Mon. Weather Rev., 123, 1128–1145, 1995.

De Lannoy, G. J. M., Reichle, R., Houser, P., Arsenault, K., Verhoest, N., and Pauwels, V.: Satellite-Scale Snow Water Equivalent Assimilation into a High-Resolution Land Surface Model, J. Hydrometeorol., 11, 352–369, https://doi.org/10.1175/2009JHM1192.1, 2010.

De Lannoy, G. J. M., Reichle, R. H., Arsenault, K. R., Houser, P. R., Kumar, S., Verhoest, N. E. C., and Pauwels, V. R. N.: Multiscale assimilation of AMSR-E snow water equivalent and MODIS snow cover fraction observations in northern Colorado, Water Resour. Res., 48, W01522, https://doi.org/10.1029/2011WR010588, 2012.

De Sève, D., Bernier, M., Fortin, J. P., and Walker, A. E.: Preliminary analysis of the snow microwave radiometry using SSM/I passive microwave data: The case of the La Grande River watershed (Québec), Ann. Glaciol., 25, 353–361, 1997.

De Sève, D., Bernier, M., Fortin, J. P., and Walker, A.: Spatio-temporal analysis of microwave radiometry of snow cover with SSM/I data in a taïga area, Eastern Snow Conference, Fredericton, Canada, 200–205, 1999.

De Sève, D., Evora, N. D., and Tapsoba, D.: Comparison of three algorithms for estimating Snow Water Equivalent (SWE) over the La Grande River watershed using SSM/I data in the context of Hydro-Québec's hydraulic power management, Int. Geosci. Remote Se., IGARSS 23–28 July 2007, https://doi.org/10.1109/IGARSS.2007.4423791, 2007.

Derksen, C., Toose, P., Lemmetyinen, J., Pulliainen, J., Langlois, A., Rutter, N., and Fuller, M.: Evaluation of passive microwave brightness temperature simulations and snow water equivalent retrievals through a winter season, Remote Sens. Environ., 117, 236–248, 2012.

Derksen, C., Lemmetyinen, J., Toose, P., Silis, A., Pulliainen, J., and Sturm, M.: Physical properties of Arctic versus subarctic snow: Implications for high latitude passive microwave snow water equivalent retrievals, J. Geophys. Res.-Atmos., 119, 7254–7270, 2014.

Dobson, M. C., Ulaby, F. T., Hallikainen, M. T., and El-Rayes, M. A.: Microwave dielectric behavior of wet soil-part II: dielectric miximg models, IEEE T. Geosci. Remote, 1, 35–46, 1985.

Dolant, C., Langlois, A., Montpetit, B., Brucker, L., Roy, A., and Royer, A.: Development of a rain-on-snow detection algorithm using passive microwave radiometry, Hydrol. Process., 30, 3184–3196, https://doi.org/10.1002/hyp.10828, 2016.

Ducharme, P.: Estimation de l'équivalent en eau du couvert nival (Plan de projet étape 3), Technical report, Plate-forme Production, Hydro-Québec publication, Varennes, QC, Canada, 2003.

Durand, M. and Margulis, S.: Feasibility test of multi-frequency radiometric data assimilation to estimate snow water equivalent, J. Hydrometeorol., 7, 443–457, 2006.

Durand, M. and Margulis, S. A.: Correcting first-order errors in snow water equivalent estimates using a multifrequency, multiscale radiometric data assimilation scheme, J. Geophys. Res., 112, D13121, https://doi.org/10.1029/2006JD008067, 2007.

Durand, M., Kim, E. J., and Margulis, S. A.: Radiance assimilation shows promise for snowpack characterization, Geophys. Res. Lett., 36, L02503, https://doi.org/10.1029/2008GL035214, 2009.

Durand, M., Kim, E., Margulis, S., and Molotch, N.: A first-order characterization of errors from neglecting stratigraphy in forward and inverse passive microwave modeling of snow, IEEE Geosci. Remote Se., 8, 730–734, 2011.

Dutra, E., Kotlarski, S., Viterbo, P., Balsamo, G., Miranda, P. M. A., Schär, C., Bissolli, P., and Jonas, T.: Snow cover sensitivity to horizontal resolution, parameterizations and atmospheric forcing in a land surface model, J. Geophys. Res., 116, D21109, https://doi.org/10.1029/2011JD016061, 2011.

Foster, J., Sun, C., Walker, J., Kelly, R., Chang, A., Dong, J., and Powell, H.: Quantifying the uncertainty in passive microwave snow water equivalent observations, Remote Sens. Environ., 94, 187–203, 2005.

Franklin, S. E.: Terrain analysis from digital patterns in geomorphometry and Landsat MSS spectral response, Photogramm. Eng. Rem. S., 53, 59–65, 1987.

Gordon, N. J., Salmond, D. J., and Smith, A. F.: Novel approach to nonlinear/non-Gaussian Bayesian state estimation, IEEE Proc.-F, 140, 107–113, 1993.

Grant, J. P., Saleh, K., Wigneron, J.-P., Guglielmetti, M., Kerr, Y. H., Schwank, M., Skou, N., and Van de Griend, A.: Calibrationof the L-MEB model over a coniferous and a deciduous forest, IEEE Trans. Geosci. Remote Sens., 46, 808–818, https://doi.org/10.1109/TGRS.2007.914801, 2008.

Grant, A., Van de Griend, A., Schwank, M., and Wigneron, J.-P.: Observations and modeling of a pine forest floor at L-band, IEEE T. Geosci. Remote, 47, 2024–2034, https://doi.org/10.1109/TGRS.2008.2010252, 2009.

Hallikainen, M.: Retrieval of Snow Water Equivalent from Nimbus-7 SMMR Data: Effect of Land-Cover Categories and Weather Conditions, IEEE J. Oceanic Eng., OE-9, 372–376, https://doi.org/10.1109/JOE.1984.1145656, 1984.

Imaoka, K., Kachi, M., Kasahara, M., Nakagawa, K., and Oki, T.: Instrument performance and calibration of AMSR-E and AMSR2, International Archives of the Photogrammetry, Remote Sensing and Spatial Information Science, ISPRS, Kyoto, Japan, 2010.

Jordan, R. E.: A one-dimensional temperature model for a snow cover: Technical documentation for SNTHERM.89, Special Rep.

91-16, U.S. Army Cold Regions Research and Engineering Laboratory, Hanover, NH, 49 pp., 1991.

Kelly, R., Chang, A. T. C., Tsang, L., and Foster, J.: A prototypeAMSR-E global snowarea and snow depth algorithm, IEEE Trans. Geosci. Rem. Sens, 41, 23, 2003.

Kelly, R., Chang, A. T. C., Tsang, L., and Foster, J.: A prototype AMSR-E global snow area and snow depth algorithm, IEEE T. Gesoci. Remote, 41, 230–242, 2010.

Klehmet, K., Geyer, B., and Rockel, B.: A regional climate model hindcast for Siberia: analysis of snow water equivalent, The Cryosphere, 7, 1017–1034, https://doi.org/10.5194/tc-7-1017-2013, 2013.

Koenig, L. S. and Forster, R. R.: Evaluation of passive microwave snow water equivalent algorithms in the depth-hoar dominated snowpack of the Kuparuk River Watershed, Alaska, USA, Remote Sens. Environ., 93, 511–527, 2004.

Kwon, Y., Yang, Z.-L., Zhao, L., Hoar, T. J., Toure, A. M., and Rodell, M.: Estimating snow water storage in North America using CLM4, DART, and snow radiance data assimilation, J. Hydrometeorol., 17, 2853–2874, 2016.

Kwon, Y., Yang, Z.-L., Hoar, T. J., and Toure, A. M.: Improving the Radiance Assimilation Performance in Estimating Snow Water Storage across Snow and Land-Cover Types in North America, J. Hydrometeorol., 18, 651–668, 2017.

Lafaysse, M., Cluzet, B., Dumont, M., Lejeune, Y., Vionnet, V., and Morin, S.: A multiphysical ensemble system of numerical snow modelling, The Cryosphere, 11, 1173–1198, https://doi.org/10.5194/tc-11-1173-2017, 2017.

Langlois, A., Royer, A., Derksen, C., Montpetit, B., Dupont, F., and Goïta, K.: Coupling of the snow thermodynamic model SNOWPACK with the Microwave Emission Model for Layered Snowpacks (MEMLS) for subarctic and arctic Snow Water Equivalent retrievals, Water Resour. Res., 48, W12524, https://doi.org/10.1029/2012WR012133, 2012.

Langlois, A. et al.: Detection of rain-on-snow (ROS) events and ice layer formation using passive microwave radiometry: a context for Peary caribou habitat in the Canadian Arctic, Remote Sens. Environ., 189, 84–95, https://doi.org/10.1109/TGRS.1982.4307521, 2017.

Larue, F., Royer, A., De Sève, D., Langlois, A., Roy, A., and Brucker, L.: Validation of GlobSnow-2 snow water equivalent over Eastern Canada, Remote Sens. Environ., 194, 264–277, https://doi.org/10.1016/j.rse.2017.03.027, 2017.

Larue, F., Royer, A., De Sève, D., Roy, A., Picard, G., Vionnet, V., and Cosme, E.: Simulation and assimilation of passive microwave data using a snowpack model coupled to a calibrated radiative transfer model over northeastern Canada, Water Resour. Res., 54, 4823–4848, https://doi.org/10.1029/2017WR022132, 2018.

Lehning, M., Bartelt, P., Brown, R. L., and Fierz, C.: A physical SNOWPACK model for the Swiss avalanche warning; Part III: meteorological forcing, thin layer formation and evaluation, Cold Reg. Sci. Technol., 35, 169–184, 2002.

Leppänen, L., Kontu, A., Vehviläinen, J., Lemmetyinen, J., and Pulliainen, J.: Comparison of traditional and optical grain size field measurements with SNOWPACK simulations in a taiga environment, J. Glaciol., 61, 151–162, 2015.

Liebe, H.: MPM – an atmospheric millimeter-wave propagation model, Int. J. Infrared Milli., 10, 631–650, 1989.

Luojus, K., Pulliainen, J., Takala, M., Lemmetyinen, J., Smolander, T., and Derksen, C.: The GlobSnow Snow Water Equivalent Product, 22 July 2014, SnowPEX ISSPI-1, College Park, Maryland, USA, 2014.

Masson, V., Le Moigne, P., Martin, E., Faroux, S., Alias, A., Alkama, R., Belamari, S., Barbu, A., Boone, A., Bouyssel, F., Brousseau, P., Brun, E., Calvet, J.-C., Carrer, D., Decharme, B., Delire, C., Donier, S., Essaouini, K., Gibelin, A.-L., Giordani, H., Habets, F., Jidane, M., Kerdraon, G., Kourzeneva, E., Lafaysse, M., Lafont, S., Lebeaupin Brossier, C., Lemonsu, A., Mahfouf, J.-F., Marguinaud, P., Mokhtari, M., Morin, S., Pigeon, G., Salgado, R., Seity, Y., Taillefer, F., Tanguy, G., Tulet, P., Vincendon, B., Vionnet, V., and Voldoire, A.: The SURFEXv7.2 land and ocean surface platform for coupled or offline simulation of earth surface variables and fluxes, Geosci. Model Dev., 6, 929–960, https://doi.org/10.5194/gmd-6-929-2013, 2013.

Mätzler, C.: Applications of the interaction of microwaveswith the natural snow cover, Remote Sens. Rev., 2, 259–387, https://doi.org/10.1080/02757258709532086, 1987.

Mätzler, C.: Passive microwave signatures of landscapes in winter, Meteorol. Atmos. Phys., 54, 241–260, 1994.

Mätzler, C., Schanda, E., and Good, W.: Towards the definition of optimum sensor specifications for microwave remote sensing of snow, IEEE T. Geosci. Remote, GE-20, 57–66, 1982.

Meissner, T. and Wentz, F.: Intercalibration of AMSR-E and WINDSAT brightness temperature measurements over land scenes, presented at the 11th Specialist Meeting on Microwave Radiometry and Remote Sensing of the Environment (MicroRad 2010), 1–4 March 2010, Washington, DC, USA, 2010.

Mo, T., Choudhury, B. J., Schmugge, T. J., Wang, J. R., and Jackson, T. J.: A model for microwave emission from vegetation-covered fields, J. Geophys. Res., 87, 11 229–11 237, 1982.

Molotch, N. P. and Bales, R. C.: Scaling snow observations from the point to the grid element: Implications for observation network design, Water Resour. Res., 41, W11421, https://doi.org/10.1029/2005WR004229, 2005.

Montpetit, B., Royer, A., Roy, A., Langlois, A., and Derksen, D.: Snow microwave emission modeling of ice lenses within a snowpack using the microwave emission model for layered snowpacks, IEEE T. Geosci. Remote, 51, 4705–4717, https://doi.org/10.1109/TGRS.2013.2250509, 2013.

Montpetit, B., Royer, A., Roy, A., and Langlois, A.: In-situ passive microwave parameterization of sub-arctic frozen organic soils, Remote Sens. Environ., 205, 112–118, 2018.

Myneni, R. B., Hoffman, S., Knyazikhin, Y., Privette, J. L., Glassy, J., Tian, Y., Wang, Y., Song, X., Zhang, Y., Smith, G. R., Lotsch, A., Friedl, M., Morisette, J. T., Votava, P., Nemani, R. R., and Running, S. W.: Global products of vegetation leaf area and fraction absorbed PAR from year one of MODIS data, Remote Sens. Environ., 83, 214–231, 2002.

Miyoshi, T.: The Gaussian Approach to Adaptive Covariance Inflation and Its Implementation with the Local Ensemble Transform Kalman Filter, Mon. Weather Rev., 139, 1519–1535, 2011.

Pampaloni, P. and Paloscia, S.: Microwave emission and plant water content: A comparison between field measurements and theory, IEEE T. Geosci. Remote, GE-24, 900–905, https://doi.org/10.1109/TGRS.1986.289705, 1986.

Pellarin, T., Kerr, Y. H., and Wigneron, J. P.: Global simulation of brightness temperatures at 6.6 and 10.7 GHz over land based on SMMR data set analysis, IEEE T. Geosci. Remote, 44, 2492–2505, 2006.

Picard, G., Brucker, L., Roy, A., Dupont, F., Fily, M., Royer, A., and Harlow, C.: Simulation of the microwave emission of multi-layered snowpacks using the Dense Media Radiative transfer theory: the DMRT-ML model, Geosci. Model Dev., 6, 1061–1078, https://doi.org/10.5194/gmd-6-1061-2013, 2013.

Pietroniro, A. and Leconte, R.: A review of Canadian Remote Sensing and Hydrology 1999–2003, Hydrol. Process., 19, 285–301, 2005.

Pulliainen, J. T., Grandel, J., and Hallikainen, M. T.: HUT snow emission model and its applicability to snow water equivalent retrieval, IEEE T. Geosci. Remote, 37, 1378—1390, 1999.

Quéno, L., Vionnet, V., Dombrowski-Etchevers, I., Lafaysse, M., Dumont, M., and Karbou, F.: Snowpack modelling in the Pyrenees driven by kilometric-resolution meteorological forecasts, The Cryosphere, 10, 1571–1589, https://doi.org/10.5194/tc-10-1571-2016, 2016.

Raleigh, M. S., Lundquist, J. D., and Clark, M. P.: Exploring the impact of forcing error characteristics on physically based snow simulations within a global sensitivity analysis framework, Hydrol. Earth Syst. Sci., 19, 3153–3179, https://doi.org/10.5194/hess-19-3153-2015, 2015.

Rott, H., Yueh, S. H., Cline, D. W., Duguay, C., Essery, R., Haas, C., Hélière, F., Kern, M. G., Malnes, E., Nagler, T., Pulliainen, J., Rebhan, H., and Thompson, A.: Cold regions hydrology high-resolution observatory for Snow and Cold Land Processes, Proc. IEEE,, 98, 752–765, https://doi.org/10.1109/JPROC.2009.2038947, 2010.

Roy, A., Royer, A., and Turcotte, R.: Analyse de l'identification de la fonte de neige printanière avec QuickSCAT dans le Sud du Québec, Canada, Can. J. Remote Sens., 36, 89–98, 2010.

Roy, A., Royer, A., Wigneron, J.-P., Langlois, A., Bergeron, J., and Cliche, P.: A simple parameterization for a boreal forest radiative transfer model at microwave frequencies, Remote Sens. Environ., 124, 371–383, 2012.

Roy, A., Picard, G., Royer, A., Montpetit, B., Dupont, F., Langlois, A., Derksen, C., and Champollion, N.: Brightness Temperature Simulations of the Canadian Seasonal Snowpack Driven by Measurements of the Snow Specific Surface Area, IEEE T. Geosci. Remote., 51, 4692–4704, 2013.

Roy, A., Royer, A., and Hall, R.: Relationship between forest microwave transmissivity and structural parameters for Canadian boreal forest, IEEE Geosci. Remote S., 11, 1802–1806, 2014.

Roy, A., Royer, A., St-Jean-Rondeau, O., Montpetit, B., Picard, G., Mavrovic, A., Marchand, N., and Langlois, A.: Microwave snow emission modeling uncertainties in boreal and subarctic environments, The Cryosphere, 10, 623–638, https://doi.org/10.5194/tc-10-623-2016, 2016.

Royer, A., Roy, A., Montpetit, B., Saint-Jean-Rondeau, O., Picard, G., Brucker, L., and Langlois, A.: Comparison of commonly-used microwave radiative transfer models for snow remote sensing, Remote Sens. Environ., 190, 247–259, 2017.

Sandells, M., Essery, R., Rutter, N., Wake, L., Leppänen, L., and Lemmetyinen, J.: Microstructure representation of snow in coupled snowpack and microwave emission models, The Cryosphere, 11, 229–246, https://doi.org/10.5194/tc-11-229-2017, 2017.

Sun, C., Walker, J. P., and Houser, P. R.: A methodology for snow data assimilation in a land surface model, J. Geophys. Res., 109, D08108, https://doi.org/10.1029/2003JD003765, 2004.

Takala, M., Luojus, K., Pulliainen, J., Derksen, C., Lemmetyinen, J., Kärnä, J.-P., Koskinen, J., and Bojkov, B.: Estimating northern hemisphere snow water equivalent for climate research through assimilation of space-borne radiometer data and ground-based measurements, Remote Sens. Environ., 115, 3517–3529, 2011.

Tapsoba, D., Fortin, V., Anctil, F., Haché, M.: Apport de la technique du krigeage avec dérive externe pour une cartographie raisonnée de l'équivalent en eau de la neige: application aux bassins de la rivière Gatineau, Can. J. Civ. Eng., 32, 289–297, 2005.

Tapsoba, D., Fortin, V., Anctil, F., and Haché, M.: Apport de la technique du krigeage avec dérive externe pour une cartographie raisonnée de l'équivalent en eau de la neige: Application aux bassins de la rivière Gatineau, Can. J. Civil Eng., 32, 289–297, 2009.

Tedesco, M. and Narvekar, P. S.: Assessment of the NASA AMSR-E SWE product, IEEE J. Sel. Top. Appl., 3, 141–159, 2010.

Tedesco, M., Kelly, R., Foster, J. L., and Chang, A. T. C.: AMSR-E/Aqua Daily L3 Global Snow Water Equivalent EASE-Grids, Version 2, NASA National Snow and Ice data center Distributed Active Archive Center, Boulder, Colorado, USA, https://doi.org/10.5067/AMSR-E/AE_DYSNO.002, 2004.

Touré, A., Goïta, K., Royer, A., Kim, E., Durand, M., Margulis, S. A., and Huizhong, Lu: A Case Study of Using a Multi-Layered Thermo-Dynamical Snow Model for Radiance Assimilation, IEEE T. Geosci. Remote, 49, 2828–2837, 2011.

Tsang, L. and Kong, J. A.: Scattering of Electromagnetic Waves, vol. 3, Advanced Topics, Wiley Intersci., Hoboken, N. J., 2001.

Tsang, L., Ding, K. H., and Wen, B.: Dense Media Radiative Transfer Theory for Dense Discrete Random Media with Particles of Multiple Sizes and Permitivities, Pr. Electromagn. Res., 6, 181–230, 1992.

Turcotte, R., Fortier-Filion, T.-C., Fortin, V., Roy, A., and Royer, A.: Simulation hydrologiques des derniers jours de la crue du printemps: le problème de la neige manquante, Hydrolog. Sci. J., 55, 872–882, 2010.

Vachon, F.: Estimation de l'équivalent en eau de la neige en milieu subarctique du Québec par télédétection micro-ondes passives, Mémoire de thèse, Université de Sherbrooke, Sherbrooke, Canada, Mai 2009.

Van Leeuwen, P. J.: Particle filtering in geophysical systems, Mon. Weather Rev., 137, 4089–4114, https://doi.org/10.1175/2009MWR2835.1, 2009.

Van Leeuwen, P. J.: Particle filters for the geosciences, Advanced Data Assimilation for Geosciences: Lecture Notes of the Les Houches School of Physics: Special Issue, June 2012, in: Advanced Data Assimilation for Geosciences, 291–318, https://doi.org/10.1093/acprof:oso/9780198723844.003.0013, 2014.

Vionnet, V., Brun, E., Morin, S., Boone, A., Faroux, S., Le Moigne, P., Martin, E., and Willemet, J.-M.: The detailed snowpack scheme Crocus and its implementation in SURFEX v7.2, Geosci. Model Dev., 5, 773–791, https://doi.org/10.5194/gmd-5-773-2012, 2012.

Wang, N., Liu, C., Ferraro, R., Wolff, D., Zipser, E., and Kummerow, C.: TRMM 2A12 land precipitation product-status and future plans, J. Met. Soc. of Japan, 87, 237–253, 2009.

Wegmüller, U. and Mätzler, C.: Rough bare soil reflectivity model, IEEE T. Geosci. Remote, 37, 1391–1395, 1999.

Xu, X., Li, J., and Tolsonv, B. A.: Progress in integrating remote sensing data and hydrologic modeling, Prog. Phys. Geog., 38, https://doi.org/10.1177/0309133314536583, 2014.

10

Exploratory studies into seasonal flow forecasting potential for large lakes

Kevin Sene, Wlodek Tych, and Keith Beven

Lancaster Environment Centre, Lancaster University, Lancaster, LA1 4YQ, UK

Correspondence: Wlodek Tych (w.tych@lancaster.ac.uk)

Abstract. In seasonal flow forecasting applications, one factor which can help predictability is a significant hydrological response time between rainfall and flows. On account of storage influences, large lakes therefore provide a useful test case although, due to the spatial scales involved, there are a number of modelling challenges related to data availability and understanding the individual components in the water balance. Here some possible model structures are investigated using a range of stochastic regression and transfer function techniques with additional insights gained from simple analytical approximations. The methods were evaluated using records for two of the largest lakes in the world – Lake Malawi and Lake Victoria – with forecast skill demonstrated several months ahead using water balance models formulated in terms of net inflows. In both cases slight improvements were obtained for lead times up to 4–5 months from including climate indices in the data assimilation component. The paper concludes with a discussion of the relevance of the results to operational flow forecasting systems for other large lakes.

1 Introduction

One of the challenges in seasonal flow forecasting is that the lead times of interest often far exceed the hydrological response time of catchments. This means that traditional approaches to data assimilation are often less effective due to the decay in information content on these longer timescales.

However, the potential for deriving operationally useful forecasts improves if there are significant storage influences. Perhaps the greatest success to date has been in snowmelt forecasting for basins with a significant winter snowpack and typically this has been based on statistical techniques or sampling of historic records for input to hydrological models; for example using an ensemble streamflow prediction approach (Day, 1985; Wood and Schaake, 2008). These techniques have also been applied more widely and other more recent developments include the use of seasonal rainfall forecasts, climate indices and ensemble Kalman filter approaches (e.g. Crochemore et al., 2017; Candogan Yossef et al., 2017; Huang et al., 2017). However, a common finding is that forecast skill may arise as much from the representation of antecedent conditions as from the meteorological inputs, with the balance depending on factors such as lead times and season, as well as location (e.g. Robertson and Wang, 2012; Greull et al., 2016; Mendoza et al., 2017).

Another situation where storage influences are important is for large lakes, and potential seasonal forecasting applications include assisting with water supply, irrigation and hydropower operations for individual lakes, and for water resources monitoring on a regional or global scale. However, some potential modelling challenges are that lake catchment areas may span several climate zones and that monitoring networks are often sparse. Also, with the exception of lake levels and outflows, the main components in the water balance – lake rainfall, tributary inflows and lake evaporation – are often difficult to measure or estimate. In some cases there may also be significant differences between lake and catchment rainfall due to influences on local climate.

For large lakes, perhaps the longest-established operational systems for seasonal forecasting of lake levels are for the Great Lakes in the USA and Canada. Both empirical techniques and physically based hydrological models are used driven by long-range climate outlooks for precipitation and air temperatures with, more recently, incorporation of ensemble outputs from regional climate models (Gronewold et al., 2011; Bollinger et al., 2017). On a smaller scale, of course, water balance models for reservoirs and regulated lakes often form part of regional water supply models typically using statistical techniques or ensemble streamflow prediction approaches (Yeh, 1985; Pagano et al., 2009). The potential for seasonal forecasting using statistical models has also been explored for the rivers downstream of two of the largest lakes in Africa – Lake Malawi (Jury, 2014) and Lake Victoria (Siam and Eltahir, 2015) – and, over longer time scales, Mulumpwa et al. (2017) derived probabilistic estimates for future levels of Lake Malawi using a univariate stochastic model.

Here we describe exploratory studies into forecasting potential for large lakes using an alternative approach. Following a brief review of the dynamic characteristics of lake response, both stochastic dynamic regression and transfer function approaches are used to explore the relationships between a range of potential predictors and lake levels and outflows. This general approach has been widely applied to real-time flood forecasting applications (e.g. Lees, 2000; Smith et al., 2014) and, in addition to the ease with which options can be explored, a key advantage is that few prior assumptions are required about the nature of those relationships (e.g. Beven, 2009; Young, 2013).

The methods are evaluated using case studies for Lake Victoria and Lake Malawi including the potential benefits of data assimilation in reducing the impacts of measurement and modelling uncertainties. As in the studies by Jury (2014) and Siam and Eltahir (2015) the use of climate indices is considered since both the El Niño Southern Oscillation and Indian Ocean Dipole are thought to have a significant influence on rainfall in east and southern Africa (e.g. Saji et al., 1999; Nicholson and Selato, 2000; Jury and Gwazantini, 2002; Manatsa et al., 2011). Since the main aim was to provide insights into possible model structures, the analyses were based primarily on historical datasets derived as part of previous water balance studies as this allowed a more detailed investigation of lake response than would be possible using contemporary datasets. By chance the periods covered included some of the most significant flood and drought periods on record allowing model performance to be evaluated under these more extreme conditions. The discussion concludes with some suggestions for how the findings could be translated into operational forecasting models.

2 Methodology

2.1 General approach

The water balance for a lake can typically be expressed as the following:

$$\frac{\mathrm{d}h}{\mathrm{d}t} = N(t) - \frac{Q_\mathrm{o}(t)}{A(t)}, \tag{1}$$

where h is the lake level, t is time, Q_o is the outflow and A the surface area. The term N is the net inflow, expressed as a volumetric flux per unit area of lake surface, which is sometimes called the net basin supply or "freewater" and, for a given time interval, is defined as

$$N = P - E + \frac{Q_\mathrm{c}}{A} = \Delta h + \frac{Q_\mathrm{o}}{A}, \tag{2}$$

where P is the rainfall on the lake surface, E the lake evaporation and Q_c is the inflow from the surrounding catchment area, again expressed in terms of unit lake surface area. All terms are averages for the selected time interval. An error term is often included to account for additional terms which normally cannot easily be quantified such as seepage and groundwater inflows at the lake bed, although for simplicity this has been omitted here.

Based on the idealised equations for fluid flow over a weir, the natural outflow from a lake is often expressed in the following form:

$$Q_\mathrm{o} = ah^b, \tag{3}$$

where a and b are empirically derived constants and it is assumed that h is defined relative to a datum value for which outflows are zero. For a rectangular weir, the theoretical estimate for b is 1.5 and in practice values can be estimated either directly from lake levels and discharge measurements or using more approximate techniques (e.g. Skaugen, 2004). Furthermore if, as is often the case, the lake area can be assumed to be constant ($A(t) = A_o$), Eq. (1) then reduces to

$$\frac{\mathrm{d}h}{\mathrm{d}t} = N(t) - \frac{a h^b(t)}{A_o}. \tag{4}$$

Some useful insights can be gained by exploring the response for a constant net inflow $N_o > 0$ and integer values of b, and for the present study, as discussed later, the case $b = 2$ is relevant in which case the solution to Eq. (4) can be expressed as

$$h = h_o \frac{(1 + ce^{-t/\tau})}{(1 - ce^{-t/\tau})}, \tag{5}$$

where τ is a time constant defined by $\tau = \frac{1}{2}\sqrt{\frac{A_o}{a N_o}}$, $c = \frac{h_1 - h_o}{h_1 + h_o}$, h_1 is the initial level and $h_o = 2N_o\tau$ is the equilibrium level (e.g. Sene, 2000). During periods of constant net inflow, lake levels therefore tend towards an equilibrium value over timescales which are a function of the net inflow itself, the area of the lake, and the outflow relationship. In contrast, during periods of heavy rainfall, a more rapid response would be expected with levels rising over much shorter timescales.

Generally the relationship between catchment rainfall and runoff is non-linear and affected by catchment antecedent conditions. However, for ungauged catchments it is often found that, on an annual basis, and sometimes a monthly basis, empirical regression relationships can be derived between flow-related parameters of interest and catchment characteristics such as the area and slope (e.g. Smakhtin, 2001). Perhaps the simplest of all such approaches is to assume a mean runoff coefficient $r = Q_c/(A_c P_c)$ in which case the net inflow is given by

$$N = P + k P_c - E\,, \qquad (6)$$

where P_c is the catchment rainfall, A_c is the catchment area and $k = r(A_c/A_o)$.

Equations (5) and (6) together provide a useful – albeit very crude – framework for considering the lake response. That is, during periods of constant net inflow, it might be assumed that the lake outflow is related to the net inflow by a non-linear relationship with a typical response timescale τ. Furthermore if – as is often the case – the variability in evaporation is much less than that in rainfall, then the variations in net inflow might be considered to be primarily a function of the lake rainfall and catchment rainfall. The extent to which these relationships are valid (or not) is explored later.

For the more general case of time-varying net inflows, Eq. (4) can be solved numerically, with the observed levels at the start of each forecast providing initial conditions. For the current exploratory studies a simple iterative solution proved to be sufficient although more computationally efficient solutions could be envisaged. Regarding the estimates for net inflows, one option would be to seek a process-based model based on lake rainfall, tributary inflows and lake evaporation estimates. However, due to the difficulties in estimating these components, and the possible interdependence between them, it is often more practicable to estimate the net inflows from the lake level and outflow terms in the water balance, as indicated by the right hand side of Eq. (2). This is because levels can usually be measured with little difficulty and outflows are often monitored closely, particularly when a lake is important for hydropower generation and/or water supply, as with the present examples.

This approach also has the advantage of avoiding some of the complexities of understanding the water balance but does require a forecasting model for net inflows and a statistical approach provides one option; for example using the following autoregressive (AR) formulation:

$$y_t = -r_1(t)\,y_{t-1} - r_2(t)\,y_{t-2} \ldots - r_n(t)\,y_{t-n} + e_t, \qquad (7)$$

where y_t is the dependent variable, r_i are the model coefficients, which can be time varying if required, n is the maximum lag time considered and e_t is a stochastic noise term. Alternatively if, as might be expected, external influences such as the lake rainfall are important, then the following linear regression formulation might be considered:

$$y_t = s_1(t)\,u_1 + s_2(t)\,u_2 + \ldots s_m(t)\,u_m + e_t\,, \qquad (8)$$

where s_i are the model coefficients and u_i are the external input values, such as rainfall or climate indices, lagged by 1, …, m time steps. In contrast the following transfer function formulation allows both serial dependence and external variables to be considered together with a pure time delay δ if required:

$$y_t = \frac{B\left(z^{-1}\right)}{A\left(z^{-1}\right)} u_{t-\delta} + \frac{D\left(z^{-1}\right)}{C\left(z^{-1}\right)} e_t\,, \qquad (9)$$

where z^{-1} is the backward shift operator ($z^{-i} y_t = y_{t-i}$). The second term represents the residuals of the transfer function input-output model via polynomials C and D and, although not used for the net inflow component, an autoregressive moving average (ARMA) model of this form (e.g. Box and Jenkins, 1970) was used in the data assimilation component described later. Here a single external input is considered but as discussed later the formulation is easily extended to multiple inputs.

The transfer function and data assimilation aspects of the models were implemented using the recursive estimation techniques available as part of the CAPTAIN Toolbox, which was developed by Lancaster Environment Centre for operation within the Matlab® programming environment. These are described in Young et al. (2007) and Young (2011) but in essence provide a range of routines for estimating model parameters and outputs. The stochastic solution techniques used inherently provide recursive estimates of parameters and uncertainty, including how both of these vary over time, as opposed to the simple estimation of posterior means and variances that many other techniques provide.

2.2 Case studies

The two lakes considered were Lake Victoria and Lake Malawi, which respectively are the first and third largest in Africa and lie within the East African Rift, which contains a number of other large lakes with both open (with outflows) and closed basins.

On a regional scale, both are economically important since the outflows are harnessed for hydropower generation and to support large-scale irrigation schemes further downstream on the White Nile and Shire rivers. More locally the livelihoods of millions of people are supported through fisheries, water supply and agriculture. To provide an indication of scale, the combined catchment and water surface area for Lake Victoria exceeds that of countries such as Uganda and Rwanda, whilst the area of Lake Malawi and its catchment is larger than Malawi itself.

For Lake Victoria the lake outlet is just north of the equator and the southernmost part of the catchment is at about 3° S, whilst Lake Malawi extends from about 14° S at the outlet to 9° S in the northernmost part of the catchment. The catchment for Lake Victoria lies mainly in Rwanda, Tanzania, Kenya and Uganda, whilst that for Lake Malawi is mainly in Malawi and Tanzania. There are also small contributing areas in the Democratic Republic of Congo (for Lake Victoria) and Mozambique (for Lake Malawi).

At their closest points, the catchments lie about 500 km apart; however, they experience markedly different climates in part due to the annual passage of the Intertropical Convergence Zone (ITCZ). For Lake Victoria, which lies fairly centrally within the zone's range, there are two main rainfall seasons and these are typically between March and May and October and December. In contrast, Lake Malawi lies towards the southernmost end of the range resulting in a single main rainfall season from November to April or May in much of the basin, although with some evidence of a temporary reduction in rainfall intensity part way through the season (Nicholson et al., 2014). The predominant climate classifications (Peel et al., 2007) for the lake catchment areas are tropical savannah for Lake Victoria and temperate (dry winters, hot summers) for Lake Malawi, with regions of arid savannah and arid steppe in the south.

Due to topographic influences, there are wide variations in annual rainfall within each basin; both lakes are also large enough for the difference between lake water surface temperatures and the surrounding land to affect the local atmospheric circulation and hence precipitation and evaporation. For example WMO (1983) notes that for Lake Malawi breezes tend to be offshore in the early morning then onshore in the afternoon, leading to "a preferential tendency for rainfall on the lake to occur in the early morning rather than the late afternoon". UNDP (1986) also notes a wind-funnelling affect in the north-western part of the lake due to local topography, which can result in annual rainfall exceeding 3000 mm in this area, in contrast to the plateau areas to the west of the lake where values are typically only 700–1000 mm. The local impacts are even more pronounced for Lake Victoria and have been the subject of several investigations, including the use of mesoscale models to study the influences on atmospheric circulation both locally (Sun et al., 2015) and regionally (Thiery et al., 2015). Lake inflows generally follow these seasonal trends, although it is worth noting that some of the lake tributaries are ephemeral with flows generally ceasing towards the end of the dry season, particularly in drier parts of the basins.

For both lakes, regular recording of lake levels began in the 1890s and some catchment rain gauge observations date back to the period 1900–1910 for Lake Victoria and the 1920s for Lake Malawi. For Lake Victoria monitoring of outflows began in about 1940, whilst for Lake Malawi the first observations began in 1948. Lake outflows have also been regulated for hydropower production from 1953 in the case of Lake Victoria and 1965 for Lake Malawi. However, the scheme designs are very different due to the nature of the topography and river channels at the outlet of each lake and some key features include the following:

- Lake Victoria – the lake outlet used to be at a spectacular natural waterfall until Owen Falls Dam was built about 3 km further downstream, drowning out the falls; hydropower generation and the lake outflow are now controlled at the dam.
- Lake Malawi – outflows are controlled at Kamuzu Barrage more than 80 km downstream from the lake outlet, which is possible since the change in elevation is only a few metres between the lake and the barrage. The main hydropower plants are in natural gorges downstream of the barrage.

However, an important point is that the operating rules for both schemes were, to a large extent, designed to mimic the response of the natural lake, and these are often represented in the form shown in Eq. (3), with values of b close to 2 (e.g. Drayton, 1984; Piper et al., 1986). Due to operational requirements, though, there are sometimes minor departures from these rules so a separate outflow record – termed the "natural flows" here – was derived in which flows were only retained when similar to those expected from the level–outflow relationships described by Eq. (3). For the Lake Victoria studies, the periods omitted only amounted to a small part of the record (which predates more recent departures) but this occurred slightly more frequently in the case of Lake Malawi, primarily in the later years of the records. During these times there is also an effect on levels although this is much less significant due to the non-linear nature of the outflow relationships.

Table 1 summarises some key characteristics of the long-term water balance for each lake based on previously published estimates. However, whilst these values are typical, it is worth noting that they can vary significantly between studies depending on the datasets, periods selected and estimation techniques used. Regarding surface areas, estimates also vary although generally the changes with levels are small, for Lake Victoria amounting to about 2 % over the historical range of observed levels (e.g. Piper et al., 1986) and for Lake Malawi by less than 1 % per metre rise or fall (Lyons et al., 2011). Areas were therefore assumed constant for these exploratory analyses, although these variations might be included in a more detailed approach.

As noted earlier, historical datasets were used and here it is worth noting two landmark hydrometeorological studies in the 1970s and early 1980s (e.g. WMO, 1982, 1983). The derived values formed the basis for a number of later studies in which new records and information were added using a wide variety of approaches; see for example Piper et al. (1986), Sene and Plinston (1994), Sutcliffe and Parks (1999), Nicholson et al. (2000) and Kizza et al. (2012, 2013) for studies on Lake Victoria, plus the citations therein, and Drayton (1984), Neuland (1984), Jury and Gwazantini (2002) and Sene et al. (2017) for Lake Malawi. The original papers should be referred to for a discussion of the methods and datasets used but, in many cases, the overall approach was similar: namely to reconstruct the rainfall, inflow and evaporation terms from gauges situated around each lake, and in some cases on islands within the lake. In some studies, rainfall–runoff or sta-

Table 1. Some key physical characteristics of Lake Victoria and Lake Malawi from various sources including Piper et al. (1986), Sutcliffe and Parks (1999) and Sene et al. (2017). All values indicative only.

Key parameters	Lake Victoria (1956–1978)	Lake Malawi (1954–1980)
Surface area (km^2)	67 000	28 750
Catchment area (km^2)	194 000	95 750
Lake rainfall (mm)	1878	1414
Catchment rainfall ($mm\,yr^{-1}$)	–	1178
Catchment runoff ($mm\,yr^{-1}$)	343	1000
Lake evaporation ($mm\,yr^{-1}$)	1595	2264
Lake outflow ($mm\,yr^{-1}$)	524	418

tistical models were also used to infill or extend tributary inflow records and – in nearly all cases – the analyses were performed on a monthly basis.

A key factor in choosing which records to use here was how well the resulting models described the overall water balance since this provides some confidence in the suitability of the individual components; although, as discussed later, over such huge areas estimates can only ever be approximate. Several of the studies cited met this criterion and, primarily on the basis of data availability and completeness, the following estimates were selected:

- Lake Victoria – monthly estimates from 1925–1978 reported by Piper et al. (1986) and subsequently updated by the Institute of Hydrology (1993) to the period 1925–1990 for lake rainfall (6–8 shoreline rain gauges), catchment rainfall (25–30 rain gauges) and tributary inflows (up to 20 river gauges) and for 1925–1992 for levels and outflows.

- Lake Malawi – monthly estimates for the period November 1954 to October 1980 reported by WMO (1983) for lake rainfall (16 shoreline rain gauges; 1 island gauge), catchment rainfall (53 rain gauges) and tributary inflows (about 21 river gauges), and for which aspects of the water balance appear in a number of the studies cited here, such as Drayton (1984) and Neuland (1984).

The numbers in brackets indicate the approximate numbers of gauges used to derive each component in the water balance in these studies, with ranges provided in some cases due to differing numbers of gauges in different periods. In the Lake Victoria studies, rainfall–runoff models were also used to estimate inflows for ungauged catchments and to help infill missing values, whilst for Lake Malawi, scaling and correlation approaches were used. For the present study, longer-term annual level records were also compiled for both lakes from these various sources dating back to the 1890s and overall catchment rainfall estimates derived from individual catchment values using a simple area-weighting approach. Regarding climate indices the following values were used: the Southern Oscillation Index (SOI; Trenberth, 1984), Niño3.4 (NINO34; Trenberth, 1997) and the Dipole Mode Index (DMI; JAMSTEC).

For the Lake Malawi records, it is worth noting that sometimes a small correction term is included to account for inflows and losses between the lake outlet and Kamuzu Barrage; however. the impacts are small when considered on a monthly basis and, as it is here, this term is often omitted. As indicated above, values are for a hydrological year of November to October but, to help comparisons with the Lake Victoria analyses, when describing results the predominant calendar year is cited; for example "1970" refers to the hydrological year 1969/70. For the Lake Victoria datasets that were used another point to note is that, to help to account for the increased rainfall over the lake surface, the lake rainfall estimates based on rain gauge observations were increased using linear scaling factors; this was to preserve an overall water balance between the start and end points of the simulation period, although with no constraints on the variability in the intervening years.

As the many citations above indicate, much has been written about the accuracy of the various gauge records available and the derived water balance components, together with issues such as how rainfall measured at the shoreline relates to average rainfall at the lake surface. For these exploratory studies, to facilitate comparisons, unless otherwise stated values for levels and other parameters were generally expressed in standardised form, based on the departure from the mean divided by the standard deviation in each time period of interest. This also helped to reduce the uncertainties in the results by focussing on the underlying signals rather than being concerned about absolute amounts, such as the estimates for lake rainfall. In forecasting applications, another consideration is that real-time observations can be used to adjust forecasts to help to account for modelling and observation errors and the potential value of this process, called data assimilation or adaptive modelling, is discussed later.

3 Results

3.1 Initial exploratory studies

Figure 1 shows some notable events in the recorded histories for both lakes. These include high levels in the late 1970s and late 1990s and – for Lake Victoria – the most extreme levels on record in the early 1960s and a prolonged period of low levels from the 1920s to the 1950s.

In contrast, for Lake Malawi, levels were unusually low up to the 1930s and several studies (e.g. Drayton et al., 1984) have suggested that this was due to a sand barrier forming at the lake outlet or in the channel(s) downstream around 1908–1915, following which levels then rose progressively until the blockage(s) cleared during that decade. However, this event falls outside the period considered here and – with the exception of the minor impacts from hydropower operations noted earlier – for both lakes variations in levels were therefore due primarily to climate influences. In particular the 1961/62 event for Lake Victoria has previously been investigated in detail with some evidence of a regional shift in climate at that time (e.g. Sutcliffe and Parks, 1999; Nicholson and Selato, 2000).

Another notable feature of the observed levels is the apparent persistence during times of falling levels and the analytical form of the water balance (Eq. 5) provides some insights into the response during these periods with, as indicated earlier, the use of a constant area and a value of $b = 2$ being reasonable approximations.

Figure 2 illustrates this response for Lake Malawi for the case of a sudden change in levels, such as might occur following a few weeks of heavy rainfall or recovery following a prolonged dry spell, and similar response curves have previously been published for Lake Victoria (e.g. Institute of Hydrology, 1993) and – using numerical simulations – for Lake Malawi (WMO, 1983; Neuland, 1984). In both cases, based on typical long-term mean values for the net inflows, the estimated time constants from Eq. (5) were in the range of 4–5 years, which may just be coincidence or is perhaps reflective of the balance between net inflows, areas and outflow characteristics required for a lake in this region to have a permanent outflow; this is a speculative point which might be worth further investigation since there are several other large lakes in the African Rift Valley.

More generally these results suggest that, in addition to monthly variations, there are longer-term aspects to the lake response related to both climate variations and the inherent time delays in response. From a forecasting perspective these can potentially be exploited and, to explore these relationships further, both time series and correlation plots were prepared on a monthly basis, in the latter case for a range of assumed lag times. Figure 3 shows one such example, for the case of the Lake Victoria datasets with zero assumed time delay between inputs and outputs. For net inflows, for both lakes, the strongest relationships were with the lake rainfall and catchment rainfall. Here the full records were considered and cross correlation coefficients were in the range 0.82–0.95 at zero time delay and about 0.5–0.8 at a lag time of 1 month for the Lake Victoria and Lake Malawi records respectively, whilst for tributary inflows the relationships were generally weaker than this. The serial dependence in values was also investigated for some records and typically the autocorrelation coefficients in net inflows were highest for a lag time of 1 month (0.4–0.7 for the two lakes), roughly halving for a lag time of 2 months and continuing to reduce at longer lag times.

Regarding climate indices, for both lake records the relationships between net inflow and the El Niño related indices were only borderline statistically significant, with the largest values for lag times of a few months. In contrast the maximum coefficients for DMI were about 0.2 for lag times of 2–4 months. Further investigation showed that for the Lake Victoria records, but not for Lake Malawi, this linkage was markedly higher for the second half of each year – and hence the second rainfall season – reaching values of about 0.27 at lag times of 2–3 months. Interestingly, when compared on a time series basis, in some periods the correspondence was unexpectedly close; for example during the 1961–1964 event in Lake Victoria the initial peak in net inflows in 1961 was preceded by a rise in DMI a few months before, with similar but smaller rises in advance of peak values for the next 3 years. However, this is just a tentative conclusion and would require further investigation if that event is of particular interest.

3.2 Net inflow forecasting

Taken together, these initial studies suggested that the following characteristics would provide a useful starting point for developing a model of the net inflows for both lakes:

- Serial correlation – a dependence on values for the past 1–2 months but probably not much beyond that
- Primary external factors – a dependence on current values for lake rainfall and catchment rainfall and possibly for the previous month
- Secondary external factors – inclusion of the DMI and possibly one or both of the indices related to the El Niño–Southern Oscillation (ENSO), at lag times of up to several months

The calibration periods used for model development were 1925–1954 for Lake Victoria and 1954–1970 for Lake Malawi, and these were chosen so that the high levels of the 1960s for Lake Victoria and for the 1970s for both lakes would fall within the validation periods, which were 1955–1990 and 1971–1980 respectively. As noted earlier, standardised values were used throughout at a monthly time step.

Considering the autocorrelation aspects first (Eq. 7), for the calibration period for the Lake Victoria net inflow record, using a simple autoregressive model the highest values achieved for R^2 were about 0.3 with a second order model. The corresponding value for the Lake Malawi record was rather better at about 0.6, again for a second order model.

Regarding external inputs (Eq. 8), several permutations were considered focussing on the use of rainfall inputs and climate indices. In terms of the R^2 performance, the differences between these various regression models were generally not large, with values typically in the range 0.81–0.94 for both lakes when using time varying parameters. However, based on the simple representation for net inflows shown in Eq. (6), the value of k was similar for both lakes suggesting that a weighted average of these two inputs might also be worth exploring, but the performance was only marginally

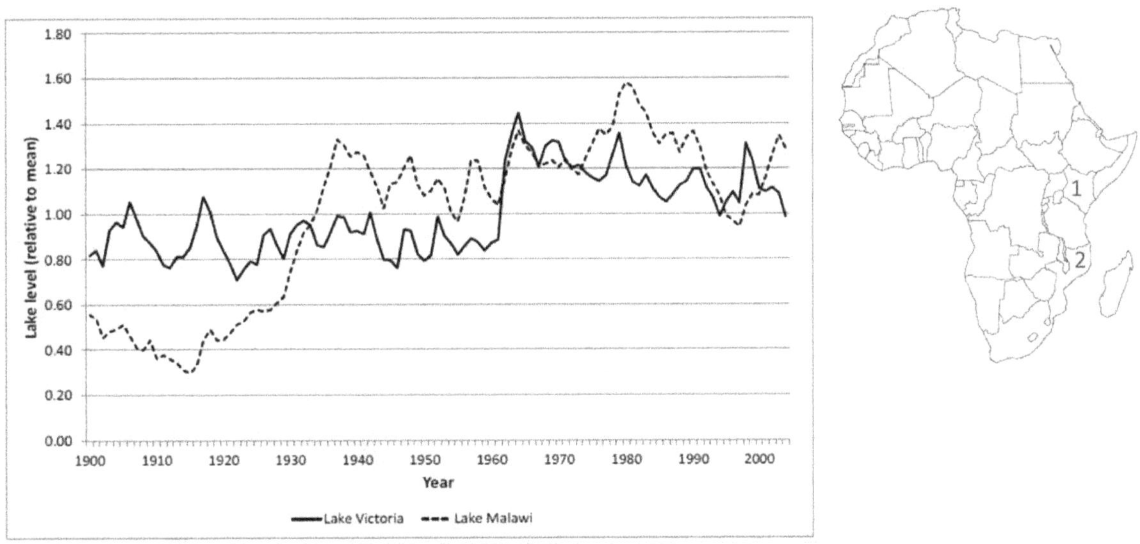

Figure 1. Annual lake level variations in Lake Victoria and Lake Malawi from 1900 to 2004 relative to the mean values in that period; values are expressed in terms of indicative depths at the lake outlet. The inset shows the locations of (1) Lake Victoria and (2) Lake Malawi.

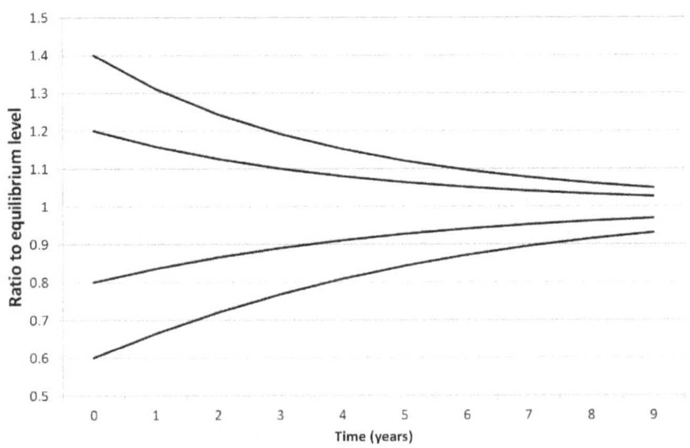

Figure 2. Examples of the response in levels for Lake Malawi from initial values equal to 0.6, 0.8, 1.2 and 1.4 times the long-term equilibrium values.

better than when using lake rainfall alone. The influence of climate indices was also small (about 0.01–0.02 in terms of R^2) due to the dominance of the rainfall terms so these were included in the data assimilation components of the models as described later.

On this basis, the decision was taken to use the lake rainfall as the only external input since – as discussed later – this would have some advantages in an operational setting. For the Lake Malawi records, best results were obtained using the latest observed lake rainfall alone, whilst – in the case of Lake Victoria – including the previous month's rainfall as well seemed worthwhile. Hence for the Lake Victoria records the expectation was that a [2 2 0] model structure might be appropriate, with a [1 1 0] structure for Lake Malawi, where the notation $[n\ m\ \delta]$ refers to the parameters in Eqs. (7) to (9).

A search of all permutations in the range [1 1 0] to [3 3 3] showed these to be in the top few options in terms of R^2 plus a range of other indicators, such as the information criterion described by Young (2011). A recursive estimation of parameter values also showed that these were reasonably stable over time and fixed parameter versions gave similar values of R^2: that is about 0.8 and 0.92 for the Lake Victoria and Malawi records respectively. Fixed parameters were therefore assumed for the forecasting runs described later and Fig. 4 shows one such example for Lake Victoria for part of the validation period. For the Lake Malawi record the validation performance was similar to that during the calibration period but slightly reduced for the Lake Victoria record. This is perhaps due to the suspected shifts in climate after the 1961 event and, from graphical comparisons, the decrease seemed to be mainly due to slight differences in timing for some years, rather than in magnitudes, as discussed further in the next section.

The estimated confidence intervals are also shown and generally encompassed both the high and low inflow observations, providing some reassurance that the main features of the response are being captured despite the simplifications of fixed parameter values and using lake rainfall values alone rather than a more complex approach. However, at this stage no noise term was included since it proved to be more convenient to include this as part of the data assimilation component.

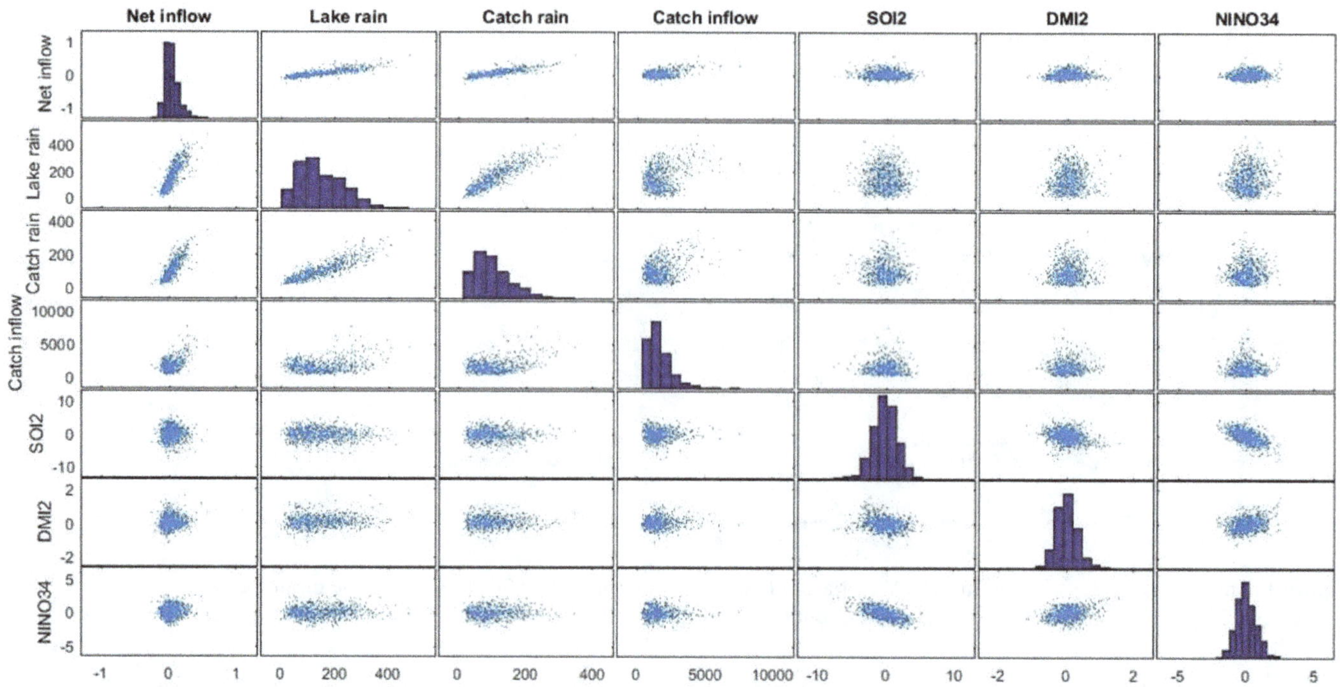

Figure 3. Example of a correlation plot for the standardised monthly records for Lake Victoria (1925–1990) assuming zero lag time between values (Catch = catchment).

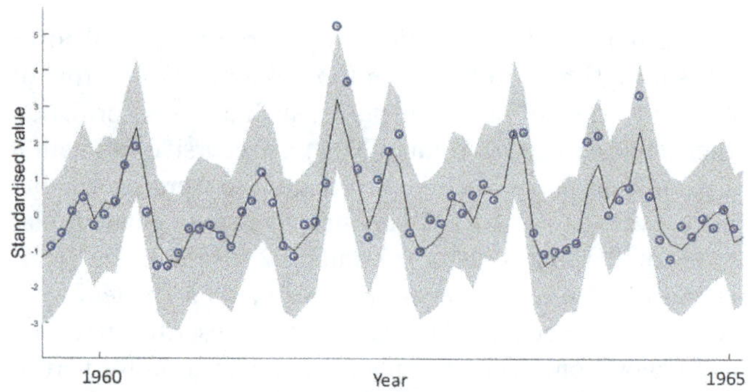

Figure 4. Example of standardised net inflow estimates from a transfer function model for Lake Victoria net inflows for part of the validation period, with fixed parameter values; observations appear as dots and the shading corresponds to twice the model standard error, which is approximately equivalent to the 95 % confidence interval. Note that, due to sampling uncertainties, the number of values within or outside the intervals varies depends on the period(s) chosen.

3.3 Forecast performance

Having developed models for the net inflows, these were expressed in recursive form for input to the water balance equation (Eq. 4). This formulation mimics how the models would be used in an operational setting in which the water balance would be solved numerically each month to derive forecasts for lake levels and outflows for the months ahead based on rainfall observations available up to the time of the forecast. To further extend forecast lead times, rainfall forecasts would ideally also be required but, for these exploratory studies, it was sufficient to use climatological estimates instead, based on the mean monthly distributions of rainfall in the calibration period. However, to provide an indication of forecast potential, the performance was also estimated assuming perfect foresight of rainfall: that is, using historical observed values beyond the forecast origin. In both cases, the first 2 years of records were ignored to allow for initialisation of the autocorrelation aspects of the models.

Figure 5 shows the estimated variations in R^2 with lead time for the validation periods for both lakes, derived by advancing the forecast origin by 1 month between each model run and then retrospectively estimating overall values at the required lead times. The differences between values for levels and outflows are primarily due to the gaps in the derived natural outflow record discussed earlier. As expected, performance decreases with increasing lead times; for example, for lake level estimates falling to about 0.8 after 7 months and 0.9 after about 3–4 months. However, for the Lake Victoria record, due to the rapid increase in levels in the early 1960s, this probably overstates the performance so the values of about 0.7–0.8 at 3–4 months and 0.4–0.6 at 7 months obtained for later years and the calibration period are more typical. This is simply an artefact of this type of performance measure since using mean values as a reference becomes less informative if a time series is non-stationary, such as exhibiting trends over time and/or quasi-step changes in values as here.

The figures also show the performance for lake levels assuming perfect foresight of rainfall, from which a speculative conclusion might be that, since many seasonal rainfall forecast products tend towards a climatological estimate at long lead times, the values for the control run or ensemble mean might therefore asymptotically tend to those estimates at longer lead times. If correct, then the differences between the climatological and perfect foresight values provide a rough indication of the potential performance gain from using seasonal rainfall forecasts, with the remaining improvements to be achieved from reducing uncertainties in the models and underlying datasets.

These results of course only give one view of performance and given the findings from previous climate studies for these lakes, and the cross correlation estimates from the present study, one possibility is that the errors in model outputs might to some extent be explained by longer-term variations in climate. From the point of view of developing data assimilation routines, this is a more attractive option than simply developing a statistical model for the residuals since there is then some underlying physical interpretation.

To test this hypothesis, regression models were developed between the residuals of the forecast outputs and the climate indices described earlier, using DMI and NINO34 as examples; similar conclusions were reached using SOI. Both single and multiple regression models were evaluated in time-varying and fixed parameter forms considering a range of possible lag times and forecast lead times. For example, for the Lake Malawi records, for the 4 month ahead forecasts, the maximum cross correlation coefficients with DMI were obtained for lag times of about 3–5 months, and at slightly longer lag times of 6–9 months for NINO34 and SOI, although the signs differed depending on the index chosen. The magnitudes of the coefficients were about 0.42 and 0.28 for DMI and NINO34 in the calibration period, and slightly lower for SOI. For the Lake Victoria record, it seemed useful to consider slightly longer lead times and, for the 6 month ahead forecasts, optimum lag times were about 6–7 months for DMI and 5–6 months for NINO34, with correlation coefficients of about 0.28 and 0.41 respectively, and again slightly less for SOI. As might be expected, these relationships were generally weaker when considering perfect foresight since these influences may already be embedded in the observed data to some extent.

Based on these results, and exploratory studies using single regression models, multiple regression models for lag times of 7 and 5 months for DMI and NINO34 were assumed for the Lake Victoria record and values of 4 and 7 months for the Lake Malawi record. In both cases, the models with time-varying parameters exhibited slightly better performance than their fixed parameter counterparts, with R^2 values of about 0.39 and 0.19 respectively for the Lake Victoria regression model and 0.44 and 0.27 for the Lake Malawi model, when using both indices together.

However, the use of R^2 only provides one view of performance and, as already noted, has some limitations. Also, lake levels are strongly seasonal and there might therefore be a case to use alternative benchmarks such as seasonally based values or metrics which incorporate lag times or moving averages as proposed by Schaefli and Gupta (2007). Indeed there are many possible measures which could be used (e.g. Jolliffe and Stephenson, 2011; Wilks, 2011) and this is a topic that would merit further research. For these exploratory studies, though, a simpler approach was adopted which was to focus on the errors in annual peak levels since these tend to occur around the same time each year and are more challenging to forecast than annual minima.

For the purpose of this exercise alone, separate relationships were also calibrated for the validation period to further explore the strength of these relationships. Figure 6 shows some example results for the fixed parameter case for the 5 month ahead forecasts of peak annual levels for the Lake Victoria record, and 4 month ahead values for Lake Malawi. The results suggest that in many – although not all – years the adjusted forecasts for maximum levels are closer to those which were subsequently observed, potentially providing a useful gain in forecast performance.

The main exception, however, was for the Lake Victoria record in the validation period (1955–1990) for which – although negative forecast errors were consistently improved – positive errors were not. Interestingly, the relationship with DMI in that period was also slightly stronger but that with NINO34 no longer significant (<0.1), which is possibly again further evidence of a shift in rainfall response. However, for both records, further investigation would be required into the optimum approach to use; for example, exploring the influence of timing errors on model performance and whether time varying parameters could be used to help represent longer-term trends and other variations.

Given these uncertainties, rather than applying these adjustments first to re-estimate the residuals, the original series were analysed further. Due to lake storage and climate influences, as might be expected the values showed some serial correlation over periods of months superimposed upon longer term variations. However, exploratory studies using an ARMA approach for the residuals – with forecasts implemented via a Kalman filter – provided mixed results. For example, for the Lake Malawi records, using the same lead time as in Fig. 6 (4 months) and climatological inputs, the forecasts for peak annual levels were only improved in a few years, and even degraded in some cases. As is common with error prediction routines, the issue here seemed to be partly due to timing differences in the residuals, although this problem seemed to be reduced to some extent when assuming perfect foresight of rainfall, again suggesting that reducing timing errors at the outset may assist with performance.

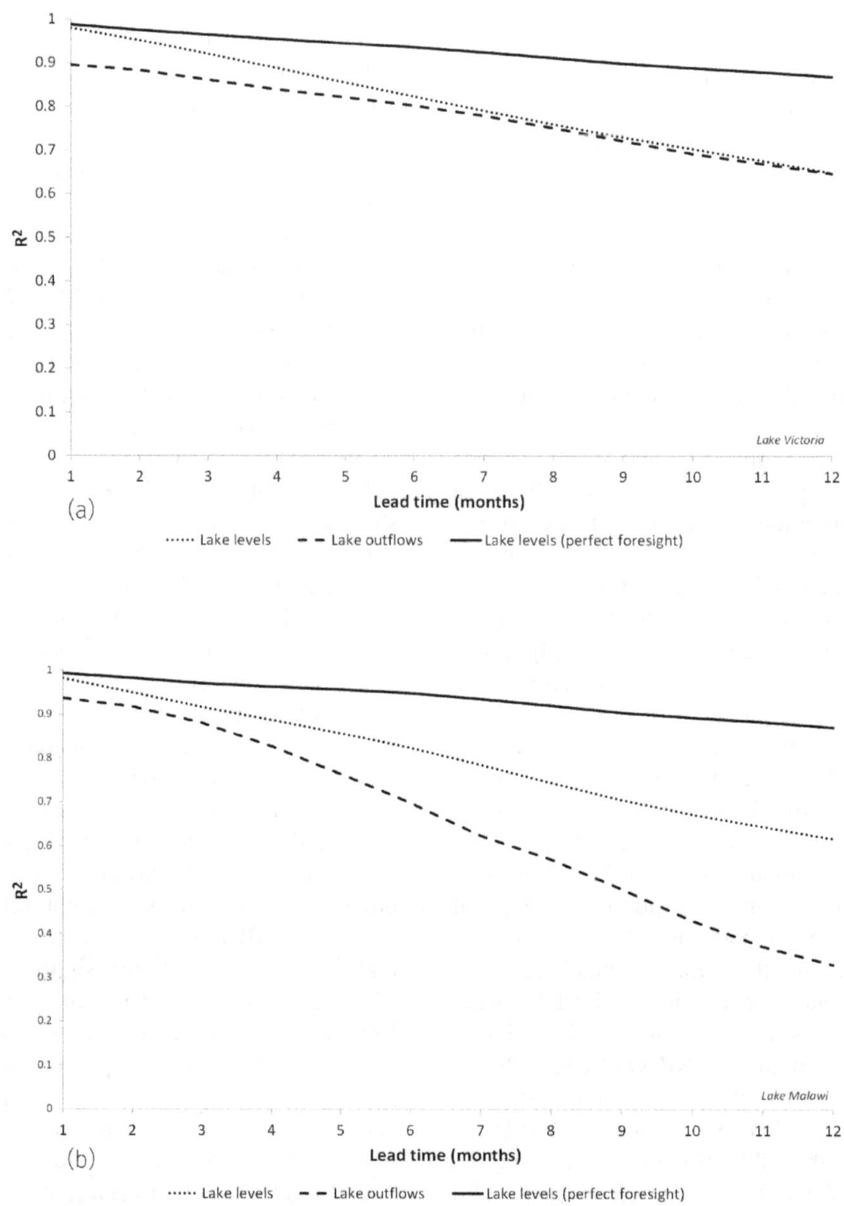

Figure 5. Variations in R^2 with lead time for levels and outflows for **(a)** Lake Victoria and **(b)** Lake Malawi using climatological rainfall inputs for the validation periods and – for levels – perfect foresight of rainfall.

Various permutations of model orders and lead times were explored and Fig. 7 shows some examples using the same structures for each lake record, namely AR(6) and MA(5) when using climatological inputs and AR(3) and MA(2) for perfect foresight inputs. The lead times used were 2 months for the Lake Victoria record and 4 and 2 months respectively for the Lake Malawi record. Again, for illustration, the results for the full record lengths are shown although the ARMA models were calibrated just for the calibration periods.

For the case of perfect foresight, some improvement was obtained in most years for the Lake Malawi record, with results more mixed for the climatological estimates, whilst the performance was similar in both cases for the Lake Victoria record, albeit using a shorter lead time for the climatological estimates. More generally, for these examples, it seemed as if the effect of these adjustments tails off significantly at lead times of more than about 2–3 months for the Lake Victoria records and 3–4 months for the Lake Malawi records. To improve the results, one option might be to optimise the model orders separately for each lake record, perhaps using a two stage approach in which the climate index adjustments are made first. Depending on the application, additional performance measures might also be considered to better un-

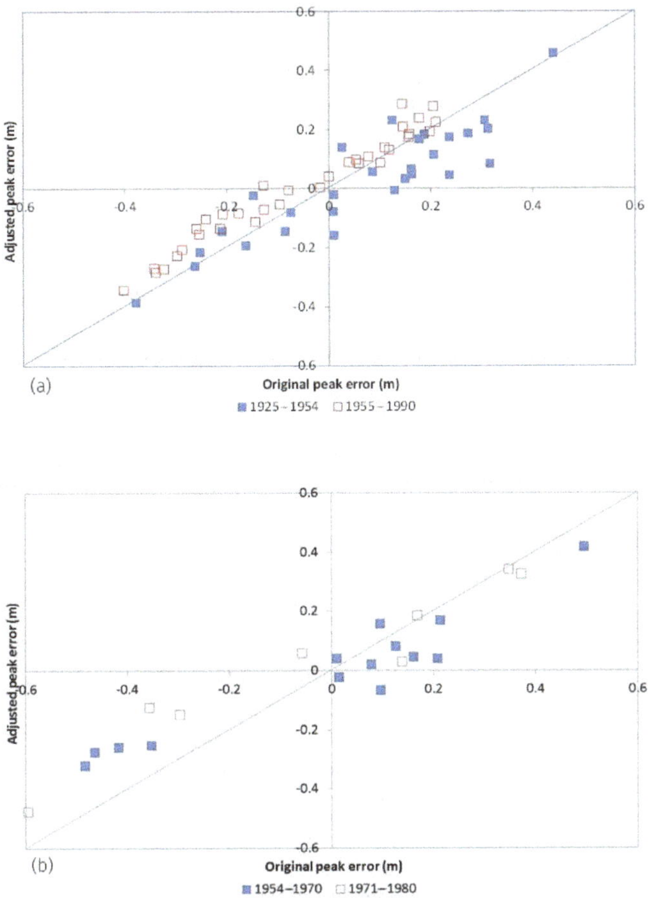

Figure 6. Illustration of the effect of including fixed parameter regression relationships with DMI and NINO34 for the model residuals on errors in annual maximum levels at lead times of **(a)** 5 months for Lake Victoria and **(b)** 4 months for Lake Malawi. Climatological rainfall inputs were used and a one-to-one trend line is shown as a guide.

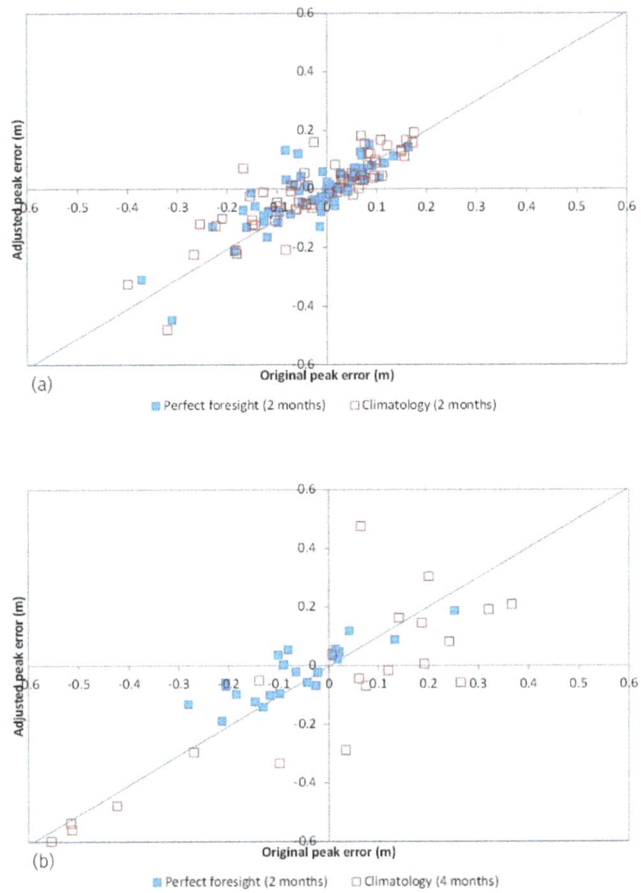

Figure 7. Illustration of the effect of an ARMA approach for the full records using (6.5) models with climatological inputs and (3.2) models with perfect foresight for **(a)** Lake Victoria and **(b)** Lake Malawi with a one-to-one trend line as a guide.

derstand where to focus effort, such as taking account of the timing differences between peak values or using threshold-based metrics. However, these issues were not pursued further since, as noted in the following section, for operational forecasting alternative inputs would be used, each with their own error characteristics and bias correction requirements.

4 Discussion and conclusions

The aim here has been to illustrate an approach to exploring possible model structures considering factors such as the choice of input variables, characteristic response times and options for data assimilation. A mixture of transfer function, regression and analytical techniques was used. In keeping with this overall approach, some aspects were only developed to the stage required to draw useful conclusions rather than providing a full solution, such as with the net inflow model and data assimilation components. In future studies, these types of analyses might then help to guide the development of more complex models for individual lakes, or groups of lakes.

For the two lakes considered here, some initial findings include the possibility of using lake rainfall alone as a model input, and the potential to use error prediction techniques that are more typical of those used for short-range flow forecasting, combined with statistical relationships incorporating climate indices. Regarding seasonal variations in levels, due to storage influences there was some evidence of forecast skill up to 3–6 months ahead solely from climatological rainfall estimates, depending on the performance measures used. For the Lake Victoria record, as found in several previous studies, there were also indications of a change in response following the extreme rainfall events of the early 1960s. In any further studies, though, for these or other lakes, the results would need to be evaluated using contemporary datasets taking account of both uncertainties in the observations and, for regulated lakes, current operating rules via an appropriate parameterisation of outflow relationships, including allowing for any changes in procedures over time.

For example, one particularly difficult decision is on the choice of rainfall records to use, and whether these should be area-averaged values (as here) or index series from representative gauges. As already noted, some measurement challenges include the spatial coverage of gauges and any local enhancement of rainfall due to the influences on atmospheric circulation from the lake. Alternatively, where rain gauge networks are sparse, satellite observations provide another possibility, although with their own measurement challenges such as the need to differentiate between locally driven convective and stratiform rainfall and land and water surfaces. With rain gauge inputs, another potential challenge is the need for data-sharing agreements when gauges are operated by more than one organisation or country.

More generally, though, estimates for lake rainfall should in principle be more accurate than for catchment rainfall since there are no topographic influences to consider – other than around the lake shoreline – and for that reason were selected here. However, given the huge areas covered, the challenges in estimating the individual components in the water balance should not be underestimated, and for forecasting purposes these are perhaps best regarded as index series themselves. The uncertainty in the estimates then cascades into the water balance estimates and hence level and outflow forecasts. For model calibration, this emphasises the importance of water balance studies in evaluating the suitability of any inputs which are proposed for operational use. Also for smaller lakes with faster response times, a weekly or even daily time step might be required to capture the main features of the response, particularly for tributary inflows. However, this brings new modelling challenges such as the need to consider the spatial variability in different components of the water balance in more detail, and the timing differences between them. Of course, if links to rainfall are not of interest, there is the option of formulating models directly in terms of the net inflows derived from levels and outflows, which are often easier to estimate.

Regarding rainfall forecasts, these offer the potential to extend lead times further through direct input of ensemble rainfall forecasts, perhaps combined with the seasonal forecasts for climate indices which are now also routinely available. For the two lakes considered, the focus on model development to date has been for Lake Victoria, including operational tools for hazardous thunderstorms (Thiery et al., 2016, 2017) and research studies on developing high-resolution models for seasonal forecasting (Argent et al., 2014). The performance of global scale models for seasonal forecasting has also been evaluated in a regional context suggesting that these offer some improvements on the consensus forecasts prepared by the Regional Climate Outlook Forums (Mwangi et al., 2014). Promising results have also been obtained for drought forecasting on seasonal timescales in southern Africa (Winsemius et al., 2014).

For the development of seasonal models, the reforecasts available from sources such as the Subseasonal to Seasonal (S2S) Prediction Project (Vitart et al., 2017) provide a valuable resource. So-called custom climate indices or predictors might also be considered based on additional meteorological and ocean parameters; for example using relationships derived from principal component analyses or a transfer function approach.

Another consideration is the modelling approach to use and this will typically depend on the operational requirement and the real-time data available, and to some extent the skills and preferences of the modelling team. For example, some additional factors to consider might possibly include artificial influences on lake inflows or outflows from hydropower or other operations, and whether modelling components are required for water quality, ecology, snowmelt and sediment transport.

The classical approach to modelling a lake water balance is to use rainfall–runoff models to estimate tributary inflows, with separate components for area-averaging of lake and catchment rainfall and the outflow response. The runoff components are typically estimated in semi-distributed or distributed form using a conceptual or physical-conceptual approach, and lake evaporation is typically estimated from local weather station records or an energy budget approach.

In contrast, if a transfer function modelling approach is adopted, the step-by-step approach illustrated here provides a powerful way to rapidly explore many options. However, this does not fully exploit the power of the stochastic techniques used, particularly regarding the use of time varying parameters, a concept from system engineering in which forecasts for the parameters themselves become part of the solution, and a way of reflecting potential long-term trends and variations in climate. In that regard, it is worth noting that the water balance itself can be solved in transfer function form with no further approximation for the case of discrete time intervals (as here) and a linear outflow relationship ($b = 1$).

For the more general non-linear case, the same approach can still be adopted but the model coefficients now become a function of the levels or – to use a phrase common in system engineering – are "state dependent" (e.g. Young, 2000). Although this would be considerably more complex than was required in this case, an overall solution could be envisaged in which the net inflow model, water balance and data assimilation components are combined using a so-called state dependent parameter (SDP) approach. Some aspects of this approach have already been illustrated by the methods described here and there is extensive literature regarding more complete solutions (e.g. Young, 2000; Sadeghi et al., 2010), including flood forecasting applications (e.g. Young and Beven, 1994; Leedal et al., 2013; Smith et al., 2014).

Competing interests. The authors declare that they have no conflict of interest.

Special issue statement. This article is part of the special issue "Sub-seasonal to seasonal hydrological forecasting". It is a result of the HEPEX workshop on seasonal hydrological forecasting, Norrköping, Sweden, 21–23 September 2015.

Acknowledgements. We would like to thank the Centre for Ecology and Hydrology for permission to use the datasets for Lake Victoria; also Michael Kizza for providing additional independently derived records for comparison. The WMO (1983) study was led by Chris Kidd and provides a valued contribution to understanding of the hydrology of Lake Malawi. The present analyses were performed whilst the lead author was an Honorary Researcher at Lancaster Environment Centre. We are grateful to Maria-Helena Ramos, Wim Thiery and Kolbjorn Engeland for their comments on the first draft of this paper.

Edited by: Maria-Helena Ramos

References

Argent, R., Sun, X., Semazzi, F., Xie, L., and Liu, B.: The Development of a Customization Framework for the WRF Model over the Lake Victoria Basin, Eastern Africa on Seasonal Timescales, Adv. Meteorol., 2015, 653473, https://doi.org/10.1155/2015/653473, 2014.

Beven, K.: Environmental Modelling: An Uncertain Future?, CRC Press, UK, 2009.

Bolinger, R. A., Gronewold, A. D., Kompoltowicz, K., and Fry, L. M.: Application of the NMME in the development of a new Regional Seasonal Climate Forecast Tool, B. Am. Meteorol. Soc., 2017, 555–564, 2017.

Box, G. E. P. and Jenkins, G. M.: Time Series Analysis: Forecasting and Control, Holden-Day, 1970 (revised 1976).

Candogan Yossef, N., van Beek, R., Weerts, A., Winsemius, H., and Bierkens, M. F. P.: Skill of a global forecasting system in seasonal ensemble streamflow prediction, Hydrol. Earth Syst. Sci., 21, 4103–4114, https://doi.org/10.5194/hess-21-4103-2017, 2017.

Crochemore, L., Ramos, M.-H., Pappenberger, F., and Perrin, C.: Seasonal streamflow forecasting by conditioning climatology with precipitation indices, Hydrol. Earth Syst. Sci., 21, 1573–1591, https://doi.org/10.5194/hess-21-1573-2017, 2017.

Day, G. N.: Extended streamflow forecasting using NWSRFS, J. Water Resour. Plan. Man., 111, 157–170, 1985.

Drayton, R. S.: Variations in the level of Lake Malawi, Hydrolog. Sci. J., 29, 1–3, 1984.

Greuell, W., Franssen, W. H. P., and Hutjes, R. W. A.: Seasonal streamflow forecasts for Europe – II. Explanation of the skill, Hydrol. Earth Syst. Sci. Discuss., https://doi.org/10.5194/hess-2016-604, in review, 2016.

Gronewold, A. D., Clites, A. H., Hunter, T. S., and Stow, C. A.: An appraisal of the Great Lakes advanced hydrologic prediction system, J. Great Lakes Res., 37, 577–583, 2011.

Huang, C., Newman, A. J., Clark, M. P., Wood, A. W., and Zheng, X.: Evaluation of snow data assimilation using the ensemble Kalman filter for seasonal streamflow prediction in the western United States, Hydrol. Earth Syst. Sci., 21, 635–650, https://doi.org/10.5194/hess-21-635-2017, 2017.

Institute of Hydrology: Review and update of the water balance of Lake Victoria in East Africa, Wallingford, Institute of Hydrology, 23 pp., IH Project Number: T04059g1, ODA Report 93/3, (Unpublished), NERC Open Access Research Archive (NORA), available at: http://nora.nerc.ac.uk/ (last access: 21 February 2017), 1993.

Jolliffe, I. T. and Stephenson, D. B.: Forecast verification, A practitioner's guide in atmospheric science, 2nd Edn., Wiley, Chichester, 2011.

Jury, M. R.: Malawi's Shire River Fluctuations and Climate, J. Hydrometeorol., 15, 2039–2049, 2014.

Jury, M. R. and Gwazantini, M. E.: Climate variability in Malawi, part 2: sensitivity and prediction of lake levels, Int. J. Climatol., 22, 1303–1312, 2002.

Kizza, M., Westerberg, I., Rodhe, A., and Ntalea, N. K.: Estimating areal rainfall over Lake Victoria and its basin using ground-based and satellite data, J. Hydrol., 464–465, 401–411, 2012.

Kizza, M., Guerrero, J.-L., Rodhe, A., Xu, C., and Ntale, H. K.: Modelling catchment inflows into Lake Victoria: regionalisation of the parameters of a conceptual water balance model, Hydrol. Res., 44, 789–808, 2013.

Leedal, D., Weerts, A. H., Smith, P. J., and Beven, K. J.: Application of data-based mechanistic modelling for flood forecasting at multiple locations in the Eden catchment in the National Flood Forecasting System (England and Wales), Hydrol. Earth Syst. Sci., 17, 177–185, https://doi.org/10.5194/hess-17-177-2013, 2013.

Lees, M. J.: Data-based mechanistic modelling and forecasting of hydrological systems, J. Hydroinform., 2, 15–35, 2000.

Lyons, R. P., Kroll, N., and Scholz, C. A.: An energy balance hydrologic model for the Lake Malawi Rift Basin, East Africa, Global Planet. Change, 75, 83–97, 2011.

Manatsa, D., Matarira, C. H., and Mukwada, G.: Relative impacts of ENSO and Indian Ocean dipole/zonal mode on east SADC rainfall, Int. J. Climatol., 31, 558–577, 2011.

Mendoza, P. A., Wood, A. W., Clark, E., Rothwell, E., Clark, M. P., Nijssen, B., Brekke, L. D., and Arnold, J. R.: An intercomparison of approaches for improving operational seasonal streamflow forecasts, Hydrol. Earth Syst. Sci., 21, 3915–3935, https://doi.org/10.5194/hess-21-3915-2017, 2017.

Mulumpwa, M., Jere, W. W. L., and Lazaro, M.: Forecasting of Lake Malawi Water Level Fluctuations using Stochastic Models, International Journal of Environmental and Ecological Engineering, 4, 2017.

Mwangi, E., Wetterhall, F., Dutra, E., Di Giuseppe, F., and Pappenberger, F.: Forecasting droughts in East Africa, Hydrol. Earth Syst. Sci., 18, 611–620, https://doi.org/10.5194/hess-18-611-2014, 2014.

Neuland, H.: Abnormal High Water Levels of Lake Malawi? – An attempt to assess the future behaviour of the lake water levels, Geojournal, 9.4, 323–334, 1984.

Nicholson, S. E. and Selato, J. C.: The influence of La Niña on African rainfall, Int. J. Climatol., 20, 1761–1776, 2000.

Nicholson, S. E., Yin, X., and Ba, M. B.: On the feasibility of using a lake water balance model to infer rainfall: an example from Lake Victoria, Hydrol. Sci. J., 45, 75–95, 2000.

Nicholson, S. E., Klotter, D., and Chavula, G. A.: detailed rainfall climatology for Malawi, Southern Africa, Int. J. Climatol., 34, 315–325, 2014.

Pagano, T. C., Garen, D. C., Perkins, T. R., and Pasteris, P. A.: Daily updating of operational statistical seasonal water supply forecasts for the Western U.S., J. Am. Water Resour. Assoc., 45, 767–778, 2009.

Peel, M. C., Finlayson, B. L., and McMahon, T. A.: Updated world map of the Köppen-Geiger climate classification, Hydrol. Earth Syst. Sci., 11, 1633–1644, https://doi.org/10.5194/hess-11-1633-2007, 2007.

Piper, B. S., Plinston, D. T., and Sutcliffe, J. V.: The water balance of Lake Victoria, Hydrol. Sci. J., 31, 25–37, 1986.

Robertson, D. E. and Wang, Q. J.: A Bayesian Approach to Predictor Selection for Seasonal Streamflow Forecasting, J. Hydrometeorol., 13, 155–171, 2012.

Sadeghi, J., Tych, W., Chotai, A., and Young, P. C.: Multi-state dependent parameter model identification and estimation for nonlinear dynamic systems, Electronic Letters, 46, 1265–1266, 2010.

Saji, N., Goswami, B., Vinayachandran, P., and Yamagata, T.: A dipole mode in the tropical Indian Ocean, Nature, 401, 360–363, 1999.

Schaefli, B. and Gupta, H. V.: Do Nash values have value?, Hydrol. Process., 21, 2075–2080, https://doi.org/10.1002/hyp.6825, 2007.

Sene, K. J.: Theoretical estimates for the influence of Lake Victoria on flows in the upper White Nile, Hydrolog. Sci. J., 45, 125–145, 2000.

Sene, K. J. and Plinston, D. T.: A review and update of the hydrology of Lake Victoria in East Africa, Hydrol. Sci. J., 39, 47–63, 1994.

Sene, K., Piper, B., Wykeham, D., McSweeney, R., Tych, W., and Beven, K.: Long-term variations in the net inflow record for Lake Malawi, Hydrol. Res., 48, 851–866, https://doi.org/10.2166/nh.2016.143, 2017.

Siam, M. S. and Eltahir, E. A. B.: Explaining and forecasting interannual variability in the flow of the Nile River, Hydrol. Earth Syst. Sci., 19, 1181–1192, https://doi.org/10.5194/hess-19-1181-2015, 2015.

Skaugen, T.: Estimating rating curves and response functions from basin geometry and flow velocity, Hydrology: Science & Practice for the 21st Century, Vol. 1, British Hydrological Society, 2004.

Smakhtin, V. U.: Low flow hydrology: a review, J. Hydrol., 240, 147–186, 2001.

Smith, P. J., Beven, K. J., Leedal, D., Weerts, A. H., and Young, P. C.: Testing probabilistic adaptive real-time flood forecasting models, J. Flood Risk Manag., 7, 265–279, 2014.

Sun, X., Xie, L., Semazzi, F., and Liu, B.: Effect of Lake Surface Temperature on the Spatial Distribution and Intensity of the Precipitation over the Lake Victoria Basin, Mon. Weather Rev., 143, 1179–1192, 2015.

Sutcliffe, J. V. and Parks, Y. P.: The Hydrology of the Nile, IAHS Special Publication no. 5, 192 pp., 1999.

Thiery, W., Davin, E. L., Panitz, H.-J., Demuzere, M., Lhermitte, S., and van Lipzig, N.: The Impact of the African Great Lakes on the Regional Climate, J. Climate, 28, 4061–4085, 2015.

Thiery, W., Davin, E. L., Seneviratne, S. I., Bedka, K., Lhermitte, S., and van Lipzig, N. P. M.: Hazardous thunderstorm intensification over Lake Victoria, Nat. Commun., 7, 12786, https://doi.org/10.1038/ncomms12786, 2016.

Thiery, W., Gudmundsson, L., Bedka, K., Semazzi, F., Lhermitte, S., Willems, P., van Lipzig, N., and Seneviratne, S. I.: The Lake Victoria Intense Storm EarlyWarning System (VIEWS), Vol. 19, EGU2017-18256, 2017.

Trenberth, K. E.: Signal versus Noise in the Southern Oscillation, Mon. Weather Rev., 112, 326–332, 1984.

Trenberth, K. E.: The Definition of El Niño, B. Am. Meteorol. Soc., 78, 2771–2777, 1997.

UNDP: Annex 2B, National Water Resources Master Plan, United Nations Development Programme Projects MLW-79-015/MLW-84-003, 1986.

Vitart, F., Ardilouze, C., Bonet, A., Brookshaw, A., Chen, M., Codorean, C., Déqué, M., Ferranti, L., Fucile, E., Fuentes, M., Hendon, H., Hodgson, J., Kang, H., Kumar, A., Lin, H., Liu, G., Liu, X., Malguzzi, P., Mallas, I., Manoussakis, M., Mastrangelo, D., MacLachlan, C., McLean, P., Minami, A., Mladek, R., Nakazawa, T., Najm, S., Nie, Y., Rixen, M., Robertson, A., Ruti, P, Sun, C., Takaya, Y., Tolstykh, M., Venuti, F., Waliser, D., Woolnough, S., Wu, T., Won, D., Xiao, H., Zaripov, R., and Zhang, L.: The Subseasonal to Seasonal (S2s) Prediction Project Database, B. Am. Meteorol. Soc., 2017, 163–173, 2017.

Wilks, D. S.: Statistical methods in the atmospheric sciences, 3rd Edn., Academic Press, Amsterdam, 2011.

Winsemius, H. C., Dutra, E., Engelbrecht, F. A., Archer Van Garderen, E., Wetterhall, F., Pappenberger, F., and Werner, M. G. F.: The potential value of seasonal forecasts in a changing climate in southern Africa, Hydrol. Earth Syst. Sci., 18, 1525–1538, https://doi.org/10.5194/hess-18-1525-2014, 2014.

WMO: Hydrometeorological Survey of the Catchments of Lake Victoria, Kyoga and Mobutu Sese Seko, World Meteorological Organisation report, Geneva, Switzerland, 1982.

WMO: A Water Resources Evaluation of Lake Malawi and the Shire River, World Meteorological Organization, Report No. MLW/77/012, Geneva, 1983.

Wood, A. W. and Schaake, J. C.: Correcting errors in streamflow forecast ensemble mean and spread, J. Hydrometeorol., 9, 132–148, 2008.

Yeh, W.: Reservoir Management and Operations Models: A State-of-the-Art Review, Water Resour. Res., 21, 1797–1818, 1985.

Young, P. C.: Stochastic, dynamic modelling and signal processing: time variable and state dependent parameter estimation, Nonlinear and nonstationary signal processing, Cambridge University Press, Cambridge, 74–114, 2000.

Young, P. C.: Recursive Estimation and Time-Series Analysis: An introduction for the student and practitioner, 2nd Edn., Springer, 2011.

Young, P. C.: Hypothetico-inductive data-based mechanistic modeling of hydrological systems, Water Resour. Res., 49, 915–935, 2013.

Young, P. C. and Beven, K. J.: Data-based mechanistic modelling and the rainfall-flow non-linearity, Environmetrics, 5, 335–363, 1994.

Young, P. C., Taylor, C. J., Tych, W., and Pedregal, D. J.: The Captain Toolbox, Centre for Research on Environmental Systems and Statistics, Lancaster University, UK, available at: www.es.lancs.ac.uk/cres/captain (last access: 18 February 2017), 2007.

11

Dissolved oxygen prediction using a possibility theory based fuzzy neural network

Usman T. Khan and Caterina Valeo

Mechanical Engineering, University of Victoria, P.O. Box 1700, Stn. CSC, Victoria, BC, V8W 2Y2, Canada

Correspondence to: Caterina Valeo (valeo@uvic.ca)

Abstract. A new fuzzy neural network method to predict minimum dissolved oxygen (DO) concentration in a highly urbanised riverine environment (in Calgary, Canada) is proposed. The method uses abiotic factors (non-living, physical and chemical attributes) as inputs to the model, since the physical mechanisms governing DO in the river are largely unknown. A new two-step method to construct fuzzy numbers using observations is proposed. Then an existing fuzzy neural network is modified to account for fuzzy number inputs and also uses possibility theory based intervals to train the network. Results demonstrate that the method is particularly well suited to predicting low DO events in the Bow River. Model performance is compared with a fuzzy neural network with crisp inputs, as well as with a traditional neural network. Model output and a defuzzification technique are used to estimate the risk of low DO so that water resource managers can implement strategies to prevent the occurrence of low DO.

1 Introduction

The City of Calgary is a major economic hub in western Canada. With a rapidly growing population, currently estimated in excess of 1 million, the city is undergoing expansion and urbanisation to accommodate the changes. The Bow River is a relatively small river that flows through the city and provides approximately 60 % of the residents with potable water (Khan and Valeo, 2015, 2016). In addition to this, water is diverted from within the city for irrigation, is used as a source for commercial and recreational fisheries, and is the source of drinking water for communities downstream of Calgary (Robinson et al., 2009; Bow River Basin Council, 2015). This highlights the importance of the Bow River, not just as a source of potable water, but also as a major economic resource.

However, urbanisation has the potential to reduce the health of the Bow River, which is fast approaching its assimilative capacity and is one of the most regulated rivers in Alberta (Bow River Basin Council, 2015). Three wastewater treatment plants (shown in Fig. 1) and numerous stormwater outfalls discharge their effluent into the river and are considered to be a major cause of water quality degradation in the river (He et al., 2015). This highlights some of the major impacts on the Bow River from the surrounding urban area. A number of municipal and provincial programmes are in place to reduce the loading of nutrients and sediments into the river such as the Total Loadings Management Plan and the Bow River Phosphorus Management Plan (Neupane et al., 2014) as well as modelling efforts – namely the Bow River Water Quality Model (Neupane et al., 2014; Golder Associates Ltd., 2004) – to predict the impact of different water management programmes on the water quality.

One of the major concerns is that low dissolved oxygen (DO) concentration has occurred on a number of occasions over the last decade in the Bow River within the city limits. DO is an indicator of overall health of the aquatic ecosystem (Dorfman and Jacoby, 1972; Hall, 1984; Canadian Council of Ministers of the Environment, 1999; Kannel et al., 2007; Khan and Valeo, 2014a, 2015), and low DO – which can be caused by a number of different factors (Pogue and Anderson, 1995; Hauer and Hill, 2007; He et al., 2011; Wen et al., 2013) – can impact various organisms in the water body. While the impact of long-term effects of low DO are largely unknown, acute events can have devastating effects on aquatic ecosystems (Adams et al., 2013). Thus, main-

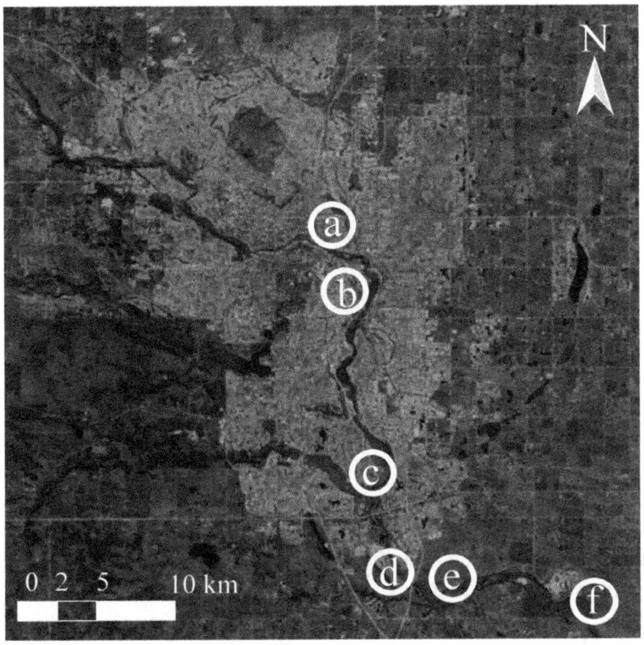

Figure 1. An aerial view of the City of Calgary, Canada, showing the locations of **(a)** flow monitoring site Bow River at Calgary (Water Survey of Canada ID: 05BH004), three wastewater treatment plants at **(b)** Bonnybrook, **(c)** Fish Creek, and **(d)** Pine Creek, and two water quality sampling sites, **(e)** Stier's Ranch and **(f)** Highwood.

taining a suitably high DO concentration, and water quality in general, is of utmost importance to the City of Calgary and downstream stakeholders, particularly as the city is being challenged to meet its water quality targets (Robinson et al., 2009).

A number of recent studies have examined the DO in the Bow River and the factors that impact its concentration. Iwanyshyn et al. (2008) found the diurnal variation in DO and nutrient (nitrate and phosphate) concentration were highly correlated, suggesting that biogeochemical processes (photosynthesis and respiration of aquatic vegetation) had a dominant impact on nutrient concentration rather than wastewater treatment effluent. Further, Robinson et al. (2009) found that the DO fluctuations in the river were primarily due to periphyton rather than macrophyte biogeochemical processes. In both studies, the seasonality of DO, nutrients, and biological concentration, and external factors (e.g. flood events) were demonstrative of the complexity in understanding river processes in an urban area, and that consideration of various inputs, outputs and their interaction is important to fully understand the system. He et al. (2011) found that seasonal variations in DO in the Bow River could be explained by a combination of abiotic factors (such as climatic and hydrometric conditions), as well as by biotic factors. The study found that while photosynthesis and respiration of biota are the main drivers of DO fluctuation, the role of nutrients was ambiguous. Neupane et al. (2014) found that organic materials and nutrients from point and non-point sources influence DO concentration in the river. The likelihood of low DO was highest downstream of wastewater treatment plants and that non-point sources have a significant impact in the open-water season. Using a physically based model, Neupane et al. (2014) predicted low DO concentration more frequently in the future in the Bow River owing to higher phosphorus concentration in the water, as well as climate change impacts.

A major issue of modelling DO in the Bow River is that rapid urbanisation within the watershed has resulted in substantial changes to land-use characteristics, sediment and nutrient loads, and to other factors that govern DO. Major flood events (like those in 2005 and 2013) completely alter the aquatic ecosystem, while new wastewater treatment plants (e.g. the Pine Creek wastewater treatment plant) added in response to the growing population further increases the stress downstream. These types of changes in a watershed increase the complexity of the system: the interaction of numerous factors over a relatively small area and across different temporal scales means that DO trends and variability in urban areas are more difficult to model and evaluating water quality in urban riverine environments is a difficult task (Hall, 1984; Niemczynowicz, 1999).

The implication of this is that the simplistic representation described in conceptual, physically based models is not suitable for complex systems, i.e. where the underlying physical mechanisms behind the factors that govern DO are still not clearly understood, in a rapidly changing urban environment. Physically based models require the parameterisation of several different variables which may be unavailable, expensive and time consuming (Antanasijević et al., 2014; Wen et al., 2013; Khan et al., 2013). In addition to this, the increase in complexity in an urban system proportionally increases the uncertainty in the system. This uncertainty can arise as a result of vaguely known relationships among all the factors that influence DO, in addition to the inherent randomness in the system (Deng et al., 2011). The rapid changes in an urban area render the system *dynamic* as opposed to *stationary*, which is what is typically assumed for many probability based uncertainty quantification methods. Thus, not only is DO prediction difficult, but it is also beset with uncertainty, hindering water resource managers from making objective decisions.

In this research, we propose a new method to predict DO concentration in the Bow River using a data-driven approach, as opposed to a physically based method that uses *possibility* theory and fuzzy numbers to represent the uncertainty rather than the more commonly used probability theory. Data-driven models are a class of numerical models based on generalised relationships, links or connections between input and output data sets (Solomatine and Ostfeld, 2008). These models can characterise a system with limited assumptions and are useful in solving practical problems, especially when there is lack of understanding of the underlying physical process, the time series are of insufficient length, or when existing models are inadequate (Solomatine et al., 2008; Napolitano et al., 2011).

1.1 Fuzzy numbers and data-driven modelling

Possibility theory is an information theory that is an extension of fuzzy set theory for representing uncertain, vague or imprecise information (Zadeh, 1978). Fuzzy *numbers* are an extension of fuzzy set theory, and express an uncertain or imprecise *quantity*. These types of numbers are particularly useful for dealing with uncertainties when data are limited or imprecise (Bárdossy et al., 1990; Guyonnet et al., 2003; Huang et al., 2010; Zhang and Achari, 2010) – in other words when epistemic uncertainty exists. This type of uncertainty is in contrast to aleatory uncertainty that is typically handled using probability theory. Possibility theory and fuzzy numbers are thus useful when a probabilistic representation of parameters may not be possible, since the exact values of parameters may be unknown or only partial information available (Zhang, 2009). Thus, the choice of using a data-driven approach in combination with possibility theory lends itself well to the constraints posed by the problem in the Bow River: the difficulty in correctly defining a physically based model for a complex urban system and the use of possibility theory to model the uncertainty in the system when probability theory based methods may be inadequate.

Data-driven models, such as neural networks, regression based techniques, fuzzy rule based systems, and genetic programming, have seen widespread use in hydrology, including DO prediction in rivers (Shrestha and Solomatine, 2008; Solomatine et al., 2008; Elshorbagy et al., 2010). Wen et al. (2013) used artificial neural networks (ANNs) to predict DO in a river in China using ion concentration as the predictors. Antanasijević et al. (2014) used ANNs to predict DO in a river in Serbia using a Monte Carlo approach to quantify the uncertainty in model predictions and temperature as a predictor. Chang et al. (2015) also used ANNs coupled with hydrological factors (such as precipitation and discharge) to predict DO in a river in Taiwan. Singh et al. (2009) used water quality parameters to predict DO and BOD in a river in India. Other studies (e.g. Heddam, 2014; Ay and Kisi, 2011) have used regression to predict DO in rivers using water temperature or electrical conductivity, amongst others, as inputs. In general, these studies have demonstrated that there is a need and demand for less complex DO models, have led to an increase in the popularity of data-driven models (Antanasijević et al., 2014), and shown that the performance of these types of models is suitable. Recent research into predicting DO concentration in the Bow River in Calgary using abiotic factors (these are non-living, physical and chemical attributes) as inputs have shown promising results (He et al., 2011; Khan et al., 2013; Khan and Valeo, 2015). The advantage of using readily available data (i.e. the abiotic inputs) in these studies is that if a suitable relationship between these factors and DO can be found, changing the factors (e.g. increasing the discharge rate downstream of a treatment plant) can potentially reduce the risk of low DO.

While fuzzy set theory based applications, particularly applications using fuzzy *logic* in neural networks, have been widely used in many fields including hydrology (Bárdossy et al., 2006; Abrahart et al., 2012), the use of fuzzy *numbers* and possibility theory based applications has been limited in comparison (Bárdossy et al., 2006; Jacquin, 2010). Some examples include maps of soil hydrological properties (Martin-Clouaire et al., 2000), remotely sensed soil moisture data (Verhoest et al., 2007), climate modelling (Mujumdar and Ghosh, 2008), subsurface contaminant transport (Zhang et al., 2009), and streamflow forecasting (Alvisi and Franchini, 2011). Khan et al. (2013) and Khan and Valeo (2015) have introduced a fuzzy number based *regression* technique to model daily DO in the Bow River using abiotic factors with promising results. Similarly, Khan and Valeo (2014a) used an autoregressive time series based approach combined with fuzzy numbers to predict DO in the Bow River. In these studies, the use of fuzzy numbers meant that the uncertainty in the system could be quantified and propagated through the model. However, due to the highly non-linear nature of DO modelling, the use of an ANN based method is of interest since these types of models are effective for modelling complex, non-linear relationships without the explicit understanding of the physical phenomenon governing the system (Alvisi and Franchini, 2011; Antanasijević et al., 2014). A fuzzy neural network method proposed by Alvisi and Franchini (2011) for streamflow prediction that uses fuzzy weights and biases in the network is further refined in this research for predicting DO concentration.

1.2 Objectives

Given the importance of DO concentration as an indicator of overall aquatic ecosystem health, there is a need to accurately model and predict DO in urban riverine environments, like that in Calgary, Canada. In this research a new data-driven method is proposed that attempts to address the issues that plague numerical modelling of DO concentration in the Bow River. The FNN method proposed by Alvisi and Franchini (2011) is adapted and extended in two critical ways. The existing method uses crisp (i.e. non-fuzzy) inputs and outputs to train the network, producing a set of fuzzy number weights and biases, and fuzzy outputs. The method is adapted to be able to handle fuzzy number inputs to produce fuzzy weights and biases, and fuzzy outputs. The advantage is that the uncertainties in the input observations are also captured within the model structure. To do this, a new method of creating fuzzy numbers from observations is presented based on a probability–possibility transformation. Second, the existing training algorithm is based on capturing a predetermined set of observations (e.g. 100, 95 or 90 %) within the fuzzy outputs. The selection of the predetermined set of observations in the original study was an arbitrary selection. A new method that exploits the relationship between possibility theory and probability theory is defined to create a more objective method of training the FNN. A consequence of this is that the resulting fuzzy number outputs from the model can then be directly used for risk analysis, specifically to quantify the risk of low DO concentration. This information is extremely valuable for managing water resources in the face of uncertainty. The impact of using fuzzy inputs and the new training criteria is evaluated by comparing results to the existing FNN method (by Alvisi and Franchini, 2011) as well as with a traditional, crisp ANN.

Following previous research for this river, two abiotic inputs (daily mean water temperature, T, and daily mean flow rate, Q) will be used to predict daily minimum DO. An advantage of using these factors is that they are routinely collected by the City of Calgary, and thus a large data set is available. Also, their use in previous studies has shown that they are good predictors of daily DO concentration in this river basin (He et al., 2011; Khan et al., 2013; Khan and Valeo, 2015). The following sections outline the background of fuzzy numbers and existing probability–possibility transformations. This is followed by the development of the new method to create fuzzy numbers from observations. Then, the new FNN method using fuzzy inputs is developed mathematically using new criteria for training, also based on possibility theory. Lastly, a method to measure the risk of low DO is described.

2 Methods

2.1 Data collection

The Bow River is 645 km long and averages a 0.4 % slope over its length (Bow River Basin Council, 2015) from its headwaters at Bow Lake in the Rocky Mountains to its confluence with the Oldman River in southern Alberta, Canada (Robinson et al., 2009; Environment Canada, 2015). The river is supplied by snowmelt from the Rocky Mountains, rainfall and discharge from groundwater. The City of Calgary is located within the Bow River Basin and the river has an average annual discharge of 90 $m^3\,s^{-1}$, and an average width and depth of 100 and 1.5 m, respectively (Khan and Valeo, 2014b, 2016).

The City of Calgary routinely samples a variety of water quality parameters along the Bow River to measure the impacts of urbanisation, particularly from three wastewater treatment plants and numerous stormwater runoff outfalls that discharge into the river. DO concentration measured upstream of the city is generally high throughout the year, with little diurnal variation (He et al., 2011; Khan et al., 2013; Khan and Valeo, 2015). The DO concentration downstream of the city is lower and experiences much higher diurnal fluctuation. The three wastewater treatment plants are located upstream of this monitoring site, and are thought to be responsible, along with other impacts of urbanisation, for the degradation of water quality (He et al., 2015).

Table 1. A summary of low DO events in the Bow River between 2004 and 2012 and the corresponding minimum acceptable DO concentration guidelines.

Year	DO < 5 $mg\,L^{-1}$,[a]	DO < 6.5 $mg\,L^{-1}$,[b]	DO < 9.5 $mg\,L^{-1}$,[c]	Total number of samples
2004	25	41	107	135
2005	1	26	133	208
2006	25	70	164	209
2007	0	27	182	211
2008	0	5	130	163
2009	0	15	85	96
2010	0	0	180	207
2011	0	0	122	204
2012	0	0	76	206
Total	51	184	1179	1639

[a] For the protection of aquatic life for 1 day (AENV, 1997); [b] for the protection of aquatic life in cold freshwater for other life (i.e. not early) stages (CCME, 1999); [c] for the protection of aquatic life in cold freshwater for early life stages (CCME, 1999).

For this research, 9 years of DO concentration data were collected from one of the downstream stations from 2004 to 2012. The monitoring station was located at Pine Creek and sampled water quality data every 30 min (from 2004 to 2005) and every 15 min (from 2006 to 2007). The station was then moved to Stier's Ranch and sampled data every hour (in 2008) and every 15 min (2009 to 2011). The monitoring site was moved further downstream to its current location (at Highwood) in 2012 where it samples every 15 min. During this period a number of low DO events have been observed in the river and are summarised below in Table 1 corresponding to different water quality guidelines.

Note that even though daily minimum DO was observed to be below 5 $mg\,L^{-1}$ on several occasions in 2004 and 2006 (in Table 1), the minimum DO was below 9.5 $mg\,L^{-1}$ only 107 and 164 days, respectively, for those 2 years. In contrast, in 2007 and 2010, no observations below 5 $mg\,L^{-1}$ were seen, yet 182 and 180 days, respectively, below the 9.5 $mg\,L^{-1}$ guideline were seen for those years. The total number of days below 9.5 $mg\,L^{-1}$ constitute approximately 90 % of all observations for those years. This shows that despite no DO events below 5 $mg\,L^{-1}$, generally speaking minimum DO on a daily basis was quite low in these 2 years. The implication of this is that only using one guideline for DO might not be a good indicator of overall aquatic ecosystem health.

A YSI sonde is used to monitor DO and T, and the sonde is not accurate in freezing water; thus, only data from the ice-free period were considered, which is approximately from April to October for most years (YSI Inc., 2015). Since low DO events usually occur in the summer (corresponding to high water temperature and lower discharge), the ice-free period data set contains the dates that are of interest for low DO modelling.

Daily mean flow rate, Q, was collected from Water Survey of Canada site "Bow River at Calgary" (ID: 05BH004) for the same period. These data are collected hourly throughout the year; thus, data where considerable shift corrections were applied (usually due to ice conditions) were removed from the analysis. The mean annual water temperature ranged between 9.23 and 13.2 °C, the annual mean flow rate was between 75 and 146 $m^3 s^{-1}$, and the mean annual minimum daily DO was between 6.89 and 9.54 $mg L^{-1}$, for the selected period.

2.2 Probability–possibility transformations

Fuzzy sets were proposed by Zadeh (1965) in order to express imprecision in complex systems, and can be described as a generalisation of classical set theory (Khan and Valeo, 2015). In classical set theory, an element x either belongs or does not belong to a set A. In contrast, using fuzzy set theory, the elements x have a degree of membership, μ, between 0 and 1 in the fuzzy set A. If μ equals 0, then x does not belong in A, and $\mu = 1$ means that it completely belongs in A, while a value $\mu = 0.5$ means that it is only a partial member of A.

Fuzzy numbers express uncertain or imprecise *quantities*, and represent the set of all possible values that define a quantity rather than a single value. A fuzzy number is defined as a specific type of fuzzy set: a *normal* and *convex* fuzzy set. Normal implies that there is at least one element in the fuzzy set with a membership level equal to 1, while convex means that the membership function increases monotonically from the lower support (i.e. $\mu = 0^L$) to the modal element (i.e. the element(s) with $\mu = 1$) and then monotonically decreases to the upper support (i.e. $\mu = 0^R$) (Kaufmann and Gupta, 1985).

Traditional representation of a fuzzy numbers has been using symmetrical, linear membership functions, typically denoted as triangular fuzzy numbers. The reason for selecting this type of membership function has to do with its simplicity: given that a fuzzy number must, by definition, be convex and normal, a minimum of three elements are needed to define a fuzzy number (two elements at $\mu = 0$ and one element at $\mu = 1$). For example, if the most credible value for DO concentration is 10 $mg L^{-1}$ ($\mu = 1$), with a support about the modal value between ($\mu = 0^L$) and 12 $mg L^{-1}$ ($\mu = 0^R$). This implies that the simplest membership function is triangular, though not necessarily symmetrical. Also, as we demonstrate below, in some probability–possibility frameworks, a triangular membership function corresponds to a uniform probability distribution – the least specific distribution in that any value is equally probable, and hence represents the most uncertainty (Dubois and Prade, 2015; Dubois et al., 2004).

However, recent research (Khan et al., 2013; Khan and Valeo, 2014a, b, 2015, 2016) has shown that such a simplistic representation may not be appropriate for hydrological data, which are often skewed and non-linear. This issue is further highlighted if the probability–possibility framework mentioned above is used: it implies that for a triangular membership function, the fuzzy number bounded by the support [8 12] $mg L^{-1}$ has a uniform probability distribution bounded between 8 and 12 $mg L^{-1}$ with a mean value of 10 $mg L^{-1}$, suggesting that values between the support are equally likely to occur. It not difficult to see that this an over-simplification of hydrological data. In many cases enough information (i.e. from observations) is available to define the membership function with more specificity, and this information should be used to define the membership function.

Multiple frameworks exist to transform a probability distribution to a possibility distribution and vice versa; a comparison of different conceptual approaches is provided in Klir and Parvais (1992), Oussalah (2000), Jaquin (2010), Mauris (2013) and Dubois and Prade (2015). However, a major issue of implementing fuzzy number based methods in hydrology is that there is no consistent, transparent and objective method to convert observations (e.g. time series data) into fuzzy numbers, or generally speaking to construct the membership function associated with fuzzy values (Abrahart et al., 2012; Dubois and Prade, 1993; Civanlar and Trussel, 1986).

A popular method (Dubois et al., 1993, 2004) converts a probability distribution to a possibility distribution by relating the area under a probability density function to the membership level (Zhang, 2009). In this framework, the possibility is viewed as the upper envelope of the family of probability measures (Jacquin, 2010; Ferrero et al., 2013; Betrie et al., 2014). There are two important considerations for this transformation, first it guarantees that something must be possible before it is probable; hence, the degree of possibility cannot be less than the degree or probability – this is known as the consistency principle (Zadeh, 1965). Second is order preservation, which means if the possibility of x_i is greater than the possibility of x_j then the probability of x_i must be greater than the probability of x_j (Dubois et al., 2004). For a discrete system, this can be represented as follows:

$$\text{if } p(x_1) > p(x_2) > \ldots > p(x_n),$$

then the possibility distribution of $x(\pi(x))$ follows the same order, that is,

$$\pi(x_1) > \pi(x_2) > \ldots > \pi(x_n).$$

The transformation is given by

For $p(x_1) > p(x_2) > \ldots > p(x_n)$:

$$\pi(x_1) = 1$$

$$f(x) = \begin{cases} \sum_{j=i}^{n} p_j, & \text{if } p_{i-1} > p_i \\ \pi(x_{i-1}), & \text{else} \end{cases} \quad (1)$$

where the x_i are elements of a fuzzy number A, $\pi(x_i)$ is the possibility of element x_i, and $p(x_i)$ is the probability of element x_i. The concept of this transformation may be more illustrative when viewed in the continuous case: for any interval $[a, b]$, the membership level μ (where $\pi(a) = \pi(b) = \mu$) is equal to the sum of the areas under the probability density function curve between $(-\infty, a)$ and (b, ∞) (Zhang et al., 2009). It is important to highlight that this particular transformation has an inverse transformation associated with, where a probability distribution can be estimated from the possibility distribution.

However, a major drawback of this transformation is that it theoretically requires a full description of the probability density function, or in the finite case, the probability associated with each element of the fuzzy number, the probability mass function. For many hydrological applications this might not be possible because the hourly time series data may not adequately fit the mould of a known class of probability density functions, or one distribution amongst many alternatives may have to be selected based on best-fit. This best-fit function may not be universal, e.g. data from one 24 h period may be best described by one class or family of probability density function, while the next day by a completely different class of density function. This means working with multiple classes of distribution functions for one application, which can be cumbersome. Also, given that each day may only have 24 data points (or fewer on days with missed samples), it is difficult to select one particular function.

In previous research by Khan and Valeo (2015), a new approach to create a fuzzy number based on observations was developed. This process used a histogram based approach to estimate the probability mass function of the observations, and then Eq. (1) was used to estimate the membership function of the fuzzy number. To create the histogram, the bin size was selected based on the extrema observations for a given day and the number of the observations. A linear interpolation scheme was then used to calculate the fuzzy number at five predefined membership levels. This method has a few shortcomings, namely that the bin-size selection was arbitrarily selected based on the magnitude and number of observations, which does not necessarily result in the optimum bin size. This lack of optimality means that the resulting histogram may either be too smooth so as not to capture the variability between membership levels, or too rough and uneven so that the underlying shape of the membership function is difficult to discern. This is a common issue with histogram selection in many applications (Shimazaki and Shinomoto, 2007). Secondly, the aforementioned transformation used by Khan and Valeo (2015) only allows one element to have $\mu = 1$ when $p(x)$ is maximum. However, there are a number of cases (e.g. bimodal distributions, or arrays when all elements are equal) where multiple elements have joint-equal maximum $p(x)$, and hence multiple elements with $\mu = 1$. This means that all elements *within* the α-cut interval $[a\ b]_{\mu=1}$ (where a and b are the minimum and maximum elements with $\mu = 1$) must by definition also have a membership level equal to 1. Thus, a method is necessary to be flexible enough to accommodate these types of issues.

In this research, a two-step procedure is proposed to create fuzzy numbers on the inputs (i.e. Q and T) using hourly (or sub-hourly) observations. First, a bin-size optimisation method is used (an extension of an algorithm proposed by Shimazaki and Shinomoto, 2007) to create histograms to represent the estimate of discretised probability density functions of the observations. This estimate of the probability distribution is then transformed to the membership function of the fuzzy number using a new numerical procedure and the transformation principles described in Eq. (1). This updated method requires no assumptions regarding the distribution of the underlying data or selection of an arbitrary bin size, has the flexibility to create different shapes of fuzzy numbers depending on the distribution of the underlying data, and allows multiple elements to have equal $\mu = 1$. The proposed algorithm is described in the proceeding section.

A new algorithm to create fuzzy numbers

Shimazaki and Shinomoto (2007) proposed a method to find the optimum bin size of a histogram when the underlying distribution of the data is unknown. The basic premise of the method is that the optimum bin size (D_{opt}) is one that minimises the error between the theoretical (but unknown) probability density and the histogram generated using the D_{opt}. The error metric used by Shimazaki and Shinomoto (2007) is the mean integrated squared error (E_{MISE}) which is frequently used for density estimation problems. It is defined as

$$E_{MISE} = \frac{1}{P} \int_0^P E\left[f_n(t) - f(t)\right]^2 dt, \quad (2)$$

where $f(t)$ is the unknown density function, $f_n(t)$ is the histogram estimate of the density function, t denotes time and P is the observation period, and $E[\bullet]$ is the expectation. In practice, E_{MISE} cannot be directly calculated since the underlying distribution is unknown, and thus an estimate of the E_{MISE} is used in its place (see C_D below). Thus, $f_n(t)$ can be found without any assumptions of the type of distribution (e.g. class, unimodality); the only assumption is that the number of events (i.e. the counts k_i) in the ith bin of the histogram follow a Poisson point process. This means that the events in two disjoint bins (e.g. the ith and $i + 1$th bins) are independent, and that mean (k) and variance (v) of the k_i in each bin are equal, due to the assumption of a Poisson process (Shimazaki and Shinomoto, 2007).

Using this property, the optimum bin size can be found as follows. Let $\boldsymbol{X}$ be the input data vector for the observation period (P), e.g. a [24 × 1] vector corresponding to hourly samples for a given day. The elements in $\boldsymbol{X}$ are binned into N bins of equal bin size D. The number of events k_i in each ith bin are then counted and the mean (k) and variance (v) of the k_i are calculated as follows:

$$k = \frac{1}{N}\sum_{i=1}^{N} k_i, \tag{3}$$

$$v = \frac{1}{N}\sum_{i=1}^{N} (k_i - k)^2. \tag{4}$$

The k and v are then used to compute the *cost function* C_D, which is defined as

$$C_\mathrm{D} = \frac{2k - v}{D^2}. \tag{5}$$

This cost function is a variant of the original E_MISE listed in Eq. (2) and is derived by removing the terms from E_MISE that are independent of the bin size D, and by replacing the unobservable quantities (i.e. $E[f(t)]$) with their unbiased estimators (details of this derivation can be found in the original paper by Shimazaki and Shinomoto, 2007). The objective then is to search for D_opt: the value of D that minimises C_D. To do this two systematic modification are made: first, C_D is recalculated at different *partitioning positions*, and secondly, the entire process is repeated for different values of N and D, until a "reliable" estimate of minimum C_D and thus D_opt is found. Using different partitioning positions means that the variability in k_i resulting from the position of the bin (rather than the size of the bin) can be quantified. Repeating the analysis at different N and D accounts for the variability due to different bin sizes. Both these techniques are ways of accounting for the uncertainty associated with estimating the histogram.

Partitioning positions are defined as the first and last points that define a bin. The most common way of defining a partitioning position is to centre it on some value a; for example, the bin defined at $[a - D/2, a + D/2]$ is centred on a and has a bin size D. Variations of this partitioning position can be found by using a moving-window technique, where the bin size D is kept constant, but the first and last points are perturbed by a small value δ: $[a - D/2 + \delta,\ a + D/2 + \delta]$, where δ ranges incrementally between 0 and D. Using these different values of δ whilst keeping D constant will result in different values of k_i and hence unique values of C_D. Thus, for a single value of D, multiple values of C_D are possible.

For this research this bin-size optimisation algorithm is implemented to determine the optimum histogram for the two input variables, Q and T. The array of daily data $\boldsymbol{X}$ (at hourly or higher frequency; see Sect. 2.1 for details regarding the sampling frequency of both inputs) for each variable was collected for the 9-year period. The bin size was calculated for each day as follows:

$$D = \frac{x_\mathrm{max} - x_\mathrm{min}}{N}, \tag{6}$$

where x_max and x_min are the maximum and minimum sampled values for $\boldsymbol{X}$, respectively, and N is the number of bins. As described above, a number of different D were considered to find the optimum C_D. This was done by selecting a number of different values of N, ranging from N_min to N_max. The minimum value N_min was set equal to 3 for all days; this is the necessary number of bins to define a fuzzy number (two elements for $\mu = 0$, and one element for $\mu = 1$). The highest value, N_max, was calculated as

$$N = \frac{x_\mathrm{max} - x_\mathrm{min}}{2r}, \tag{7}$$

where $2r$ is the measurement resolution of the device used to measure either Q or T, set at twice the accuracy (r) of the device. The rationale for this decision is that as N increases, D necessarily decreases (as per Eq. 6). However, D cannot be less than the measurement resolution; this constraint (i.e. $N \leq N_\mathrm{max}$) ensures that the optimum bin size is never less than what the measurement devices can physically measure. For this research, the accuracy for T is listed as $\pm 0.1\,^\circ\mathrm{C}$, and thus, the resolution ($2r$) is $0.2\,^\circ\mathrm{C}$ (YSI Inc., 2015). For Q all measurements below $99\,\mathrm{m^3\,s^{-1}}$ have an accuracy of $\pm 0.1\,\mathrm{m^3\,s^{-1}}$ and thus, a resolution of $0.2\,\mathrm{m^3\,s^{-1}}$, while measurements above $99\,\mathrm{m^3\,s^{-1}}$ have an accuracy of $\pm 1\,\mathrm{m^3\,s^{-1}}$, and thus a resolution of $2\,\mathrm{m^3\,s^{-1}}$. This is based on the fact that all data provided by the Water Survey of Canada is accurate to three significant figures (Lane, 1999). Note that for the case where x_min equals x_max (i.e. no variance in the daily observed data) then $D = 2r$, which means that the only uncertainty considered is due to the measurement.

Once the N_max is determined, the bin size D was calculated for each N between N_min and N_max. Then, starting at the largest D (i.e. $D = (x_\mathrm{max} - x_\mathrm{min})/N_\mathrm{min}$), the cost function C_D is calculated at the *first* partitioning positing, where the first bin is centred at x_min, $[(x_\mathrm{min} - D/2)\ (x_\mathrm{min} + D/2)]$, and the Nth bin is centred on x_max, $[(x_\mathrm{max} - D/2),\ (x_\mathrm{max} + D/2)]$. Then, C_D is calculated at the *next* partitioning positing, where the first bin is $[(x_\mathrm{min} - D/2 + \delta)\ (x_\mathrm{min} + D/2 + \delta)]$, and the Nth bin is $[(x_\mathrm{max} - D/2 + \delta),\ (x_\mathrm{max} + D/2 + \delta)]$. The value of δ ranged between 0 and D at $(D/100)$ intervals. Thus, for this value of D, 100 values of C_D were calculated since 100 different partitioning positions were used. The mean value of these C_D was used to define the final cost-function value for the given D.

This process is then repeated for the next N between N_min and N_max, using the corresponding D at 100 different partitioning positions, and so on until the smallest D (at N_max). This results in $[N_\mathrm{max} - N_\mathrm{min}]$ values of mean C_D: the value of D corresponding to the minimum value of C_D is considered to be the optimum bin size D_{opt}. This D_{opt} is then used to construct the optimum histogram of each daily observation. This histogram can be used to calculate a discretised probability density function ($p(x)$), where for each x (an element of $\boldsymbol{X}$), the $p(x)$ is calculated by dividing the number of events in each bin by the total number of elements in $\boldsymbol{X}$. x and $p(x)$ can then be used to calculate the possibility distribution using the transformation described in Eq. (1).

First, the $p(x)$ are ranked from highest to lowest, and the x corresponding to the highest $p(x)$ has a membership level of 1. Then the $\pi(x)$ values for the remaining x are calculated using Eq. (1). For cases where multiple elements have equal $p(x)$, the highest $\pi(x)$ is assigned to each x. For example, if $p(x_i) = p(x_j)$, and $x_i > x_j$, then $\pi(x_j) = \pi(x_i)$. This means that in some cases, for each calculated membership level, $\pi(x)$, there exists an α-cut interval $[a, b]_{\mu=\pi(x)}$ where all the elements between a and b have equal $p(x)$ and hence equal $\pi(x)$. By definition of α-cut intervals, all values of x within the interval $[a, b]$ have *at least* a possibility of $\pi(x)$. A special case of this occurs when multiple x have joint-equal maximum $p(x)$, meaning that multiple elements have a membership level of $\mu = 1$. Thus, an α-cut interval is created for the $\mu = 1$ case, creating a trapezoidal membership function, where the modal value of the fuzzy number is defined by an interval rather than a single element.

Once all the $\pi(x)$ are calculated for each element x in $\boldsymbol{X}$, a *discretised* empirical membership function of the fuzzy number $\boldsymbol{X}$ can be constructed using the calculated α-cut intervals. That is, the fuzzy number is defined by a number of intervals at different membership levels. The upper and lower limits of the intervals at higher membership levels define the extent of the limits of the intervals at lower membership levels. This way the constructed fuzzy numbers maintain convexity (similar to a procedure used by Alvisi and Franchini, 2011), where the widest intervals have the lowest membership level. For example, the interval at $\mu = 0.2$ will contain the interval $\mu = 0.4$, and this interval will contain the interval at $\mu = 0.8$.

In creating this discretised empirical membership function this way (rather than assuming a shape of the function) means that this function best reflects the possibility distribution of the observed data. However, it also means that all fuzzy numbers created using this method are not guaranteed to be defined at the same $\pi(x)$, nor have an equal number of $\pi(x)$ intervals used to define the fuzzy number. Thus, direct fuzzy arithmetic between multiple fuzzy numbers using the extension principle is not possible since it requires each fuzzy number to be defined at the same α-cut intervals (Kaufmann and Gupta, 1985). Thus, linear interpolation is used to define each fuzzy number at a pre-set α-cut interval using the empirical $\pi(x)$ calculated using the transformation. To select the pre-set α-cut intervals it is illustrative to see the impact of selecting two extreme cases: (i) if only two levels are selected (specifically $\mu = 0$ and 1) the constructed fuzzy number will reduce to a triangular fuzzy number. As discussed above there are important implications of using triangular membership functions that make it undesirable for hydrological data; (ii) if a large number of intervals (e.g. 100 intervals between $\mu = 0$ and 1) are selected, there is a risk that the number of pre-set intervals is much larger than the empirical $\pi(x)$, which means not enough data (empirical α-cut levels intervals) to conduct interpolation, leading to equal interpolated values at multiple α-cut levels. For this research, results (discussed in the following section) of the bin-size optimisation showed that most daily observations for T and Q resulted in 2 to 10 unique $p(x)$ values. Based on this, six pre-set α-cut intervals were selected: 0, 0.2, 0.4, 0.6, 0.8 and 1. The empirical $\pi(x)$ can then be converted to a standardised function at pre-defined membership levels using linear interpolation.

2.3 Fuzzy neural networks

2.3.1 Background on artificial neural networks

Artificial neural networks (ANNs) are a type of data-driven model that are defined as a massively parallel distributed information processing system (Elshorbagy et al., 2010; Wen et al., 2013). ANN models have been widely used in hydrology when the complexity of the physical systems is high owing partially to an incomplete understanding of the underlying process and the lack of availability of necessary data (He et al., 2011; Kasiviswanathan et al., 2013). Further, ANNs arguably require fewer data and do not require an explicit mathematical description of the underlying physical process (Antanasijević et al., 2014), making them a simpler and practical alternative to traditional modelling techniques.

A multilayer perceptron (MLP) is a type of feedforward ANN and is one of the most commonly used in hydrology (Maier et al., 2010). A trained MLP network can be used as a universal approximator with only one hidden layer (Hornik et al., 1989). This means that models are relatively simple to develop and theoretically have the capacity to approximate any linear or non-linear mapping (ASCE Task Committee on Application of Artificial Neural Networks in Hydrology, 2000; Elshorbagy et al., 2010; Napolitano et al., 2011; Kasiviswanathan et al., 2013). Further, the popularity of MLPs has meant that subsequent research has continued to use MLPs (He and Valeo, 2009; Napolitano et al., 2011) and thus form a reference for the basis of comparing ANN performance (Alvisi and Franchini, 2011).

In the simplest case, an MLP consists of an input layer, a hidden layer, and an output layer as shown in Fig. 2. Each layer consists of a number of neurons (or nodes) that each receive a signal, and on the basis of the strength of the signal, emit an output. Thus, the final output layer is the synthesis and transformation of all the input signals from both the input and the hidden layer (He and Valeo, 2009).

The number of neurons in the input (n_I) and output (n_O) layers corresponds to the number of variables used as the

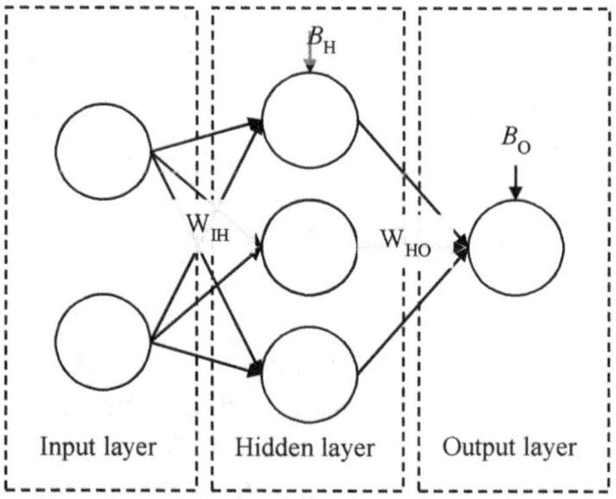

Figure 2. An example of a three-layer multilayer perceptron feed-forward ANN, with two input neurons, the hidden-layer neurons, and one output neuron. WIH are the weights between the input and hidden layers, WHO are the weights between the hidden and output layers, BH are the biases in the hidden layer, and BO is the bias in the output layer.

input and the output, respectively, and the number of neurons in the hidden layer (n_H) are selected based on the relative complexity of the system (Elshorbagy et al., 2010). A typical MLP is expressed mathematically as follows:

$$y_i = f_{HID}\left(\mathbf{W}_{IH}\boldsymbol{x}_i + \boldsymbol{B}_H\right), \tag{8}$$

$$z_i = f_{OUT}\left(\mathbf{W}_{HO}\boldsymbol{y}_i + \boldsymbol{B}_O\right), \tag{9}$$

where $\boldsymbol{x}_i$ is the ith observation (an $n_I \times 1$ vector) from a total of n observations, $\mathbf{W}_{IH}$ is an $n_H \times n_I$ matrix of weights between the input and hidden layers, $\boldsymbol{B}_H$ is a vector ($n_H \times 1$) of biases in the hidden layer, and $\boldsymbol{y}_i$ is the ith output (an $n_H \times 1$ vector) of the input signal through the hidden-layer transfer function, f_{HID}. Similarly, $\mathbf{W}_{HO}$ is an $n_O \times n_H$ matrix of weights between the hidden and output layers, $\boldsymbol{B}_O$ is an $n_O \times 1$ vector of biases in the output layer, and f_{OUT} the final transfer function to generate the ith modelled output z_i (an $n_O \times 1$ vector).

The values of all the weights and biases in the MLP are calculated by training the network by minimising the error – typically mean squared error (E_{MSE}) (He and Valeo, 2009) – between the modelled output and the target data (i.e. observations). The Levenberg–Marquardt algorithm (LMA) is one of the most common training algorithms (Alvisi et al., 2006). In LMA, the error between the output and target is back-propagated through the model using a gradient method where the weights and biases are adjusted in the direction of maximum error reduction. The LMA is well suited for problems that have a relatively small number of neurons. To counteract potential over-fitting issues, an early-stopping procedure is used (Alvisi et al., 2006; Maier et al., 2010), which is a form of regularisation where the data are split into three subsets (for training, validation and testing). The training is terminated when the error on the validation subset increases from the previous iteration.

Most ANNs have a deterministic structure without a quantification of the uncertainty corresponding to the predictions (Alvisi and Franchini, 2012; Kasiviswanathan and Sudheer, 2013). This means that users of these models may have excessive confidence in the forecasted values and misinterpret the applicability of the results (Alvisi and Franchini, 2011). This lack of uncertainty quantification is one reason for the limited appeal of ANN among water resource managers (Abrahart et al., 2012; Maier et al., 2010). Without this characterisation, the results produced by these models have limited value (Kasiviswanathan and Sudheer, 2013).

In this research, two methods are proposed to quantify the uncertainty in MLP modelling to predict DO in the Bow River. First, the uncertainty in the input data (daily mean water temperature and daily mean flow rate) is represented through the use of fuzzy numbers. These fuzzy numbers are created using the probability–possibility transformation discussed in the previous section. Second, the *total uncertainty* (as defined by Alvisi and Franchini, 2011) in the weights and biases of an MLP are quantified using a new possibility theory based FNN. The total uncertainty represents the overall uncertainty in the modelling process, and not of the individual components (e.g. randomness in observed data). The following section describes the proposed FNN method.

2.3.2 FNN with fuzzy inputs and possibility based intervals

Alvisi and Franchini (2011) proposed a method to create a FNN, where the weights and biases, and by extension the output, of the neural network are fuzzy numbers rather than crisp (non-fuzzy) numbers. These fuzzy numbers quantify the total uncertainty of the calibrated parameters. Most fuzzy set theory based applications of ANN in hydrology have used fuzzy logic, e.g. the widely used Adaptive Neuro-Fuzzy Inference System, where automated IF-THEN rules are used to create *crisp* outputs (Abrahart et al., 2010; Alvisi and Franchini, 2011). Thus, the advantage of fuzzy outputs (as developed by Alvisi and Franchini, 2011) is that it provides the uncertainty of the predictions in addition to the uncertainty of the parameters. This uncertainty quantification can be used to by end users to assess the value of the model output.

In their FNN, the MLP model presented in Eqs. (5) and (6) is modified to predict an interval rather than a single value for the weights, biases and output, corresponding to an α-cut interval (at a defined membership level μ). This is repeated for several α-cut levels, thus building a discretised fuzzy number at a number of membership levels. This is done by using a stepwise, constrained optimisation approach:

$$\left[y_i^L y_i^U\right] = f_{HID}\left(\left[\mathbf{W}_{IH}^L \mathbf{W}_{IH}^U\right]\boldsymbol{x}_i + \left[\boldsymbol{B}_H^L \boldsymbol{B}_H^U\right]\right), \tag{10}$$

$$\left[z_i^L z_i^U\right] = f_{OUT}\left(\left[\mathbf{W}_{HO}^L \mathbf{W}_{HO}^U\right] \times \left[y_i^L y_i^U\right] + \left[\boldsymbol{B}_O^L \boldsymbol{B}_O^U\right]\right). \tag{11}$$

where all the variables are as described as before, and the superscripts U and L represent the upper and lower limits of the α-cut interval, respectively. The constraints are defined so that the upper and lower limits of each weight and bias (in both layers) minimise the width of the predicted interval:

$$\min\left(\sum_{i=1}^{n}\left(z_i^{\mathrm{L}}-z_i^{\mathrm{U}}\right)\right),$$
$$\frac{1}{n}\sum_{i=1}^{n}(\delta_i)\geq P_{\mathrm{CI}}, \tag{12}$$
$$\delta_i=\begin{cases}1, & \text{if } z_i^{\mathrm{L}}<t_i<z_i^{\mathrm{U}}\\ 0, & \text{otherwise,}\end{cases}$$

where t is that target (observed data) and P_{CI} is a predefined percentage of data. Alvisi and Franchini (2011) defined P_{CI} to be 100 % at $\mu=0$, 99 % at $\mu=0.25$, 95 % at $\mu=0.5$ and 90 % at $\mu=0.75$. This algorithm was built starting at $\mu=0$ and moving to higher membership levels to maintain convex membership functions of the generated fuzzy numbers by using the results of the previous optimisation as the upper and lower limit constraints for the proceeding optimisation. Lastly, at $\mu=1$, the interval collapses to a singleton, represent the crisp results from non-fuzzy ANN. Therefore, these α-cut intervals of the FNN output quantify the uncertainty around the crisp prediction, within which is expected to contain P_{CI} percentage of data.

In this research, this method is modified in two ways. First, the inputs x are also fuzzy numbers, which means that Eqs. (10) and (11) are revised as follows:

$$\left[y_i^{\mathrm{L}}y_i^{\mathrm{U}}\right]=f_{\mathrm{HID}}\left(\left[\mathbf{W}_{\mathrm{IH}}^{\mathrm{L}}\mathbf{W}_{\mathrm{IH}}^{\mathrm{U}}\right]\times\left[\boldsymbol{x}_i^{\mathrm{L}}\boldsymbol{x}_i^{\mathrm{U}}\right]+\left[\boldsymbol{B}_{\mathrm{H}}^{\mathrm{L}}\boldsymbol{B}_{\mathrm{H}}^{\mathrm{U}}\right]\right), \tag{13}$$
$$\left[z_i^{\mathrm{L}}z_i^{\mathrm{U}}\right]=f_{\mathrm{OUT}}\left(\left[\mathbf{W}_{\mathrm{HO}}^{\mathrm{L}}\mathbf{W}_{\mathrm{HO}}^{\mathrm{U}}\right]\times\left[y_i^{\mathrm{L}}y_i^{\mathrm{U}}\right]+\left[\boldsymbol{B}_{\mathrm{O}}^{\mathrm{L}}\boldsymbol{B}_{\mathrm{O}}^{\mathrm{U}}\right]\right). \tag{14}$$

Note that now the input vector is represented by its upper and lower limits. The major impact on this is that the training algorithm for the FNN needs to accommodate this fuzzy α-cut interval, which requires the implementation of fuzzy arithmetic principles (Kaufmann and Gupta, 1985). The cost function for the optimisation remains unchanged.

The second modification of the original algorithm is related to the selection of the percent of data included in the predicted interval (P_{CI}). In the original, the selection is arbitrary and end-users of this method may be interested in the events that are not included in the selected P_{CI}. Thus, a full spectrum of possible values for a given prediction is required. Thus, the Alvisi and Franchini (2011) approach is further refined by utilising the same relationship between probability and possibility that was used to define the input fuzzy numbers, giving a more objective means of designing FNNs with fuzzy weights, biases and output.

In the adopted possibility–probability framework, the interval $[a\ b]_\alpha$ created by the α-cut at $\mu=\alpha$ implies that

$$[p(x<a)+p(x>b)]=\alpha. \tag{15}$$

This can be used to calculate the probability

$$[p(a<x<b)]=(1-\alpha). \tag{16}$$

Table 2. Selected values for P_{CI} for the FNN optimisation.

μ	P_{CI} (%)
1.00	0.00
0.80	20.00
0.60	40.00
0.40	60.00
0.20	80.00
0.00	99.50

This means that there is a probability of $(1-\alpha)$ that the random variable x will fall within the interval $[a\ b]_\alpha$. In other words, the α-cuts of a possibility distribution (at any μ) correspond to the $(1-\alpha)$ confidence interval of the probability distribution of the same variable (Serrurier and Prade, 2013). This principle is used to select the different P_{CI} for the optimisation constraints rather than the predetermined P_{CI} selected by Alvisi and Franchini (2011). These are shown in Table 2.

Note that for practical purposes, P_{CI} was selected as 99.50 % at $\mu=0$ to prevent over-fitting. The implication of this selection is that at $\mu=0$, *nearly all* the observed data should fall within this predicted FNN interval, reflecting the highest uncertainty in the prediction. The uncertainty decreases as μ increases. For the $\mu=1$ case the values of the weights and biases were determined to be the mid-point of the interval at $\mu=0.8$ to maintain convexity of the produced fuzzy numbers, and the difficulty in finding an interval containing 0 % of the data.

2.3.3 Network architecture and implementation

For this research a three layer, feedforward MLP architecture was selected to model minimum daily DO (the output) using fuzzified daily flowrate (Q) and fuzzified daily water temperature (T) as the inputs. The three layers consist of an input layer, an output layer, and a hidden layer (with 5 neurons based on a trial-and-error search procedure). This architecture was selected for three reasons: it is one of the most commonly used in hydrology (Maier et al., 2010), it can be used as a universal approximator (Hornik et al., 1989), and as reference for comparing performance with previous research (He and Valeo, 2009; Napolitano et al., 2011). In particular, a previous study modelling minimum DO in the Bow River used a three-layer MLP feedforward network (see He et al., 2011). Two transfer functions are required for FNN implementation: the hyperbolic tangent sigmoid function was selected for f_{HID}, and a pure linear function for f_{OUT}. Both function selections follow Alvisi and Franchini (2011), Wen et al. (2013) and Elshorbagy et al. (2010), and are described as follows:

$$f_{\mathrm{HID}}=\frac{e^x-e^{-x}}{e^x+e^{-x}} \tag{17}$$

$$f_{\mathrm{OUT}} = x. \tag{18}$$

The LMA method was used to train the network, minimising E_{MSE}. The input and output data were pre-processed before training, validating and testing: the data were normalised so that all input and output data fell within the interval $[-1\ 1]$. Furthermore, the data were randomly divided into training, validation and testing subsets, following a 50–25–25 % split.

This FNN optimisation algorithm was implemented in MATLAB (version 2015a). First, the built-in MATLAB Neural Network Toolbox was used to estimate the value of weights and biases using the midpoint of the interval at $\mu = 1$. The results from this were used as the constraints to solve the FNN optimisation (Eqs. 12 to 14) at subsequent lower membership levels. The Shuffled Complex Evolution algorithm (commonly known as SCE-UA, Duan et al., 1992) was used to find the optimisation solution. The optimisation is run such that the intervals at higher membership levels govern the upper and lower bounds of the predicted interval in order to preserve the convexity of fuzzy numbers. The same process and network architecture was used to run the original FNN method (proposed by Alvisi and Franchini, 2011) using crisp inputs for comparison purposes. For this case, further refinement of the optimised solution was conducted using the built-in MATLAB minimisation function, f_{mincon}. Note that for the crisp inputs, values of fuzzified daily flowrate (Q) and fuzzified daily water temperature (T) at $\mu = 1$ were used to enable direct comparison. This option allows for the closest comparison between the two approaches that have completely distinct applications. Other options for the crisp inputs (e.g. mean daily value, or maximum daily value) may also be selected for the existing FNN case.

2.4 Risk analysis using defuzzification

Risk analyses for complex systems is challenging for a number of reasons, including an insufficient understanding of the failure mechanisms (Deng et al., 2011). The use of imprecise information (e.g. fuzzy numbers) is an effective method of conducting a risk analysis (Deng et al., 2011). However, communicating uncertainty is an important, yet difficult task, and many different frameworks exist to do so; water quality indices (Sadiq et al., 2007; Van Steenbergen et al., 2012) are one example. Since water resource managers often prefer to use probabilistic measures (rather than possibilistic ones), it is important to convert the possibility of low DO to a comparable probability for effective communication of risk analysis. Note that the linguistic parameters (e.g. "most likely") that are often used to convey risk or uncertainty (Van Steenbergen et al., 2012) have a probability based meaning – in this case "most likely" is a measure of *likelihood.*

In this research, a *defuzzification* procedure is used to convert the possibility of low DO to a probability measure, to represent the risk of observing low DO (below a given threshold) in the Bow River. This method uses the inverse of the transformation described in Eq. (1); however, instead of calculating the probability of one element, $p(x)$, which is of limited value in most applications, it is generalised to calculated $P(\{X < x\})$, as follows (from Khan and Valeo, 2014a, 2016): for any x in the support (defined as the α-cut interval at $\mu = 0$) of a fuzzy number $[a\ b]$, we have the corresponding μ and the paired value x' which shares the same membership level. The value μ is the sum of the cumulative probability between $[a, x]$ and $[x', b]$, labelled P_{L} and P_{R}, respectively:

$$\mu(x) = P_{\mathrm{L}} + P_{\mathrm{R}}, \tag{19}$$

where P_{L} represents the cumulative probability between a and x which is assumed to equal the probability $P(\{X < x\})$, since the fuzzy number defines any values to less than a to be impossible (i.e. $\mu = 0$). Given the fact that the fuzzy number is not symmetrical, the lengths of the two intervals $[a, x]$ and $[x', b]$ can be used to establish a relationship between P_{L} and P_{R}. Then, P_{L} can be estimated as

$$P\{X < x\} = P_{\mathrm{L}} = \frac{\mu}{1 + \frac{(b-x')}{(x-a)}}. \tag{20}$$

Thus, Eq. (20) gives the probability that the predicted minimum DO for a given day is below the threshold value x. For example, if the lowest acceptable DO concentration for the protection of aquatic life for cold water ecosystems ($6.5\ \mathrm{mg\,L^{-1}}$, Canadian Council of Ministers of the Environment, 1999) is selected, then this transformation can be used to calculate the probability that the predicted fuzzy DO will be below $6.5\ \mathrm{mg\,L^{-1}}$.

3 Results and discussion

3.1 Probability–possibility transformation using bin-size optimisation

The bin-size optimisation and the probability–possibility transformation algorithms were applied to the collected Q and T data for the 9-year period. The constructed fuzzy numbers were then used to calibrate the FNN model. This section compares the results of constructing a discretised probability distribution with and without the bin-size optimisation algorithm and its impact on the resulting membership function of the fuzzy number. The comparison is illustrated through five examples each for Q and T as a means of illustrating the advantages of using the proposed approach.

Figure 3 shows sample results of converting hourly Q observations to fuzzy numbers for five cases. The left-most column in the figure shows the raw data, i.e. the observations sampled over the course of 24 h. The resulting histogram based probability functions are shown for both the optimised (D_{opt}, illustrated with circles) and original ($D_{\mathrm{orig}} = 2r$; see Sect. "A new algorithm to create fuzzy numbers" for the definition, illustrated with squares) bin sizes in the second column. The third and fourth columns in Fig. 3 show the resulting discretised empirical membership function using each of

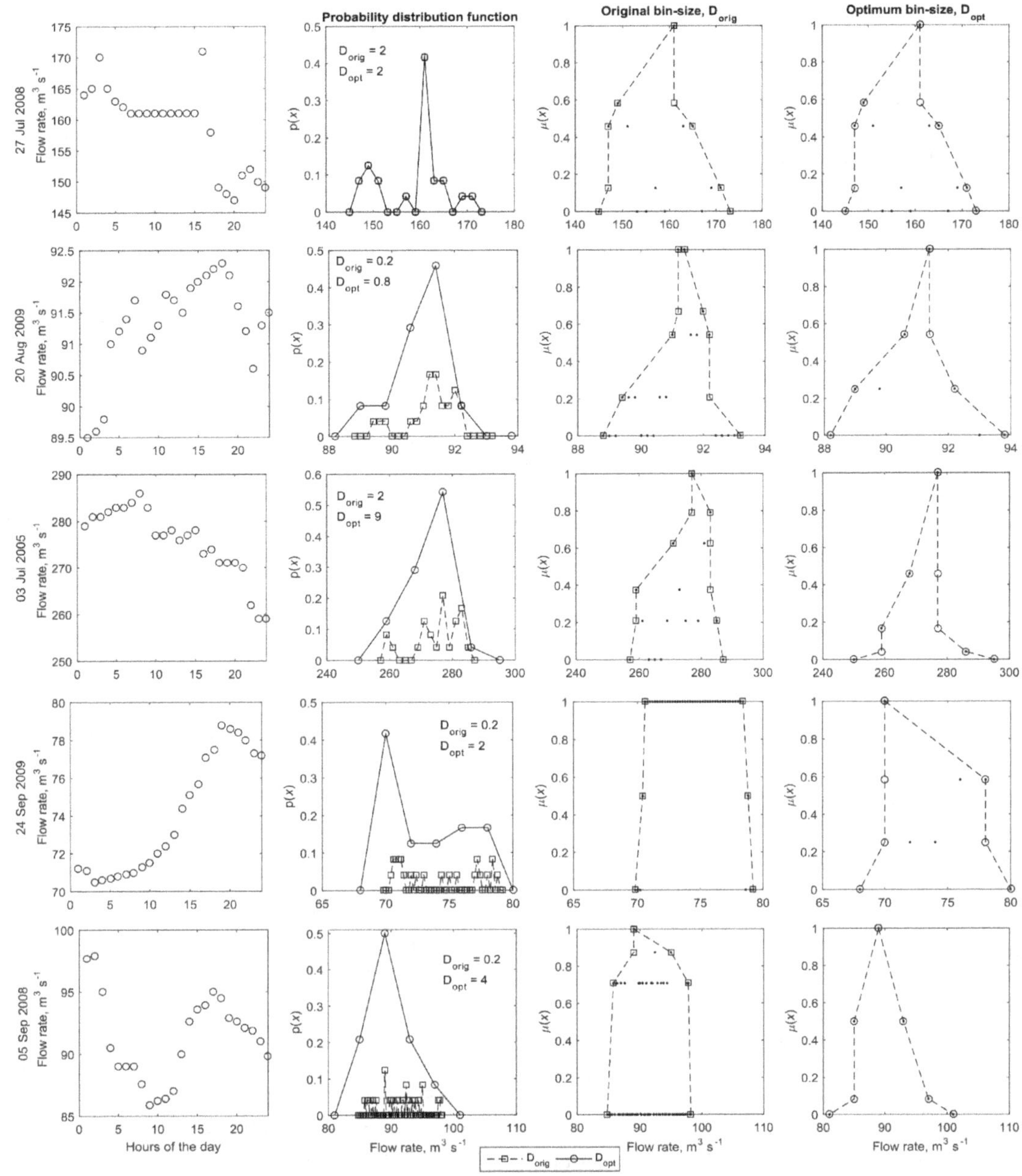

Figure 3. Sample results of probability–possibility transformation for flow rate, Q.

the histograms. The five examples selected here represent a full spectrum of results for the bin-size optimisation. The first row shows an example of when the optimum result was equal to the measurement resolution ($D_{\text{orig}} = D_{\text{opt}} = 2$), followed by cases where the D_{opt} was 4, 4.5, 10 and 20 times greater than the initial bin size.

The example in the first row illustrates cases where the bin-size optimisation algorithm calculates an optimum bin size, corresponding to the minimum cost function C_{D}, which is equal to the instrument measurement resolution. Thus, the resulting probability distributions for both cases are equivalent, as are the membership functions. In most cases, this occurred when the calculated minimum C_{D} would result in a D_{opt} smaller than $D_{\text{orig}} = 2r$, and since this is not physically feasible (measurable), the algorithm did not consider any bin sizes below $2r$. Of note in this example is that the transformation of the probability distribution results in five empirical membership levels. Only one element was found to have a membership level equal to 1 (at $Q = 161\ \text{m}^3\,\text{s}^{-1}$). Thus, the α-level cut at this level is a simple singleton: $[161]_{\mu=1}$. The next membership level was calculated as 0.58; again the resulting α-cut level only has one element at $Q = 149$ (which is less than the modal value). However, at this level the upper and lower limits of α-cuts at higher membership levels define the upper and lower limits of α-cuts at lower levels. Thus, using the information from the α-level cut at $\mu = 1$, the level at $\mu = 0.58$ was defined as $[149\ 161]_{\mu=0.58}$. The next membership level calculated was 0.46, and four elements had equal membership levels, ranging between 147 and 165. The α-cut interval at this level was defined as $[147\ 165]_{\mu=0.46}$. Note that in this case, this interval captures both the intervals at higher membership levels within its limits, i.e. the lower limit is less than the lower limits of higher intervals, and the upper limit is greater than then upper limits at higher levels. The next membership level was calculated to be 0.125, and three elements between 157 and 171 were assigned this value. However, the lower limit at $\mu = 0.46$ (the next higher membership level) was 147, which is less than 157, and thus, for the α-cut level at this membership, the interval is then revised to $[147\ 171]_{\mu=0.125}$ rather than $[157\ 171]_{\mu=0.125}$ to maintain convexity. Again, the reason here is that if something is possible at $\mu = 0.46$, it must be possible (by definition) at $\mu = 0.125$. The last membership level found for this particular example was $\mu = 0$, with six elements sharing this value, ranging from 145 to 173, resulting in an α-cut level of $[145\ 173]_{\mu=0}$. Together, these five membership levels define a discretised membership function of the fuzzy number for Q on 27 July 2008. Following this, linear interpolation was conducted to find the elements corresponding to the six predefined membership levels of $\mu = 0$, 0.2, 0.4, 0.6, 0.8 and 1. The results are not explicitly shown in the figure for clarity, but can are essentially located on the dashed line in the last column on the corresponding membership levels.

The second row in Fig. 3 shows the results for 20 August 2009, where the optimum bin size was found to be 4 times higher than the original bin size ($D_{opt} = 0.8$ vs. $D_{orig} = 0.2$). The impact of this change is clearly evident in the distribution functions in the second column. The original histogram is multi-modal, and with multiple candidates as the modal value (where $\mu = 1$), whereas the post-optimisation histogram is considerably smoother, with a definitive modal value at $Q = 91.4\,m^3\,s^{-1}$. The impact of this increase in bin size is that the resulting membership function is defined at four membership levels (0, 0.25, 0.54 and 1), whereas the original function was defined at six levels, including an interval (rather than singleton) at $\mu = 1$. This decrease in membership levels in this case has a consequence of smoothing out the membership function, as can be seen by comparing the shapes of the functions in columns 3 and 4. The overall impact of this smoothing out of both the distribution and the membership functions is that the heightened specificity of the original function at $\mu = 0.54$ and above is reduced to a more generalised shape.

Since the objective of the bin-size algorithm was to reduce the error between the histogram created using D_{opt} and the unknown theoretical distribution, the density function plotted in Fig. 3 represents the closest distribution to the unknown distribution. Hence, the membership function generated using this optimum distribution better reflects the underlying phenomenon than the membership function generated using D_{orig}. Thus, in comparing columns 3 and 4 for the second row, the smoother membership function representing D_{opt} is preferred. Linear interpolation is then performed on this membership function to get values of Q at the six predefined membership levels.

Similar results can be seen in the third row in Fig. 3, where the optimised bin size is 4.5 times greater than the original bin size ($D_{opt} = 9$ vs. $D_{orig} = 2$). Again, the original histogram is extremely uneven, whereas the post-optimisation histogram is considerably smoother, with a definitive modal value at $Q = 277\,m^3\,s^{-1}$. The overall impact of this smoothing is that the specificity of the function at $\mu = 0.6$ and higher of the original function is reduced to a more general shape in the optimised function.

The fourth row shows a different phenomenon, where instead of smoothing out the original membership function, the combined bin-size optimisation and transformation algorithm creates a membership function with more specificity. In this case D_{opt} is 10 times higher than D_{orig}, and the consequence of this increase is the smoother probability density function with one clear modal value (at $Q = 70\,m^3\,s^{-1}$). In contrast, the original histogram had six elements with joint-equal $p(x)$, resulting in a membership function that is shaped similarly to a uniform distribution (column 3) and defined with only three membership levels. This means that all values are considered equally *possible* and represents maximum vagueness. However, using the optimised value, this is no longer the case and the modal value is assigned a membership level of 1, and the remaining elements are defined at three other membership levels. This suggests that this modal value is more possible (since it has a higher possibility), and this is reflected in the observations. This example illustrates that the method can not only generalise the data to smoother functions (as shown in the first three examples), but can also be more specific when the underlying data demonstrate this; however, this is not captured by the non-optimised bin-width distribution function.

The last example for Q in Fig. 3 is an example of a case where the number of membership levels for both the original and optimised membership functions are equal (four in this case); however, the bin size is 20 times greater for the optimised case. In this case, an optimum bin size was found that did not change the specificity of the membership function, i.e. it is still defined with the same number of intervals but at different membership levels. In this case, the probability for D_{orig} is extremely uneven but smoothed out to a unimodal function with the D_{opt}. The final membership function for D_{opt} is defined more generally (smoothly), especially at higher membership levels compared to the one defined by D_{orig}. This example again demonstrates the utility of the new coupled optimisation–transformation method to create fuzzy numbers for data where the underlying distribution is unknown.

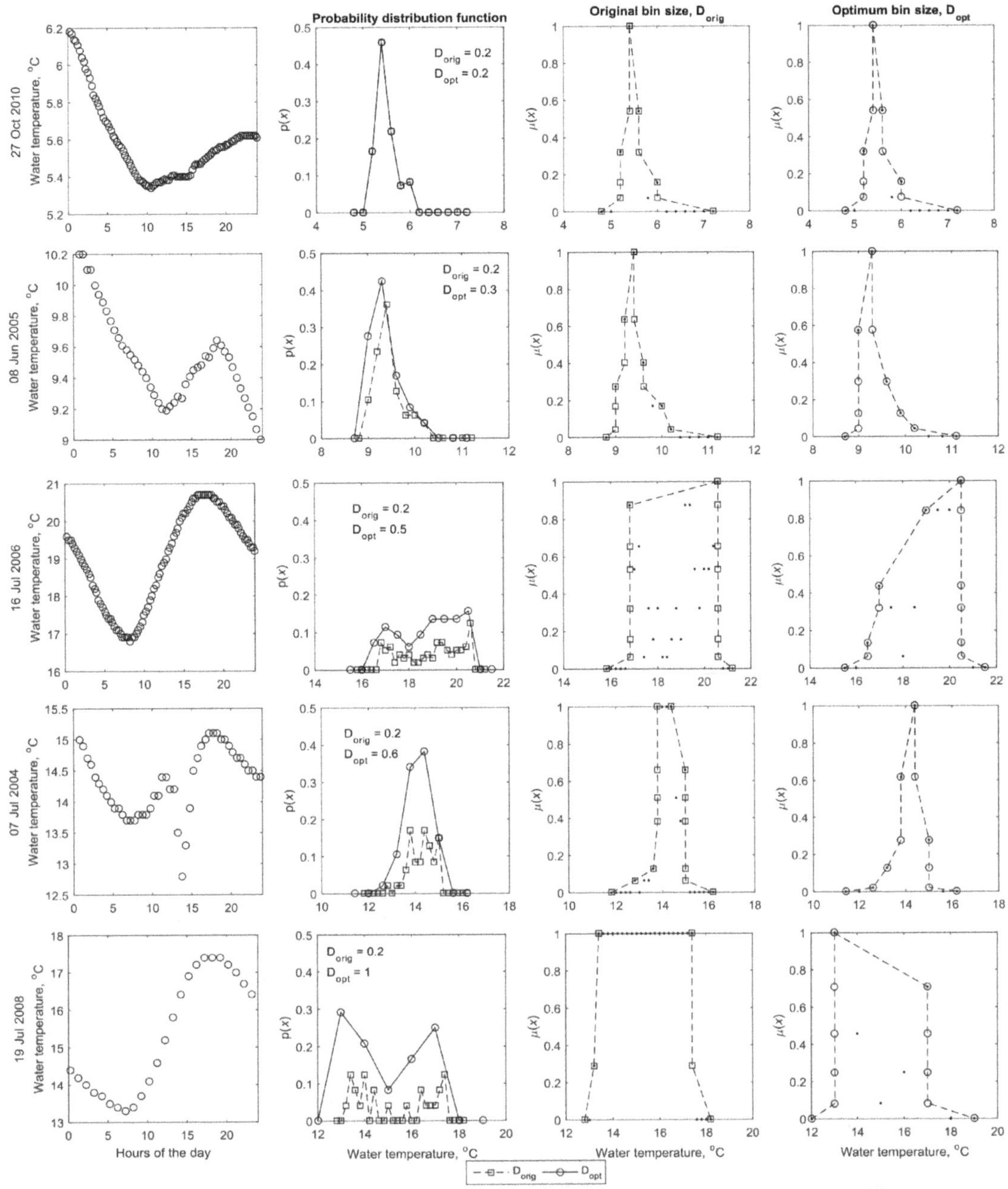

Figure 4. Sample results of probability–possibility transformation for water temperature, T.

Figure 4 shows similar results for the five water temperature examples, where the D_{opt} was equal to the D_{orig} (the first example in the top row) or increased by a factor of 1.5, 2.5, 3 or 5. The first example shows a case with very little T variation over a given day and the water temperature falls between 5.2 and 6.2 °C for the entire day. This lack of variability is responsible for the minimal bin-size selection D_{opt}: a unimodal distribution is best constructed using smaller bin sizes for these cases. The second example shows another case where D_{opt} is only slightly greater than the original, resulting in a somewhat smoother probability function and a slightly smoother membership function.

A major difference between the T and Q data is that the former is strongly diurnal, increasing after sunrise in the morning, peaking in late afternoon, and then decreasing through the night. This temporal trend is seen for all examples in Fig. 4, but most significantly in the bottom three examples. A major implication for this in developing a probability density function for these data is that the resulting shape will have a tendency to be bimodal. This means that the resulting membership functions might be trapezoidal or

Table 3. The E_{MSE} and E_{NSE} for each subset of the fuzzy neural network using the method proposed (using fuzzy inputs) and using the original method (using crisp inputs).

	E_{MSE} $(mg\,L^{-1})^2$		E_{NSE}	
	Proposed	Original	Proposed	Original
Train	1.52	1.55	0.52	0.51
Validation	1.19	1.18	0.49	0.49
Test	1.09	1.10	0.54	0.54

near-trapezoidal (and hence most vague) in shape, which is clearly demonstrated in the functions created using D_{orig} in the bottom three examples. However, in each case the optimised bin size creates a smoother probability distribution with a clearer modal value, resulting in membership functions that are no longer trapezoidal.

Thus, without using the bin-size optimisation algorithm, there is a risk that the resulting membership functions will be too vague and will not represent the information that can be gained from the observations. It is worth nothing that for these three examples, if linear interpolation is used on the original membership function, the resulting interpolated fuzzy numbers will all have equal intervals (due to the trapezoidal shape), transferring no useful information to the final fuzzy number.

Overall, the above examples illustrate the advantages of using the coupled methods of bin-size optimisation and probability–possibility transformation to create fuzzy numbers for the FNN application. The applicability of this method is not necessarily restricted to this application and can be applied whenever there is a need to construct fuzzy numbers from observed data. The utility of the first component, bin-size optimisation to estimate the density function is that in cases where either not enough information is available to define a probability distribution, or if the data do not follow the mould of a known density function, or if assumptions about the class of distribution cannot be made, the optimum bin size can be calculated to define an empirical distribution for the probability–possibility transformation. The advantage of the second component, the algorithm to construct the possibility distribution (i.e. the membership function of the fuzzy number) is that it provides a consistent, transparent and objective method to convert observations (e.g. time series data) into fuzzy numbers – which has been cited as a major hurdle in implementing fuzzy number based applications in the literature (Abrahart et al., 2010; Dubois and Prade, 1993; Civanlar and Trussel, 1986). A noteworthy component of this algorithm is that the fuzzy numbers do not reduce to the simple, triangular shaped functions that are widely used, but rather the functions better represent the information from the observations.

3.2 Training the fuzzy neural network

Once the observations of the abiotic input parameters (Q and T) were converted to fuzzy numbers, the FNN training algorithm was run using five neurons in the hidden layer, to predict daily minimum DO in the Bow River. First, the values of the fuzzy numbers at $\mu = 1$ were used to train the crisp network. This was done in order to have initial estimates of the 10 $\mathbf{W}_{IH}$ (5 for each input), 5 $\boldsymbol{B}_{IH}$, 5 $\mathbf{W}_{HO}$, and 1 $\boldsymbol{B}_O$. These initial estimates were used to provide the upper and lower limits of the constraints for the proceeding optimisation algorithm. Once these estimates were calculated, the optimisation algorithm was used to calculate the fuzzy weights and biases using fuzzy inputs, and was started from $\mu = 0$, moving sequentially to higher membership levels until $\mu = 0.8$. The final level (at $\mu = 1$) was calculated using the midpoint of the intervals estimated at $\mu = 0.8$. The total optimisation time for the proposed method was 13 h, whereas the existing method with crisp inputs was 8 h, using a 2.40 GHz Intel® *Xeon* microprocessor (with 4 GB RAM).

The E_{MSE} and the Nash–Sutcliffe model efficiency coefficient (E_{NSE}; Nash and Sutcliffe, 1970) for the training, validation and testing scenarios for $\mu = 1$ for both methods are shown in Table 3. The E_{MSE} for each data set are low, between 11 and 16 % of the mean annual minimum DO seen in the Bow River for the study period. The E_{NSE} values are approximately equal to 0.5 for each subset, which is higher than E_{NSE} values in the literature for water quality parameters when modelled daily (see Moriasi et al. (2007) for a survey of results) and is considered to be "satisfactory" by their standards. In comparing the two methods, it is obvious that including additional information (in the form of fuzzy inputs) does not decrease performance, as the metrics are nearly identical for both methods. This shows that the proposed method has successfully incorporated input data uncertainty in the model architecture. These model performance metrics highlight that in general, predicting minimum DO using abiotic inputs and a data-driven approach is an effective technique.

The results of the optimisation component of the algorithm are summarised in Table 4, which shows the percentage of data (P_{CI}) captured within the resulting α-cut intervals for each of the three data subsets. The performance for each of the data sets (i.e. train, validation and test) for both methods is nearly identical (on an interval-by-interval basis): the exact amount of data captured within the intervals, as required by the constraints, except for the $\mu = 0.8$ interval. At this interval, the amount of coverage decreases (i.e. lower performance) as the membership level increases, which is unavoidable when the width of the uncertainty bands decrease. As required by Eq. (12), the amount of data within the interval has to be greater-than or equal-to the limit defined by P_{CI} (as per Alvisi and Franchini, 2011) which is true for all training data. This means that a solution to satisfy the constraints with a lower amount of data (e.g. reducing the 29.91 % for

Table 4. Percentage of data captured within each α-cut interval for the three subsets of data.

	Percent captured, P_{CI} (%)					
	Proposed method			Existing method		
μ	Train	Validation	Test	Train	Validation	Test
1.00	–	–	–	–	–	–
0.80	29.91	28.54	28.78	20.02	14.39	18.05
0.60	39.93	40.98	40.24	40.05	35.85	40.49
0.40	59.95	66.10	64.15	60.07	60.73	61.95
0.20	79.98	80.49	82.93	80.10	79.51	82.44
0.00	99.39	98.78	99.02	99.51	98.54	99.02

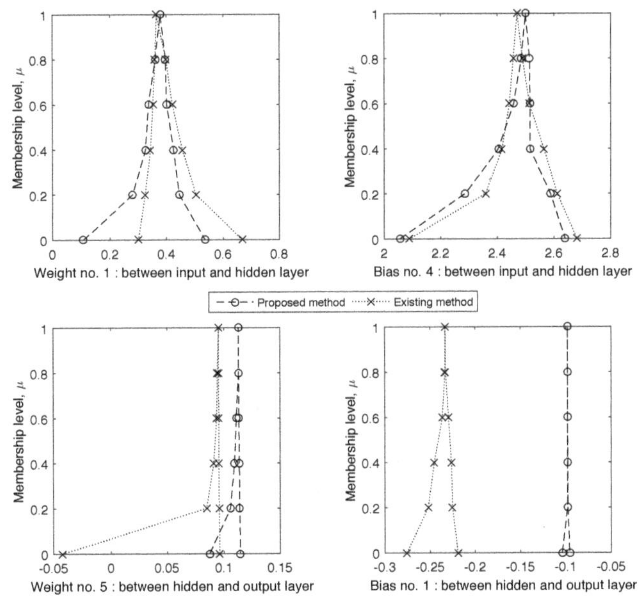

Figure 5. Sample plots of the produced membership functions for the weights and biases of the fuzzy neural network for both the proposed and existing methods.

the $\mu = 0.8$ interval for the proposed method) would either result in non-minimal intervals (though this is unlikely) or that the constraints on the values of the intervals could not be maintained. This latter issue will be discussed in detail with Fig. 5 below. Lastly, as mentioned above, the performance of non-training data sets for both methods decrease as the interval get narrow: this can be seen best by the inability for both methods to capture the exact amount of data required at the $\mu = 0.8$ interval for the validation and testing data sets. These results are similar to the testing data set in Alvisi and Franchini (2011). This comparison again demonstrates the ability of the proposed method with fuzzy inputs to function in similar manner to the original algorithm that used crisp inputs.

A sample of the fuzzy weights and biases produced through the optimisation are shown in Fig. 5. Note that the membership functions are assumed to be piecewise linear (following similar assumptions made in Alvisi and Franchini, 2011; Khan et al., 2013; Khan and Valeo, 2015), i.e. that the intervals at each membership levels can be joined to create a fuzzy number. This can be confirmed by the fact that each of the weights and biases are *convex* where intervals at lower levels are wider than intervals at higher levels, and are *normal* with at least one element with $\mu = 1$. Note that each *weight* and *bias* has a non-linear membership function, i.e. none of the functions produced follow the typical triangular functions and are not necessarily symmetric about the modal value. The shapes of the fuzzy weights and biases for the proposed and existing methods are generally the same for the input-hidden layer; however, differences can be seen for the hidden-output layer plots. Since the existing method uses crisp inputs, it requires the produced weights and biases to represent the uncertainty in the data, to produce output intervals wide enough to capture the set amount of observations. This is reflected in hidden-output layer plots where the lower limit of the membership function for Weight #5 is highly skewed, which enables this method to capture the low DO events. Similarly, Bias #1 in the hidden-output layer has been translated to a lower value, to produce fuzzy DO outputs that capture the low DO observations.

The figure demonstrates that enough α-cut levels (i.e. six levels equally spaced between 0 and 1) have been selected to *completely* define the shape of the membership functions. At a smaller number of levels, e.g. two levels, one at $\mu = 0$ and one at $\mu = 1$, the fuzzy number collapses to a triangular fuzzy number, which is not desirable for this research, as discussed in previous sections. When only two levels are selected, the figures demonstrate that significant differences exist between those simple functions and the ones generated using six membership levels: the decrease in the width of the intervals with an increase in membership level is not linear as is in triangular shaped function. Similarly, a higher number of intervals e.g. 100 intervals, equally spaced between 0 and 1, could be selected. The risk in selecting many intervals is that as the membership level increases (closer to 1) the intervals become narrower as a consequence of convexity. This will result in numerous closely spaced intervals, with essentially equal upper and lower bounds, making the extra information redundant. This is demonstrated in the sample membership functions in Fig. 5 for $\mathbf{W}_{IH}$ number 5 and $\boldsymbol{B}_O$ (for the proposed method) where the intervals at the higher membership levels collapse to a singleton, or are extremely narrow. Thus, defining more uncertainty bands between the existing levels would not add more detail but would merely replicate the information already calculated.

Connecting this back to the results in Table 4, these two particular weights and biases show why the percentage of data calculated at $\mu = 0.8$ (for training) cannot be improved by further optimisation. At some point, if the intervals at $\mu = 0.8$ for the various weights and biases collapse to a single element, no further refinement in the model is possible (since all the constraints are met) and the minimum interval width of the predicted DO whilst capturing *at least* the P_{CI} amount

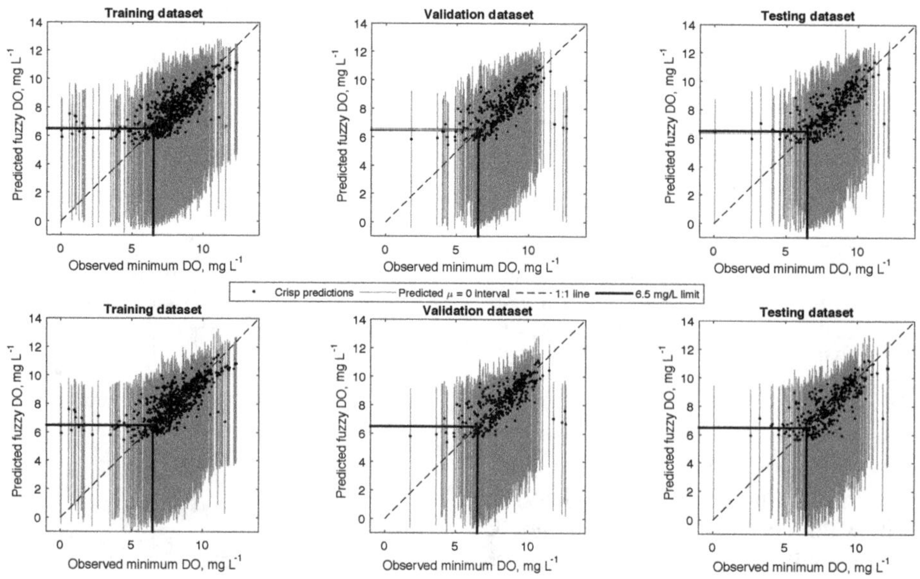

Figure 6. A comparison of the predicted and observed minimum DO at the $\mu = 0$ interval (grey line) and at $\mu = 1$ (black dots) for the proposed (top row panels) and existing (bottom row panels) methods.

of data has been reached. It is worth emphasizing here that the uncertainty represented by these fuzzy number weights and biases is not the uncertainty of the particular weight or bias, but is the total forecasting uncertainty defined by the quantifying bands around the crisp predicted value.

Tables 3 and 4 and Fig. 5 demonstrate the overall success of the proposed approach to calibrate an FNN model as compared to a crisp ANN, as well as an FNN that uses crisp inputs. The optimisation algorithm is defined based on the principles of possibility theory (i.e. defining the amount of data to include in each interval) and is a transparent, repeatable and objective (not arbitrary) method to create the fuzzy numbers for the FNN model.

The observed versus crisp predictions (black dots) and fuzzy predictions at $\mu = 0$ (grey lines) for daily minimum DO for the three different data subsets (training, validation and testing) are shown in Fig. 6. The figure shows that *nearly all* (specifically, 99.4, 98.8 and 99.0 % of the training, validation, and testing subsets, respectively, for the proposed method, with similar results for the crisp input FNN method) of the observations fall within the $\mu = 0$ interval, since the observed values (black dots) tend to fall inside the grey lines. This figure also highlights one of the major advantages of the FNN over a simple non-fuzzy ANN: almost all of the fuzzy results intersect the 1 : 1: line, whereas many of the crisp results are quite far from that line, especially at low DO values (which is marked at 6.5 mg L^{-1} in the figure). In other words, while the fuzzy number prediction may not predict the observed value exactly, it provides at least some *possibility* of the observed value within its various α-cut intervals, but the crisp results do not provide this additional information. This figure illustrates that E_{NSE} (listed in Table 2 for the $\mu = 1$ case only) is not representative of the entire fuzzy number predictions, since it does not capture the performance at different membership levels. Thus, there is a need to develop an equivalent performance metric when comparing crisp observations to fuzzy number predictions.

Figure 6 also demonstrates the benefit of the FNN approach as compared to the crisp ANN approach with respect to predicting low DO (i.e. when DO is less than 6.5 mg L^{-1}). Both the FNN methods predict more of the low DO events within its intervals as compared to the crisp method. The figure demonstrates that both the crisp ($\mu = 1$) and fuzzy predictions tend to overpredict the low DO events (since they fall above the 1 : 1: line), but the fuzzy intervals are closer to the observations (i.e. they intersect the 1 : 1 line for the majority of low DO events), and therefore predict some *possibility* (even if it is a low probability) of the low DO events occurring. Thus, generally speaking, the ability of the FNN to capture *nearly all* of the data within its predicted intervals guarantees that most of the low DO events will be successfully predicted. This is a major improvement over conventional methods used to predict low DO. In comparing the two FNN methods, both methods give similar results: the average width of predicted low DO intervals for the 9-year period (at $\mu = 0$) is 8.84 mg L^{-1} for the proposed method, and 8.60 mg L^{-1} for the existing method. The impact of the width of the predicted intervals is discussed later.

Trend plots of observed minimum DO and predicted fuzzy minimum DO for the years 2004, 2006, 2007 and 2010 are illustrated in Figs. 7 and 8. These results are shown only for the proposed method for clarity; the difference between the existing method (using crisp inputs) and the proposed method (using fuzzy inputs) is discussed later. These years were se-

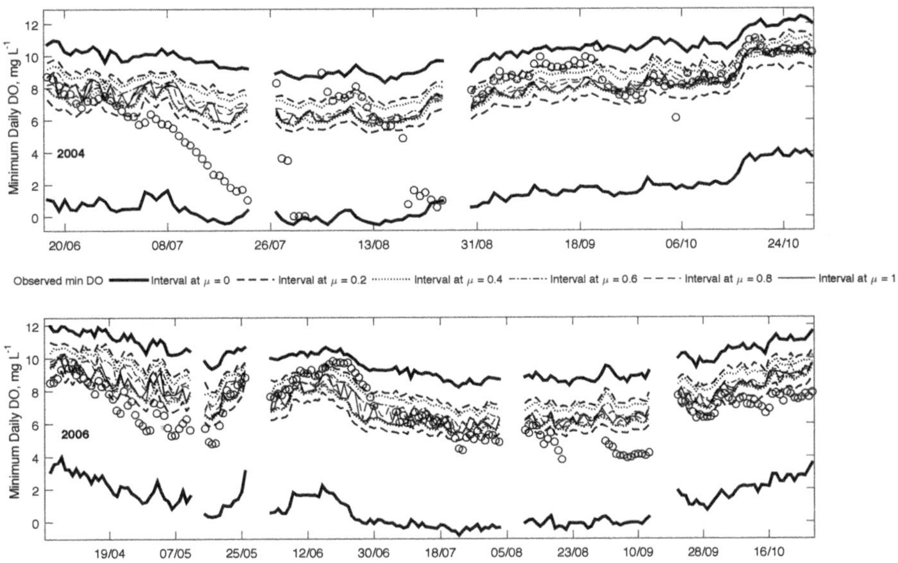

Figure 7. A comparison of the observed and predicted minimum DO trends for 2004 (top panel) and 2006 (bottom panel).

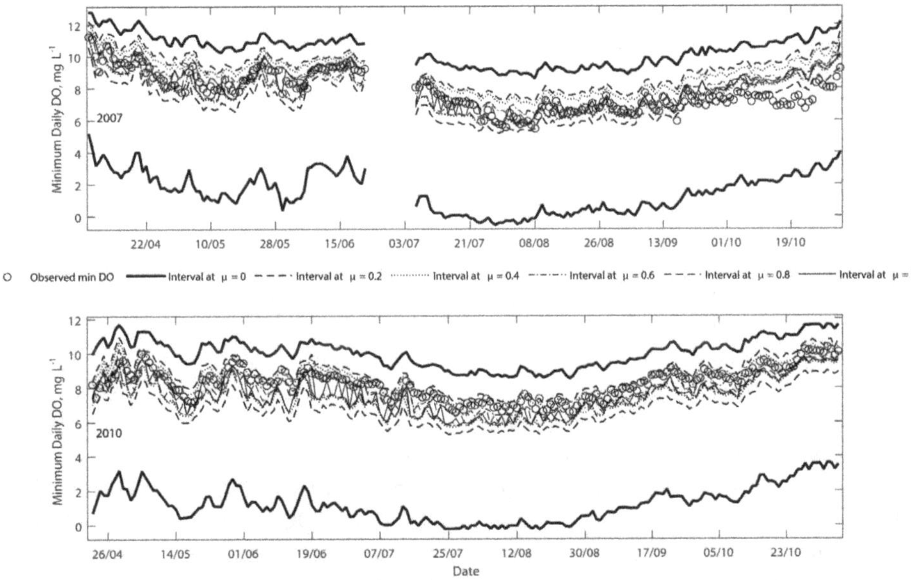

Figure 8. A comparison of the observed and predicted minimum DO trends for 3 sample years: 2007 (top panel) and 2010 (bottom panel).

lected due to the high number of low DO occurrences in each year (as listed in Table 1), and highlight the utility of the proposed method to predict minimum DO using abiotic factors in the absence of a complete understanding of the physical mechanisms that govern DO in the Bow River. Note that for each year, 50 % of the data are training data, 25 % are validation data and 25 % are testing data. However, for clarity this difference is not individually highlighted for each data point in these figures.

In Figs. 7 and 8, the predicted minimum DO at equivalent membership levels (e.g. 0^L or 0^R) at different time steps are joined together, creating bands representative of the predicted fuzzy numbers calculated at each time step. In doing so, it is apparent that all the observed values fall within the $\mu = 0$ interval for the years 2006, 2007 and 2010, and all but one observation in 2004. The width of each band represents the amount of uncertainty associated with each membership level. For example, the bands are the widest at $\mu = 0$, meaning the results have the most vagueness associated with it.

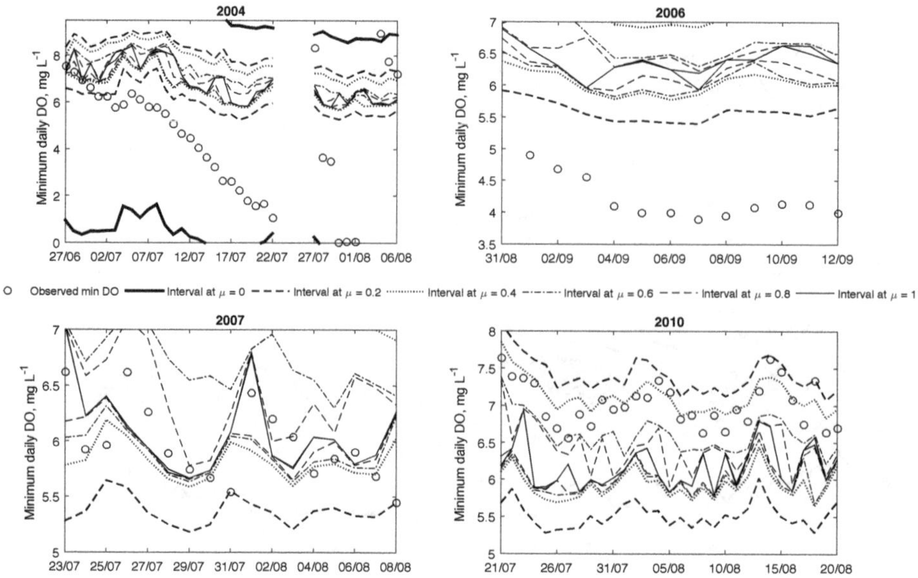

Figure 9. Zoomed-in views of the trend plots for 4 sample years corresponding to important periods with low DO occurrences.

Narrower bands are seen as the membership level increases until $\mu = 1$. This reflects a decrease in vagueness, increase in credibility, or less uncertainty of the predicted value, as the membership level increases. Note that the majority of the predictions at $\mu = 1$ are single elements but some predictions are α-cut intervals (e.g. $[a\ b]_{\mu=1}$). This means that when not enough information is available, the fuzzy prediction collapse to trapezoidal membership functions.

In each of the years shown, the majority of the observations tend to fall within the $\mu = 0.2$ interval or higher, with only the low DO (i.e. $< 5\,\mathrm{mg\,L^{-1}}$) falling within the $\mu = 0$ and $\mu = 0.2$ bands. This suggests that the low DO events are predicted with less certainty compared to the occasions when DO concentration is high. Also note that the interval at $\mu = 0$ is highly skewed towards the lower limit ($\mu = 0^{\mathrm{L}}$), i.e. the modal value is not at the centre of the interval. This shows that the FNN has been trained to capture these low DO events, but predicts them with lower credibility. Compared to the crisp results (i.e. those at $\mu = 1$), for these low DO events, the proposed method provides some possibility of low DO, whereas the crisp results do not predict a possibility of low DO. Thus, the ability to capture the full array of DO observations within different intervals is an advantage of the proposed method over existing methods.

The trend plot for 2004 shows that observed DO decreases rapidly from late June to late July, followed by a few days of missing data and near-zero observations, before increasing to higher concentrations. Details of this trend are shown in Fig. 9 which shows magnified versions of important periods for each year. The reason for this rapid decrease in 2004 is unclear and may be related to problems with the real-time monitoring device which was in its first year of operation that year. However, it demonstrates that the efficacy of data-driven methods is dependent on the quality of the data. Since the proposed method was calibrated to capture *nearly all* the observations (including outliers like those seen in 2004) within the least certain band at $\mu = 0$, the resulting network predicts results to include these outliers, but at low credibility levels. As the data length increases (i.e. more data are added and the FNN is subsequently updated), the number of these types of outliers included within the $\mu = 0$ band will decrease because the optimisation algorithm (Eqs. 12 to 14) searches for the smallest width of the interval whilst including 99.5 % of the data. Thus, with more data, it is expected that these extreme events (i.e. the outliers seen in 2004) will no longer be captured within the $\mu = 0$ band.

The time series plot for 2006 shows that all the observations fall within the predicted intervals, and that the predicted trend generally follows the observed trend. The majority of the 25 low DO events ($< 5\,\mathrm{mg\,L^{-1}}$) occur from mid-July and continue occasionally until mid-September. Details of some of these low DO events are plotted in Fig. 9. Figure 7 demonstrates these low DO events are captured between $\mu = 0$ and 0.2 intervals, similar to the 2004 case, meaning that the credibility of these predictions is the lowest. However, unlike the 2004 case, Fig. 9 demonstrates that in 2006 the predicted intervals tend to follow the same trend as the observations for these low DO events, even if it is predicting them at a low credibility.

In contrast to the results from 2004 and 2006, the majority of observations are captured at higher membership levels (i.e. greater than $\mu = 0.2$) in 2007 as shown in Fig. 8. That is, only a limited number of observations are captured within the lowest credibility band. More importantly, 26 out of the 27 *low DO* ($< 6.5\,\mathrm{mg\,L^{-1}}$) events are captured at a membership level greater than 0.2^{L}, meaning that the low DO predictions in 2007 for the 6.5 mg L^{-1} guideline are predicted with higher credibility than the 2004 and 2006 cases. Another difference for the results from this year is that many of the low DO observations for 2007 are more evenly scattered around the $\mu = 1$ predictions (as seen in Fig. 9), in contrast to the 2004 and 2006 cases, where the low DO events were always predicted to be closer to lower bounds of the intervals.

The trend plot for 2010 is shown in Fig. 8, and it is clear that all observations fall within the $\mu = 0.2$ interval or at higher α-level intervals, meaning that the predictions capture the observations with higher certainty. This is likely due to the lack of DO events below 5 or 6.5 mg L^{-1} in 2010. Also of note for this year is that *all* observations are less than 10 mg L^{-1}, and about 90 % of all observations are below the 9.5 mg L^{-1} guideline (as listed in Table 1). The trend plot again illustrates that the FNN generally reproduces the overall trend of observed minimum DO. This can be seen in a period in early May where DO falls from a high of 10 mg L^{-1} to a low of 7 mg L^{-1}, and all the predicted intervals replicate the trend. This is an indication that the two abiotic input parameters are suitable parameters for predicting minimum DO in this urbanised watershed.

Figure 9 shows details of a low DO (< 9.5 mg L^{-1}) event in 2010 in late July through to late August. The bulk of low DO events are captured between the $\mu = 0.6^R$ and 0.2^R intervals – demonstrating that these values are predicted with higher credibility than the other low DO cases in 2004 and 2006, and are predicted closer to the upper end of the interval. All of the low DO (< 9.5 mg L^{-1}) observations in this plot are underpredicted by the crisp method (though not with the FNN method, since they are captured within a fuzzy interval). This shows that the crisp ANN results tend to overpredict extremely low DO events (i.e. < 5 mg L^{-1}) while underpredicting the DO < 9.5 mg L^{-1} events.

The analysis of the trend plots for these four sample years shows that the proposed FNN method is extremely versatile in capturing the observed daily minimum DO in the Bow River using Q and T as inputs. The crisp case (at $\mu = 1$) cannot capture the low DO events (as shown in Figs. 7 and 8); however, the FNN is able to capture these low DO events. Generally speaking, the training method selected for the FNN has been successful in creating nested intervals to represent the predicted fuzzy numbers. The widths of the predicted intervals correspond to the certainty of the predictions (i.e. larger intervals for more uncertainty). The utility of this method is further demonstrated in Sect. 3.3, where the risk of low DO is estimated using a possibility–probability (i.e. defuzzification) technique.

Figure 10 shows a comparison of the predicted minimum DO trends from both the proposed FNN method (solid black line) and the existing FNN method (dashed black line), along with the observed data (circles) for each membership level for the 2009 data. These figures show that despite the use of more data for the inputs (i.e. fuzzy numbers versus crisp numbers), both methods are optimised to show similar results (due to the optimisation algorithm requiring a specific amount of data being captured at each level). This shows that the optimisation algorithm developed in this manuscript for fuzzy inputs successfully mimics the original algorithm developed by Alvisi and Franchini (2011) that only used crisp inputs. Thus, when modelling a complex system, such as the minimum daily DO in the Bow River, the uncertainty in the inputs can also be quantified and propagated through the data-driven model, by using the proposed method. This is a major advantage over the original model (Alvisi and Franchini, 2011) that only allowed crisp inputs to be used. Note that as per Table 4, both methods approximately capture the same amount of data at each interval, however as Fig. 10 indicates this does not necessarily mean that the predicted intervals are exactly the same for both methods. Both methods predict unique intervals, with the overall result being that P_{CI} amount of data is captured within each interval. Note that at the interval $\mu = 1$, the existing method using crisp inputs by design predicts this interval as a singleton (thus the interval width will always be zero), whereas the proposed method has an additional feature of predicting an interval for the $\mu = 1$ level. An instance of this can be seen near the end of September where an interval is predicted rather than a singleton.

Figure 11 compares with width of predicted intervals at four selected membership levels for both methods, generally showing mixed results. As discussed above the existing method does not predict an interval for $\mu = 1$, thus, a comparison cannot be made and is not included. At the $\mu = 0.8$ level, the average width of the intervals for the existing method is close to 0, whereas for the proposed method is 0.36 mg L^{-1}. This is consistent with the results shown in Table 4 and Fig. 5, which demonstrate the narrow interval at higher membership levels. Annual comparisons of the remaining intervals show that the intervals at $\mu = 0$ are larger for the proposed method compared to the existing method for all but one year (2009). However, at $\mu = 0.2$, 0.4 and 0.6, the width of the intervals is smaller for the proposed method for a majority of years, however the overall differences are not statistically significant ($p > 0.05$ using the two-sample Kolmogorov–Smirnov test). The results demonstrate that while both method can achieve the optimisation objectives whilst respecting the constraints (i.e. P_{CI}), it is reflected differently in the predicted fuzzy intervals. Generally, the predicted fuzzy numbers using the proposed method have a larger support (at $\mu = 0$) signifying that the increased uncertainty due to fuzzy inputs into the model are propagated through the model. Whereas, the lack of inclusion of uncertain input data in the existing method results in a slightly narrower average support. In essence, the proposed FNN model is modelling a more complex system (because of the inclusion of input uncertainty) whereas the existing method models the system by assuming lower complexity (by ignoring the input uncertainty).

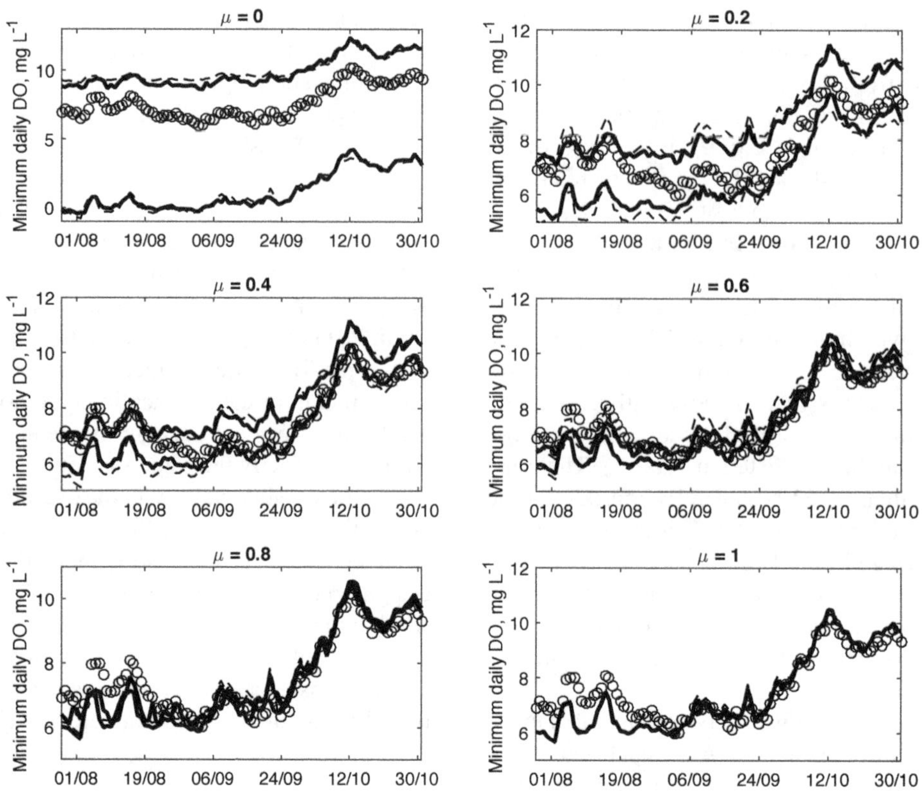

Figure 10. Comparison of predicted trends of the proposed (solid black line) and existing (dashed black line) methods shown for 2009 for each membership level. Observations are shown as black circles.

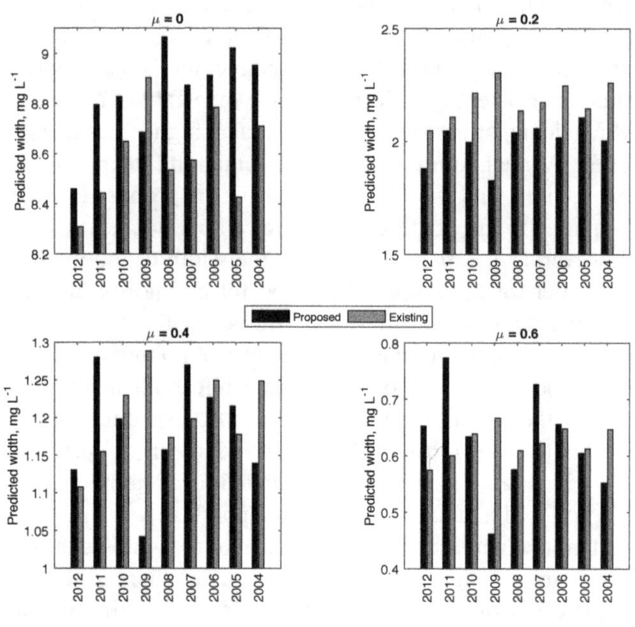

Figure 11. A comparison of average annual interval widths of predicted fuzzy numbers using the proposed and existing FNN methods for four selected membership levels.

3.3 Risk analysis for low DO events

The utility of the FNN method is illustrated through an analysis of the ability of the proposed model to predict low DO events, and then a possibility–probability transformation is used to assess the risk of these low DO events. The number of occasions when observed DO was below any of the three guidelines used for this research are summarised in Table 1. The FNN model was cable to capture 100 % of all low DO events (i.e. below 5, or 6.5 or 9.5 $\mathrm{mg\,L^{-1}}$) within the predicted intervals. In comparison, the crisp ANN network (i.e. at $\mu = 1$) did not predict DO to be less than 5 $\mathrm{mg\,L^{-1}}$ on any of the 51 occasions. Similarly, it predicted DO to be less than the more conservative limit of 6.5 $\mathrm{mg\,L^{-1}}$ in only 53 % of the 184 occurrences. For the last case, the 9.5 $\mathrm{mg\,L^{-1}}$ limit, the ANN method still trailed the FNN method, by predicting 96 % of these low DO events. This illustrates that not only can the FNN method capture more low DO events within its predicted intervals, it also performs exceptionally better for the highest risk case (DO $< 5\,\mathrm{mg\,L^{-1}}$). In general, more days were correctly identified when there was a risk of low DO using FNN rather than the typical ANN approach. This is one of the major advantages of using a fuzzy number based uncertainty analysis component for low DO prediction.

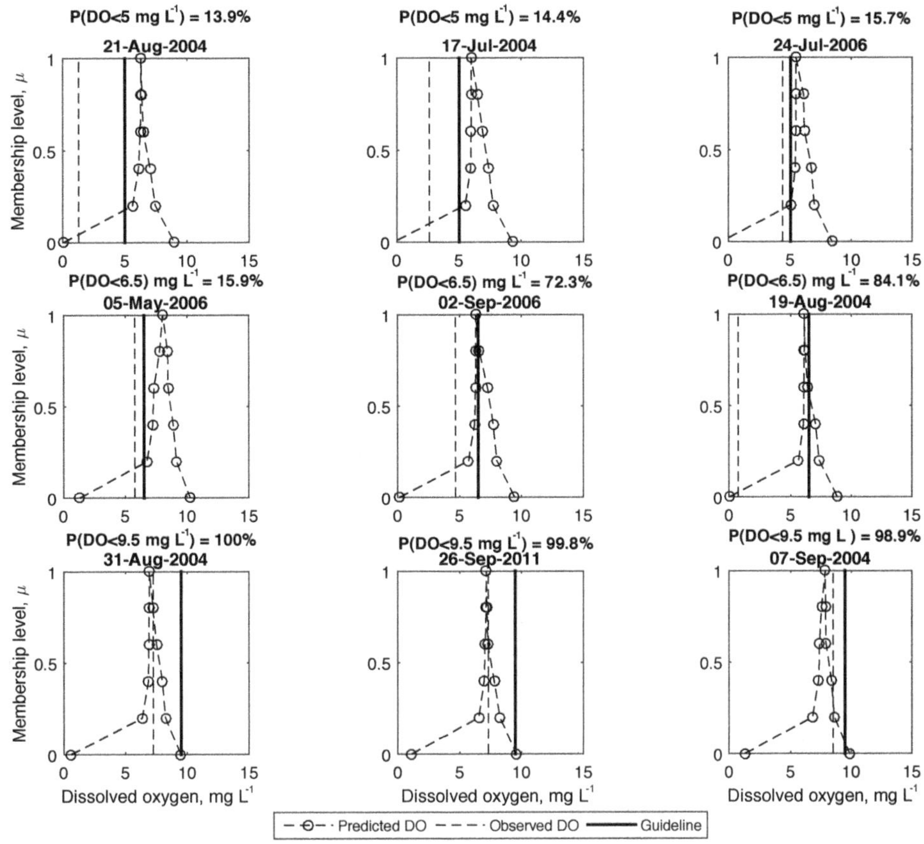

Figure 12. Sample plots of low DO events and the corresponding risk of low DO calculated using a possibility–probability transformation for the 5 $mg\,L^{-1}$ (top panels), 6.5 $mg\,L^{-1}$ (middle panels), and 9.5 $mg\,L^{-1}$ (bottom panels) guidelines.

Once all the low DO events were identified, the inverse transformation (defuzzification) described in Sect. 2.4 was used to estimate the probability of low DO. The primary reason for converting from possibility to probability is to improve the communication of the risk of low DO. For each low DO event (i.e. at 5, or 6 or 9.5 $mg\,L^{-1}$), the predicted membership function was used to determine the possibility of low DO, i.e. identify the membership level where the membership function intersects either of the low DO guidelines (some examples of low DO events are shown in Fig. 12). Once these were identified, the defuzzification technique was used to predict the probability of low DO (e.g. $P(\{DO < 5\,mg\,L^{-1}\})$).

For the first case, $P\{(DO < 5\,mg\,L^{-1}\})$, the probability ranged between 11.5 and 16.6 % for the 51 events, with a median value of 14 %. This means that on days when DO was observed to be below 5 $mg\,L^{-1}$, the FNN results identified the possibility of low DO and the probability of DO to be below the 5 $mg\,L^{-1}$ guideline of ~ 14 %. Thus, the FNN method predicts a probability of low DO (even if it is relatively small) on days when the crisp ANN does not predict a low DO event. This value can be used as a threshold by water resource managers for estimating the risk of low DO. For example, if forecast water temperature and flow rate are used to predict minimum fuzzy DO using the calibrated model, and if the risk of low DO reaches 14 %, the event can be flagged. Appropriate defence mechanisms can then be implemented to prevent the occurrence of low DO.

For the 184 cases where DO was observed to be less than 6.5 $mg\,L^{-1}$, the probability–possibility transformation estimated the risk of low DO to be between 13.7 and 92.9 %, with a median value of 73.4 %. Compared to the first case, the probability of low DO for this threshold is higher and more variable. The low probabilities are associated with predictions of low DO at lower credibility levels at the lower limit of the intervals (i.e. L), whereas the higher probabilities are associated with predictions corresponding to the upper limits of the intervals (R). For 43 out of the 184 low DO events, the probability of low DO was less than 21 % – these events correspond to predictions of low DO at low credibility levels at the lower limits. For the majority of events (107 out of 184), the risk was high, more than 65 %. It is worth noting that the crisp network only predicted 53 % of these low DO events, and of those correctly identified, the majority were over-predicted.

For the last, most conservative case, the probability of predicting DO to be less than 9.5 mg L^{-1} (1179 events) varied between 21.9 and 100 %, with a median value of 98.1 %. Only 46 out of the 1179 events had a probability of less than 70 %; the majority of events had a high risk of low DO: more than 80 % of the events had a risk of low DO of more than 90 %. This shows that the FNN can predict, with high probability, the events where minimum daily DO is observed to be below the 9.5 mg L^{-1} limit.

It is worth noting here that the proposed FNN model was designed to only include data from the April to October each year, corresponding to the ice-free period (as defined in Sect. 2.1). This implies that the analysis has been conducted on the time period that is most critical or susceptible to low DO. Thus, as the proposed FNN model predicts, there is possibility of low DO on most days (as shown in the trend plots in Figs. 7 and 8). However, the consistency principle (Zadeh, 1978) implies that an event must be possible before it is probable. Thus, a possibility to predict low DO does not imply that it will occur with a high probability. In fact, nearly all the possibility of low DO events occurs at low membership levels (i.e. $\mu < 0.2$) implying a low possibility – and the skewed nature of the results deem the probability to be low as well. For example, for the DO < 5 mg L^{-1} case, the proposed FNN model predicted 1367 days where low DO was predicted but not observed, however, on average the probability of low DO for 98 % of these events was below the threshold criteria (14 %) mentioned above. Thus, the number of "false alarms" predicted by the proposed method is very low. Similarly, for the DO < 6.5 and 9.5 mg L^{-1} cases, each had ~94 % of the low DO cases to fall below their respective threshold criteria. This shows that while FNN model correctly predicts a possibility of low DO for the majority of the days (corresponding to the typical low DO conditions), the risk of predicting a "false alarm" is low. Lastly, it should be noted that the wide intervals predicted at $\mu = 0$ are a function of the rapidly decreasing low DO value seen in 2004 (discussed in Sect. 3.2) that are likely due to instrument error. With the inclusion of new data as it becomes available, and as the model parameters are updated, it is expected that these outliers will be part of the 0.5 % of P_{CI} not included in the predicted intervals, resulting in narrower bands of predictions.

The predicted membership functions of minimum DO for nine examples are shown Fig. 12, along with the observed minimum DO (the vertical dashed line). Three samples are shown for each low DO guideline: 5, 6.5, or 9.5 mg L^{-1}; along with the associated risk of low DO calculated using the defuzzification technique. Note that the membership functions of the predicted fuzzy numbers show that each is uniquely shaped, *convex* and *normal*, highlighting the fact that the proposed optimisation algorithm successfully produces nested intervals at each membership levels (as it does for the weights and biases shown in Fig. 5). For the predicted fuzzy DO, the intervals are largest at $\mu = 0$, which decrease in size as the membership level increases. The shape of the membership functions are not triangular shaped as assumed in many fuzzy number based applications. This is of significance because it shows that the amount of uncertainty (or credibility) of the model output does not change linearly with the magnitude of DO, which has important implications regarding the risk of low DO.

For the 5 mg L^{-1} guideline, the intersection of the membership function and the guideline occurs at low possibility levels (between $\mu = 0^L$ and 0.2^L), meaning that the corresponding probability will be low as well, as illustrated by the probability values shown in the figure. This again highlights that the risk of low DO (< 5 mg L^{-1}) is predicted to be low by the FNN mostly due to the fact that the observations are captured at low membership levels. Note that the crisp ANN results (at $\mu = 1$) always overpredict low DO, as shown in these three examples. The observed value falls within the predicted interval for each case, also at low membership levels.

The examples for the 6.5 mg L^{-1} guideline (second row in Fig. 12) show that the intersection between the membership function and the guidelines occurs between $\mu = 0.4$ and 0.6 on 26 July 2006, between $\mu = 0.6$ and 0.8 on 8 August 2007, and at about $\mu = 0.6$ on 29 September 2004. This illustrates the broader trend with the 6.5 mg L^{-1} guideline (which was discussed earlier and had a large range of risk predictions), which is that for the full data set, the possibility of low DO (< 6.5 mg L^{-1}) occurs at every interval, with the majority occurring at higher intervals. This is in contrast to the 5 mg L^{-1} guideline where the possibility of low DO only occurs between $\mu = 0$ and 0.2.

The last row in Fig. 12 show sample low DO results for the 9.5 mg L^{-1} guideline. As discussed above, more than 80 % of these events had a high (more than 90 %) risk of low DO. In the first example, on 23 September 2004, the guideline intersects the membership function at $\mu = \sim 0.2^R$, corresponding to a ~97 % risk of low DO. The 6 August 2008 has a low DO prediction of 100 % – this is because the predicted fuzzy number is entirely below the guideline limit. A similar result can be seen for the last example. These examples also illustrate that had only a triangular membership function been used (i.e. the fuzzy numbers defined at two membership levels), the probability of low DO could not be quantified as specifically as it has been here. The slight changes in membership function shapes between intervals impact the final probability, and a linear function would have not captured these changes.

These examples are meant to illustrate the potential utility of the data-driven and abiotic input parameter DO model, which can be used to assess the risk of low DO. Given that it is a data-driven approach, the model can be continually updated as more data are available, further refining the predictions. Various combinations of input values can be used to predict fuzzy minimum DO and defuzzification technique can be used to *quantify* the risk of low DO given the input values. The utility of this method is that a water-resource manager can use forecasted water temperature data and expected flow rates to *quantify the risk* of low DO events in the Bow River, and can plan accordingly. For example, if the risk of low DO reaches a specific numerical threshold or trigger, different actions or strategies (e.g. increasing flow rate in the river by controlled release from the upstream dams) can be implemented. The quantification of the risk to specific probabilities means that the severity of the response can be tuned to the severity of the calculated risk.

4 Conclusions

A new method to predict DO concentration in an urbanised watershed is proposed. Given the lack of understanding of the physical system that governs DO concentration in the Bow River (in Calgary, Canada), a data-driven approach using fuzzy numbers is proposed to account for the uncertainty. Further, the model uses abiotic (non-living, physical and chemical attributes) factors as inputs to the model. Specifically, water temperature and flow rate were selected, which are routinely monitored, and thus a large data set is available.

The data-driven approach proposed is a modification of an existing fuzzy neural network method that quantifies the total uncertainty in the model by using fuzzy number weights and biases. The proposed model refines the exiting model by (i) using possibility theory based intervals to calibrate the neural network (rather than arbitrarily selecting confidence intervals), and (ii) using fuzzy number inputs rather than crisp inputs. This research also proposes a new two-step method to construct these fuzzy number inputs using observations. First a bin-size optimising algorithm is used to find the optimum histogram (as an estimate of the underlying but unknown probably density function of the observations). Then a probability–possibility transformation is used to determine the shape of the fuzzy number membership function.

The results demonstrate that the network training algorithm proposed can be successfully implemented. Model results demonstrate that low DO events are better captured by the fuzzy network as compared to a non-fuzzy network. A defuzzification technique is then used to calculate the risk of low DO events. Generally speaking, the method demonstrates that a data-driven approach using abiotic inputs is a feasible method for predicting minimum daily DO. Results from this research can be implemented by water resource managers to assess conditions that lead to and quantify the risk of low DO.

Author contributions. Usman T. Khan conducted all aspects of this research and wrote the manuscript. Caterina Valeo assisted in manuscript preparation and revisions.

Acknowledgements. The authors would like to thank S. Alvisi from the Università degli Studi di Ferrara for providing the MATLAB code for the original Fuzzy Neural Network model. The authors would also like to thank: the Natural Sciences and Engineering Research Council of Canada; the Ministry of Advanced Education, Innovation and Technology – Government of British Columbia; and the University of Victoria, for funding this research. The authors are grateful for the City of Calgary and Environment Canada for providing the data used in this research. Lastly, the authors would like to acknowledge the comments of two anonymous reviewers, whose feedback greatly helped improve this manuscript.

Edited by: D. Solomatine

References

Abrahart, R. J., Anctil, F., Coulibaly, P., Dawson, C. W., Mount, N. J., See, L. M., Asaad Y. Shamseldin, A. Y., Solomatine, D. P., Toth, E., and Wilby, R. L.: Two decades of anarchy? Emerging themes and outstanding challenges for neural network river forecasting, Prog. Phys. Geogr., 36, 480–513, doi:10.1177/0309133312444943, 2012.

Adams, K. A., Barth, J. A., and Chan, F.: Temporal variability of near-bottom dissolved oxygen during upwelling off central Oregon, J. Geophys. Res.-Oceans, 118, 4839–4854, doi:10.1002/jgrc.20361, 2013.

AENV – Alberta Environment: Alberta water quality guideline for the protection of freshwater aquatic life: Dissolved oxygen, Catalogue #: ENV-0.94-OP, Standards and Guidelines Branch, Alberta Environment, Edmonton, Alberta, Canada, 42–56, 1997.

Alvisi, S. and Franchini, M.: Fuzzy neural networks for water level and discharge forecasting with uncertainty, Environ. Model. Softw., 26, 523–537, doi:10.1016/j.envsoft.2010.10.016, 2011.

Alvisi, S., and Franchini, M.: Grey neural networks for river stage forecasting with uncertainty, Phys. Chem. Earth, 42, 108–118, doi:10.1016/j.pce.2011.04.002, 2012.

Alvisi, S., Mascellani, G., Franchini, M., and Bárdossy, A.: Water level forecasting through fuzzy logic and artificial neural network approaches, Hydrol. Earth Syst. Sci., 10, 1–17, doi:10.5194/hess-10-1-2006, 2006.

Antanasijević, D., Pocajt, V., Perić–Grujić, A., and Ristić, M.: Modelling of dissolved oxygen in the Danube River using artificial neural networks and Monte Carlo Simulation uncertainty analysis, J. Hydrol., 519, 1895–1907, doi:10.1016/j.jhydrol.2014.10.009, 2014.

ASCE Task Committee on Application of Artificial Neural Networks in Hydrology.: Artificial Neural Networks in Hydrology. II: Hydrologic Applications, J. Hydrol. Eng., 5, 124–137, doi:10.1061/(ASCE)1084-0699(2000)5:2(124), 2000.

Ay, M. and Kisi, O.: Modeling of dissolved oxygen concentration using different neural network techniques in Foundation Creek, El Paso County, Colorado, J. Environ. Eng.-ASCE, 138, 654–662, doi:10.1061/(ASCE)EE.1943-7870.0000511, 2011.

Bárdossy, A., Bogardi, I., and Duckstein, L.: Fuzzy regression in hydrology, Water Resour. Res., 26, 1497–1508, doi:10.1029/WR026i007p01497, 1990.

Bárdossy, A., Mascellani, G., and Franchini, M.: Fuzzy unit hydrograph, Water Resour. Res., 42, W02401, doi:10.1029/2004WR003751, 2006.

Betrie, G. D., Sadiq, R., Morin, K. A., and Tesfamariam, S.: Uncertainty quantification and integration of machine learning techniques for predicting acid rock drainage chemistry: A probability bounds approach, Sci. Total Environ., 490, 182–190, doi:10.1016/j.scitotenv.2014.04.125, 2014.

Bow River Basin Council – Profile of the Bow River Basin: http://wsow.brbc.ab.ca/index.php?option=com_content&view=article&id=259&Itemid=83, last access: 3 September 2015.

CCME – Canadian Council of Ministers of the Environment: Canadian water quality guidelines for the protection of aquatic life: Dissolved oxygen (freshwater), Canadian Council of Ministers of the Environment, Winnipeg, MB, Canada, 1–6, 1999.

Chang, F. J., Tsai, Y. H., Chen, P. A., Coynel, A., and Vachaud, G.: Modeling water quality in an urban river using hydrological factors–Data driven approaches, J. Environ. Manage., 151, 87–96, doi:10.1016/j.jenvman.2014.12.014, 2015.

Civanlar, M. R. and Trussell, H. J.: Constructing membership functions using statistical data, Fuzzy Sets Syst., 18, 1-13, doi:10.1016/0165-0114(86)90024-2, 1986.

Deng, Y., Sadiq, R., Jiang, W., and Tesfamariam, S.: Risk analysis in a linguistic environment: a fuzzy evidential reasoning-based approach, Expert Syst. Appl., 38, 15438–15446, doi:10.1016/j.eswa.2011.06.018, 2011.

Dorfman, R. and Jacoby, H. D.: An illustrative model of a river basin pollution control, in: Models for Managing Regional Water quality, edited by: Dorfman, R., Jacoby, H. D., and Thomas Jr., H. A., Harvard University Press, Cambridge, MA, USA, 84–141, doi:10.4159/harvard.9780674419216.c3, 1972.

Duan, Q., Sorooshian, S., and Gupta, V.: Effective and efficient global optimisation for conceptual rainfall-runoff models, Water Resour. Res., 28, 1015–1031, doi:10.1029/91WR02985, 1992.

Dubois, D. and Prade, H.: Fuzzy sets and probability: misunderstandings, bridges and gaps, in: Proceedings of the Second IEEE International Conference on Fuzzy Systems, 28 March–1 April 1993, San Francisco, USA, 1059–1068, doi:10.1109/FUZZY.1993.327367, 1993.

Dubois, D. and Prade, H.: Possibility Theory and its Applications: Where do we stand?, in: Springer Handbook of Computational Intelligence, edited by: Kacprzyk, J. and Pedrycz, W., Springer-Verlag, Berlin, Heidelberg, 31–60, doi:10.1007/978-3-662-43505-2_3, 2015.

Dubois, D., Prade, H., and Sandri, S.: On possibility/probability transformations, in: Fuzzy logic, edited by: Lowen, R. and Roubens, M., Springer Netherlands, 103–112, doi:10.1007/978-94-011-2014-2_10, 1993.

Dubois, D., Foulloy, L., Mauris, G., and Prade, H.: Probability–possibility transformations, triangular fuzzy sets, and probabilistic inequalities, Reliab. Comput., 10, 273–297, doi:10.1023/B:REOM.0000032115.22510.b5, 2004.

Elshorbagy, A., Corzo, G., Srinivasulu, S., and Solomatine, D. P.: Experimental investigation of the predictive capabilities of data driven modeling techniques in hydrology – Part 1: Concepts and methodology. Hydrol. Earth Syst. Sci., 14, 1931–1941, doi:10.5194/hess-14-1931-2010, 2010.

Environment Canada: Wateroffice Hydrometric Station Meta Data, https://wateroffice.ec.gc.ca/station_metadata/stationList_e.html, last access: 3 September 2015.

Ferrero, A., Prioli, M., Salicone, S., and Vantaggi, B.: A 2-D Metrology-Sound Probability–Possibility Transformation, IEEE T. Instrum. Meas., 62, 982–990, doi:10.1109/TIM.2013.2246910, 2013.

Golder Associates Ltd.: Bow River Impact Study – Phase 1: Model Development and Calibration, City of Calgary Wastewater, Utilities and Environmental Protection, Calgary, Alberta, 134 pp., 2004.

Guyonnet, D., Bourgine, B., Dubois, D., Fargier, H., Come, B., and Chiles, J.-P.: Hybrid approach for addressing uncertainty in risk assessments, J. Environ. Eng.-ASCE, 129, 68–78, doi:10.1061/(ASCE)0733-9372(2003)129:1(68), 2003.

Hall, M. J.: Urban Hydrology, Elsevier Applied Science, Essex, England, 299 pp., 1984.

Hauer, F. R. and Hill, W. R.: Temperature, light and oxygen, in: Methods in stream ecology, edited by: Hauer, F. R. and Lamberti, G. A., Academic Press, San Diego, CA, USA, 103–117, doi:10.1016/B978-012332908-0.50007-3, 2007.

He, J. and Valeo, C.: Comparative study of ANNs versus parametric methods in rainfall frequency analysis, J. Hydrol. Eng., 14, 172–184, doi:10.1061/(ASCE)1084-0699(2009)14:2(172), 2009.

He, J., Chu, A., Ryan, M. C., Valeo, C., and Zaitlin, B.: Abiotic influences on dissolved oxygen in a riverine environment, Ecol. Eng., 37, 1804–1814, doi:10.1016/j.ecoleng.2011.06.022, 2011.

He, J., Ryan, M. C., and Valeo, C.: Changes in Water Quality Characteristics and Pollutant Sources along a Major River Basin in Canada, in: Environmental Management of River Basin Ecosystems, edited by: Ramkumar, M., Kumaraswamy, K., and Mohanraj, R., Springer International Publishing, Cham, Switzerland, 525–548, doi:10.1007/978-3-319-13425-3_25, 2015.

Heddam, S.: Generalized regression neural network-based approach for modelling hourly dissolved oxygen concentration in the Upper Klamath River, Oregon, USA, Environ. Technol., 35, 1650–1657, doi:10.1080/09593330.2013.878396, 2014.

Hornik, K., Stinchcombe, M., and White, H.: Multilayer feedforward networks are universal approximators, Neural Networks, 2, 359–366, doi:10.1016/0893-6080(89)90020-8, 1989.

Huang, Y., Chen, X., Li, Y. P., Huang, G. H., and Liu, T.: A fuzzy-based simulation method for modelling hydrological processes under uncertainty, Hydrol. Process., 24, 3718–3732, doi:10.1002/hyp.7790, 2010.

Iwanyshyn, M., Ryan, M. C., and Chu, A.: Separation of physical loading from photosynthesis/respiration processes in rivers by mass balance, Sci. Total Environ., 390, 205–214, doi:10.1016/j.scitotenv.2007.09.038, 2008.

Jacquin, A. P.: Possibilistic uncertainty analysis of a conceptual model of snowmelt runoff. Hydrol. Earth Syst. Sci., 14, 1681–1695, doi:10.5194/hess-14-1681-2010, 2010.

Kannel, P. R., Lee, S., Lee, Y. S., Kanel, S. R., and Khan, S. P.: Application of water quality indices and dissolved oxygen as indicators for river water classification and urban impact assessment, Environ. Monit. Assess., 132, 93–110, doi:10.1007/s10661-006-9505-1, 2007.

Kasiviswanathan, K. S. and Sudheer, K. P.: Quantification of the predictive uncertainty of artificial neural network based river flow forecast models, Stoch. Environ. Res. Risk A., 27, 137–146, doi:10.1007/s00477-012-0600-2, 2013.

Kasiviswanathan, K. S., Cibin, R., Sudheer, K. P., and Chaubey, I.: Constructing prediction interval for artificial neural network rainfall runoff models based on ensemble simulations, J. Hydrol., 499, 275–288, doi:10.1016/j.jhydrol.2013.06.043, 2013.

Kaufmann, A. and Gupta, M. M.: Introduction to Fuzzy Arithmetic: Theory and Applications, Van Nostrand Reinhold Company, New York, NY, USA, 351 pp., 1985.

Khan, U. T. and Valeo, C.: Predicting Dissolved Oxygen Concentration in Urban Watersheds: A Comparison of Fuzzy Number Based and Bayesian Data-Driven Approaches, in: Proceedings of the International Conference on Marine and Freshwater Environments, 6–8 August 2014, St John's, Canada, 1–10, 2014a.

Khan, U. T. and Valeo, C.: Peak flow prediction using fuzzy linear regression: Case study of the Bow River, in: Proceedings of the International Conference on Marine and Freshwater Environments, 6–8 August 2014, St Johns, NFLD, Canada, 1–10, 2014b.

Khan, U. T. and Valeo, C.: A new fuzzy linear regression approach for dissolved oxygen prediction, Hydrolog. Sci. J., 60, 1096–1119, doi:10.1080/02626667.2014.900558, 2015.

Khan, U. T. and Valeo, C.: Short-term peak flow rate prediction and flood risk assessment using fuzzy linear regression, J. Environ. Inform., in press, 2016.

Khan, U. T., Valeo, C., and He, J.: Non-linear fuzzy-set based uncertainty propagation for improved DO prediction using multiple-linear regression, Stoch. Environ. Res. Risk A., 27, 599–616, doi:10.1007/s00477-012-0626-5, 2013.

Klir, G. J. and Parviz, B.: Probability-possibility transformations: a comparison, Int. J. Gen. Syst., 21, 291–310, doi:10.1080/03081079208945083, 1992.

Lane, R. J.: The Water Survey of Canada: Hydrometric Technician Career Development Program Lesson Package No. 10.1 – Principles of Discharge Measurement, available at: http://publications.gc.ca/site/eng/463990/publication.html (last access: 10 June 2016), 1999.

Maier, H. R., Jain, A., Dandy, G. C., and Sudheer, K. P.: Methods used for the development of neural networks for the prediction of water resource variables in river systems: current status and future directions, Environ. Model. Softw., 25, 891–909, doi:10.1016/j.envsoft.2010.02.003, 2010.

Martin-Clouaire, R., Cazemier, D. R., and Lagacherie, P.: Representing and processing uncertain soil information for mapping soil hydrological properties, Comput. Electron. Agr., 29, 41–57, doi:10.1016/S0168-1699(00)00135-6, 2000.

Mauris, G.: A review of relationships between possibility and probability representations of uncertainty in measurement, IEEE T. Instrum. Meas., 62, 622–632, doi:10.1109/TIM.2012.2218057, 2013.

Moriasi, D. N., Arnold, J. G., Van Liew, M. W., Bingner, R. L., Harmel, R. D., and Veith, T. L.: Model evaluation guidelines for systematic quantification of accuracy in watershed simulations, T. ASABE, 50, 885–900, doi:10.13031/2013.23153, 2007.

Mujumdar, P. P. and Ghosh, S.: Modeling GCM and scenario uncertainty using a possibilistic approach: Application to the Mahanadi River, India, Water Resour. Res., 44, W06407, doi:10.1029/2007WR006137, 2008.

Napolitano, G., Serinaldi, F., and See, L.: Impact of EMD decomposition and random initialisation of weights in ANN hindcasting of daily stream flow series: an empirical examination, J. Hydrol., 406, 199–214, doi:10.1016/j.jhydrol.2011.06.015, 2011.

Nash, J. E. and Sutcliffe, J. V.: River flow forecasting through conceptual models: Part I. A discussion of principles, J. Hydrol., 10, 282–290, doi:10.1016/0022-1694(70)90255-6, 1970.

Neupane, A., Wu, P., Ghanbarpour, R. and Martin, N.: Bow River Phosphorus Management Plan: Water Quality Modeling Scenarios, Alberta Environment and Sustainable Resource Development, Edmonton, AB, Canada, 23 pp., 2014.

Niemczynowicz, J.: Urban hydrology and water management – present and future challenges, Urban Water, 1, 1–14, doi:10.1016/S1462-0758(99)00009-6, 1999.

Oussalah, M.: On the probability/possibility transformations: a comparative analysis, Int. J. Gen. Syst. 29, 671–718, doi:10.1080/03081070008960969, 2000.

Pogue, T. R. and Anderson, C. W.: Processes controlling dissolved oxygen and pH in the upper Willamette River and major tributaries, Oregon, 1994, Water Resources Investigations Report 95-4205, US Geological Survey, Portland, OR, USA, 77 pp., 1995.

Robinson, K. L., Valeo, C., Ryan, M. C., Chu, A., and Iwanyshyn, M.: Modelling aquatic vegetation and dissolved oxygen after a flood event in the Bow River, Alberta, Canada, Can. J. Civil Eng., 36, 492–503, doi:10.1139/L08-126, 2009.

Sadiq, R., Rodriguez, M. J., Imran, S. A., and Najjaran, H.: Communicating human health risks associated with disinfection by-products in drinking water supplies: a fuzzy-based approach, Stoch. Environ. Res. Risk A., 21, 341–353, doi:10.1007/s00477-006-0069-y, 2007.

Serrurier, M. and Prade, H.: An informational distance for estimating the faithfulness of a possibility distribution, viewed as a family of probability distributions, with respect to data, Int. J. Approx. Reason., 54, 919–933, doi:10.1016/j.ijar.2013.01.011, 2013.

Shimazaki, H. and Shinomoto, S.: A method for selecting the bin size of a time histogram, Neural Comput., 19, 1503–1527, doi:10.1162/neco.2007.19.6.1503, 2007.

Shrestha, D. L. and Solomatine, D. P.: Data-driven approaches for estimating uncertainty in rainfall-runoff modelling, Int. J. River Basin Manage., 6, 109–122, doi:10.1080/15715124.2008.9635341, 2008.

Singh, K. P., Basant, A., Malik, A., and Jain, G.: Artificial neural network modelling of the river water quality – a case study, Ecol. Model., 220, 888–895, doi:10.1016/j.ecolmodel.2009.01.004, 2009.

Solomantine, D. P. and Ostfeld, A.: Data-driven modelling: some past experiences and new approaches, J. Hydroinform., 10, 3–22, doi:10.2166/hydro.2008.015, 2008.

Solomatine, D. P., See, L. M., and Abrahart, R. J.: Data-driven modelling: concepts, approaches and experiences, in: Practical Hydroinformatics, edited by: Abrahart, R. J., See, L. M., and Solomatine, D. P., Springer, Berlin, Heidelberg, Germany, 17–30, doi:10.1007/978-3-540-79881-1_2, 2008.

Van Steenbergen, N., Ronsyn, J., and Willems, P.: A non-parametric data-based approach for probabilistic flood forecasting in support of uncertainty communication, Environ. Model. Softw., 33, 92–105, doi:10.1016/j.envsoft.2012.01.013, 2012.

Verhoest, N. E. C., De Baets, B., Mattia, F., Satalino, G., Lucau, C., and Defourny, P.: A possibilistic approach to soil moisture retrieval from ERS synthetic aperture radar backscattering under soil roughness uncertainty, Water Resour. Res., 43, W07435, doi:10.1029/2006WR005295, 2007.

Wen, X., Fang, J., Diao, M., and Zhang, C.: Artificial neural network modelling of dissolved oxygen in the Heihe River, Northwestern China, Environ. Monit. Assess., 185, 4361–4371, doi:10.1007/s10661-012-2874-8, 2013.

YSI Inc.: YSI 5200: A Multiparameter Monitor and Control Specifications, https://www.ysi.com/File_Library/Documents/Specification_Sheets/W45-01-5200A.pdf, last access: 3 September 2015.

Zadeh, L. A.: Fuzzy sets, Inform. Control, 8, 338–353, doi:10.1016/S0019-9958(65)90241-X, 1965.

Zadeh, L. A.: Fuzzy sets as a basis for a theory of possibility, Fuzzy Sets Syst., 1, 3–28, doi:10.1016/0165-0114(78)90029-5, 1978.

Zhang, K.: Modeling uncertainty and variability in health risk assessment of contaminated sites, PhD thesis, Civil Engineering, University of Calgary, Canada, 290 pp., 2009.

Zhang, K. and Achari, G.: Correlations between uncertainty theories and their applications in uncertainty propagation, in: Safety, reliability and risk of structures, infrastructures and engineering systems, edited by: Furuta, H., Frangopol, D. M., and Shinozuka, M., Taylor and Francis Group, London, UK, 1337–1344, doi:10.1201/9781439847657-c20, 2010.

Zhang, K., Li, H., and Achari, G.: Fuzzy-stochastic characterisation of site uncertainty and variability in groundwater flow and contaminant transport through a heterogeneous aquifer, J. Contam. Hydrol., 106, 73–82, doi:10.1016/j.jconhyd.2009.01.003, 2009.

PERMISSIONS

All chapters in this book were first published in HESS, by Copernicus Publications; hereby published with permission under the Creative Commons Attribution License or equivalent. Every chapter published in this book has been scrutinized by our experts. Their significance has been extensively debated. The topics covered herein carry significant findings which will fuel the growth of the discipline. They may even be implemented as practical applications or may be referred to as a beginning point for another development.

The contributors of this book come from diverse backgrounds, making this book a truly international effort. This book will bring forth new frontiers with its revolutionizing research information and detailed analysis of the nascent developments around the world.

We would like to thank all the contributing authors for lending their expertise to make the book truly unique. They have played a crucial role in the development of this book. Without their invaluable contributions this book wouldn't have been possible. They have made vital efforts to compile up to date information on the varied aspects of this subject to make this book a valuable addition to the collection of many professionals and students.

This book was conceptualized with the vision of imparting up-to-date information and advanced data in this field. To ensure the same, a matchless editorial board was set up. Every individual on the board went through rigorous rounds of assessment to prove their worth. After which they invested a large part of their time researching and compiling the most relevant data for our readers.

The editorial board has been involved in producing this book since its inception. They have spent rigorous hours researching and exploring the diverse topics which have resulted in the successful publishing of this book. They have passed on their knowledge of decades through this book. To expedite this challenging task, the publisher supported the team at every step. A small team of assistant editors was also appointed to further simplify the editing procedure and attain best results for the readers.

Apart from the editorial board, the designing team has also invested a significant amount of their time in understanding the subject and creating the most relevant covers. They scrutinized every image to scout for the most suitable representation of the subject and create an appropriate cover for the book.

The publishing team has been an ardent support to the editorial, designing and production team. Their endless efforts to recruit the best for this project, has resulted in the accomplishment of this book. They are a veteran in the field of academics and their pool of knowledge is as vast as their experience in printing. Their expertise and guidance has proved useful at every step. Their uncompromising quality standards have made this book an exceptional effort. Their encouragement from time to time has been an inspiration for everyone.

The publisher and the editorial board hope that this book will prove to be a valuable piece of knowledge for researchers, students, practitioners and scholars across the globe.

LIST OF CONTRIBUTORS

P. Licznar
Faculty of Environmental Engineering, Wroclaw University of Technology, Wrocław, Poland

C. De Michele
Department of Civil and Environmental Engineering, Politecnico di Milano, Milan, Italy

W. Adamowski
Institute of Environmental Engineering, John Paul II Catholic University of Lublin, Stalowa Wola, Poland

Heiko Apel, Lars Gerlitz, Katy Unger-Shayesteh, Sergiy Vorogushyn and Abror Gafurov
GFZ German Research Centre for Geoscience, Section 5.4 Hydrology, Potsdam, Germany

Zharkinay Abdykerimova
Hydro-Meteorological Service of Kyrgyzstan, Bishkek, Kyrgyzstan

Marina Agalhanova
Hydro-Meteorological Service of Turkmenistan, Ashgabat, Turkmenistan

Azamat Baimaganbetov
Hydro-Meteorological Service of Kazakhstan, Almaty, Kazakhstan

Nadejda Gavrilenko
Hydro-Meteorological Service of Uzbekistan, Tashkent, Uzbekistan

Olga Kalashnikova
CAIAG Central Asian Institute for Applied Geoscience, Bishkek, Kyrgyzstan

Ricardo Zubieta and Jhan Carlo Espinoza
Subdirección de Ciencias de la Atmósfera e Hidrósfera (SCAH), Instituto Geofísico del Perú (IGP), Lima, Peru
Programa de Doctorado en Recursos Hídricos, Universidad Nacional Agraria La Molina, Peru

Augusto Getirana
Hydrological Sciences Laboratory, NASA Goddard Space Flight Center, Greenbelt, MD, USA
Earth System Science Interdisciplinary Center, College Park, MD, USA

Waldo Lavado-Casimiro
Programa de Doctorado en Recursos Hídricos, Universidad Nacional Agraria La Molina, Peru
Servicio Nacional de Meteorología e Hidrología (SENAMHI), Lima, Peru

Luis Aragon
Programa de Doctorado en Recursos Hídricos, Universidad Nacional Agraria La Molina, Peru

Natthachet Tangdamrongsub, Shin-Chan Han and In-Young Yeo
School of Engineering, University of Newcastle, Callaghan, New South Wales, Australia

Mark Decker
ARC Centre of Excellence for Climate System Science, University of New South Wales, Sydney, New South Wales, Australia

Hyungjun Kim
Institute of Industrial Science, the University of Tokyo, Tokyo, Japan

Charlotte Marie Emery
LEGOS, Université de Toulouse, CNES, CNRS, IRD, UPS, Toulouse, France
JPL, Pasadena, CA, USA

Adrien Paris
LEGOS, Université de Toulouse, CNES, CNRS, IRD, UPS, Toulouse, France
GET, Université de Toulouse, UPS, CNRS, IRD, Toulouse, France
LMI OCE IRD/UNB, Campus Darcy Ribeiro, Brasilia, Brazil
CLS, Ramonville-Saint-Agne, France

Sylvain Biancamaria and Stéphane Calmant
LEGOS, Université de Toulouse, CNES, CNRS, IRD, UPS, Toulouse, France

Aaron Boone
Meteo France CNRS, CNRM UMR 3589, Toulouse, France

Pierre-André Garambois
ICUBE – UMR 7357, Fluid Mechanics Team, INSA, Strasbourg, France

Joecila Santos da Silva
CESTU, Universidade do Estado do Amazonas, Manaus, Brazil

Tsun-Hua Yang, Chin-Cheng Tsai and Jui-Yi Ho
Taiwan Typhoon and Flood Research Institute (TTFRI), National Applied Research Laboratories (NARLabs), Taipei, 10093, Taiwan

Gong-Do Hwang
Taiwan Typhoon and Flood Research Institute (TTFRI), National Applied Research Laboratories (NARLabs), Taipei, 10093, Taiwan
Department of Atmospheric Science, National Taiwan University, Taipei, 10617, Taiwan

Jason M. Hunter, Holger R. Maier, Eloise R. Foale, Naomi A. Grosvenor, Nathan P. Harders and Tahali C. Kikuchi-Miller
School of Civil, Environmental and Mining Engineering, The University of Adelaide, Adelaide, 5005 SA, Australia

Matthew S. Gibbs
School of Civil, Environmental and Mining Engineering, The University of Adelaide, Adelaide, 5005 SA, Australia
Department of Environment and Water, Adelaide, 5001 SA, Australia

Louise Arnal
Department of Geography and Environmental Science, University of Reading, Reading, RG6 6AB, UK
European Centre for Medium-Range Weather Forecasts, Shinfield Park, Reading, RG6 9AX, UK

Elisabeth Stephens and Jessica Neumann
Department of Geography and Environmental Science, University of Reading, Reading, RG6 6AB, UK

Fredrik Wetterhall, Blazej Krzeminski and Florian Pappenberger
European Centre for Medium-Range Weather Forecasts, Shinfield Park, Reading, RG6 9AX, UK

Christel Prudhomme
European Centre for Medium-Range Weather Forecasts, Shinfield Park, Reading, RG6 9AX, UK
Department of Geography, Loughborough University, Loughborough, LE11 3TU, UK
NERC Centre for Ecology & Hydrology, Wallingford, OX10 8BB, UK

Hannah L. Cloke
Department of Geography and Environmental Science, University of Reading, Reading, RG6 6AB, UK
Department of Meteorology, University of Reading, Reading, RG6 6BB, UK
Department of Earth Sciences, Uppsala University, Uppsala, 752 36, Sweden
Centre of Natural Hazards and Disaster Science, CNDS, Uppsala, 752 36, Sweden

Fanny Larue
CARTEL, Université de Sherbrooke, Québec, Canada
Centre d'Études Nordiques, Québec, Canada
IREQ, Hydro-Québec, Québec, Canada

Alain Royer
CARTEL, Université de Sherbrooke, Québec, Canada
Centre d'Études Nordiques, Québec, Canada

Danielle De Sève
IREQ, Hydro-Québec, Québec, Canada

Emmanuel Cosme
Institut des Géosciences de l'Environnement, IGE, UGA-CNRS, Grenoble, France

Alexandre Roy
CARTEL, Université de Sherbrooke, Québec, Canada
Centre d'Études Nordiques, Québec, Canada
Département de Géographie, Université de Montréal, Québec, Canada
Université du Québec à Trois-Rivière, Québec, Canada

Kevin Sene, Wlodek Tych and Keith Beven
Lancaster Environment Centre, Lancaster University, Lancaster, LA1 4YQ, UK

Usman T. Khan and Caterina Valeo
Mechanical Engineering, University of Victoria, Stn. CSC, Victoria, BC, V8W 2Y2, Canada

Index

www.ingramcontent.com/pod-product-compliance
Lightning Source LLC
LaVergne TN
LVHW060537210926
848485LV00005B/1215